Basics of Human Andrology

Anand Kumar • Mona Sharma
Editors

Basics of Human Andrology

A Textbook

Editors
Anand Kumar
Department of Reproductive Biology
All India Institute of Medical Sciences
New Delhi, India

Mona Sharma
Department of Reproductive Biology
All India Institute of Medical Sciences
New Delhi, India

ISBN 978-981-10-9943-4 ISBN 978-981-10-3695-8 (eBook)
DOI 10.1007/978-981-10-3695-8

Softcover reprint of the hardcover 1st edition 2017

Printed on acid-free paper

This Springer imprint is published by Springer Nature
The registered company is Springer Nature Singapore Pte Ltd.
The registered company address is: 152 Beach Road, #21-01/04 Gateway East, Singapore 189721, Singapore

Dedicated to The Mother & Sri Aurobindo
&
All India Institute of Medical Sciences

Foreword

Because of high muscular strength and apparent physique, man gives an impression of robustness. Therefore, male reproductive health has been a neglected area for physicians. Little do we realise that males are actually the weaker sex, and they too need care. They die faster because of their sex hormones and mind. They die in wars and adventures apart from hormone created morbidity. With increase in longevity, reproductive and sexual health needs more attention than ever before.

I hope this book will lay the knowledge base for professionals who wish to delve in male reproductive health.

I enjoyed reading this book and I wish that you enjoy it too.

MC Misra, MS, FACS, FAMS, FRCS
Director, All India Institute of Medical Sciences,
New Delhi, India

About the Editors

Professor Anand Kumar, MD, FAMS, has been the Head of the Department of Reproductive Biology at All India Institute of Medical Sciences (AIIMS), New Delhi. He was a postdoctoral fellow at the University of Hamburg and WHO fellow at the University of Bonn. He was a visiting professor at the University of Hamburg and the University of Science, Malaysia. He has over 100 research publications and several books to his name.

Dr. Mona Sharma, MD, DNB, MAMS, is an assistant professor in the Department of Reproductive Biology at All India Institute of Medical Sciences (AIIMS), New Delhi. She is a motivated teacher and an enthusiastic researcher in the field of andrology and reproductive genetics. She has presented her research at various scientific conferences and has numerous publications to her name.

Preface

Male fertility parameters are declining globally. The Y chromosome is shortening over generations and may eventually disappear in course of evolution. Colleagues have shown a higher proportion of X chromosome-bearing sperm in the semen and have suggested that there is a preferential elimination of Y chromosome-bearing sperm during their passage through the male genital tract. However, male embryos are better protected in womb. At birth, for 100 females 106 males are born. Males in each age group are more prone to morbidity and mortality, and at the end, elderly females outnumber men in most parts of the world. Males retain their fertility till the ripe old age, whereas females lose theirs in the middle phase of life. Males are biologically driven to be aggressive and polygamous probably to amplify gene diversity. Male sexuality and reproduction have always remained an enigma.

Truth is variously presented by writers in their own unique way; any one book cannot serve all and be appreciated universally. Each book serves a particular group of readers. Therefore, many books may be needed to state the same knowledge keeping corresponding readers in mind. We too present here the current knowledge of andrology in our own unique way for our potential readers.

This book addresses various aspects of male reproduction ranging from mind to testis. This book has been designed for all students and practitioners across disciplines of the medical science. We have kept the language simple so that an inquisitive person with a background of biology too may read it. The scope of our book is reflected by wide range of topics covered by our contributors.

This book has 30 chapters ranging from chromosome to yoga. The basis of maleness lies in the Y chromosome. Reproductive functions depend upon the development of male organs from embryo to manhood. Testis, the male gonad, produces hormones and sperm; latter are ejaculated in semen secreted by accessory sex glands. The testicular events are under neuroendocrine regulation which coordinates reproductive life from puberty to andropause and to the end of life.

Biology is as important as psychology in the control of reproduction. Behaviors are rooted in the brain. Various brain areas and neural circuits regulate male behaviors. Brain sexual polymorphism is the basis of homosexuality and transgenders. Neurophysiology has always been complex to understand. But, we have attempted to present it in a simpler way.

Reproductive organs receive systemic influences, too. This book describes roles of metabolic, immune and thyroid status in reproduction. The book has chapters on male reproductive pathophysiology. Principles of diagnosis and management are also included. The last section deals with contraception and yoga. The traditional wisdom of yoga has been used for millennia to enhance sexual and reproductive experience.

Learning objectives are given in the beginning of each chapter to provide a quick overview of the content. Concepts are developed using flow charts, tables and illustrations. At the end of each chapter, a few key questions are given to the reader for self-evaluation.

We hope that this book will serve basic medical scientists, urologists, nephrologists, surgeons, andrologists, endocrinologists, gynaecologists, nurses, councellors and also the students of biological sciences who want to study reproduction in human male.

We convey our sincere thanks to Dr. Manish Jain, Dr. Bodhana Dhole, Ms. Amanpreet Kaur Kalsi and artist Mr. Ramchandra Pokale for their painstaking efforts in making this book in its present form.

Your suggestions regarding this book will be welcomed. Wish you a happy reading!

Anand Kumar
Mona Sharma

Contents

Contributors

Editorial Assistants

Manish Jain, PhD Department of Reproductive Biology, All India Institute of Medical Sciences, New Delhi, India

Bodhana Dhole, PhD Department of Reproductive Biology, All India Institute of Medical Sciences, New Delhi, India

Amanpreet Kaur Kalsi, MSc Department of Reproductive Biology, All India Institute of Medical Sciences, New Delhi, India

Authors

Mohit Agrawal, MD Department of Microbiology, Mahatma Gandhi University of Medical Sciences and Technology, Jaipur, India

Sweety Agrawal, MD Department of Endocrinology and Metabolism, All India Institute of Medical Sciences, New Delhi, India

†**A.C. Ammini, MD, DM** Department of Endocrinology and Metabolism, All India Institute of Medical Sciences, New Delhi, India

P.K. Anish, MD Department of Psychiatry, All India Institute of Medical Sciences, New Delhi, India

A.S. Ansari, PhD Centre for Advanced Studies, Department of Zoology, University of Rajasthan, Jaipur, India

Imtiyaz Alam, MSc Centre for Advanced Studies, Department of Zoology, University of Rajasthan, Jaipur, India

Rony Benson, MD Department of Radiotherapy, All India Institute of Medical Sciences, New Delhi, India

Rakesh Kumar Chadda, MD, FAMS, FRC Department of Psychiatry, All India Institute of Medical Sciences, New Delhi, India

Bodhana Dhole, PhD Department of Reproductive Biology, All India Institute of Medical Sciences, New Delhi, India

Mohd Ashraf Ganie, MD, DM Department of Endocrinology and Metabolism, All India Institute of Medical Sciences, New Delhi, India

Ravinder Goswami, MD, DM, FASc, FNASc, FAMS Department of Endocrinology and Metabolism, All India Institute of Medical Sciences, New Delhi, India

Devendra K. Gupta, MS, MCh, FAMS, FCS, FRCS, DSc Department of Pediatric surgery, All India Institute of Medical Sciences, New Delhi, India

Surabhi Gupta, PhD Department of Reproductive Biology, All India Institute of Medical Sciences, New Delhi, India

Ashutosh Halder, MD, DNB, DM, FAMS, MRCOG-I Department of Reproductive Biology, All India Institute of Medical Sciences, New Delhi, India

Kunhi Parambath Haresh, MD, DNB, MNAMS Department of Radiotherapy, All India Institute of Medical Sciences, New Delhi, India

Venkateswaran Iyer, MD Department of Pathology, All India Institute of Medical Sciences, New Delhi, India

Manish Jain, PhD Department of Reproductive Biology, All India Institute of Medical Sciences, New Delhi, India

Vandana Jain, MD Department of Pediatrics, All India Institute of Medical Sciences, New Delhi, India

Amanpreet Kaur Kalsi, MSc Department of Reproductive Biology, All India Institute of Medical Sciences, New Delhi, India

Amit Kumar, MA, DYED Department of Orthopaedics, All India Institute of Medical Sciences, New Delhi, India

Anand Kumar, MD, FAMS Department of Reproductive Biology, All India Institute of Medical Sciences, New Delhi, India

Prashant Kumar, MSc, MPhil Department of Reproductive Biology, All India Institute of Medical Sciences, New Delhi, India

N.K. Lohiya, PhD, FNASc, FAMS, FIAES, FEMSI Centre for Advanced Studies, Department of Zoology, University of Rajasthan, Jaipur, India

Manmohan Misro, PhD Reproductive Biomedicine, National Institute of Health and Family Welfare, New Delhi, India

Tarala D. Nandedkar, PhD, FNASc National Institute for Research in Reproductive Health, Mumbai, India

Sobia Nisar, MD Department of Geriatric Medicine, All India Institute of Medical Sciences, New Delhi, India

Sk Hammadur Rahaman, MD Department of Endocrinology and Metabolism, All India Institute of Medical Sciences, New Delhi, India

Nishant Raizada, MD, DM Department of Endocrinology and Metabolism, All India Institute of Medical Sciences, New Delhi, India

Ashok Rattan, MD, MAMS Department of Microbiology, Mahatma Gandhi University of Medical Sciences and Technology, Jaipur, India

Amit Satapathy, MD Department of Pediatrics, All India Institute of Medical Sciences, New Delhi, India

Mona Sharma, MD, DNB, MAMS Department of Reproductive Biology, All India Institute of Medical Sciences, New Delhi, India

Shilpa Sharma, MS, MCh, PhD Department of Pediatric surgery, All India Institute of Medical Sciences, New Delhi, India

Skand Shekhar, MD Saint Peter's University Hospital, Rutgers Robert Wood Johnson Medical School, New Jersey, NJ, USA

Manjit Singh, MD Additional Medical Superintendent (Retired), Safdarjung Hospital & Vardhman Mahavir Medical College, New Delhi, India

Shivaprakash Somasundaram, MD, DM Department of Endocrinology and Metabolism, All India Institute of Medical Sciences, New Delhi, India

Part I

Anatomical, Developmental and Genetic Basis of Maleness

The Testis

1

Bodhana Dhole and Anand Kumar

Learning Objectives
- Internal structure of testis
- Development and regulation of Sertoli Cells
- Blood-testis barrier
- Description of the peritubular myoid cells
- Development and regulation of the Leydig Cells
- Description of testicular macrophages

1.1 Introduction

Testes are the principal male reproductive organs which are responsible for production of sperm and hormones required for development and maintenance of male sexual characteristics. Testes are paired ovoid extra-abdominal organs. The testis weighs 10.5–15 g and has an approximate length of 4.5 cm and 2.5 cm in breadth. Testicular volume in a normal adult human varies between 12 and 17 ml (Fattini et al. 1998). The gross anatomy of testis includes an outer tough fibrous capsule and two morphologically and functionally distinct compartments, namely, (1) the tubular compartment and (2) the interstitial compartment (see Fig. 1.1c).

B. Dhole, PhD • A. Kumar, MD, FAMS (✉)
Department of Reproductive Biology, All India Institute of Medical Sciences, New Delhi, India
e-mail: anandkumarrepbiol@hotmail.com

A. Kumar, M. Sharma (eds.), *Basics of Human Andrology*,
DOI 10.1007/978-981-10-3695-8_1

1.2 Testicular Capsule

The testicular capsule is composed of three layers-tunica vaginalis, tunica albuginea and tunica vasculosa from outside to inside. The tunica vaginalis is formed prior to the descent of foetal testes into the scrotum from abdomen. Tunica vaginalis is a thin layer composed of mesothelial cells. The tunica vaginalis layer consists of a visceral and a parietal lamina. The space between the visceral and parietal lamina forms the cavity of tunica vaginalis. Tunica albuginea is a dense, bluish white covering and is composed of fibroblasts and collagen fibres. The external part of the tunica albuginea is covered by the visceral layer of tunica vaginalis (see Fig. 1.1a). Tunica vasculosa consists of blood vessels and loose connective tissue. Tunica vasculosa layer lies in close approximation to the underlying seminiferous tubules and interstitial tissue (Glass and Mundy 2005).

1.3 The Tubular Compartment

The testis is divided by septa into 250–300 lobules. The lobules differ in size and are largest and longest in the centre. Each lobule is composed of one to four seminiferous tubules. Each seminiferous tubules is a coiled tube with average dimension being 70–80 cm in length and 0.12–0.3 mm in diameter (Glass and Mundy 2005). Each seminiferous tubules towards the end forms straight tubule which becomes continuous with the rete testis. Rete testis is a network of interconnecting tubes in the testicular hilum or mediastinum. The rete testis is lined by squamous or cuboidal epithelial cells that are nonsecretory (Walker 2003). Rete testis carries sperm from seminiferous tubules to vasa efferentia or deferens. During transit through the rete testis, most of the seminal fluid is reabsorbed, and sperm become concentrated.

Seminiferous tubule consists of a basement membrane surrounded by elastic fibres and myoid cells. It is lined by a layer of seminiferous epithelium, also known as germinal epithelium. Seminiferous tubules are populated by germ cells in close association with the Sertoli cells (see Fig. 1.1b; Kerr et al. 2006). In this chapter, we will describe the Sertoli cells and peritubular myoid cells surrounding the seminiferous tubules. The details of germ cell and their relationship with Sertoli cells are described in the chapter, Spermatogenesis.

1.3.1 The Sertoli Cells

Sertoli cells were first described by an Italian physiologist Enrico Sertoli in 1865. Sertoli cells are 'nursing' cells with their base attached to the basement membrane of the seminiferous tubule and their apex extending towards the lumen. Sertoli cells are derived from the coelomic epithelium of the genital ridge (Karl and Capel 1998). Sertoli cells express the SRY (sex determining region Y) gene and produce

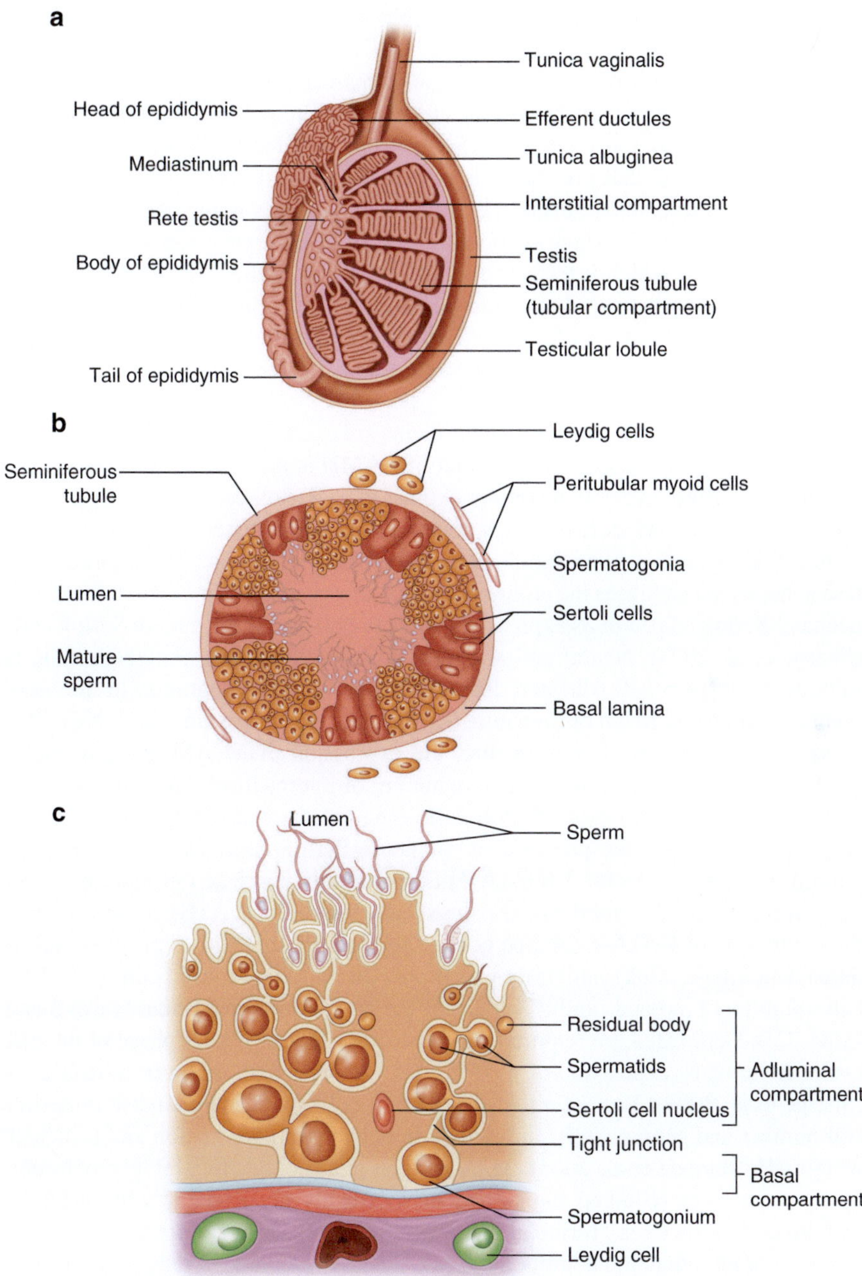

Fig. 1.1 Schematic diagram showing detailed structure of the testis. (**a**) Longitudinal section of testis showing internal structure. (**b**) Cross section of a seminiferous tubule. (**c**) Schematic diagram showing different compartments in the seminiferous tubule

anti-Mullerian hormone (AMH). AMH inhibits the development of the Müllerian or paramesonephric duct in the male embryo.

Two distinct types of Sertoli cells, the immature and the mature type, differ extensively in their morphological characteristics and biochemical activity (Sharpe et al. 2003). Only immature Sertoli cells proliferate. In primates, including humans, Sertoli cell proliferation occurs over two distinct periods. The first proliferative period starts during foetal life and continues in the neonatal life and then stops. The second stage of immature Sertoli cell expansion is in the peri-pubertal life after which it terminally differentiates into the mature stage. In rats, as the neonatal and peri-pubertal life overlaps, therefore it is difficult to identify two distinct periods of proliferation. The mature Sertoli cells lose their ability to divide. These mature Sertoli cells have a large nucleus with a prominent nucleolus (Sharpe et al. 2003).

In Sertoli cells, different genes and proteins are expressed maximally during different development stages. During foetal life, AMH is maximally expressed in the Sertoli cells. High levels of AMH are expressed in Sertoli cells from foetal life to prepubertal stage and decrease as spermatogenesis progresses. In mature Sertoli cells, AMH expression is negligible or absent (Arambepola et al. 1998). The aromatase gene, which catalyses the aromatization of androgens, is expressed in foetal and neonatal Sertoli cells, but its expression is downregulated in the mature Sertoli cells (Sharpe et al. 2003). Neural cell adhesion molecule (NCAM), a glycoprotein, is important for germ cell adhesion during their intra-tubular migration to the basement membrane. In foetal or immature Sertoli cells, maximum amount of NCAM is expressed. When Sertoli cells mature, the expression of NCAM is significantly decreased. In humans, cytokeratin 18, a marker of intermediate filaments, is exclusively expressed in immature Sertoli cells. The presence of cytokeratin 18 in the adult testes indicates the presence of immature Sertoli cells. On the other hand, Globin transcription factor 1 (GATA-1) is exclusively expressed in mature Sertoli cells in mice. GATA-1 inhibits the expression of AMH, and AMH in turn inhibits the expression of GATA-1 (Sharpe et al. 2003). $p27^{Kip1}$ and $p21^{Cip1}$ are two cyclin-dependent kinase (Cdk) inhibitors which play an important role in cell cycle. The Cdk inhibitors bind and inhibit activity of the cyclin-dependent kinases Cdk2 and Cdk4. Cdk2 and Cdk4 are required for cells to pass from G1 to S phase of the cell cycle. In humans, expression of $p27^{Kip1}$ coincides with maturation of Sertoli cells (Sharpe et al. 2003). Loss of either $p27^{Kip1}$ or $p21^{Cip1}$ causes an increase in Sertoli cell number and subsequently an increase in daily sperm production and testicular weight (Holsberger et al. 2003). M2A antigen is a monomeric sialoglycoprotein. The protein lacks N-linked sugars and is extensively glycosylated with O-linked carbohydrate structures. Terminal sialic acid residues on M2A are important for reactivity with monoclonal antibody to M2A. M2A is expressed only by immature and not mature Sertoli cells in humans (Sharpe et al. 2003). In human testis, androgen receptor expression is first observed in Sertoli cells at the age of 5 months. High levels of androgen receptors are expressed from the age of 8 years through adulthood (Chemes et al. 2008). Table 1.1 summarizes the factors expressed by immature and mature Sertoli cells.

Table 1.1 Proteins expressed by immature and mature Sertoli cells

Protein	Immature Sertoli cell (prior to puberty)	Mature Sertoli cell (from puberty onwards)
AMH	(+)	(−)
Aromatase	(+)	(−)
NCAM	(+)	(−)
Cytokeratin 18	(+)	(−)
M2A	(+)	(−)
GATA-1	(−)	(+)
$p27^{Kip1}$ and $p21^{Cip1}$	(−)	(+)
Androgen receptor	(−)	(+)

1.3.1.1 Regulation of Sertoli Cells

Follicle stimulating hormone (FSH) increases the rate of proliferation of Sertoli cells. An increase in FSH receptors on Sertoli cell increases their proliferation (Walker and Cheng 2005). However, FSH is not responsible for cessation of Sertoli cell proliferation and initiation of differentiation. The 3, 5, 3′-L-triiodothyronine (T_3), testosterone and retinoic acid inhibit the proliferation of immature Sertoli cells of rat testis (Buzzard et al. 2003). Hypothyroidism delays the differentiation of Sertoli cells, resulting in a significant increase in adult testis size and sperm production. Hyperthyroidism causes early cessation of Sertoli cell proliferation, decrease in testis size and sperm production (van Haaster et al. 1993). T_3 acts by binding to thyroid hormone receptors (TR), $TR\alpha_1$ and $TR\alpha_2$, present on the Sertoli cells (Buzzard et al. 2000; Jannini et al. 1998). T_3, testosterone and retinoic acid increase the expression of $p27^{Kip1}$ and $p21^{Cip1}$ proteins. Accumulation of these inhibitory proteins is responsible for cessation of Sertoli cell proliferation (Holsberger et al. 2003). T_3 also regulates certain Sertoli cell proteins associated with its maturational status. T_3 increases Sertoli cell androgen receptor-mRNA in Sertoli cell cultures and decreases AMH-mRNA in Sertoli cells (Arambepola et al. 1998). T_3 downregulates NCAM production in Sertoli cell-germ cell cocultures (Laslett et al. 2000). In mouse Sertoli cell line, TM4 and T_3 decrease transcription of the aromatase gene (Catalano et al. 2003).

1.3.2 The Blood-Testis Barrier

In the mammalian testis, the blood-testis barrier (BTB) is composed of specialized junctions between adjacent Sertoli cells near the basement membrane of seminiferous epithelium (Cheng and Mruk 2012). BTB divides the germinal epithelium into basal and adluminal compartments (see Fig. 1.1c). The basal compartment is populated by the spermatogonial stem cells and primary spermatocytes. The adluminal compartment is closer to the tubular lumen and is populated by the secondary spermatocytes, spermatids and mature sperm. The unique functional compartmentalization of BTB creates a special micro-environment where germ cell meiosis and further development occur (Mruk and Cheng 2004).

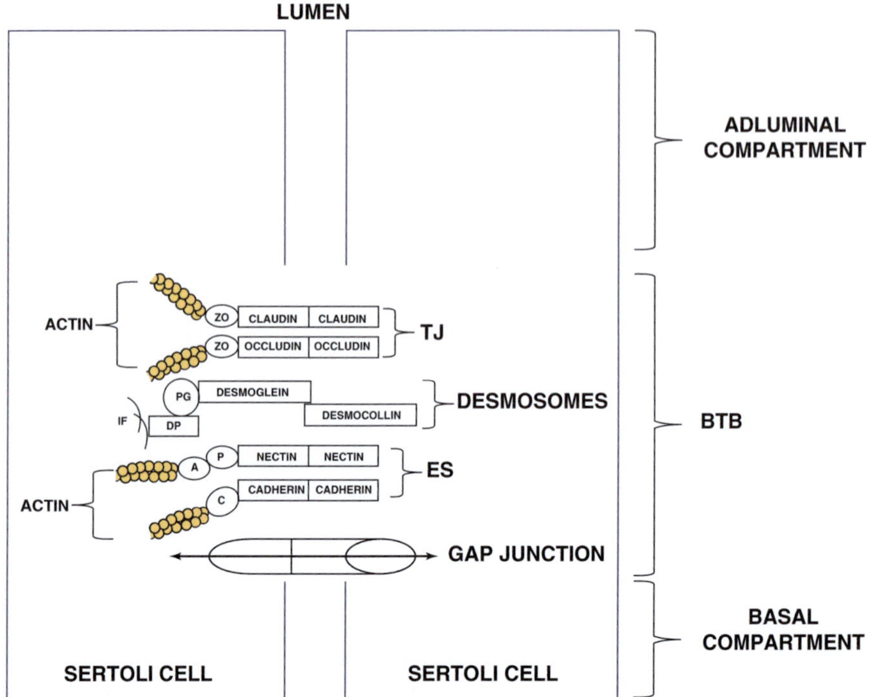

Fig. 1.2 Schematic representation of BTB between adjacent Sertoli cells. *ZO* zonula occludens, *PG* plakoglobulin, *DP* desmoplakin, *IF* intermediate filaments, *P* ponsin, *A* afadin, *C* catenin, *TJ* tight junctions, *ES* ectoplasmic specialization

BTB is formed by tight junctions (TJs) between two adjacent Sertoli cells. TJs coexist with basal ectoplasmic specialization (ES), gap junctions and desmosome-like junctions (Su et al. 2011). See Fig. 1.2 for the structural organization of BTB.

TJs are points of contacts between plasma membrane of adjacent Sertoli cells. There is no intracellular space between the two cells at TJs. The most abundant TJ proteins are claudins, occludins, coxsackievirus and adenovirus receptor (CAR)/junctional adhesion molecule (JAM) protein complexes. All of the junctional proteins are transmembrane proteins (Mruk and Cheng 2010). The TJ proteins also bind to scaffolding proteins such as zonula occludens (ZO), multi-PDZ domain protein 1 (MUPP1), membrane-associated guanylate kinase (MAGUK) and cingulin on the cytoplasmic side of plasma membrane. The C-terminal of ZO-1 and ZO-3 bind to actin, thereby connecting TJ proteins with actin (Hartsock and Nelson 2008).

Claudins have four membrane spanning domains and two extracellular domains. Androgens regulate the levels of claudins in TJs (Hartsock and Nelson 2008). In azoospermic genetically engineered mice, with reduced levels of androgen receptors on Sertoli cells, there is a marked decrease in claudin 3-mRNA levels. The gene encoding claudin 3 has an androgen response element in its promoter region suggesting it to be a direct transcriptional target of androgen receptors. Loss of androgen receptors increases the permeability of BTB by disrupting the TJs between adjacent Sertoli cells (Meng et al. 2005). Claudin levels are also regulated by cytokines such as transforming growth factor beta 3 (TGF-β3) and tumor necrosis factor alpha (TNF-α). Both these cytokines suppress claudin levels in the testis (Lui and Cheng 2007).

Knockout of occludin gene in mice does not affect barrier function. The exact function of occludins in the TJs is not clearly known. Occludins are highly phosphorylated proteins which are deactivated by dephosphorylation of their specific serine and threonine residues (Mruk and Cheng 2010). Both TGF-β3 and TNF-α cytokines downregulate the expression of occludins in the TJs (Lui and Cheng 2007).

JAMs are immunoglobulin-like proteins capable of homophilic and heterophilic interactions. JAMs play an important role in adhering cells to the basement membrane by binding to integrins. Integrins are transmembrane receptors which bind to proteins on extracellular matrix (ECM). Knockout of JAM-A gene in mice disrupts the interaction between the Sertoli cell and ECM (Mruk and Cheng 2010).

CAR, a membrane receptor, is an important component of TJs. CAR mediate homophilic interactions and heterophilic interactions with JAM. Like JAMs, CAR also binds and activates the integrin-mediated signalling pathway (Mruk and Cheng 2010).

ES is a surface modification of Sertoli cell at specific sites of cell-cell junctions. ES formed between adjacent Sertoli cells in the region of BTB are referred to as basal ES. Two multi-protein transmembrane complexes, afadin/ponsin/nectin complex, and catenin/p120/cadherin complex anchor onto the cytosolic actin filaments. The multi-protein complexes project into the intercellular space (ICS) where they join similar complexes from the plasma membrane of the opposing Sertoli cells (Lee and Cheng 2004).

Nectins are integral membrane proteins belonging to immunoglobulin-like protein superfamily. Nectins bind to afadin protein at their cytoplasmic domain. The binding of nectin to afadin is mediated by ponsin, an adaptor protein. Afadin are actin-binding proteins which link the nectin proteins to actin filaments via their C-terminal domain (Lee and Cheng 2004).

Cadherins are calcium-dependent transmembrane proteins abundantly found in ES. The cadherin proteins are expressed by Sertoli cells (Lee and Cheng 2005). Cadherin levels in ES are regulated by both testosterone and cytokines. Sertoli cell cultures treated with TGF-β2 increase internalization of cadherins in clathrin-coated vesicles. Similarly, testosterone enhances rate of cadherin endocytosis, it also stimulates its recycling back to the Sertoli cell surface (Kopera et al. 2010). Cadherins bind to α, β and γ catenins and p120 proteins. β and γ catenins bind to cadherins at their conserved intracellular cytoplasmic domain. β and γ catenins connect to actin filaments via actin-binding α catenin protein thereby linking cadherins to the underlying cytoplasmic actin filaments. p120 binds to cadherins at sites which are different from β and γ catenin binding site (Lee and Cheng 2005).

Desmosomes are intercellular adhesive junctions between adjacent Sertoli cells in the region of BTB (Kopera et al. 2010). Structurally, desmosomes are divided into three distinct regions, namely, (1) extracellular core region or desmoglea, (2) outer dense plaque (ODP) and (3) inner dense plaque (IDP). Extracellular core region separates the two connected cells by a distance of 34 nm. The core region is primarily composed of calcium-dependent transmembrane glycoproteins, desmogleins and desmocollins belonging to cadherin superfamily. Both the proteins have five extracellular domains, a single transmembrane domain and a cytoplasmic anchor on the cytoplasmic side of plasma membrane (Delva et al. 2009). The ODP comprises of the cytoplasmic anchor of desmosomal cadherins attached to armadillo family proteins. Armadillo proteins, including plakoglobin and plakophilins, have distinct central domain and C- and N-terminal domains and bind directly to the cytoplasmic anchor of the desmosomal cadherins. Finally, the IDP consists of plakin family of linker proteins interacting with keratin intermediate filaments. Desmoplakin, a plakin family protein, has globular N- and C-terminals connected by a central, α-helical domain. The N-terminal head region of desmoplakin binds to armadillo proteins and serves as a key linker between intermediate filaments and desmosomal proteins (Delva et al. 2009). When desmoglein-2 and desmocollin-2 are simultaneously knocked down using interfering RNA, it results in a decrease in ZO-1 levels by 30%, decrease in their internalization and dislocalization from the cytoplasmic side of plasma membrane to the cytosol and also the dislocalization of CAR proteins and cSrc, a protein kinase from Sertoli cell surface to the cytosol. Thus, knockdown of both desmosomal cadherins disrupts barrier function of TJ proteins (Lie et al. 2010).

Gap junctions are cluster of channels formed between two adjacent Sertoli cells. Gap junctions are composed of proteins belonging to the connexin superfamily. Each connexin protein has four transmembrane domains which are connected by extracellular loops. Each loop has three conserved cysteine residues that form the intra-connexin disulphide bond. Each connexin protein has an N- and C-terminal cytoplasmic domain and a cytoplasmic loop connecting the second and third transmembrane domains (Mese et al. 2006).

The primary function of the BTB is to restrict the passage of nutrients and steroids across the Sertoli cell into the adluminal compartment. Therefore, the maturing sperm develop in a unique immunologically privileged region. Specific transporters located along the basolateral and apical membranes of the Sertoli cells regulate the movement of molecules in or out of the lumen. Many ions, organic solutes and peptides are transported into and out of the testis via specific transporters. Each transporter carries specific molecules in and out of the lumen (Mital et al. 2011).

The BTB acts as an immunological barrier as it conceals germ cell antigens and prevents entry of leukocytes and antibodies into the adluminal compartment. Although most of the auto-antigenic germ cells are sequestered behind the BTB, preleptotene spermatocytes and spermatogonia, located within the basal compartment, express antigens that can evoke an immune response (Mital et al. 2011).

1.3.3 Peritubular Myoid Cells

In all mammals, peritubular myoid cells surround the testicular seminiferous tubules (see Fig. 1.1c). In humans, myoid cells organize to form multiple layers. Peritubular myoid cells contain large amount of actin filaments. The arrangement of these actin filaments changes as the testes develop. The myoid cells also contain other cytoskeletal proteins, such as myosin, desmin or vimentin and α-actin (Maekawa et al. 1996). The contractile movement of the myoid cells help in the transport of sperm and testicular fluid into the seminiferous tubule. Myoid cells secrete a number of extracellular matrix components such as type I and IV collagens, fibronectin, proteoglycans and growth factors such as TGFβ, insulin-like growth factor-1 (IGF-I) and activin-A (Maekawa et al. 1996). Myoid cells also express androgen receptors. If the androgen receptors are specifically knocked out from the mice peritubular myoid cells, it results in reduced testicular weight, azoospermia and infertility (Welsh et al. 2009).

1.4 The Interstitial Compartment

The interstitial compartment occupies one-third of the total testicular volume and contains the androgen-producing Leydig cells, blood vessels, lymphatics and nerves. In addition, collagen, elastic fibres and a large number of macrophages are found in the interstitial space.

1.4.1 The Leydig Cells

Leydig cells are the principal steroidogenic hormone-producing cells of the testis. During embryonic life, Leydig cells develop from mesenchymal stem cells and produce androgens required for foetal masculinization. Mammalian Leydig cells are classified into two distinct groups- foetal Leydig cells (FLCs) and adult Leydig cells (ALCs). Though both FLCs and ALCs are derived from the mesenchymal stem cells in the testis interstitium, each has different morphology, hormonal environment and physiological functions. Androgens produced by FLCs are primarily required for normal masculine differentiation, whereas in adults, their function is maintenance of male sexual characteristics (Huhtaniemi and Pelliniemi 1992).

FLCs are only present during the foetal period and undergo apoptosis after birth. Post-birth, ALCs populate the interstitial compartment of testis. The precursors of the FLCs first appear during the 10th week of gestation in humans (Huhtaniemi and Pelliniemi 1992). The initial development and differentiation of FLCs are independent of gonadotropins because pituitary luteinizing hormone (LH) and testicular LH receptors develop much later in foetal life (Migrenne et al. 2001). First, the nucleus of the FLCs starts to grow in size followed by accumulation of cytoplasmic organelles such as mitochondria, endoplasmic reticulum and lipid droplets. Initially, FLCs are elongated but later become polygonal having a diameter of about 20 μm. By the 14th week of gestation, FLCs fill up the majority of the interstitial space. The relative number of FLCs starts to decrease around the age of 16 weeks because of the fast

Table 1.2 Features of different stages of adult Leydig cells (ALCs)

Characteristic features	Progenitor Leydig cells	Newly formed ALCs	Immature ALCs	Mature ALCs
LH receptors	(+)	(+)	(+)	(+)
Steroidogenic enzymes-CYP11A1, CYP17A1 and 3β-HSD	(+)	(+)	(+)	(+)
5α-reductase	(−)	(−)	(+)	(+) [low]

growth of the testis, even though their absolute number remains constant until the 24th week. Steroidogenesis in FLCs starts at around the foetal age of 8 weeks and peaks during weeks 12–14 (Huhtaniemi and Pelliniemi 1992).

Postnatally, mesenchymal stem cells first differentiate into progenitor cells. The differentiation of mesenchymal cells into progenitor cells does not require LH. Both the mesenchymal stem cells and progenitor cells are spindle shaped. Progenitor Leydig cells, unlike mesenchymal cells, express LH receptors and steroidogenic enzymes such as CYP11A1, CYP17A1 and 3-beta hydroxysteroid dehydrogenase (3β-HSD). Spindle-shaped progenitor cells transform into polygonal-shaped ALCs. The newly formed ALCs are smaller than the immature and mature adult Leydig cells (Mendis-Handagama and Ariyaratne 2001). Cytoplasmic organelles in the newly formed ALCs are sparse, and the nucleus appears large and prominent. The newly formed ALCs contain little or no cytoplasmic lipid droplets in them. These ALCs increase in size to become the immature ALCs. The cytoplasm of ALCs is rich in lipid droplets and testosterone-metabolizing enzymes such as 5α-reductase. Therefore, the immature ALCs produce mainly 5α-reduced androgens instead of testosterone. Finally, the immature ALCs differentiate into mature ALCs. Mature ALCs are larger than immature ALCs and have fewer number of cytoplasmic lipid droplets. The mature ALCs secrete maximum amount of testosterone. The mature ALCs have large number of LH receptors on their plasma membrane and are highly responsive to circulating LH. Loss of testosterone-metabolizing enzymes occurs during this time which contributes to the increased testosterone-secreting capacity of the mature ALCs (Mendis-Handagama and Ariyaratne 2001). See Table 1.2 for characteristic features of different stages of Leydig cells.

1.4.1.1 Regulation of Leydig Cell Functions

Leydig cell functions are classically known to be regulated by LH. In recent years, physiologically low levels of oxygen concentration and thyroid hormones have emerged as important regulators of Leydig cell development and function.

Though, not enough information is available on the role of physiological hypoxia on testicular functions, but it has been suggested that physiological hypoxia limits the action of LH on Leydig cells. In the tumor Leydig cell-derived MA-10 cells, cobalt chloride ($CoCl_2$)-induced hypoxia decreases the basal human chorionic gonadotropin hormone (hCG), and 8- Br-cAMP, a 3′,5′-cyclic adenosine monophosphate analogue, stimulated progesterone production. However, both induced hypoxia and 8-Br-cAMP stimulate vascular endothelial growth factor production (Kumar et al. 2012). The decrease in progesterone production is possibly due to the decrease in mRNA expression of CYP11A1 or increase in reactive oxygen species (ROS; Kumar et al. 2014). In seminiferous tubules, physiological hypoxia regulates spermatogenesis and sperm production.

Thyroid hormones regulate Leydig cell development and steroidogenesis. T_3 enhances the proliferation and accelerates differentiation of mesenchymal stem cells into Leydig cells (Ariyaratne et al. 2000). T_3 stimulates steroid production in Leydig cells and their derived cell lines (Manna et al. 1999; Maran et al. 2000).

1.4.2 Testicular Macrophages

Testicular macrophages possibly arise from circulating blood monocytes (Svingen and Koopman 2013). Testicular macrophages appear in the testicular interstitial compartment close to the time of birth. The number of testicular macrophages increases drastically when ALCs appear and germ cell meiosis starts (Frungieri et al. 2002). Macrophages constitute 25% of the interstitial cells. Testicular macrophages reside exclusively in the interstitial compartment. However, when germ cells die, testicular macrophages enter the seminiferous tubule to phagocytose the dead germ cells. Testicular macrophages have indented nucleus, numerous lysosomes, polymorphic residual bodies, cytoplasmic vesicles and microvilli. Testicular macrophages secrete inflammatory cytokines such as interleukin 1 (IL-1) and TNF-α (Cohen et al. 1999). Testicular macrophages develop specialized cell-cell junctional complexes with the Leydig cells. In the adult testis, macrophages secrete 25-hydroxycholesterol, which enters ALCs and is used in testosterone synthesis (Svingen and Koopman 2013). When Leydig cell population is completely erased from the rat testis, testicular macrophage number also reduces (Wang et al. 1994).

Key Questions

- What are the differences in protein expression between mature and immature Sertoli cells?
- Describe the hormonal regulation of Sertoli cell development.
- Describe the structural components of blood-testis barrier.
- Describe the stages of Leydig cell development.
- What is the role of thyroid hormones in regulating Leydig cell development and function?
- What are the functions of peritubular myoid cells and macrophages?

References

Arambepola NK, Bunick D, Cooke PS. Thyroid hormone and follicle-stimulating hormone regulate Mullerian-inhibiting substance messenger ribonucleic acid expression in cultured neonatal rat Sertoli cells. J Endocrinol. 1998;139:4489–95.

Ariyaratne HBS, Mills N, Mason IJ, Mendis-Handagama SML. Effects of thyroid hormone on Leydig cell regeneration in the adult rat following ethane dimethane sulphonate treatment. Biol Reprod. 2000;63:1115–23.

Buzzard JJ, Morrison JR, O'Bryan MK, Song Q, Wreford NG. Developmental expression of thyroid hormone receptors in the rat testis. Biol Reprod. 2000;62:664–9.

Buzzard JJ, Wreford NG, Morrison JR. Thyroid hormone, retinoic acid, and testosterone suppress proliferation and induce markers of differentiation in cultured rat Sertoli cells. Endocrinology. 2003;144:3722–31.

Catalano S, Pezzi V, Chimento A, Giordano C, Carpino A, Young M, et al. Triiodothyronine decreases the activity of the proximal promoter (PII) of the aromatase gene in the mouse Sertoli cell line, TM4. Mol Endocrinol. 2003;17:923–34.

Chemes HE, Rey RA, Nistal M, Regadera J, Musse M, Gonzalez-Peramato P, et al. Physiological androgen insensitivity of the fetal, neonatal and early infantile testis is explained by the ontogeny of the androgen receptor expression in Sertoli cells. J Clin Endocrinol Metab. 2008;93:4408–12.

Cheng CY, Mruk DD. The blood-testis barrier and its implications for male contraception. Pharmacol Rev. 2012;64:16–64.

Cohen PE, Nishimura K, Zhu L, Pollard JW. Macrophages: important accessory cells for reproductive function. J Leukoc Biol. 1999;66:765–72.

Delva E, Tucker DK, Kowalczyk AP. The desmosome. Cold Spring Harb Perspect Biol. 2009;1(2):a002543.

Fattini CA, Rezende GL, Nogueira Filho JR, Fukumoto WK, Simões MJ, Ambrósio JD. Effect of age on the volume and weight of human testis. Rev Chil Anat. 1998;16:185–90.

Frungieri MB, Calandra RS, Lustig L, Meineke V, Köhn FM, Vogt HJ, et al. Number, distribution pattern, and identification of macrophages in the testes of infertile men. Fertil Steril. 2002;78:298–306.

Glass J, Mundy AR. Testes and epididymis. In: Healy JC, Glass J, Mundy AR, editors. Gray's anatomy. The anatomical basis of clinical practice. 29th ed. Edinburgh: Elsevier Churchill Livingstone; 2005. p. 1305–10.

Hartsock A, Nelson WJ. Adherens and tight junctions: structure, function and connections to the actin cytoskeleton. Biochim Biophys Acta. 2008;1778:660–9.

Holsberger DR, Jirawatnotai S, Kiyokawa H, Cooke PS. Thyroid hormone regulates the cell cycle inhibitor p27Kip1 in postnatal murine Sertoli cells. Endocrinology. 2003;144:3732–8.

Huhtaniemi I, Pelliniemi LJ. Fetal Leydig cells: cellular origin, morphology, life span, and special functional features. Pro Soc Exp Biol Med. 1992;201:125–40.

Jannini EA, Crescenzi A, Rucci N, Screponi E, Carosa E, De Matteis A, et al. Ontogenetic pattern of thyroid hormone receptor expression in the human testis. J Clin Endocrinol Metab. 1998;85:3453–7.

Karl J, Capel B. Sertoli cells of the mouse testis originate from the coelomic epithelium. Dev Biol. 1998;203:323–33.

Kerr JB, Loveland KL, O'Bryan MK, de Kretser DM. Cytology of the testis and intrinsic control mechanism. In: Neil JD, editor. Knobil and Neill's physiology of reproduction. 3rd ed. New York: Elsevier; 2006. p. 827–947.

Kopera IA, Bilinska B, Cheng CY, Mruk DD. Sertoli-germ cell junctions in the testis: a review of recent data. Philos Trans R Soc Lond B Biol Sci. 2010;365(1546):1593–605.

Kumar A, Rani L, Dhole B, Chaturvedi PK. Oxygen as a regulator of MA-10 cell functions: effect of cobalt chloride on VEGF production. Andrologia. 2012;44(Suppl 1):615–20.

Kumar A, Rani L, Dhole B. Role of oxygen in the regulation of Leydig tumour derived MA-10 cell steroid production: the effect of cobalt chloride. Syst Biol Reprod Med. 2014;60:112–8.

Laslett AL, Li LH, Jester Jr WF, Orth JM. Thyroid hormone downregulates neural cell adhesion molecule expression and affects attachment of gonocytes in Sertoli cell-gonocyte cocultures. Endocrinology. 2000;141:1633–41.

Lee NP, Cheng CY. Ectoplasmic specialization, a testis-specific cell-cell actin-based adherens junction type: is this a potential target for male contraceptive development? Hum Reprod Update. 2004;10:349–69.

Lee NP, Cheng CY. Protein kinases and adherens junction dynamics in the seminiferous epithelium of the rat testis. J Cell Physiol. 2005;202:344–60.

Lie PPY, Cheng CY, Mruk DD. Crosstalk between desmoglein-2/desmocollin-2/Src kinase and coxsackie and adenovirus receptor/ZO-1 protein complexes regulates blood-testis barrier dynamics. Int J Biochem Cell Biol. 2010;42:975–86.

Lui W, Cheng CY. Regulation of cell junction dynamics by cytokines in the testis – a molecular and biochemical perspective. Cytokine Growth Factor Rev. 2007;18:299–311.

Maekawa M, Kamimura K, Nagano T. Peritubular myoid cells in the testis: their structure and function. Arch Histol Cytol. 1996;59:1–13.

Manna PR, Tena-Sempere M, Huhtaniemi IT. Molecular mechanisms of thyroid hormone-stimulated steroidogenesis in mouse Leydig tumor cells. J Biol Chem. 1999;274:5909–18.

Maran RR, Arunakaran J, Aruldhas MM. T_3 Directly stimulates basal and modulates LH induced testosterone and oestradiol production by rat Leydig cells in vitro. Endocr J. 2000;47:417–28.

Mendis-Handagama SMLC, Ariyaratne HBS. Differentiation of the adult Leydig cell population in the postnatal testis. Biol Reprod. 2001;65:660–71.

Meng J, Holdcraft RW, Shima JE, Griswold MD, Braun RE. Androgens regulate the permeability of the blood-testis barrier. Proc Natl Acad Sci USA. 2005;102:16696–700.

Meşe G, Richard G, White TW. Gap junctions: basic structure and function. J Invest Dermatol. 2006;127:2516–24.

Migrenne S, Pairault C, Racine C, Livera G, Géloso A, Habert R. Luteinizing hormone-dependent activity and luteinizing hormone-independent differentiation of rat fetal Leydig cells. Mol Cell Endocrinol. 2001;172:193–202.

Mital P, Hinton BT, Dufour JM. The blood-testis and blood-epididymis barriers are more than just their tight junctions. Biol Reprod. 2011;84:851–8.

Mruk DD, Cheng CY. Sertoli-Sertoli and Sertoli-Germ cell interactions and their significance in germ cell movement in the seminiferous epithelium during spermatogenesis. Endocr Rev. 2004;25:747–806.

Mruk DD, Cheng CY. Tight junctions in the testis: new perspectives. Philos Trans R Soc Lond B Biol Sci. 2010;365:1621–35.

Sharpe RM, McKinnell C, Kivlin C, Fisher JS. Proliferation and functional maturation of Sertoli cells, and their relevance to disorders of testis function in childhood. Reprod. 2003;125:769–84.

Su L, Mruk DD, Lee WM, Cheng CY. Drug transporters and blood – testis barrier function. J Endocrinol. 2011;209:337–51.

Svingen T, Koopman P. Building the mammalian testis: origins, differentiation, and assembly of the component cell populations. Genes Dev. 2013;27:2409–26.

van Haaster LH, de Jong FH, Docter R, de Rooij DG. High neonatal triiodothyronine levels reduce the period of Sertoli cell proliferation and accelerate tubular lumen formation in the rat testis, and increase serum inhibin levels. Endocrinology. 1993;133:755–60.

Walker WH, Cheng J. FSH and testosterone signalling in Sertoli cells. Reprod. 2005;130:1470–626.

Walker WH. Molecular mechanisms controlling Sertoli cell proliferation and differentiation. Endocrinology. 2003;144:3719–21.

Wang J, Wreford NGM, Lan HY, Atkins R, Hedger MP. Leukocyte population of the adult rat testis following removal of Leydig cells by treatment with ethane dimethane sulfonate and subcutaneous testosterone implants. Biol Reprod. 1994;51:551–61.

Welsh M, Saunders PTK, Atanassova N, Sharpe RM, Smith LB. Androgen action via testicular peritubular myoid cells is essential for male fertility. FASEB J. 2009;23:4218–30.

The Prostate Gland

2

Mona Sharma, Surabhi Gupta, Bodhana Dhole, and Anand Kumar

Learning Objectives

- Development and gross anatomical features
- Lobes and zones
- Microstructure and functional regulation
- Components of prostatic fluid
- Disorders of prostate gland

2.1 Introduction

The name 'prostate' is originally derived from a Greek word 'prohistani' (means 'to stand in front of'). The prostate is the largest accessory sex gland of males. It is a musculo-glandular, exocrine gland that secretes alkaline fluid which constitutes about 20–30 % volume of the seminal fluid. Changes in the prostatic fluid composition and/or secretion affect sperm functions and may lead to male infertility. The gland is often associated with disorders of elderly, benign prostatic hyperplasia (BPH) and carcinoma.

M. Sharma, MD, DNB, MAMS (✉) • S. Gupta, PhD • B. Dhole, PhD
A. Kumar, MD, FAMS
Department of Reproductive Biology, All India Institute of Medical Sciences, New Delhi, India
e-mail: dr.mona18sharma@gmail.com

A. Kumar, M. Sharma (eds.), *Basics of Human Andrology*,
DOI 10.1007/978-981-10-3695-8_2

2.2 Development

The ventral division of cloaca which is the terminal part of hindgut, forms the urogenital sinus. During ninth–tenth week of development, the mesenchyme surrounding the urogenital sinus interact with endoderm of proximal part of urogenital sinus which later forms the proximal part of urethra. As a result of these interactions, the initial outgrowths arise from the lateral aspect of the endodermal tube. The outgrowths form the outer glandular zone of prostate. The subsequent outgrowths arise from its dorsal wall which forms the internal glandular zone. The outgrowths develop into five distinct groups of epithelial buds by the end of the 11th week and are completed by the 16th week. According to the classification given by Lowsley, five groups of epithelial buds give rise to five lobes, namely, the median, right and left lateral and posterior and anterior lobes (Lowsley 1912). These lobes of prostate gland are recognized till the 20th week of gestation. With an advance in gestational age, only three lobes are recognizable–two lateral lobes and a median lobe (Standring 2005). The epithelial buds branch and rebranch ending into complex ductal system that meets the differentiating mesenchymal cells (see Fig. 2.1a–e). The mesenchymal cells develop around the tubules by the 16th week and become denser at the periphery to form the prostatic capsule (Grayhack and Kozlowski 1996).

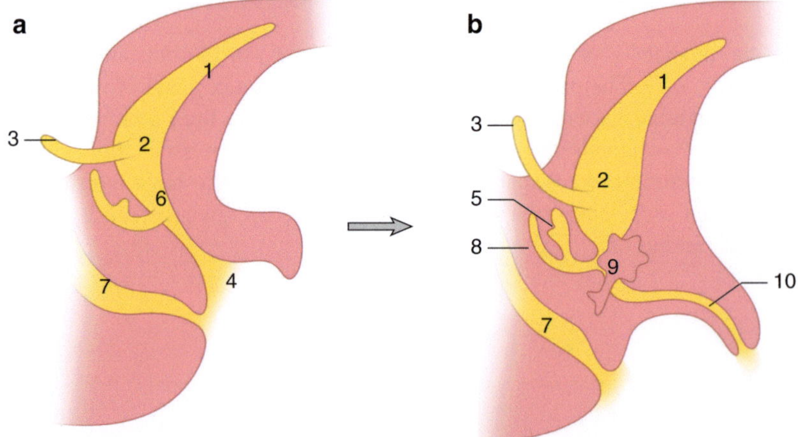

Fig. 2.1 Development of prostate gland (**a**–**e**): the endodermal outgrowths from the prostatic urethra into the surrounding mesenchyme form the gland primordium which further proliferate and enlarges. *1* allantois, *2* urinary bladder, *3* ureter, *4* definitive urogenital sinus, *5* seminal vesicle, *6* pelvic part of urogenital sinus, *7* anorectal canal, *8* ductus deferens, *9* prostate, *10* penile urethra, *11* endodermal tube, *12* glandular outgrowths, (**f**) prostatic glands, (**g**) enlarged view of tubuloalveolar gland

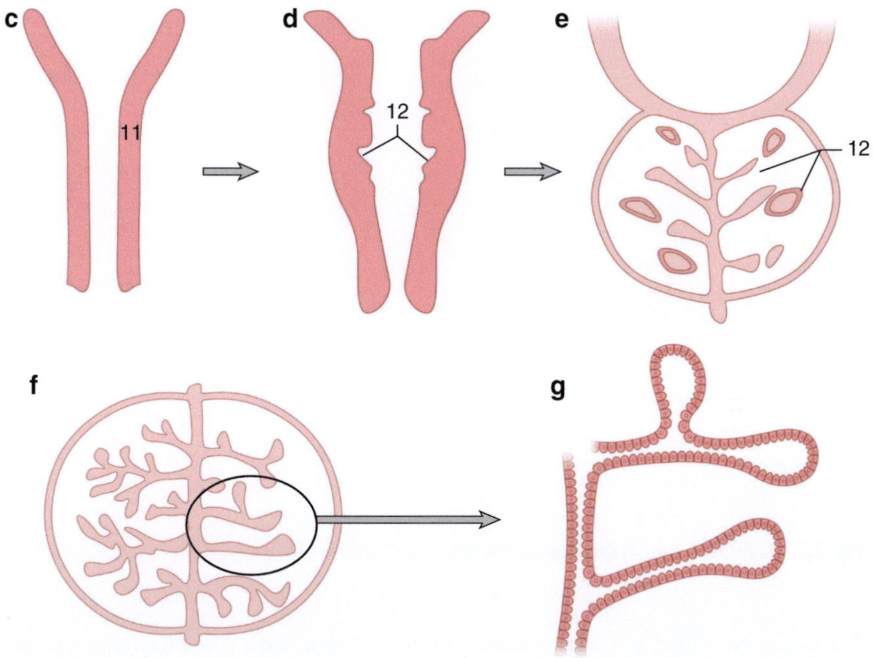

Fig. 2.1 (continued)

2.3 Gross Anatomical Features

The normal adult gland is about the size of a walnut and resembles a cone measuring approximately 4 cm (transverse) × 3 cm (vertical) and 2 cm (anteroposterior) and weighs approximately 20 grams in adults. The gland comprises of glandular and stromal elements, tightly fused within a capsule (Grayhack and Kozlowski 1996; Lee et al. 2011).

The prostate is positioned subperitoneally below the pelvic diaphragm. It lies posterior to the symphysis pubis, anterior to the rectum and inferior to the urinary bladder surrounding its neck (Lee et al. 2011). Being pyramidal in shape, it has a base superiorly, a neck inferiorly along with anterior, posterior and two inferolateral surfaces. The anterior surface lies behind the pubic arch (see Fig. 2.2). The urethra pierces the prostate near the middle of the base and exits on the anterior surface above and in front of its apical portion. The inferolateral surfaces are related to the lateral pelvic wall. The posterior surface of prostate and seminal vesicles is separated from the rectum by a thin layer of connective tissue called 'Denonvilliers' fascia' (Hammerich et al. 2009). This fascia forms a surgical plane of excision for rectal cancers. The rectum along with the fascia is separated anteriorly from seminal vesicles in males and vagina in females (Decker and du Plessis 1986).

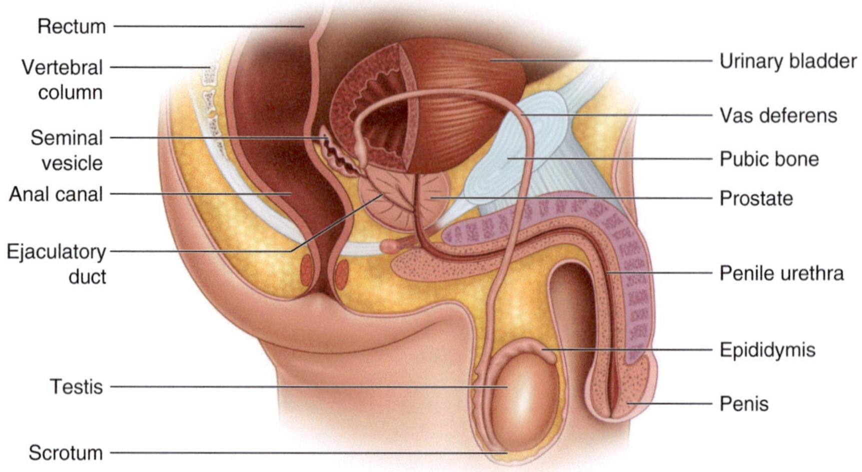

Fig. 2.2 Sagittal section of the male pelvis showing relations of the prostate

The ejaculatory duct traverses the base on its posterior surface and terminates adjacent to the seminal colliculus, also known as verumontanum (Grayhack and Kozlowski 1996).

2.4 Anatomical Lobes and Zones

The anterior lobe lies in front of prostatic urethra and continuous with lateral lobes on either side. The median lobe is situated behind the urethra and in front of ejaculatory ducts. The posterior lobe lies behind the median lobe and below ejaculatory ducts. Another simplified and clinically important way of describing prostatic morphology has been proposed. There are four prostatic zones, namely, the central zone (CZ), the transitional zone (TZ), the peripheral zone (PZ), and the periurethral gland or anterior fibromuscular stroma (AFMS) (McNeal 1981).

The cone-shaped CZ occupies approximately 25 % of the glandular prostate. CZ surrounds ejaculatory ducts and extends from the neck of the bladder to verumontanum. This zone incorporates median and posterior lobes. A major proteolytic enzyme of the seminal fluid, pepsinogen II, is produced by the cells of the central zone.

TZ represents only 5 % of the gland. TZ surrounds the distal part of preprostatic urethra near the apex of central zone and ejaculatory ducts. It incorporates part of anterior and median lobes. It is the exclusive site of BPH and, less commonly, adenocarcinoma.

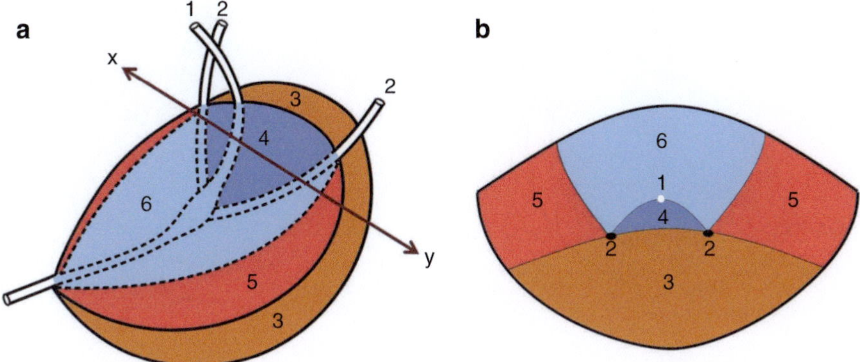

Fig. 2.3 Schematic diagram showing lobes of prostate gland in superolateral view (**a**) and transverse section (**b**) across plane *x–y*. *1* urethra, *2* ejaculatory ducts, *3* posterior lobe, *4* median lobe, *5* lateral lobes, *6* anterior lobe

PZ constitutes 70 % of the glandular prostate. This zone encloses the central and transition zones and preprostatic urethra except anteriorly. It extends distally to enclose lower part of prostatic urethra below verumontanum. The PZ contains the lateral lobes and the remaining part of the posterior lobe (De Krester et al. 1982). In this zone, carcinoma, chronic prostatitis and postinflammatory atrophy are relatively more common. With age, CZ undergoes atrophic changes and TZ enlarges due to BPH which compresses PZ (Standring 2005).

AFMS comprises less than 1 % of the glandular prostate; it fills the space between PZs anterior to preprostatic urethra and incorporates the major part of the anterior lobe. It is predominantly fibromuscular with little or no glandular structures (Grayhack and Kozlowski 1996; Lee et al. 2011; see Figs. 2.3 and 2.4).

2.5 Microstructure

The human prostate is composed of 30–50 tubuloalveolar glands and fibromuscular stroma. It is enclosed by a strong fibrous capsule which sends fibromuscular septa within the gland thereby dividing the gland into smaller regions or lobules. The glands are elongated with sac-like ends and have irregular size (see Fig 2.1g). The glands are arranged in inner mucosal layer, middle submucosal layer and outermost layer which have the main prostatic glands. The glands in the mucosal layer open directly into the urethra, whereas glands of outer two layers open into the prostatic sinuses on the posterior wall of urethra on either side of urethral crest. Each prostatic gland opens into the urethra through a prostatic duct. The characteristic feature seen in the lumen of these glands is concretions formed by aggregations of dead epithelial cells and secretory precipitations. The fibromuscular stroma consists of smooth muscle cells, contractions of which help in forcing the secretions into the urethra during ejaculation (Standring 2005; Ross and Pawlina 2006). There are different types of

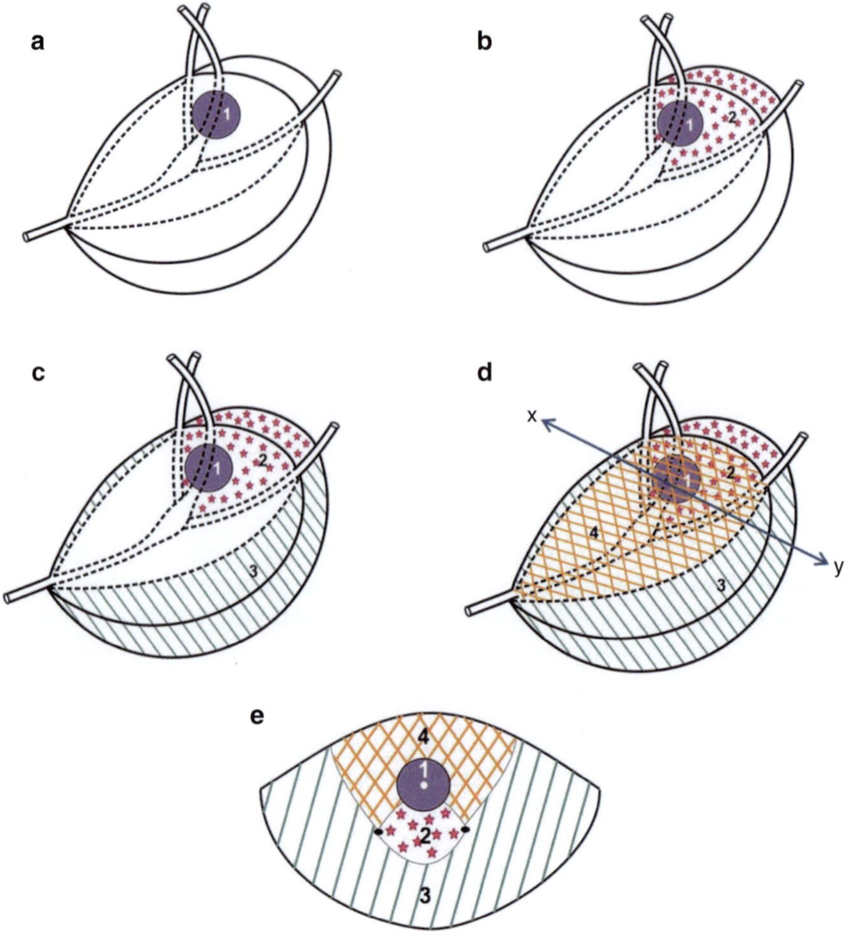

Fig. 2.4 Schematic diagram showing zones of prostate gland in superolateral view (**a**–**d**) and transverse section (**e**) across plane *x*–*y*. *1* transition zone (part of anterior and median lobes), *2* central zone (part of median and posterior lobes), *3* peripheral zone (lateral and remaining posterior lobes), *4* anterior fibromuscular stroma (anterior lobe)

cells in the glandular epithelium: secretory tall columnar or luminal cells, basal cells, stem cells, amplifying cells and neuroendocrine cells (see Figs 2.5 and 2.6).

2.5.1 Luminal Cell or Secretory Cell

The secretory cell (SC) is the predominant cell type and opposes the lumen. SC is 20 μm in height and has a round nucleus located at the base. The basal cytoplasm contains free ribosomes, rough endoplasmic reticulum and mitochondria, while the apical cytoplasm has lysosomes and dense bodies. The dense bodies contain a yellow pigment, lipofuschin. In elderly men, these pigments are found in the base or

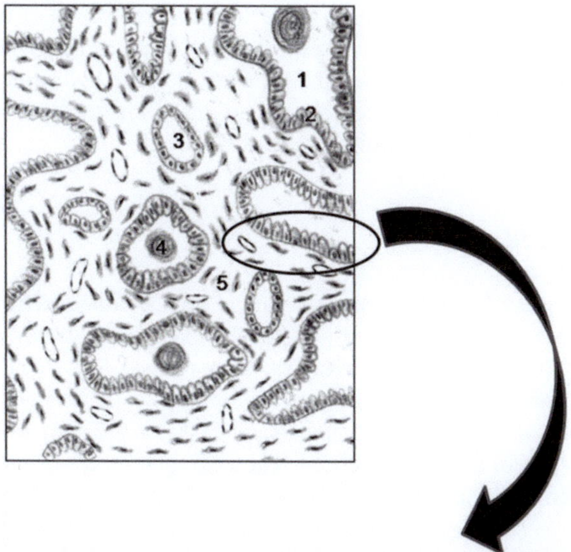

Fig. 2.5 Schematic diagram showing microstructure of prostate gland. *1* glandular acinus, *2* glandular epithelium, *3* duct of prostatic gland, *4* prostatic concretions, *5* fibromuscular stroma

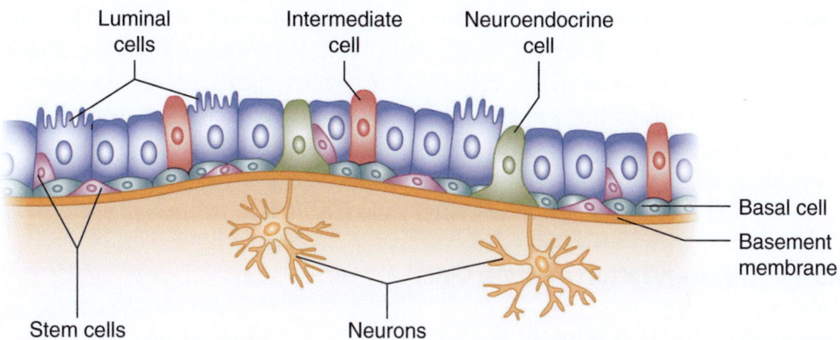

Fig. 2.6 Schematic diagram showing cells of prostatic epithelium

supranuclear areas of glands. Villiform projections are present in the apical cytoplasm, and they extend into the lumen (see Fig. 2.6). However, these projections are not found in areas associated with secretory granules or cells that have released their contents (Grayhack and Kozlowski 1996). SCs have comparatively limited proliferating capacity, as suggested by their high mitotic index, i.e. the ratio of the mitotic phase-arrested cell to the total number of cells (Ikeda et al. 2000). These cells express nuclear androgen receptor, cytokeratins 8 and 18 as well as the cell-surface marker CD57 (Schalken 2005). They are androgen-dependent cells and secrete fluid, prostate-specific antigen (PSA) and prostate acid phosphatase (PAP) into the glandular lumen. The activity of acid phosphatase is mainly localized within the secretory vacuoles and lysosomes, while that of amino peptidase and PSA are in the apical cell border and cytoplasm, respectively (Grayhack and Kozlowski 1996).

Androgen deprivation decreases the number of SCs. Following damage and infection, with testosterone replacement, the luminal compartment rapidly regenerates itself from the basal compartment (Lee et al. 2012).

2.5.2 Basal Cell

Basal cells (BCs) are found between the luminal cells and the underlying basement membrane (see Fig. 2.6). They are polygonal in shape and have large irregular-shaped nuclei. BCs lack secretory vesicles. They possess mitochondria, endoplasmic reticulum, free ribosomes, Golgi apparatus and pinocytic vesicles. Their specific markers are cytosolic cytokeratins 5 and 14 and CD44 on the cell surface (Grayhack and Kozlowski 1996; Long et al. 2005). The androgen receptors are expressed at very low level in the basal layer.

2.5.3 Stem Cell

A stem-like cell population in the basal layer is responsible for the development of epithelial cells in the prostate (see Fig. 2.6; Long et al. 2005; Peehl 2005). The stem cells self-renew by a symmetrical division, i.e. the daughter cells demonstrate similar growth and duplication potential as the mother cell. Following asymmetric division of the stem cell, the two daughter cells differ in characteristics, i.e. one daughter cell retains stem cell properties and the other becomes differentiated daughter cell. The markers which identify stem cells are CD133, CD44, $\alpha_2\beta_1$ integrin, breast cancer-reserved protein (BCRP-1), ATP-binding cassette subfamily G member 2 (ABCG-2), telomerase reverse transcriptase (TERT) and CK5/CK14 (Mimeault et al. 2008).

2.5.3.1 Differentiation of Stem Cells

Linear differentiation model

According to this model, stem cell found in the basal layer undergoes asymmetric division to give rise to stem cell that has unlimited self-renewal capacity and progenitor cells which further divide and differentiate to intermediate cells. The intermediate cell subsequently differentiates into either luminal or neuroendocrine cells (Taylor et al. 2010). This model is derived on the basis of studies on mouse prostatic tissue.

Bidirectional differentiation model

Bidirectional differentiation model is developed based on the study of prostatic cells and tissue of mouse as well as human. The model is supported by the existence of intermediate cells that express both BC- and SC-specific markers, e.g. cytokeratins 8 and/or 18 and cytokeratins 5 and/or 14 in both developing and adult stages of prostate development. In this model, one population of stem cell maintains its own cell number through self-renewal, and another is committed to further division and differentiation, giving rise to intermediate cells. This intermediate cell differentiates to a variety of lineage-specific progenitors which further differentiate to basal, luminal and neuroendocrine cells (Wang et al. 2001).

Independent lineage model
The third independent lineage model suggests that stem cells are also present among the luminal cells. This is based on the presence of a typical cell castrate-resistant Nkx3.1-expressing cells (CARNs). CARNs cells have the capacity to self-replenish in vivo or undergo division to give rise to a variety of progenitors of specific cell lineage, i.e. basal, luminal and neuroendocrine cells. The basal stem cells self-replenish or differentiate into basal cell through lineage-specific progenitors. The basal and/or luminal stem cells can be multipotent and generate the opposing lineage with basal stem cells giving rise to SCs and luminal stem cells giving rise to BCs (Taylor et al. 2010).

2.5.4 Intermediate or Amplifying Cells

Intermediate cells are proliferating cells which express markers present in both the BCs and SCs. These cells express luminal marker cytokeratins 8 and 18 in the basal layer, whereas some cells in the luminal layer express basal cell cytokeratins 5 and 14 (see Fig. 2.6; Long et al. 2005).

2.5.5 Neuroendocrine Cells

Neuroendocrine cells (NECs) are highly specialized epithelial cells and are androgen-independent cells which are distributed all over the basal layer. They are of two types: (a) the 'open' cell type that has long extensions that connects to the lumen and (b) the 'closed' type that do not have direct contact with the lumen. NECs possess uneven dendritic processes that extend between adjacent cells (see Fig. 2.6). The major markers are chromogranin A, serotonin, bombesin, neuron-specific enolase and calcitonin (Vashchenko and Abrahamsson 2005). Although its function is still not clearly understood, these cells regulate growth, differentiation and homeostatic regulation of the secretory processes (Abrahamsson and diSant'Agnese 1993).

2.6 Regulation of Prostatic Function

2.6.1 Androgens

Testosterone synthesized by the testis and adrenals, diffuses into the prostatic epithelium and is rapidly converted to dihydrotestosterone (DHT) by the enzyme 5α-reductase. DHT has fivefold higher affinity for the androgen receptor than testosterone. In the resting state, unligated androgen receptors in the cytoplasm are bound to heat shock proteins (HSPs), viz. HSP90, 70, 56 and 23, which stabilize the tertiary structure of androgen receptor in a confirmation that facilitates androgen binding. Upon binding of androgens to the androgen receptor, the HSPs dissociate leading to androgen receptor dimerization and subsequent translocation into the nucleus, where the dimerized androgen receptor binds to the androgen response elements (AREs) in the promoter/enhancer regions of target genes. The activated DNA-bound androgen receptor dimer complex then recruits co-regulatory proteins (co-activators or co-repressors) and the general transcription machinery. The co-activators, members of the p160 family that have histone acetyltransferase activity, such as NCOA1-3, TIP60, etc., alter the chromatin structure to a form which is more accessible for the transcription machinery. In contrast, the co-repressors, e.g. silencing mediator of retinoid and thyroid (SMRT) hormone receptors and nuclear receptor co-repressor (NCOR), mediate chromatin condensation leading to transcription silencing. Androgen receptor-regulated target genes include PSA, PAP, cyclin-dependent kinase 8 (CDK8) and the p85 catalytic subunit of phosphatidylinositol 3-kinase (PIK3R1) and RAB4A–a member of the Ras oncogene family of proteins. While PSA and PAP are major secretory proteins of the prostate, CDK8 is involved in cell cycle regulation; PI3KR1 activates signalling pathways important for cell proliferation and survival, and RAB4A is involved in receptor trafficking and vesicle fusion. Collectively, the activation of these genes by androgens leads to cell growth, survival and increased production of the secretory proteins (Feldman and Feldman 2001; Taplin 2007).

2.6.2 Estrogens

Estrogen action is mediated in two different ways: endocrine effects acting via pituitary to indirectly lower androgens and local paracrine effects directly targeting the prostate tissue itself. The paracrine effect is mediated by the two types of estrogen receptors- alpha (ERα) and beta (ERβ), which are localized mainly in stromal cells and epithelium, respectively. ERβ complexes with endothelial nitric oxide synthase (eNOS) and hypoxia inducible factor (HIF) which leads to chromatin remodelling and induction of transcription. This activation of ERβ has anti-proliferative effects that balance the proliferative action of androgen on the prostatic epithelia. In contrast, the activation of ERα leads to aberrant proliferation, inflammation and the development of premalignant lesions, though its molecular mechanism has not been worked out. Therefore, estrogens also play a role in the aetiology of prostatic diseases (Harkonen and Makela 2004; Prins and Korach 2008; Ellem and Risbridger 2009).

2.6.3 Prolactin

Prolactin (PRL), a hypophyseal peptide hormone, belongs to the growth hormone family. It has been reported to be locally produced in the prostate. Prolactin receptors (PRLR) are also found in the prostate. As mentioned earlier, in the developing prostate, prolactin promotes ductal morphogenesis. PRL has a physiological role in stimulating citrate production by regulating the metabolic genes, namely, mitochondrial aspartate aminotransferase, pyruvate dehydrogenase and mitochondrial aconitase via protein kinase C (PKC) signalling pathway (Costello and Franklin 2002). PRL acts as a strong mitogen and survival factor for prostate epithelium, an effect mediated by autocrine signalling through PRLR-Jak2-Stat5a/b pathway. Hence, it has a role during prostate cancer progression, and the inhibition of this pathway leads to apoptosis (Goffin et al. 2011).

2.6.4 Oxytocin

Oxytocin (OXT) is a neurohypophysial nonapeptide hormone with sequence similar to vasopressin, also known as antidiuretic hormone. In normal prostate, oxytocin is found at a concentration ranging from 0.5 to 30 nM. Both oxytocin and its receptor are expressed by the epithelial as well as stromal cells of benign and normal prostate. Its secretion is under regulation of androgens and estrogen, with increased estrogen/androgen ratio stimulating the greatest secretion. OXT inhibits proliferation of stromal and epithelial cells both directly and by regulating the local concentration of androgens. Oxytocin increases the expression and activity of the 5α-reductase enzyme, thus affecting the androgen metabolism. It also increases prostatic muscular tone and contractile activity (Thackare et al. 2006; Nicholson and Whittington 2007; Assinder 2008).

2.6.5 Thyroid Hormones

The existence of prostate-thyroid axis has been reported in the rodents (Mani Maran et al. 1998). Recently, a prospective study showed that men with hypothyroid state had a lower risk of developing prostate cancer (Mondul et al. 2012). The thyroid hormone, 3,5,3′-L-triiodothyronine (T_3), stimulates PSA expression at both mRNA and protein level in the presence of androgens. A functional thyroid hormone response element (TRE) is located in the 5′-promoter region (−198 to −183 upstream from the start codon) of the PSA gene. However, T3 has only a minimal effect on PSA expression in the absence of androgens, and it is yet to be explored how androgen-androgen receptor-ARE complex interacts with T3-TR-TRE complex in modulating PSA expression (Zhu and Young 2001).

2.6.6 Growth Factors

Besides hormones, normal prostate development and homeostasis between epithelial and stromal cells are also regulated by various growth factors. A complex interaction exists between the growth factors themselves, and, in turn, they also are regulated either by androgen or by other factors. Hence, prostate is very sensitive to any alteration, either up- or downregulation, in the expression of growth factors and/or their receptors.

Insulin-like growth factor (IGF) family includes polypeptide growth factors that have amino acid sequence and functional homology to insulin. IGFs are produced by the stromal cells and act on epithelial cells in a paracrine manner in response to androgen stimulation resulting in enhanced proliferation of the prostatic cells (Reynolds and Kyprianou 2006).

Epidermal growth factor (EGF) and transforming growth factor alpha (TGF α) are structurally and functionally related polypeptides that signal through the same cell surface receptor, viz. EGF receptor (EGFR), a transmembrane tyrosine kinase. EGF is an important activator of normal prostate growth, and its expression is positively regulated by androgens. It is even found in large quantities in the prostatic fluid. However, TGF-α is expressed chiefly in the stroma, while its receptor is expressed by the epithelial cells suggesting that it works in a paracrine /juxtacrine manner in the normal prostate. The upregulation of both EGF and TGF-α is associated with the development of prostate cancer as they exert a mitogenic effect on epithelial cells (Russell et al. 1998).

TGF-β family includes TGF-β1, TGF-β2 and TGF-β3. They are expressed during prostate development and in the adult prostate, both normal and malignant. TGF-β is produced by both stromal and epithelial cells, but its receptors are found only on the stromal cells. It regulates prostatic growth by inhibiting cell proliferation and inducing apoptosis (Russel et al. 1998).

The fibroblast growth factor (FGF) family includes nine structurally related heparin-binding peptides, of which three are implicated in prostate cancer, viz. bFGF, aFGF and KGF. bFGF is produced by both stromal and epithelial cells of the prostate, but its receptor is present only on stromal cells in normal prostate (Sherwood et al. 1992). In contrast, KGF is secreted by the stromal cells while its receptor is present on the epithelial cells. It upregulates epithelial cell proliferation in a paracrine manner (McGarvey and Sterans 1995). Role of aFGF in the development of rat prostate has been shown, but it is not detected in human prostate (Russel et al. 1998).

In a nutshell, the growth factors IGF, EGF, TGF-β and FGF are predominantly stimulators of proliferation, while TGF-β predominantly inhibits prostatic growth.

2.6.7 Bone Morphogenetic Proteins

Bone morphogenetic proteins (BMPs) belong to the TGF-β superfamily and regulate growth, differentiation and apoptosis in many tissues in addition to the bone. The expression of BMPs (viz. BMP-2, -3, -4 and -6) by the normal human prostate and human prostate cancer cell lines has been reported. The higher expression of BMP-6 in malignant prostate tissue indicates that it contributes to prostate carcinogenesis (Shimasaki et al. 2004). In contrast, the expression of BMP-7 is strongly downregulated in prostate cancer tissue compared to normal prostate luminal epithelium (Buijs et al. 2007). BMP-4 and BMP-7 inhibit branching morphogenesis during prostate gland development (Prins and Putz 2008).

2.7 Components of Prostatic Fluid

2.7.1 Prostate-Specific Antigen

Prostate-specific antigen (PSA) is an androgen-regulated serine protease. It is a member of human tissue kallikrein gene family, located on chromosome 19q13.4. PSA is produced by the secretory epithelial cells in the ducts and acini of the prostate and is present at a concentration of 0.5–2.0 mg/ml in the seminal fluid. Its physiological function is to liquefy the seminal coagulum formed just after ejaculation by cleaving the semenogelins.

The transcription of the PSA gene is positively regulated by androgen receptor that attaches to specific DNA sequences called ARE. There are three regions upstream of the PSA gene which contain the consensus ARE: region −156 to −170 in the PSA gene proximal promoter, an androgen responsive region from −365 to –400 and a PSA distal enhancer about 4.2 kb upstream of the transcription start site. Following transcription and translation, PSA is secreted as an inactive proenzyme (proPSA) having 244 amino acids. In the lumen, proPSA is converted to a 237-amino acid long, active PSA by prostatic protease, viz. human kallikrein-2. About 30 % of the PSA in the seminal fluid is in the active form, whereas 5 % remains bound with protein C inhibitor. The remaining PSA undergoes internal cleavage to form catalytically inactive PSA. Both the active and cleaved PSA can enter into the circulation. In the peripheral blood, 70 % to 90 % PSA remains bound to protease inhibitors, chiefly alpha 1 anti-chymotrypsin while 10 % to 30 % is the inactive, cleaved PSA which circulates as free PSA (Diamandis 2000; Balk et al. 2003). This ratio of free to total PSA (fPSA /tPSA, termed the PSA index) is lower in many patients with prostate cancer and can discriminate between normal and malignant prostate.

2.7.2 Prostatic Acid Phosphatase

Prostatic acid phosphatase (PAP) is a dimeric glycoprotein enzyme. PAP is synthesized by the columnar secretory epithelia of the prostate. The enzyme is expressed at puberty and is androgen regulated. Two forms of PAP are known–cellular (cPAP) and secreted (sPAP); both are transcribed from the same gene but undergo different post-translational modifications. Interestingly, though PAP shows only 50 % sequence similarity with the lysosomal acid phosphatase, the amino acids in the active site of both enzymes are conserved. PAP catalyses the hydrolysis of various phosphate esters, viz. phosphorylcholine and phosphocreatine which are energy-rich compounds present in the seminal fluid. The presence of high amount of PAP in seminal fluid (~ 1 mg/ml) suggests a role in fertility, perhaps in increasing sperm mobility, but this role of sPAP is controversial (Kong and Byun 2013). Many studies show that higher levels of sPAP are found in seminal plasma of azoospermic men and also an inverse correlation between seminal PAP levels, and sperm concentration has been observed in oligospermic men (Kong and Byun 2013). Normal levels of serum sPAP vary from 1 to 3 ng/ml, while prostate cancer patients show elevated levels which increase as the disease progresses. In contrast, levels of cPAP are inversely related with progression of prostate cancer (Hassan et al. 2010; Muniyan et al. 2013).

2.7.3 Citric Acid

Citrate level in the normal prostatic fluid is about 400–1500 times the level found in blood plasma. Though its precise function in semen is not known, it is possibly required to maintain osmotic/electrolytic equilibrium in the semen. In normal mammalian cells, citrate is crucial in the citric acid cycle or Kreb's cycle for ATP generation. However, in the peripheral zone of the prostate, increased accumulation of zinc inhibits mitochondrial aconitase activity which oxidizes citrate to isocitrate. Thus, Kreb's cycle in these cells gets truncated, and citrate is secreted in the prostatic fluid. Therefore, the prostatic epithelial cells are also known as 'citrate-producing cells'. In prostate cancer, a decrease in the accumulation of zinc occurs. This 70–90 % reduced tissue levels of zinc do not inhibit aconitase activity; consequently, citrate is oxidized via the Kreb's cycle. Therefore, the malignant cells are referred to as 'citrate-oxidizing cells', and the concentration of citrate in the prostatic fluid is markedly decreased from 100 mM in normal to 8 mM in prostate cancer (Costello and Franklin 2009).

2.7.4 Zinc

High levels of zinc in the prostatic epithelial cells help to carry out the major physiological functions of citrate production and secretion. High concentration of zinc in the seminal plasma confers its bactericidal activity. Decrease in the levels of zinc and citrate in the normal prostatic cells is an important factor in the development of malignancy. Decreased accumulation of zinc in the prostatic epithelial cells (1000 ug/ml in normal vs. 150 ug/ml in cancerous) may be due to the downregulation of specific zinc uptake transporters proteins, viz. hZIP1. hZIP1 belongs to the ZIP (Zrt/Irt-like Proteins) family of transporters. The hormones, prolactin and testosterone, promote accumulation of zinc in prostate epithelial cells. However, the exact mechanism of zinc accumulation and its regulation in prostatic cells remain to be explored (Franklin et al. 2005). Zinc ions in the prostatic fluid are important for semen coagulation as they trigger a conformational change in the semenogelin proteins which form an insoluble protein complex of the coagulum. The normal concentration of zinc in seminal fluid is about 2.4 mM which is 100-fold higher than that in blood plasma.

2.7.5 Spermine

Spermine is a basic aliphatic polyamine secreted by the prostate gland. Other cellular polyamines are known to stimulate protein and RNA synthesis by binding non-covalently to nucleic acids and structures containing nucleic acid, e.g. ribosomes. Spermine also has strong affinity for phosphate ions, nucleic acids or phospholipids. When semen is kept undisturbed at room temperature, a transparent yellow crystal of spermine phosphate is formed. This is due to the hydrolysis of seminal phosphorylcholine, produced mainly in the epididymis, by acid phosphatase resulting in the release of inorganic phosphate ions. Finally, the phosphate ions react with spermine to form spermine phosphate. The enzyme diamine oxidase, present in the seminal plasma, oxidizes spermine to an aldehyde product that is responsible for the unique odour of semen (Folk et al. 1980).

2.7.6 Prostatic Inhibin

Prostatic inhibin is also known as β-inhibin, β-micro seminoprotein, immunoglobulin binding factor, prostatic inhibin peptide and prostatic secretory protein. It is made up of 94 amino acids and is synthesized and secreted by prostatic epithelium. It is a follicle stimulating hormone (FSH)-suppressing, non-glycosylated, cysteine-rich polypeptide (10.7 kDa). It is not related to dimeric pituitary inhibin but is a single-chain polypeptide of 94 amino acids which constitutes about 20 % of total seminal plasma proteins. Unlike PSA and PAP, prostatic inhibin expression is not dependent on androgens. Its concentration is 100 times more in the prostate than in testis. In contrast to PSA, lower levels of prostatic inhibin are found in the prostate, serum as well as urine of patients with prostate cancer. It has been implicated as a tumor growth suppressor, and silencing of its gene expression has been associated with promoter methylation of EZH2, a polycomb group member protein (Garde et al. 1992; Haiman et al. 2013).

2.8 Disorders of Prostate Gland

2.8.1 Benign Prostatic Hyperplasia (BPH)

BPH is a non-cancerous enlargement of the prostate. It is a very common urological problem among ageing men and leads to lower urinary tract symptoms. Although increase in both stromal and epithelial elements of the prostate occurs in BPH, the major increase in prostate volume is due to increased number of stromal smooth muscle cells. The exact pathogenesis of BPH is unknown, but a role of DHT in causing pathological prostate growth has been demonstrated. Since the enzyme 5-alpha reductase is responsible for conversion of testosterone to DHT in the prostate, inhibitors of this enzyme are used as drugs for reducing the prostate volume. Two subtypes of the enzyme are known with type-2 being the predominant type in the prostate. However, it has been seen that non-selective inhibitors of 5-alpha reductase (dutasteride) have higher efficacy than selective inhibitor of type-2 isozyme (finasteride) in suppressing the enzyme activity. Alpha-1 receptor blockers target the alpha-1A receptors present on the stromal smooth muscle cells thereby relaxing these muscles and, hence, reducing the dynamic component of bladder outflow obstruction. Both 5-alpha reductase inhibitor and alpha-1 blockers are used either alone or in combination for therapy depending on the disease status (Prabhav and Bairy 2009; Briganti et al. 2009).

2.8.2 Prostate Cancer

Prostate cancer is currently the most common cause of cancer death in men. The exact cause of developing prostate cancer are not known though ageing, ethnicity and heredity are important factors involved in the initiation and development of this cancer. Ageing is considered the most prominent risk factor, with majority of cases being diagnosed in men between 60 and 70 years of age. The role of various genes such as NKX3.1, Myc, ERG, PTEN and EZH2 has been implicated in prostate cancer (Shen and Abate-Shen 2010).

In recent years, the role of xenobiotics such as cadmium, lead, organochlorine pesticides and polychlorinated biphenyls have been implicated with prostate cancer risk in agricultural and general population. (Dich and Wiklund 1998; Settimi et al. 2003; Mink et al. 2008). The carcinogenic effect of the pesticide methyl bromide is due to its methyl group which binds to DNA to form DNA adducts, O6- and N7-methylguanine, which ultimately result in a G:C to A:T transition mutation. These gene mutations represent the early steps in prostate carcinogenesis (Cockburn et al. 2011). Another toxicant, lead, is a potential risk factor for prostate cancer. Its action is possibly mediated by increasing reactive oxygen species generation and/or by reducing the accumulation of zinc, which acts as an inhibitor of cellular growth (Siddiqui et al. 2002). Animal studies have shown that phthalates also may cause proliferative and inflammatory disorders of the rat prostate (Scarano et al. 2009). The other details of prostate cancer are given in the chapter, Male Reproductive Cancers.

2.8.3 5α-Reductase 2 Deficiency or Pseudovaginal Perineoscrotal Hypospadias

5α-reductase 2 deficiency is a result of decreased DHT production due to mutations in the 5α-reductase 2 gene (SRD5A2 gene) which is located on chromosome 2 (2p23). It is an autosomal recessive disorder of male sexual development with more than 50 mutations reported in the gene. The affected males have normal internal reproductive structures; however, they possess external genitalia similar to females, i.e. ambiguous genitalia. The prostate is underdeveloped with fibrous connective tissue, smooth muscle and no identifiable epithelial tissue, which suggests atrophic epithelium or lack of epithelial differentiation. However, these individuals undergo partial virilization of the external genitalia at puberty (Azzouni et al. 2012).

Key Questions

- Name the different zones of the prostate gland with their clinical relevance.
- Write briefly about the different cell specific markers of the prostatic epithelial cells.
- Name the key components of prostatic fluid.
- Explain briefly the physiological importance of the prostatic fluid components in semen.
- Discuss the regulation of prostate function by androgens.
- Write a short note on the physiological role of growth factors produced by the prostate gland.

References

Abrahamsson PA, di Sant'Agnese PA. Neuroendocrine cells in the human prostate gland: Minireview. J Androl. 1993;14:307–9.

Assinder SJ. Oxytocin increases 5a-reductase activity of human prostate epithelial cells, but not stromal cells. Prostate. 2008;68(2):115–21.

Azzouni F, Godoy A, Li Y, Mohler J. The 5 alpha-reductase isozyme family: a review of basic biology and their role in human diseases. Adv Urol. 2012;79:1197–205.

Balk SP, Ko YJ, Bubley GJ. Biology of prostate-specific antigen. J Clin Oncol. 2003;21:383–91.

Briganti A, Capitanio U, Suardi N, Gallina A, Salonia A, Bianchi M, et al. Benign prostatic hyperplasia and its aetiologies. Eur Urol Suppl. 2009;8:865–71.

Buijs JT, Rentsch CA, van der Horst G, Overveld PG, Wetterwald A, Schwaninger R, et al. BMP7, a putative regulator of epithelial homeostasis in the human prostate, is a potent inhibitor of prostate cancer bone metastasis in vivo. Am J Pathol. 2007;171(3):1047–57.

Cockburn M, Mills P, Zhang X, Zadnick J, Goldberg D, et al. Prostate cancer and ambient pesticide exposure in agriculturally intensive areas in California. Am J Epidemiol. 2011;173(11):1280–8.

Costello LC, Franklin RB. Prostatic fluid electrolyte composition for the screening of prostate cancer: a potential solution to a major problem. Prostate Cancer Prostatic Dis. 2009;12(1):17–24.

Costello LC, Franklin RB. Testosterone and prolactin regulation of metabolic genes and citrate metabolism of prostate epithelial cells. Horm Metab Res. 2002;34(8):417–24.

De Krester DM, Temple-Smith PD, Kerr JB. Anatomical and functional aspects of male reproductive organs, ch 7-Prostate. In: Bandhauer K, Fricj J, editors. Disturbances in male infertility. Berlin/Heidelberg/New York: Springer; 1982. p. 98–114.

Decker GAG, du Plessis DJ. The large bowel, anal canal and ischiorectal fossa. In: Le Mcgregor's synopsis of surgical anatomy, 12th ed. Indian Edition, Bombay: Varghese publishing house. 1986. p. 41–78.

Diamandis EP. Prostate-specific antigen: a cancer fighter and a valuable messenger? Clin Chem. 2000;46:896–900.

Dich J, Wiklund K. Prostate cancer in pesticide applicators in Swedish agriculture. Prostate. 1998;34(2):100–12.

Ellem SJ, Risbridger GP. The dual, opposing roles of estrogen in the prostate. Ann N Y Acad Sci. 2009;1155:174–86.

Feldman BJ, Feldman D. The development of androgen-independent prostate cancer. Nat Rev Cancer. 2001;1:34–45.

Folk JE, Park MH, Chung SI, Schrode J, Lester EP, Cooper HL. Polyamines as physiological substrates for transglutaminases. J Biol Chem. 1980;255:3695–700.

Franklin RB, Milon B, Feng P, Costello LC. Zinc and zinc transporters in normal prostate function and the pathogenesis of prostate cancer. Front Biosci. 2005;10:2230–9.

Garde SV, Sheth AR, Lohiya NK, Shah MG. Prostatic inhibin peptide: occurrence, localization and its hormonal modulation in prostates of primates and rodents. J Biosci. 1992;17(1):67–85.

Goffin V, Hoang DT, Bogorad RL, Nevalainen MT. Prolactin regulation of the prostate gland: a female player in a male game. Nat Rev Urol. 2011;8:597–607.

Grayhack JT, Kozlowski JM. Benign Prostatic Hyperplasia. In: Gillenwater JY, Grayhack JT, Howards SS, Duckett JW, editors. Adult and Pediatrics Urology, 3rd ed. vol. 2; C.V.Mosby, St.Louis.1996. p. 1501–74.

Haiman CA, Stram DO, Vickers AJ, Wilkens LR, Braun K, Valtonen-Andre C, et al. Levels of beta-microseminoprotein in blood and risk of prostate cancer in multiple populations. J Natl Cancer Inst. 2013;105(3):237–43.

Hammerich KH, Ayala GE, Wheeler TM. Anatomy of the prostate gland and surgical pathology of prostate cancer. In: Hricak H, Scardino PT, editors. Prostate cancer. Cambridge/New York: Cambridge University Press; 2009. p. 1–14.

Harkonen PL, Makela SI. Role of estrogens in development of prostate cancer. J Steroid Biochem Mol Biol. 2004;92:297–305.

Hassan MI, Aijaz A, Ahmad F. Structural and functional analysis of human prostatic acid phosphatase. Expert Rev Anticancer Ther. 2010;10:1055–68.

Ikeda K, Pant B, Mishiro A, Ozawa K, Masujima T, Sugiyama M. A convenient method for the evaluation of anti-tumor agents affecting the cell cycle. J Biosci Bioeng. 2000;90:574–6.

Kong HY, Byun J. Emerging roles of human prostatic acid phosphatase. Biomol Ther (Seoul). 2013;21(1):10–20.

Lee CH, Akin-Olugbade O, Kirschenbaum A. Overview of prostate anatomy, histology, and pathology. Endocrinol Metab Clin North Am. 2011;40:565–75.

Lee SO, Tian J, Huang CK, Ma Z, Lai KP, Hsiao HM, et al. Suppressor role of androgen receptor 1 in proliferation of prostate basal epithelial and progenitor cells. J Endocrinol. 2012;213:173–82.

Long RM, Morrissey C, Fitzpatrick JM, Watson RWG. Prostate epithelial cell differentiation and its relevance to the understanding of prostate cancer therapies. Clin Sci. 2005;108:1–11.

Lowsley OS. The development of the human prostate gland with reference to the development of other structures at the neck of the urinary bladder. Am J Anat. 1912;13:299–349.

Mani Maran RR, Subramanian S, Rajendiran G, Sunil N, Archunan G, Arunakaran J, Govindarajulu P, et al. Prostate-thyroid axis: stimulatory effects of ventral prostate secretions on thyroid function. Prostate. 1998;36(1):8–13.

McGarvey TW, Sterans ME. Keratinocyte growth factor and receptor mRNA expression in benign and malignant human prostate. Exp Mol Pathol. 1995;63:52–62.

McNeal JE. The zonal anatomy of the prostate. Prostate. 1981;2(1):35–49.

Mimeault M, Mehta PP, Hauke R, Batra SK. Functions of normal and malignant prostatic stem/progenitor cells in tissue regeneration and cancer progression and novel targeting therapies. Endocr Rev. 2008;29:234–52.

Mink PJ, Adami HO, Trichopoulos D, Britton NL, Mandel JS. Pesticides and prostate cancer: a review of epidemiologic studies with specific agricultural exposure information. Eur J Cancer Prev. 2008;17(2):97–110.

Mondul AM, Weinstein SJ, Bosworth T, Remaley AT, Virtamo J, Albanes D. Circulating thyroxine, thyroid-stimulating hormone, and hypothyroid status and the risk of prostate cancer. PLoS One. 2012;7(10):e47730.

Muniyan S, Chaturvedi NK, Dwyer JG, LaGrange CA, Chaney WG, Lin MF. Human prostatic acid phosphatase: structure, function and regulation. Int J Mol Sci. 2013;14:10438–64.

Nicholson HD, Whittington K. Oxytocin and the human prostate in health and disease. Int Rev Cytol. 2007;263:253–86.

Peehl DM. Primary cell cultures as models of prostate cancer development. Endocr Relat Cancer. 2005;12:19–47.

Prabhav T, Bairy L. Pharmacotherapy of benign prostatic hyperplasia. JPBS. 2009;22:6–11.

Prins GS, Putz O. Molecular signaling pathways that regulate prostate gland development. Differentiation. 2008;76(6):641–59.

Prins GS, Korach KS. The role of estrogens and estrogen receptors in normal prostate growth and disease. Steroids. 2008;73(3):233–44.

Reynolds RA, Kyprianou N. Growth factor signalling in prostatic growth: significance in tumour development and therapeutic targeting. Br J Pharmacol. 2006;147:144–52.

Ross MH, Pawlina W. Male reproductive system. In: Histology, a text and atlas. 5th ed. Baltimore: Lippincott Williams and Wilkins; 2006. p. 728–71.

Russel PJ, Bennett S, Stricker P. Growth factor involvement in progression of prostate cancer. Clin Chem. 1998;44(4):705–23.

Scarano WR, Toledo FC, Guerra MT, de Campos SG, Junior LA, Felisbino SL, et al. Long-term effects of developmental exposure to di-n-butyl-phthalate (DBP) on rat prostate: proliferative and inflammatory disorders and a possible role of androgens. Toxicology. 2009;262(3):215–23.

Schalken J. Androgen-receptor mediated growth of prostate cancer. Eur Urol Suppl. 2005;4:4–11.

Settimi L, Masina A, Andrion A, Axelson O. Prostate cancer and exposure to pesticides in agricultural settings. Int J Cancer. 2003;104(4):458–61.

Shen MM, Abate-Shen C. Molecular genetics of prostate cancer: new prospects for old challenges. Genes Dev. 2010;24:1967–2000.

Sherwood ER, Fong C-J, Lee C, Kozlowsski JM. Basic fibroblast growth factor: a potential mediator of stromal growth in the human prostate. Endocrinology. 1992;130:2955–63.

Shimasaki S, Kelly Moore R, Otsuka F, Erickson GF. The bone morphogenetic protein system in the mammalian reproduction. Endocr Rev. 2004;25:72–101.

Siddiqui MK, Srivastava S, Mehrotra PK. Environmental exposure to lead as a risk for prostate cancer. Biomed Environ Sci. 2002;15(4):298–305.

Standring S. Prostate. In: Gray's anatomy. 39th ed. Elsevier/Churchill Livingstone; 2005. p. 1301–1304.

Taplin ME. Drug insight: role of the androgen receptor in the development and progression of prostate cancer. Nat Clin Pract Oncol. 2007;4(4):236–44.

Taylor RA, Toivanen R, Risbridger GP. Stem cells in prostate cancer: treating the root of the problem. Endocr Relat Cancer. 2010;17:273–85.

Thackare H, Nicholson HD, Whittington K. Oxytocin: its role in male reproduction and new potential therapeutic uses. Hum Reprod Update. 2006;12(4):437–48.

Vashchenko N, Abrahamsson PA. Neuroendocrine differentiation in prostate cancer: implications for new treatment modalities. Eur Urol. 2005;47:147–55.

Wang Y, Hayward WS, Thayer MCKA, Cunha GR. Cell differentiation lineage in the prostate. Differentiation. 2001;68:270–9.

Zhu W, Young CY. Androgen-dependent transcriptional regulation of the prostate-specific antigen gene by thyroid hormone 3,5,3'-L-triiodothyronine. J Androl. 2001;22(1):136–41.

3 Seminal Vesicles

Surabhi Gupta, Mona Sharma, and Anand Kumar

Learning Objectives

- Gross anatomy and development
- Maturation and involution
- Microstructure and function
- Hormonal regulation
- Disorders

3.1 Introduction

Berengario da Carpi, an *Italian* anatomist gave the first ever report on seminal vesicles in 1521. He regarded them as storage organs for semen. Various morphological studies have suggested that the seminal vesicles along with the ampulla of the ductus deferens and the ejaculatory ducts function as a single unit known as ampullo-vesiculoductal complex (Riva et al. 1989).

S. Gupta, PhD (✉) • M. Sharma, MD, DNB, MAMS
A. Kumar, MD, FAMS
Department of Reproductive Biology, All India Institute of Medical Sciences, New Delhi, India
e-mail: surabhi72@rediffmail.com

A. Kumar, M. Sharma (eds.), *Basics of Human Andrology*,
DOI 10.1007/978-981-10-3695-8_3

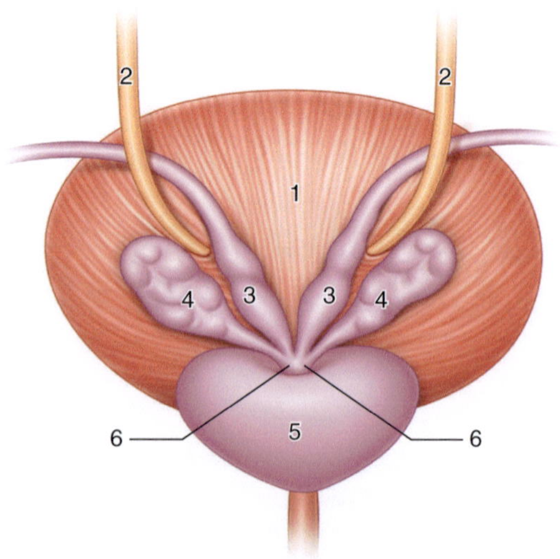

Fig. 3.1 Schematic diagram (posterosuperior view) showing anatomical relations of seminal vesicles: *1* urinary bladder, *2* ureters, *3* ductus deferens, *4* seminal vesicles, *5* prostate gland, *6* ejaculatory ducts

3.2 Gross Anatomical Features

The seminal vesicles or glandulae vesiculosae or vesicular glands are a pair of simple tubular glands located between the bladder anteriorly and rectum posteriorly (Frandson et al. 2009). The vesicles are blind pouches and are rounded on their most superior aspects and taper to their inferior aspects, where they constrict to ultimately form short ducts. Each seminal vesicle is a single-coiled tube of approximately 5 cm length with irregular folds in the lumen. The full unfolded length of seminal vesicle is approximately 10–15 cm. They descend inferomedially while lying on the posterior surface of the urinary bladder (see Fig. 3.1). The dilated portion or ampulla of the vas deferens lies along the medial margin of each seminal vesicle. The veins of the prostatic venous plexus are related lateral to it. The duct of seminal vesicle joins the distal portion of the vas deferens and becomes the ejaculatory duct, which drains into the prostatic urethra (Standring 2005a).

3.3 Development

Developmentally, when the testes begin to form during intrauterine life, the mesonephric or Wolffian duct differentiates into the male genital system under the influence of testosterone. Initially, the seminal vesicle and ductus/vas deferens form a common swelling which later during fourth month, gets separated into two structures. Mesonephric duct ultimately forms the epididymis, ductus deferens, seminal vesicle, and ejaculatory duct. (see Fig. 3.2, Standring 2005b).

The growth and differentiation of the seminal vesicle continues until birth, and the interior of the gland forms mucosal folds and ridges, as well as smooth muscle

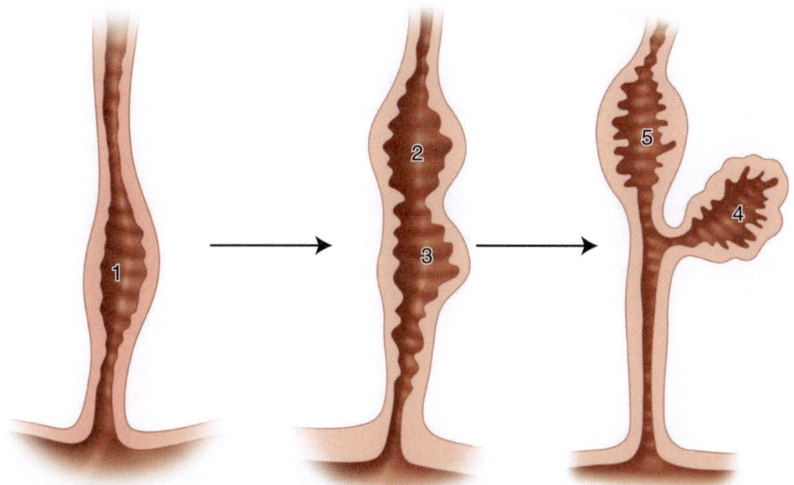

Fig. 3.2 Schematic diagram showing development of seminal vesicles. Mesonephric duct forming a common swelling (*1*) which later divides into two swellings (*2*, *3*); the swelling that grows laterally forms the seminal vesicle (*4*) whereas the other one forms ductus deferens (*5*)

Table 3.1 Maturation stages of seminal vesicles

Age (years)	Neonatal	1	1–5	8–14	15
Weight (gm)	0.05	0.08	0.09	0.1–0.15	1.5
Length (mm)	10	15	17	20–25	61
Width (mm)	3.3	4	4	4	6.6

cells differentiate in two layers within the seminal vesicle wall. The distal end of the vas deferens develops into the ampulla. The prostatic tissue surrounds the terminal portion of the mesonephric duct and forms the ejaculatory duct.

3.4 Maturation and Involution

Until puberty, growth of seminal vesicles proceeds slowly (see Table 3.1, Aumuller and Riva 1992). In children, the mucosal epithelium of the seminal vesicles consists of glandular and basal cells. The glandular weight increases during puberty. The gland develops typical mucosal folds projecting into the lumen. Connective tissue and smooth muscle cells increase in amount. Following this, the glandular epithelium becomes secretory. After 45 years of age, the regressive changes start with respect to the glandular and muscular contents of the vesicles. The smooth muscle content of the organ is reduced because of atrophic changes. There is decrease in secretion by the glandular cells, degeneration of the connective tissue, and decrease in number of mucosal folds. The overall volume of seminal vesicles is reduced (Aumuller and Riva 1992).

3.5 Microstructure

Seminal vesicles are the elongated coiled tubes having mucosal folds which projects into the lumen. The wall of seminal vesicle is composed of three coats: an internal or mucous coat, middle or muscular coat, and an external connective tissue coat or adventitia (see Fig. 3.3).

The mucous coat consists of interconnected folds which significantly increase the surface area. The folds are the primary folds which further branch into secondary folds. The folds are lined by pseudostratified columnar epithelium which consists of tall, columnar secretory cells and short, round basal cells that are the stem cells. The height and activity of the columnar cells depend on the blood testosterone levels. The lumen of the seminal vesicle is large and fluid secretions are stored there in between ejaculations.

The lamina propria is a loose connective tissue layer which extends into the folds. It consists of smooth muscle cells and elastic fibres. It is surrounded by inner circular and outer longitudinal layers of smooth muscles which are not clearly distinguished because of extensive mucosal folds. During ejaculation, smooth muscle contractions discharge the fluid produced by the seminal vesicles into the ejaculatory ducts and sperm out of the urethra. The smooth muscle layer is surrounded by connective tissue of the adventitia (Ross and Pawlina 2011).

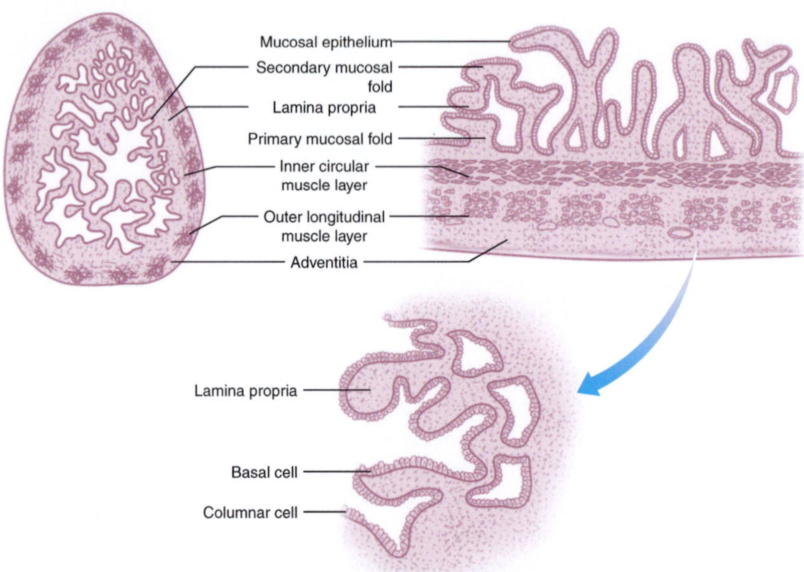

Fig. 3.3 Schematic diagram showing microstructure of seminal vesicle

3.6 Functions

The main function of the seminal vesicles is to produce components of the seminal plasma. The seminal plasma acts as a natural diluent and transport vehicle for the sperm. It also assists sperm passage through the male and female reproductive tract by eliciting smooth muscle contractions in both. Though it appears in the later fractions of ejaculation, seminal vesicle secretions form the largest portion (~65–75%) of a typical human ejaculate (Drabovich et al. 2014). The normal secretion of the seminal vesicle is alkaline which is important for protecting the sperm from the acidic vaginal environment. It has a whitish yellow colour due to the presence of fluorescent flavins (Burgos 1974).

The important constituents of seminal plasma secreted by the seminal vesicle include fructose, semenogelin I and II, prostaglandins, citric acid, ascorbic acid, ergothioneine, potassium ions, and inorganic phosphorus.

The seminal plasma has a high concentration of reducing substances belonging to two groups: one made up of molecules such as ergothioneine and ascorbic acid, the other being carbohydrates. The most important free carbohydrate secreted by seminal vesicles is fructose with the others being inositol, glucose, ribose, fucose, and sorbitol which are secreted in small amounts (Mann 1964). Fructose is found at a concentration of 200 mg/dl in the semen. It originates from the blood glucose and three different metabolic pathways for its biosynthesis have been proposed (Mann and Lutwak-Mann 1981). These are (i) glycogenolysis, (ii) direct phosphorylation to form glucose-6-phosphate, and (iii) non-phosphorylative pathway via formation of sorbitol (see Fig. 3.4).

Fructose is the primary energy source for the sperm following deposition of sperm in the female genital tract. Since hypoxic conditions are present in the vagina, fructose undergoes anaerobic degradation or fructolysis to produce lactic acid and

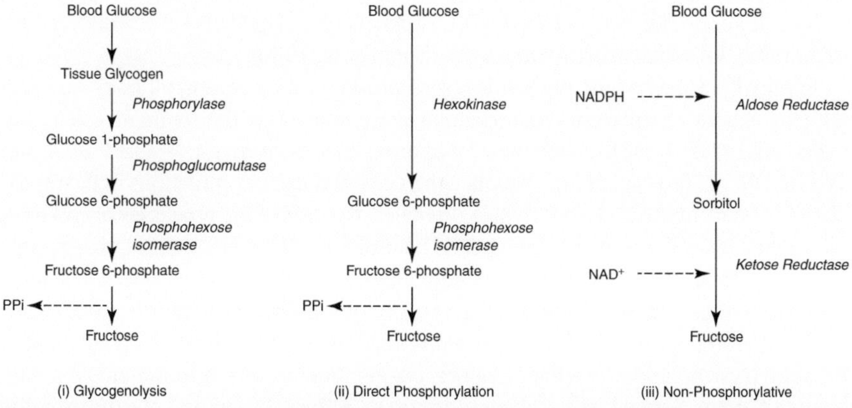

Fig. 3.4 Flow chart showing biosynthesis of fructose from blood glucose by seminal vesicles

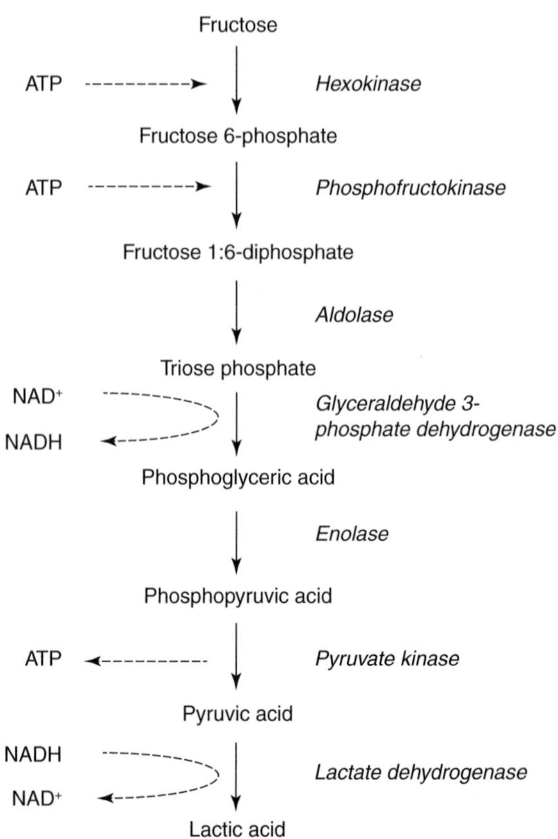

Fig. 3.5 Flow chart showing fructolysis by mammalian sperm

adenosine triphosphate (ATP; see Fig. 3.5). A positive correlation exists between the rate of fructolysis and the degree of sperm motility (Peterson and Freund 1976). In fact, if sperm are immobilized with a spermicidal agent, they immediately and irreversibly lose their fructolytic ability (Mann et al. 1980).

Seminal plasma has the highest levels of prostaglandins among all body fluids in humans. About 15 different prostaglandins are produced by the seminal vesicle; the main being PGE-1, PGE-2, and their 19-hydroxylated derivatives: 19-OH PGE-1 and 19-OH PGE-2. Prostaglandins stimulate the smooth muscle contractions in the reproductive tracts of both male and female. In males, they affect the processes of erection, ejaculation as well as testicular and penile contractions. While in females, these seminal fluid prostaglandins affect the cervical mucus, vaginal secretions, and induce uterine contractions, thereby helping the sperm to move in the female reproductive tract.

Another important product of seminal vesicles is semenogelins which are needed for semen coagulation. Details of semen coagulation are given in the chapter, The Human Semen. Semenogelins account for up to 30% of all seminal plasma proteins (Drabovich et al. 2014). Though the precise physiological importance of semen coagulation is not known, it results in coating the sperm surface with many molecules which may help in the natural process of fertilization.

The components of seminal plasma may not be absolutely essential for successful fertilization as sperm taken from epididymis or even testis can be used for fertilizing an egg. However, under natural course, the seminal fluid and its constituents optimize the conditions for sperm transport through the female reproductive tract and hence, increase the chances of successful fertilization.

3.7 Hormonal Regulation

The development as well as secretory function of the seminal vesicle is under androgen control. The secretory activity can serve as a measure of androgen supply to the organ (Mann and Lutwak-Mann 1981). Castration results in cessation of the secretory activity within a few hours and start of the involution process in the seminal vesicles and other accessory sex organs (Aumuller and Seitz 1990). Exogenous testosterone given to the castrated animals can fully restore the involuted organs in both size and secretory function. This reversible regression of the accessory sex organs occurs naturally in seasonally active males, such as deer and sheep. In humans, involution of the gland can result from anti-androgen treatment also.

Male pseudo-hermaphrodites with deficiency of the 5-alpha reductase 2 enzyme have low semen volume and high seminal viscosity (see chapter, Prostate for details; Cai et al. 1994). Since seminal vesicles are responsible for producing about 65–75% of the total semen, the low semen volume indicates hypofunction of the seminal vesicles.

Both testosterone and luteinizing hormone (LH) increase the weight of seminal vesicles when given to prepuberal rats (Foreman and Weeks 1978). The action of LH maybe through testosterone or direct but no research has been done to delineate this.

Intraperitoneal injection of oxytocin in mouse reduces fructose concentration in the seminal vesicle. However, circulating levels of glucose and testosterone, both of which are required for fructose synthesis, remain unaltered indicating a direct effect of oxytocin (Kumar and Farooq 1994).

The proliferative actions of androgens are enhanced by prolactin in the mouse seminal vesicles by increasing its weight as well as DNA levels. This effect of prolactin was not seen in castrated mice; however, androgenic stimulation was enhanced if prolactin and testosterone were administered simultaneously. The mechanism for this is not well understood but prolactin neither increases androgen accumulation nor enhances conversion of testosterone to dihydrotestosterone (DHT) (Thomas and Keenan 1976; Keenan et al. 1981).

3.8 Disorders

Congenital abnormalities of the seminal vesicle are rare and include agenesis and cysts, while the acquired diseases include tumors and inflammation.

3.8.1 Seminal Vesicle Agenesis

Defects in the embryonic development of the mesonephric duct or CFTR gene mutations can lead to congenital seminal vesicle agenesis. Seminal vesicle agenesis always occurs in association with unilateral or bilateral agenesis or ectopic presence of the vas deferens (Wu et al. 2005). Some cases also present with unilateral renal agenesis, possibly arising if the insult to the mesonephric ducts occurred prior to 7 weeks of gestation, i.e. before ureteral budding (Arora et al. 2007). About 80–95% of men with CFTR gene mutations show bilateral agenesis of vas deferens or seminal vesicles (Kavoussi and Costabile 2012). Infertility is the major symptom associated with seminal vesicle agenesis even though testicular spermatogenesis is intact (Bouzouita et al. 2014).

3.8.2 Seminal Vesicle Cyst

Seminal vesicle cysts can be congenital or acquired. Congenital cysts are mostly solitary but may also be associated with upper urinary tract abnormalities or polycystic kidney disease, an autosomal dominant genetic disorder. They are usually detected in men in their twenties or thirties, possibly coinciding with the onset of sexual activity. About two-thirds of the patients with seminal vesicle cysts also have unilateral renal agenesis or dysplasia. This may reflect maldevelopment of the distal mesonephric duct and improper ureteral budding (Arora et al. 2007). Inflammation and obstruction of ejaculatory ducts, caused by urinary infection or calculi, are associated with acquired seminal vesicle cysts.

3.8.3 Seminal Vesicle Tumors

Though primary tumors of the seminal vesicle are rare, secondary involvement of seminal vesicles in the malignant spread of carcinoma of prostate, bladder, or rectum or from lymphoma has been reported (Kavoussi and Costabile 2012). Most primary tumors of seminal vesicle are carcinomas. The glycoprotein CA-125 can be used as a tumor marker in these patients as its serum levels correlate well with the clinical course of the disease (Thiel and Effert 2002).

3.8.4 Inflammation of Seminal Vesicle

Seminal vesiculitis or inflammation of the seminal vesicles can be caused by bacteria commonly found in urogenital infection. Often vesiculitis is seen in combination with prostatitis and can result in abnormal semen parameters (Behre et al. 2010). Recently, low expression of semenogelin I has been seen in association with seminal vesiculitis pointing to an antibacterial role for semenogelin (Liu et al. 2016).

Key Questions

- At which stage of life do the development, maturation, and involution of seminal vesicles take place?
- What are the key components of seminal plasma produced by the seminal vesicles?
- Describe the physiological functions of fructose, prostaglandins, and semenogelin in the seminal plasma.
- Discuss the hormonal regulation of seminal vesicles.

References

Arora SS, Breiman RS, Webb EM, Westphalen AC, Yeh BM, Coakley FV. CT and MRI of congenital anomalies of the seminal vesicles. AJR Am J Roentgenol. 2007;189:130–5.

Aumuller G, Riva A. Morphology and functions of the human seminal vesicle. Andrologia. 1992;24:183–96.

Aumuller G, Seitz J. Protein secretion and secretory processes in male accessory sex glands. Int Rev Cytol. 1990;121:127–231.

Behre HM, Nieschlag E, Weidner W, Wieacker P. Diseases of the seminal ducts. In: Nieschlag E, Behre HM, Nieschlag S, editors. Andrology: male reproductive health and dysfunction. Heidelberg/New York: Springer; 2010. p. 263–78.

Bouzouita A, Kerkeni W, Abouda H, Khrouf M, Elloumi H, Mnif N, et al. Seminal vesicle agenesis: an uncommon cause of azoospermia. Can Urol Assoc J. 2014;8:E266–9.

Burgos MH. Biochemical and functional properties related to sperm metabolism and fertility. In: Brandes D, editor. Male accessory sex organs: structure and function in mammals. New York: Academic Press; 1974. p. 151–60.

Cai LQ, Fratianni CM, Gautier T, Imperato-McGinley J. Dihydrotestosterone regulation of semen in male pseudohermaphrodites with 5 alpha-reductase-2 deficiency. J Clin Endocrinol Metab. 1994;79:409–14.

Drabovich AP, Saraon P, Jarvi K, Diamandis EP. Seminal plasma as a diagnostic fluid for male reproductive system disorders. Nat Rev Urol. 2014;11:278–88.

Foreman D, Weeks J. Decreased fructose concentrations in coagulating glands of prepuberal rats treated with testosterone and LH. Horm Res. 1978;9:225–35.

Frandson RD, Wilke WL, Fails AD. Anatomy and physiology of farm animals. 7th ed. Wiley-Blackwell: Ames; 2009.

Kavoussi PK, Costabile RA. Surgery of the scrotum and seminal vesicles. In: McDougal WS, Wein AJ, Kavoussi LR, Novick AC, Partin AW, Peters CA, Ramchandani P, editors. Campbell-Walsh urology tenth edition review. Philadelphia: Saunders; 2012. p. 190–4.

Keenan EJ, Klase PA, Thomas JA. Interaction between prolactin and androgens in the accessory sex organs of male mice. J Endocrinol. 1981;90:323–30.

Kumar A, Farooq A. Effect of oxytocin on the concentration of fructose in the accessory glands of mouse. Life Sci. 1994;55:PL19–24.

Liu B, Song Z, Su S, Tang M, Xu A, Song N, et al. Abnormal expression of Sg I is closely related to seminal vesiculitis. Urology. 2016;88:227 e9–e14.

Mann T. The biochemistry of semen and of the male reproductive tract. London/New York: Methuen/Wiley; 1964.

Mann T, Lutwak-Mann C. Male reproductive function and semen: themes and trends in physiology, biochemistry, and investigative andrology. Berlin/New York: Springer; 1981.

Mann T, Jones R, Sherins RJ. Fructolysis in human spermatozoa under normal and pathological conditions. J Androl. 1980;1:229–33.

Peterson RN, Freund M. Metabolism of human spermatozoa. In: Hafez ESE, editor. Human semen and fertility regulation in men. Saint Louis: Mosby; 1976. p. 176–86.

Riva A, Usai E, Scarpa R, Cossu M, Lantini MS. Fine structure of the accessory glands of the human male genital tract. In: Motta PM, editor. Developments in ultrastructure of reproduction: progress in clinical and biological research. New York: Liss; 1989. p. 233–40.

Ross MH, Pawlina W. Histology: a text and atlas: with correlated cell and molecular biology. 6th ed. Philadelphia: Wolters Kluwer/Lippincott Williams & Wilkins Health; 2011.

Standring S. Accessory glandular structures. In: Standring S, editor. Gray's anatomy: the anatomical basis of clinical practice. Edinburgh/New York: Elsevier Churchill Livingstone; 2005a. p. 1319.

Standring S. Development of the urogenital system. In: Standring S, editor. Gray's anatomy: the anatomical basis of clinical practice. Edinburgh/New York: Elsevier Churchill Livingstone; 2005b. p. 1373–95.

Thiel R, Effert P. Primary adenocarcinoma of the seminal vesicles. J Urol. 2002;168:1891–6.

Thomas JA, Keenan EJ. Prolactin influences upon androgen action in male accessory sex organs. Adv Sex Horm Res. 1976;2:425–70.

Wu HF, Qiao D, Qian LX, Song NH, Feng NH, Hua LX, et al. Congenital agenesis of seminal vesicle. Asian J Androl. 2005;7:449–52.

Genital Ducts and Other Accessory Sex Glands

4

Surabhi Gupta and Anand Kumar

Learning Objectives

- Development and anatomical features
- Microstructure and functions
- Hormonal regulation

4.1 Introduction

Male accessory sex glands include prostate, seminal vesicles, Cowper's glands or bulbourethral glands, and glands of Littre or periurethral glands. All these glands contribute to the semen composition. The duct system includes rete testis, efferent ductules, epididymis, and vas deferens.

Two of the male accessory sex glands, prostate and seminal vesicles, have already been described in the previous chapters. In this chapter, discussions on rete testis, epididymis, vas deferens, Cowper's gland, and glands of Littre are focused.

4.2 Development

The prostate and Cowper's glands have an endodermal origin, while seminal vesicles, vas deferens, and epididymis originate from the mesoderm. It has been suggested that this difference in embryologic origin may lead to differential predisposition of these organs to diseases.

S. Gupta, PhD (✉) • A. Kumar, MD, FAMS
Department of Reproductive Biology, All India Institute of Medical Sciences,
New Delhi, India
e-mail: surabhi72@rediffmail.com

A. Kumar, M. Sharma (eds.), *Basics of Human Andrology*,
DOI 10.1007/978-981-10-3695-8_4

During the 10th week of development, the epithelial buds from the urogenital sinus give rise to the prostate and Cowper's glands in response to dihydrotestosterone (DHT). In contrast, the seminal vesicles, vas deferens, and epididymis develop from the mesonephric ducts (Wolffian ducts) in response to testosterone (Chughtai et al. 2005).

4.3 Gross Anatomical Features

The epididymis is a long, coiled tube which rests on the backside of each testicle. It continues as a long muscular tube called vas deferens which joins with the seminal vesicle duct to form the ejaculatory duct. This ejaculatory duct transverses the central zone of the prostate and opens into the posterior wall of prostatic urethra at either side of the verumontanum (see Fig. 4.1).

Cowper's glands are paired glands located inside the urogenital diaphragm, just below the prostate on either sides of the urethra. Resembling a pea in shape and size, they are entrapped within the muscles forming the urogenital sinus and have short ducts that drain into the bulbar urethra. The glands of Littre are very small urethral glands situated in the submucous tissue lining the penile urethra (see Fig. 4.1).

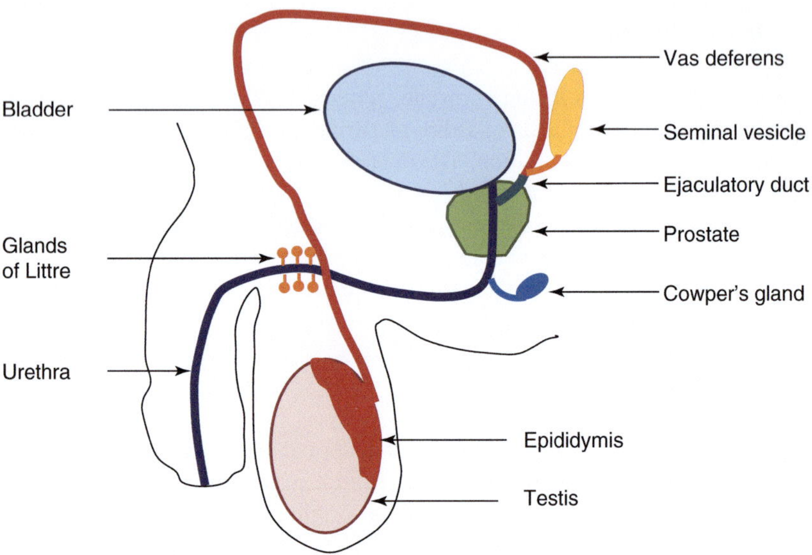

Fig. 4.1 Schematic diagram showing male accessory sex organs in humans

4.4 Microstructure

The seminiferous tubules of the testis are continuous with the excurrent ducts which comprise of straight tubules (tubuli recti) in the beginning followed by the rete testis. They lead to about 12–18 efferent ducts (ductuli efferentes) which connect the rete testis to the epididymis.

The epithelium lining the lumen of this duct system changes from cuboidal epithelium in the excurrent ducts to pseudostratified columnar epithelial cells in the epididymis (see Fig. 4.2). The epithelium of the efferent ducts contains groups of cuboidal epithelial cells alternating with groups of columnar epithelial cells giving it an uneven surface (see Fig. 4.2). The columnar epithelial cells have cilia which are important for transporting the testicular fluid and sperm from the testis to epididymis. In contrast, the cuboidal cells exhibit microvilli and are involved in absorbing most of the fluid produced by Sertoli cells in the seminiferous tubules.

As these efferent ducts transition into the epididymis, the epithelial lining becomes pseudostratified columnar epithelium consisting of tall columnar principal cells with long, non-motile stereocilia and short basal cells near the base of the

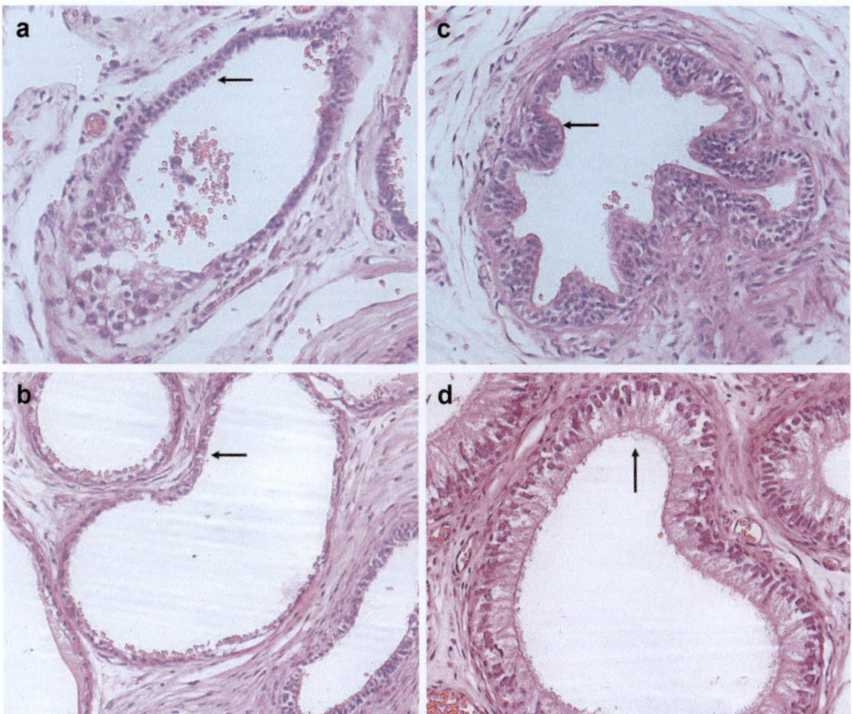

Fig. 4.2 Hematoxylin- and eosin-stained sections showing epithelium of (**a**) straight tubules, (**b**) rete testis, (**c**) efferent duct, and (**d**) epididymis

epithelium (see Fig. 4.2). The columnar epithelial cells have important absorptive and secretory functions. This epithelial layer of the epididymis is surrounded by a thin layer of smooth muscle cells and connective tissue (see Fig. 4.3).

The epididymis is continuous with the vas deferens; however, unlike the lumen of the epididymis, the vas deferens has a narrow and irregular lumen with longitudinal mucosal folds (see Fig. 4.3). Here also, the lumen is lined by pseudostratified columnar epithelium with stereocilia and a thin lamina propria consisting of compact collagen fibers and a fine network of elastic fibers. This is surrounded by three smooth muscle layers – a thin inner longitudinal layer, a thick middle circular layer, and another thin outer longitudinal layer. These muscle layers are surrounded by the adventitia which has abundant blood vessels and nerves (see Fig. 4.3).

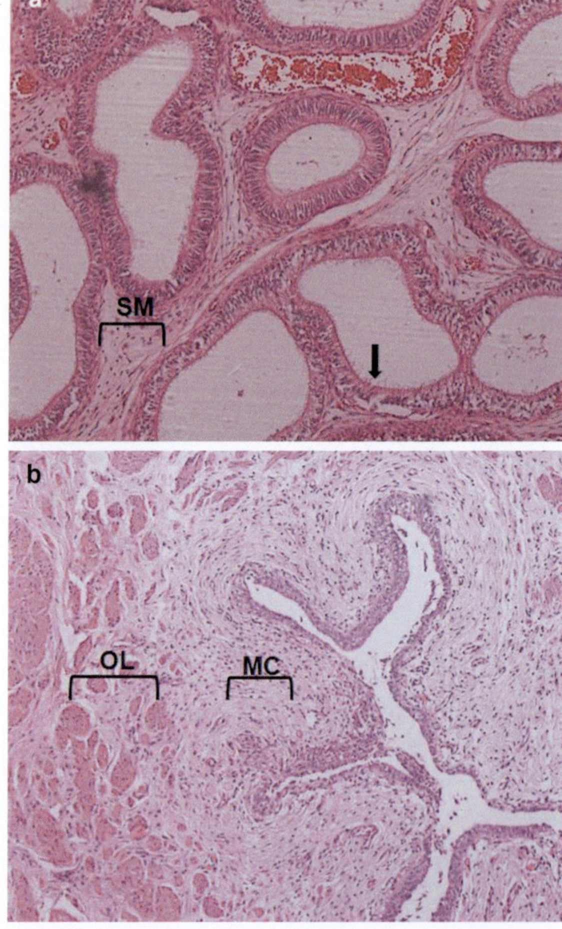

Fig. 4.3 Hematoxylin- and eosin-stained cross sections of (**a**) epididymis and (**b**) vas deferens. *Arrow* shows stereocilia in the epididymal epithelium; *SM* Smooth muscle layer of epididymis, *OL* outer longitudinal muscle layer, *MC* middle circular muscle layer of vas deferens.

Cowper's glands are compound tubuloalveolar glands radiating from a central duct lined by pseudostratified epithelium. They have a thin connective tissue capsule surrounding the simple columnar epithelium (Chughtai et al. 2005). Histologically, they are positive for mucin and smooth muscle actin, variable for prostate-specific antigen (PSA) and CK903, while they stain negative for prostatic acid phosphatase (PAP) and carcinoembryonic antigen (CEA) (Elgamal et al. 1994).

4.5 Functions

4.5.1 Epididymis

The epididymis is the site where the sperm produced by the seminiferous tubules are stored and also undergo maturation by acquiring motility and the capability to fertilize the egg. The stereocilia of the principal cells continue to absorb the excess testicular fluid which was not absorbed during the sperm passage through the efferent ducts. The principal cells are also responsible for clearing the abnormal or degenerating sperm by phagocytosis and also the remaining residual bodies that were not removed by the Sertoli cells in the testis.

Free L-carnitine, glycerophosphocholine (GPC), and neutral α-glucosidase are produced specifically by the epididymis and are used clinically as epididymal markers. Of these, neutral α-glucosidase shows the highest specificity and sensitivity for epididymal disorders (Cooper et al. 1990). Another isoform of α-glucosidase, an acidic form, is produced by the prostate and is present in the seminal plasma. However, selective inhibition of the acidic form by sodium dodecyl sulfate (SDS) allows the specific measurement of the neutral form (WHO 2010). The lower reference limit for neutral α-glucosidase is 20 mU per ejaculate (Cooper et al. 1991).

Both L-carnitine and acetyl-L-carnitine are present in high amounts in the epididymis and are also found in the ejaculated seminal fluid. L-carnitine is a highly polar, small quaternary amine. It plays an important role in sperm metabolism by providing energy to sperm for motility and maturation.

Many proteins secreted by the epididymis interact with and modify the sperm surface during their transit through the epididymis. Some of them have been shown to be important for sperm maturation and fertilization capability but the function of many is not known yet. One of them, P34H, acts as a receptor for the zona protein ZP3 while another one, SPAM1/PH20, a sperm surface hyaluronidase, possibly works as an enzyme for zona penetration and as a zona receptor (Evans et al. 2003). Another epididymal protein, HE6, is a candidate target for male contraception since male mice lacking HE6 counterpart are infertile. However, the exact function, potential ligand(s), or signal transduction pathways used by HE6 have not been elucidated (Kirchhoff et al. 2006). The role of other epididymal proteins in sperm maturation is described in the chapter, The Sperm.

4.5.2 Vas Deferens

The vas deferens transports the mature sperm along with seminal plasma constituents secreted by other accessory sex organs into the penile urethra.

4.5.3 Cowper's Glands

The clear secretion of the Cowper's glands contains mucins which are expelled even before ejaculation and serve to neutralize the acidity of the residual urine in the urethra. It also serves as lubricant during the initial phase of ejaculation. Another physiological role of the Cowper's gland secretions is believed to be immunoprotection of the genitourinary tract since they contain high amounts of immunoglobulin A (IgA).

4.6 Hormonal Regulation

Androgens are the major regulators of the development and function of all accessory sex glands. Epididymal functions are modulated by androgens as well as other hormones and testicular factors. Testicular testosterone reaches the epididymis in two forms – (a) bound to androgen-binding protein via the efferent ducts and rete testis or (b) through the circulation. A receptor-mediated process allows the bound testosterone to enter the principal cells of the epididymis while free testosterone is taken up by passive diffusion. Testosterone is converted to DHT, the key active androgen, inside the cells by 5α-reductase. Removal of testosterone by orchiectomy results in regression of the epididymis with reduction of both luminal diameter and epithelial cell height. This regression can be reversed by administration of exogenous androgen (Hamzeh and Robaire 2009). Androgens also regulate the metabolic and resorptive activity of the epididymal epithelial tissue. The synthesis of many of the secretory proteins including insulin-like growth factor-1 (IGF-1) is also under androgen control (Robaire and Hamzeh 2011).

Both type 1 and type 2 isozymes of 5α-reductase are highly expressed in the epididymis. Inhibitors of 5α-reductase modulate the expression of genes which create the optimal microenvironment for sperm maturation in the epididymal lumen of adult rats. This effect is accompanied by a significant decline in the percentage of progressively motile and morphologically normal sperm in the cauda epididymis (Robaire and Henderson 2006).

Fluid resorption by epithelial cells of epididymis is estrogen dependent and is driven by the absorption of Na^+ ions as well as the androgen-dependent secretion of other components (Cooper and Yeung 2010). Mutant mice lacking ERα show morphological and functional changes in the ducts and inhibition of fluid resorption which results in fluid accumulation and luminal dilation of the ducts (Hess et al. 2000; Joseph et al. 2010).

Key Questions

- What are the embryonic origins of the epididymis, vas deferens, and Cowper's gland?
- Describe the key histological features of the excurrent duct system.
- What are the important functions of epididymis and Cowper's gland?
- Name the markers which are used clinically to assess epididymal dysfunction.

References

Chughtai B, Sawas A, O'Malley RL, Naik RR, Ali Khan S, Pentyala S. A neglected gland: a review of Cowper's gland. Int J Androl. 2005;28:74–7.

Cooper TG, Jockenhovel F, Nieschlag E. Variations in semen parameters from fathers. Hum Reprod. 1991;6:859–66.

Cooper TG, Weidner W, Nieschlag E. The influence of inflammation of the human male genital tract on secretion of the seminal markers alpha-glucosidase, glycerophosphocholine, carnitine, fructose and citric acid. Int J Androl. 1990;13:329–36.

Cooper TG, Yeung CH. Physiology of sperm maturation and fertilization. In: Nieschlag E, Behre HM, Nieschlag S, editors. Andrology: male reproductive health and dysfunction. Heidelberg/New York: Springer; 2010. p. 61–85.

Elgamal AA, Van de Voorde W, Van Poppel H, Lauweryns J, Baert L. Immunohistochemical localization of prostate-specific markers within the accessory male sex glands of Cowper, Littre, and Morgagni. Urology. 1994;44:84–90.

Evans EA, Zhang H, Martin-DeLeon PA. SPAM1 (PH-20) protein and mRNA expression in the epididymides of humans and macaques: utilizing laser microdissection/RT-PCR. Reprod Biol Endocrinol. 2003;1:54.

Hamzeh M, Robaire B. Effect of testosterone on epithelial cell proliferation in the regressed rat epididymis. J Androl. 2009;30:200–12.

Hess RA, Bunick D, Lubahn DB, Zhou Q, Bouma J. Morphologic changes in efferent ductules and epididymis in estrogen receptor-alpha knockout mice. J Androl. 2000;21:107–21.

Joseph A, Hess RA, Schaeffer DJ, Ko C, Hudgin-Spivey S, Chambon P, et al. Absence of estrogen receptor alpha leads to physiological alterations in the mouse epididymis and consequent defects in sperm function. Biol Reprod. 2010;82:948–57.

Kirchhoff C, Obermann H, Behnen M, Davies B. Role of epididymal receptor HE6 in the regulation of sperm microenvironment. Mol Cell Endocrinol. 2006;250:43–8.

Robaire B, Hamzeh M. Androgen action in the epididymis. J Androl. 2011;32:592–9.

Robaire B, Henderson NA. Actions of 5alpha-reductase inhibitors on the epididymis. Mol Cell Endocrinol. 2006;250:190–5.

WHO. WHO laboratory manual for the examination and processing of human semen. 5th ed. Geneva: WHO Press; 2010.

Neurovascular Supply and Lymphatic Drainage of Male Reproductive Organs

5

Mona Sharma and Anand Kumar

Learning Objectives

- Arterial supply
- Draining venous tributaries
- Formation and supply of autonomic plexuses
- Lymphatic drainage

5.1 Introduction

The anatomical details of male reproductive organs also include their neurovascular supply and lymphatic drainage. Male reproductive organs are supplied by arterial branches of the abdominal aorta, and veins from these organs drain into the inferior vena cava. The nerve supply to male reproductive organs comes from various autonomous nerve plexuses. The lymphatics from these organs drain into lymph nodes present around the major blood vessels supplying them. We shall now discuss the anatomical details of the neurovascular supply and lymphatic drainage of male reproductive organs.

M. Sharma, MD, DNB, MAMS (✉) • A. Kumar, MD, FAMS
Department of Reproductive Biology, All India Institute of Medical Sciences, New Delhi, India
e-mail: dr.mona18sharma@gmail.com

A. Kumar, M. Sharma (eds.), *Basics of Human Andrology*,
DOI 10.1007/978-981-10-3695-8_5

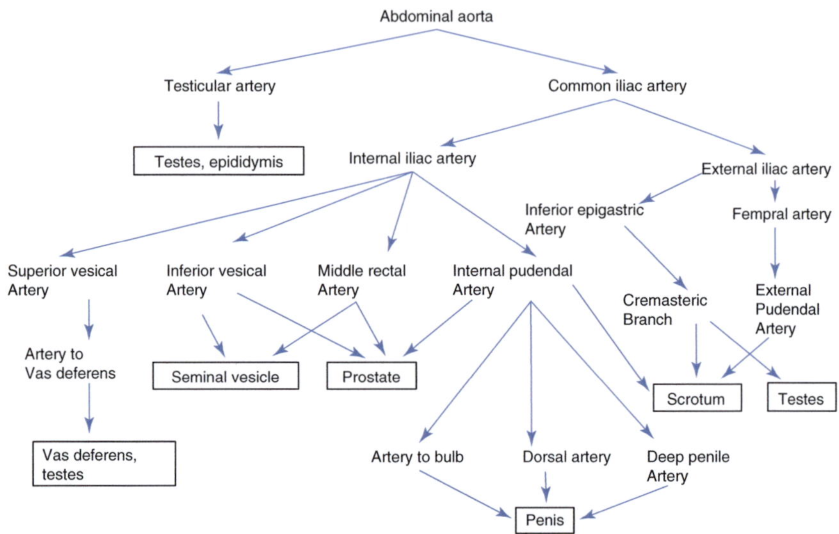

Fig. 5.1 Flowchart showing branches of the abdominal aorta supplying male reproductive organs

5.2 Arterial Supply

The arterial supply to male reproductive organs comes from the branches of the abdominal aorta. The aortic branches are described as anterior, lateral, and dorsal branches. The anterior and lateral branches supply the organs, whereas the dorsal branches supply mainly the body wall. The gonadal arteries are the lateral branches, and the common iliac arteries are terminal branches of the abdominal aorta which supply male reproductive organs (see Figs. 5.1, 5.2, 5.3, and 5.4).

In males, gonadal arteries are the testicular arteries arising from the lateral aspect of the aorta below the origin of renal arteries. Each artery passes obliquely downward to reach the deep inguinal ring, then passes through the inguinal canal, and finally comes out of the superficial inguinal ring. The artery further descends and divides into several branches which supply the testis and epididymis (Standring 2005a). The abdominal aorta divides into common iliac arteries. Each common iliac artery divides into the external and internal iliac artery. The external iliac artery gives the inferior epigastric artery which further gives a cremasteric branch that supplies the testes and scrotum. Then, external iliac artery continues as femoral artery. The femoral artery gives a branch external pudendal artery that supplies the scrotum (Standring 2005a, b).

The internal iliac artery supplies pelvic organs including reproductive organs. The internal iliac artery divides into the anterior and posterior trunk. The branches of the anterior trunk which supply male reproductive organs are superior vesical, inferior vesical, middle rectal, and internal pudendal. The internal pudendal arteries give scrotal branches that supply the scrotum and provide arterial supply to the penis. Each internal pudendal artery gives the following branches: artery to the bulb of the penis which supplies the urethral bulb, the posterior portion of the corpus cavernosum and the bulbourethral gland, the deep penile/corpus

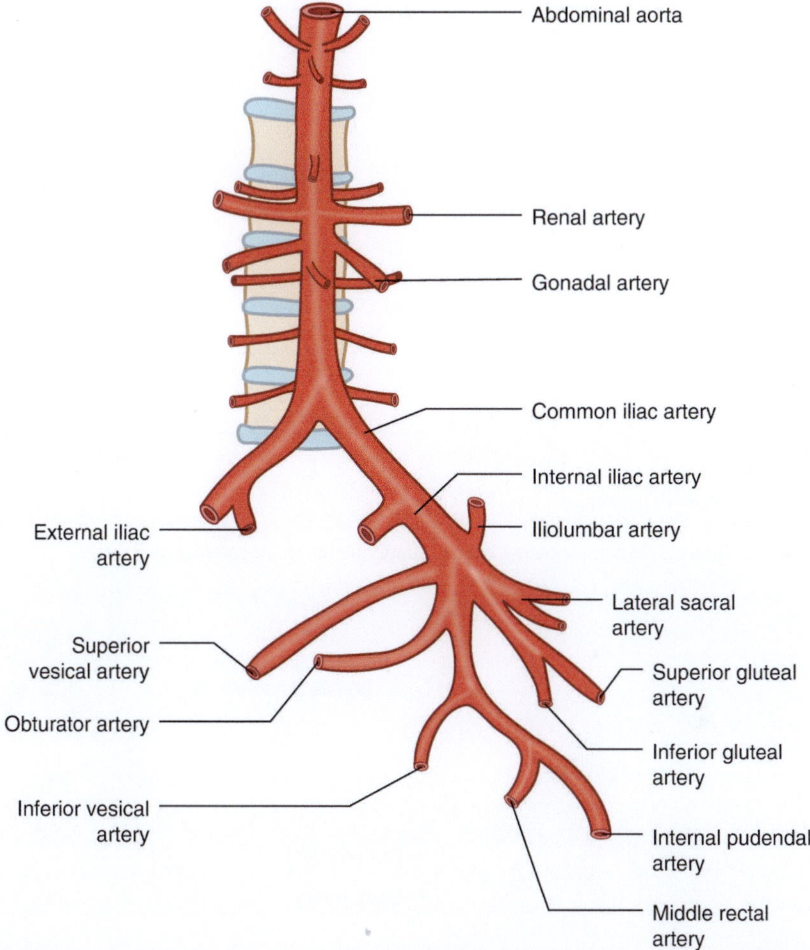

Fig. 5.2 Schematic diagram showing branches of the abdominal aorta supplying male reproductive organs

cavernosal artery which supplies the corpus cavernosum, and, finally, the dorsal artery of the penis which supplies the penile skin, glans penis, and corpus spongiosum (Standring 2005c). The deep penile artery runs in the middle within the corpus cavernosum and gives perpendicular branches called helicine arteries that open in the cavernous sinusoidal space (Krane et al. 1989). During erection, the deep penile artery dilates to twice its diameter as compared to its flaccid state.

The inferior vesical, internal pudendal, and middle rectal arteries supply the prostate (Standring 2005d). All the branches perforate the prostate gland along a posterolateral line extending from its base to apex. The inferior vesical and middle rectal arteries supply the seminal vesicles. Artery to vas deferens is a branch usually from the superior vesical artery; it may also be the branch from the inferior vesical artery. Artery to vas deferens supplies the vas deferens and testes (Standring 2005e).

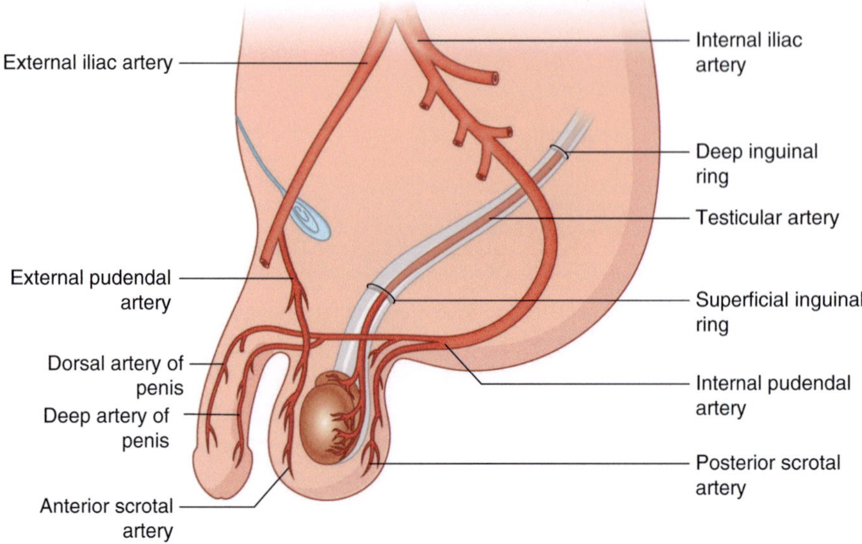

Fig. 5.3 Schematic diagram showing arteries supplying the testis, scrotum, and penis

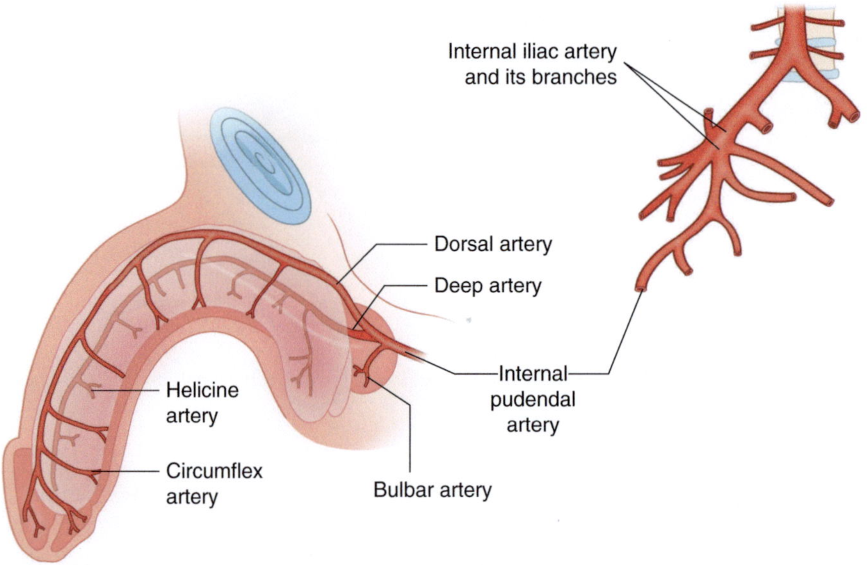

Fig. 5.4 Schematic diagram showing arteries supplying the penis

5.3 Venous Drainage

The veins draining the areas below diaphragm open into the inferior vena cava which conveys blood to the right atrium. The veins draining male reproductive organs and opening into the inferior vena cava are the right testicular vein and common iliac veins. Testicular veins after draining the testis and epididymis form the pampiniform plexus. On reaching the inguinal canal, plexus drains into 3–4 veins. In the abdomen, these veins join and form two veins which ultimately form a single vein. The left testicular vein drains into the left renal vein, whereas the right testicular vein drains into the inferior vena cava (Standring 2005a).

The penis is drained via the superficial and deep venous systems. The skin of the penis is drained by the superficial system of veins, which gives rise to a single superficial dorsal penile vein that eventually drains into the external pudendal, saphenous, and femoral veins; and finally into the external iliac vein. The deep venous system drains the corpora cavernosa, the corpus spongiosum, and penile urethra. The subtunical venules drain the three corporal bodies. The venules interdigitate through cavernous spaces and form subtunical plexus. The plexus drains into emissary veins which penetrate the tunica albuginea. Emissary veins then drain into circumflex veins. The circumflex veins drain into a single deep dorsal vein. The deep dorsal vein empties into the internal pudendal veins and ultimately into the prostatic venous plexus (Standring 2005c). The draining veins form a plexus along the anterolateral aspect of the prostate. The prostatic plexus ultimately drains into the vesical plexus. The veins from seminal vesicles and vas deferens also drain into the vesical venous plexus from where the veins drain into the inferior vena cava through inferior vesical veins (Standring 2005d, e). The venous blood coming from prostate may drain from internal iliac veins into the internal vertebral venous plexus through lateral sacral veins. This may be the route of prostate cancer spread to vertebral bodies or to the skull bones. The veins from the scrotum drain into internal and external pudendal veins and into inferior epigastric veins (Standring 2005b; see Figs. 5.5, 5.6, 5.7, and 5.8).

5.4 Nerve Supply

The branches of autonomic plexuses supply the male reproductive organs. The thoracic spinal segments (T10–T12) form aortic and renal plexuses which give origin to testicular nerves. These nerves pass through the spermatic cord and supply testes (Standring 2005a).

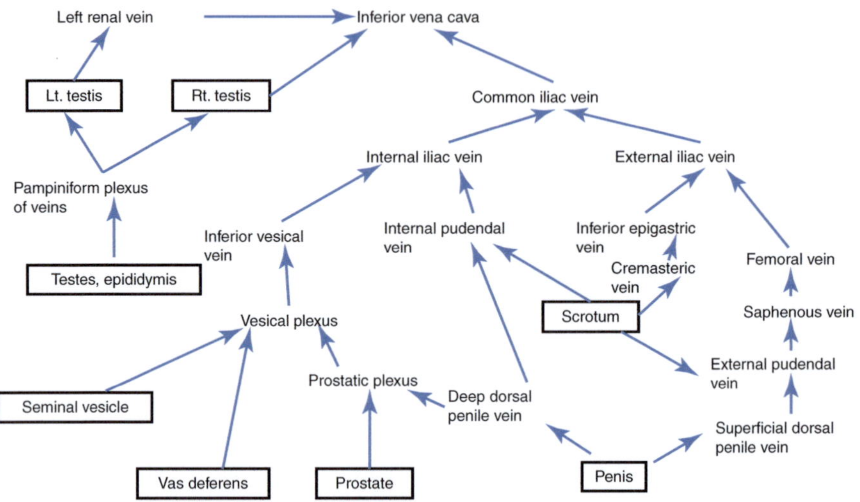

Fig. 5.5 Flowchart showing venous drainage of male reproductive organs

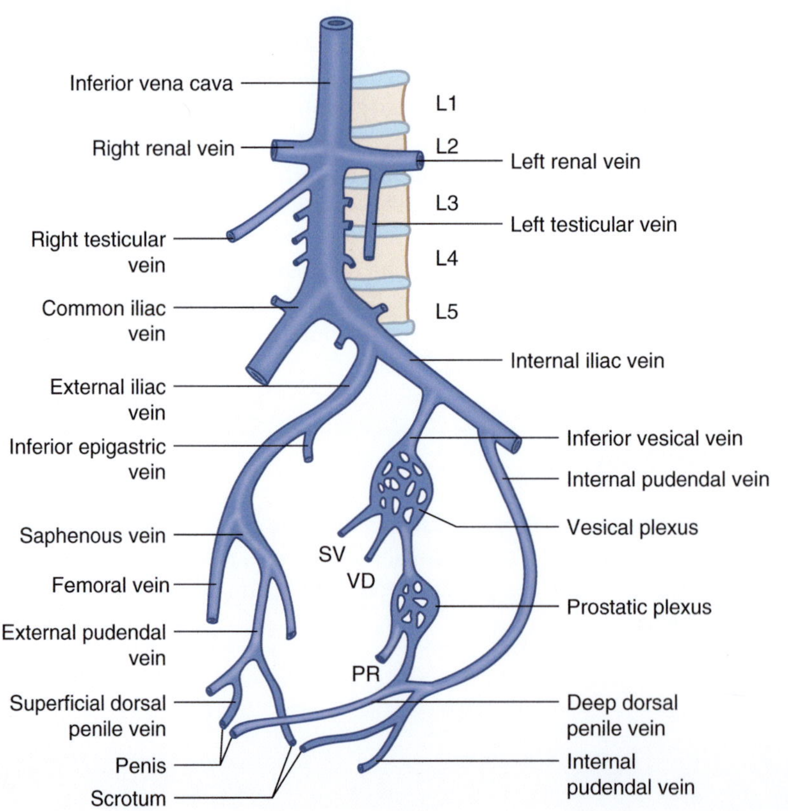

Fig. 5.6 Schematic diagram showing venous drainage of male reproductive organs. *L1–L5* lumbar vertebrae, *SV* seminal vesicle, *VD* vas deferens, *PR* prostate

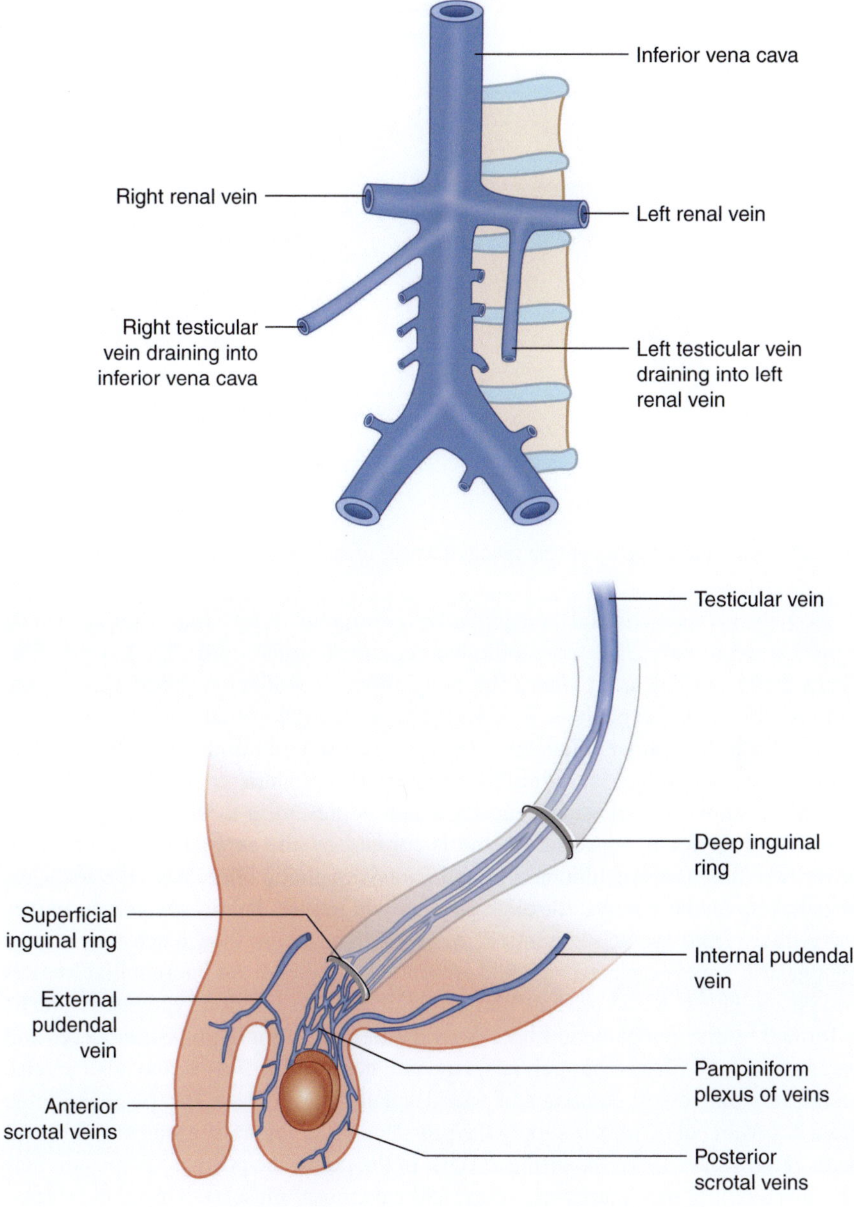

Fig. 5.7 Schematic diagram showing venous drainage of the testis and scrotum

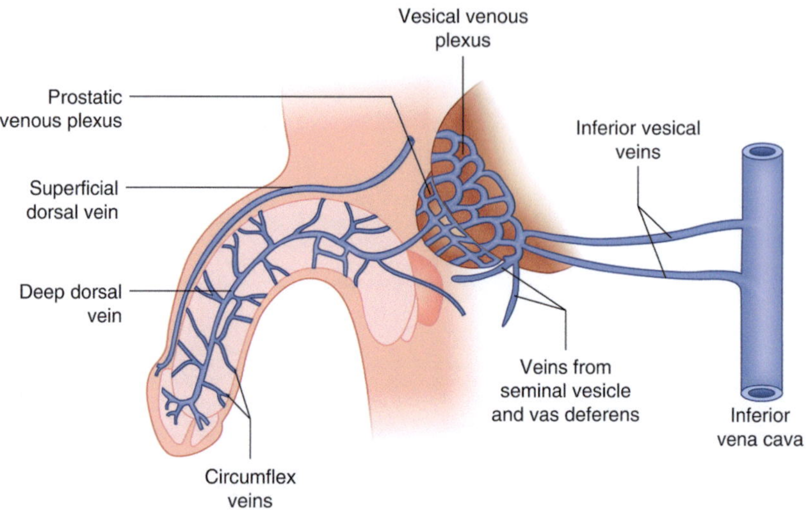

Fig. 5.8 Schematic diagram showing venous drainage of the penis

Both the sympathetic and parasympathetic components give innervations to male reproductive organs. The thoracolumbar region of spinal cord (T_{11}–L_2) provides sympathetic outflow from where the preganglionic cholinergic fibers pass to the sympathetic chain ganglia (see Fig. 5.9). The postganglionic adrenergic fibers then pass through the lumbar splanchnic nerves to reach the inferior mesenteric plexus which is present around the origin of the inferior mesenteric artery. The fibers then pass to the superior hypogastric plexus which is present in front of aortic bifurcation. The superior hypogastric plexus is formed by the sympathetic fibers from lower two lumbar splanchnic nerves and parasympathetic fibers from the branches of pelvic splanchnic nerves through hypogastric nerves. Pelvic splanchnic nerves carry fibers from sacral segments 2, 3, and 4. The fibers then reach the inferior hypogastric or pelvic plexus which is present just medial to the internal iliac vessels and posterolateral to the prostate and seminal vesicles. Inferior hypogastric plexus is formed by the sympathetic fibers from the left and right hypogastric nerves and parasympathetic fibers from pelvic splanchnic nerves. The plexus may also receive branches from lowest lumbar and sacral splanchnic nerves. The postganglionic branches from pelvic plexus supply the prostate, vas deferens, seminal vesicles, and penis (Hollinshed 1966; Standring 2005f). In the penis, the postganglionic adrenergic fibers supply penile arteries, veins, and cavernosal muscles. This causes vasoconstriction and muscle contraction leading to detumescence following orgasm. Penile innervation is also regulated by the supraspinal centers and spinal parasympathetic pathways. The important supraspinal centers for sexual function and penile erection are the medial preoptic area (MPOA) and paraventricular nucleus (PVN) of the hypothalamus and the hippocampus (Sachs and Meisel 1988; Marson et al. 1993). The outflow fibers from MPOA project through the medial forebrain bundle, midbrain tegmental region, and substantia nigra. The projecting nerve fibers then enter the parasympathetic sacral and sympathetic thoracolumbar spinal centers controlling erection (Mc Kenna 1998).

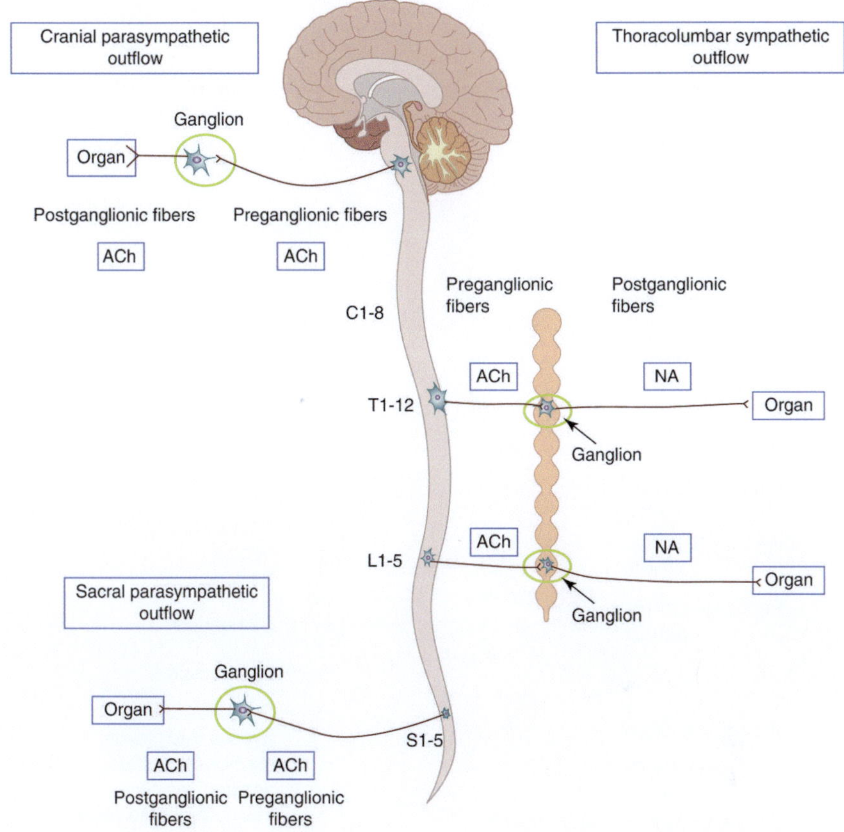

Fig. 5.9 Schematic diagram showing neurons and neurotransmitters of the autonomic nervous system. *Ach* acetylcholine, *NA* noradrenaline, *C1–C8* cervical segment, *T1–T12* thoracic segments, *L1–L5* lumbar segments, *S1–S5* sacral segments of the spinal cord

The parasympathetic fibers after originating from the intermediolateral gray matter of sacral 2 to sacral 4 spinal cord segments leave the spinal cord as preganglionic pelvic splanchnic nerves and run in the retroperitoneal space toward the prostate and perineal membrane. These preganglionic cholinergic fibers join the pelvic or inferior hypogastric plexus. The details of formation of this plexus are already discussed in the previous section. From this plexus, the postganglionic cholinergic fibers arise which form the cavernous nerves. Cavernous nerves supply the erectile tissue of the penis. The rich cholinergic innervations to the vessels lead to vasodilatation and muscle relaxation. This causes increased blood flow into the penis and penile erection (Standring 2005c). The penis is also innervated by somatic nerves. The somatic sensations from the glans, skin over the shaft of the penis, and urethra are carried by dorsal nerves of the penis which join the fibers of the pudendal nerve. The pudendal nerve fibers then enter the sacral 2 to 4 spinal cord segments. Fibers then ascend in the spinal cord and synapse in the thalamus and eventually terminate in the contralateral primary sensory cortex. The somatomotor supply to the penis comes from the sacral 2 to 4 spinal segments. The nerve fibers travel in sacral nerves to join the pudendal nerve. Pudendal nerve then

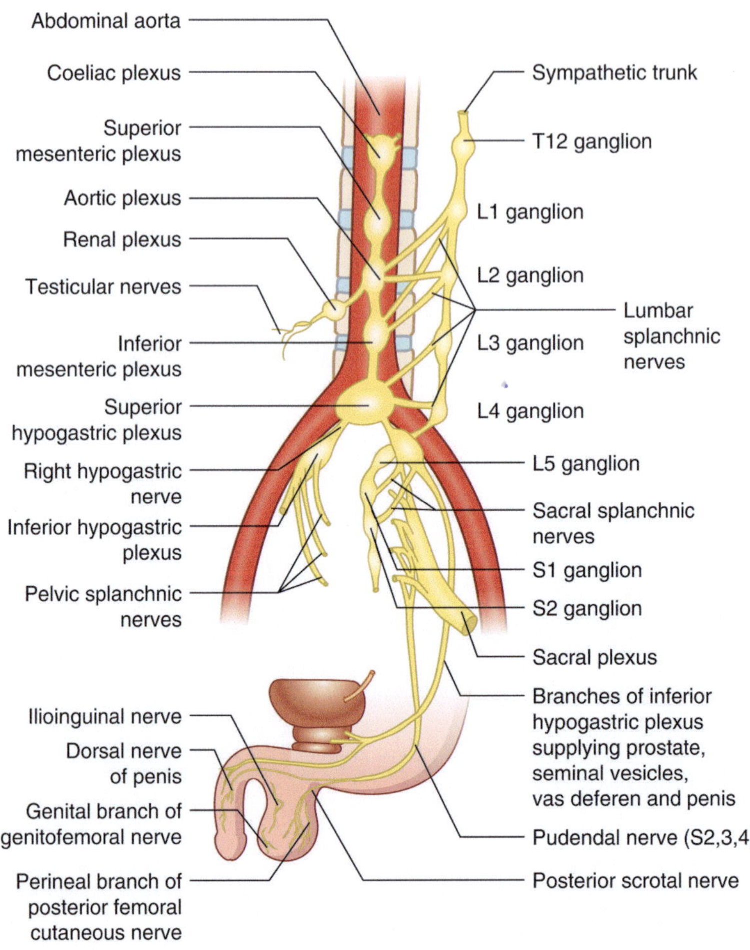

Fig. 5.10 Schematic diagram showing nerve supply of male reproductive organs. *T12* 12th thoracic ganglion, *L1–L5* lumbar ganglia, *S1–S2* sacral ganglia, *S2*, *S3*, *S4* sacral 2–4 spinal nerves

innervates the bulbospongiosus and ischiocavernosus muscles (Goldstein 1988). The contraction of ischiocavernosus muscle produces erection, whereas the rhythmic contraction of bulbospongiosus facilitates ejaculation (Mc Kenna 1998).

Pudendal nerve is formed by the anterior division of sacral 2, 3, and 4 spinal nerves. The pudendal nerve gives posterior scrotal nerves which supply the scrotum. The scrotum is also innervated by the ilioinguinal nerve (L1), genital branch of genitofemoral nerve (L1, L2), and perineal branch of posterior femoral cutaneous nerve (S2, S3) (Standring 2005b, see Fig. 5.10).

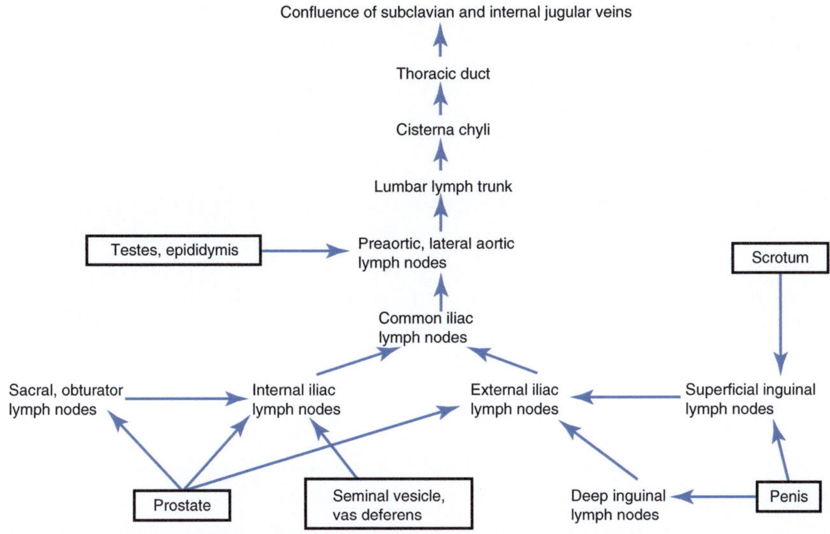

Fig. 5.11 Flowchart showing lymphatic drainage of male reproductive organs

5.5 Lymphatic Drainage

The lymphatic system consists of lymphoid organs like lymph nodes, spleen, thymus, tonsils, and lymphatic vessels. Lymph is the extravasated tissue fluid having plasma proteins and cells. Lymph is first collected by the lymphatic capillaries which drain them to the collecting vessels that carry lymph to the regional lymph nodes. The lymph nodes are connected to the lymphatic trunks which ultimately drain lymph into venous system. The lymph nodes are arranged around the corresponding vessels. The major part of lymph from male reproductive organs is drained to the internal iliac group of lymph nodes; some lymphatics also drain into the external iliac, sacral, and obturator, inguinal, and aortic group of lymph nodes (see Fig. 5.11).

The lymphatics from testes ascend in the spermatic cord and end in lateral aortic and preaortic group of lymph nodes (Standring 2005a). The skin of the penis is drained by the lymphatics accompanying the external pudendal vessels to the superficial group of inguinal lymph nodes. The lymphatics from the glans penis drain into the deep group of inguinal lymph nodes and external iliac lymph nodes. The lymphatics from the penile part of the urethra and the erectile tissue of the penis drain into the internal iliac lymph nodes (Standring 2005c). The lymphatic vessels from the scrotum drain into the superficial group of inguinal lymph nodes (Standring 2005b). Prostate lymphatics from the anterior surface drain into the internal iliac lymph nodes; lymphatics from the posterior surface drain into the external iliac lymph nodes; and lymphatics also drain into the sacral and obturator group of lymph nodes (Standring 2005d). Lymphatics from seminal vesicles and vas deferens drain into the iliac group of lymph nodes (De Krester et al. 1982; Standring 2005e; see Fig. 5.11).

Key Questions

- Name the branches of the abdominal aorta.
- Discuss briefly the arterial supply of testis.
- Name the arteries supplying the penis.
- Discuss briefly the venous drainage of testis.
- Describe the superficial and deep venous systems of the draining penis.
- Draw a flowchart showing lymphatic drainage of male reproductive organs.
- Write briefly the formation of the superior and inferior hypogastric plexus.
- Name the structures innervated by the pudendal nerve.
- Discuss the neural pathway mediating penile erection.

References

De Krester DM, Temple-Smith PD, Kerr JB. Anatomical and functional aspects of male reproductive organs, ch 7-Prostate. In: Bandhauer K, Fricj J, editors. Disturbances in male infertility. Berlin/Heidelberg/New York: Springer; 1982. p. 98–114.

Goldstein I. Evaluation of penile nerves. In: Tanagho EA, Lue TF, McClure RD, editors. Contemporary management of impotence and infertility. Baltimore: Williams and Wilkins; 1988. p. 70–83.

Hollinshed WH. Anatomy for surgeons, vol 2: the thorax, abdomen and pelvis. NewYork: Harper and Row; 1966.

Krane RJ, Goldstein I, Saenz de Tejada I. Impotence. N Engl J Med. 1989;321:1648–59.

Marson L, Platt KB, Mc Kenna KE. Central nervous system innervations of penis as revealed by the transneuronal transport of pseudorabies virus. Neurosci. 1993;55(1):263–80.

Mc Kenna KE. Central control of penile erection. Int J Impot Res. 1998;10(1):S25–34.

Sachs B, Meisel R. The physiology of male sexual behavior. New York: Raven press; 1988. p. 1393–423.

Standring S. Testes and epididymis. In: Gray's anatomy. 39th ed. Edinburgh/New York: Elsevier/Churchill Livingstone; 2005a. p. 1306–10.

Standring S. Spermatic cord and scrotum. In: Gray's anatomy. 39th ed. Edinburgh/New York: Elsevier/Churchill Livingstone; 2005b. p. 1313–4.

Standring S. Penis. In: Gray's anatomy. 39th ed. Edinburgh/New York: Elsevier/Churchill Livingstone; 2005c. p. 1315–7.

Standring S. Prostate. Gray's Anatomy. 39th ed. Edinburgh/New York: Elsevier/Churchill Livingstone; 2005d. p. 1301–4.

Standring S. Vas deferens and ejaculatory ducts. In: Gray's anatomy. 39th ed. Edinburgh/New York: Elsevier/Churchill Livingstone; 2005e. p. 1311–4.

Standring S. Overview of large intestine. In: Gray's anatomy. 39th ed. Edinburgh/New York: Elsevier/Churchill Livingstone; 2005f. p. 1177–86.

Development of the Male Gonad

6

Shilpa Sharma, Anand Kumar, and Devendra K. Gupta

Learning Objectives

- Normal sexual development
- Gonads and sex differentiation
- Testicular descent
- Stages of errors in development

6.1 Introduction

Sexual development is a complex process which begins with fertilization and stops at gamete formation. The phenotypic gender as first apparent at birth seems to be a simple phenomenon, but, in fact, the differences between a male and female are strikingly complex (Wilhelm et al. 2007). The purpose of sexual differentiation is to provide the needful anatomy and physiology for reproduction. The sexual fate is determined following gonadal development. All secondary sexual dimorphisms pursue from the gonadal differentiation and their acquirement of endocrine function.

S. Sharma, MS, MCh, PhD (✉) • D.K. Gupta, MS, MCh, FAMS, FCS, FRCS, DSc
Department of Pediatric Surgery, All India Institute of Medical Sciences, New Delhi, India
e-mail: drshilpas@gmail.com

A. Kumar, MD, FAMS
Department of Reproductive Biology, All India Institute of Medical Sciences,
New Delhi, India

A. Kumar, M. Sharma (eds.), *Basics of Human Andrology*,
DOI 10.1007/978-981-10-3695-8_6

6.2 Normal Sexual Development

According to the Jost Paradigm, sex is determined through:

(a) Chromosomal or genetic sex: influenced by the presence of Y chromosome
(b) Gonadal sex or primary sex: influenced by the presence of testis-determining factor (TDF)
(c) Phenotypic sex or secondary sexual characters: influenced by hormones produced by gonads (Jost 1979)

The chromosomal sex determines the gonadal sex. The gonadal sex in turn decides the fate of internal ducts and type of phenotypic appearance that leads to the gender identity.

6.2.1 Chromosomal or Genetic Sex

The male chromosomal pattern is 46, XY, while the female pattern is 46, XX. During initial development, all embryos are similar with regard to sexual differentiation having bipotential gonads, Mullerian and Wolffian ducts; and bipotential external genitalia.

6.2.2 Gonadal Sex

The gonadal/genital ridge develops during the first week of gestation and becomes a bipotential gonad. The gonadal differentiation is guided by the TDF. During second month of gestation, in the presence of TDF, the indifferent gonad undergoes testicular development. On the contrary, it develops into an ovary in absence of this factor.

6.2.3 Phenotypic Sex

The gonad exerts paracrine effect on the internal duct and influences its development. In the absence of testicular tissue, female internal duct or Mullerian duct develops, and development of female phenotype occurs. Similarly, in the presence of testicular tissue along with adequate testosterone and Mullerian inhibiting substance (MIS), male internal duct or Wolffian duct develops, and phenotype becomes masculine. MIS is produced by Sertoli cells of testes during eighth week of gestation. In a male fetus with normal testicular functions, testosterone stimulates development of Wolffian duct, and MIS represses Mullerian duct development. As a result, the epididymis, vas deferens, and seminal vesicles are formed (see Fig. 6.1; Wilhelm et al. 2007). In contrast to testes, the follicular cells of ovary are not active in steroid production before puberty.

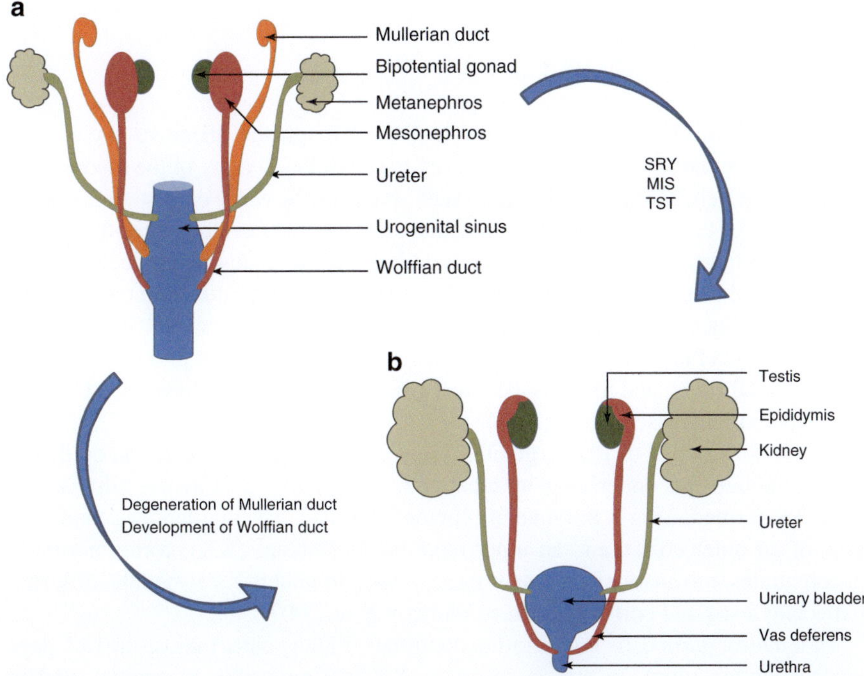

Fig. 6.1 Schematic diagram showing development of the male internal genitalia from the Wolffian duct (**a**) and degeneration of Mullerian duct (**b**); *MIS* Mullerian inhibiting substance, *SRY* sex-determining region Y, *TST* testosterone

The external genitalia are bipotential till 7 weeks of intrauterine life. Without androgen effect, the external genitalia appear phenotypically female. In the absence of MIS, normal vagina is formed.

If dihydrotestosterone (DHT) is sufficient from seventh to eighth week of gestation until birth and fetal response is normal, the bipotential genitalia develop into male type (Siitera and Wilson 1974). The genital tubercle forms the glans. The urethral folds from both sides fuse in midline to form urethral meatus at the glans tip. The scrotum is formed by fusion of two genital swellings. The prostate develops from urogenital sinus. The development of prostate gland has been described in the chapter, The Prostate Gland.

6.3 Gonads and Sex Differentiation

Various genes including some autologous ones are involved in sexual differentiation. The sex-determining region Y (SRY) gene gives rise to a protein product; the presence of which in the somatic cells leads to male development. This step is the primary sex determination. Testosterone production then determines the further sexual differentiation. The gonads are derived from the following embryonic structures:

(a) Mesothelium
(b) Underlying mesenchyme
(c) Primordial germ cells (PGCs)

PGCs originate in the epiblast and migrate through the primitive streak. The PGCs migrate during the fourth week along the dorsal mesentery of the hindgut and arrive at the genital ridge in the fifth week. At the fifth week of gestation, on the medial side of the mesonephros, a thickened mesothelium develops. Proliferation of the mesothelial cells and condensation of underlying mesenchyme produces a bulge called genital ridge. The factors involved in increased proliferation and migration of cells from adjacent mesonephros include hepatocyte growth factor (HGF), platelet-derived growth factor (PDGF), and neurotrophin-3 (NT-3). PGCs lose their motility in the genital ridge, and they invade genital ridges during the sixth week. PGCs are thus bipotential (Hilscher et al.1974).

Just before arrival of PGCs, genital ridge epithelium proliferates, and epithelial cells penetrate the underlying mesenchyme. Finger-like primitive epithelial sex cords are formed. This is the state of bipotential gonad. The bipotential gonad consists of an outer cortex and an inner medulla. In females (XX), cortex normally differentiates into an ovary and medulla regresses. In males (XY), medulla differentiates into testis and cortex regresses (Wilhelm et al. 2007).

In females, germ cells proliferation continues till they enter meiosis at 13.5 days postconception (dpc). In female embryos, XX PGCs develop as oocytes, and in male embryos, XY PGCs develop as prospermatogonia (Palmer and Burgoyne 1991b). Certain processes occur only in males after SRY expression onset (Capel et al. 1999). Sertoli cells get polarized and reside within the testis. The size of gonads increases.

The female gonad continues to appear indifferent and does not express SRY. The primitive sex cords are ill-defined. The in-growth of coelomic and mesonephric epithelial cells is followed by their condensation in the more cortical regions. The mesenchymal cells secrete outer membrana propria. The mesenchymal cells give rise to the granulosa cells of follicles, while the oogonia give rise to oocytes. During the 20th–25th week, primordial follicles give rise to the primary follicles. Each follicle contains a single germ cell. The Mullerian duct differentiates into the oviducts, uterus, cervix, and upper vagina, while the Wolffian duct degenerates in females.

Under the influence of SRY gene, there is proliferation of primitive sex cords that penetrate deeper into the medulla to form the medullary or testis cords. With further differentiation, tubules of rete testis are formed. A dense layer of fibrous connective tissue or tunica albuginea is formed, which separates the testis cords from surface epithelium. During the fourth month, testis cords become continuous with those of the rete testis. The testis cords are composed of two types of cells – Sertoli cells and primitive germ cells. Both cell types are derived from the surface epithelium of the gonad. Leydig cells develop shortly after the beginning of the differentiation of testis cords and are derived from the original mesenchyme of the genital ridge. The mesenchyme is present between the testis cords. Leydig cells produce testosterone.

Testes are able to influence differentiation of the genital ducts and external genitalia by testosterone; the secretion of which starts around the eighth week.

At puberty, testis cords acquire lumen and transform into the seminiferous tubules. The seminiferous tubules join rete testis which enters efferentes ductules. The efferent ductules link rete testis and the Wolffian duct, which becomes the ductus deferens (Wilhelm et al. 2007).

6.3.1 Descent of Testes

During intrauterine life, testes descend from the initial intra-abdominal site at tenth thoracic vertebral level to scrotal location at around 33–35 weeks of gestation or just before birth in humans. In abdomen, the mesonephric gonadal complex is held in position by cranial suspensory ligament and gubernaculum (see Fig. 6.2). Cranial suspensory ligament runs from cranial portion of the gonad to the diaphragm. Gubernaculum is mesenchymal in origin and is attached to the caudal portion of the gonad and extends to the peritoneal floor where it is attached to the fascia between

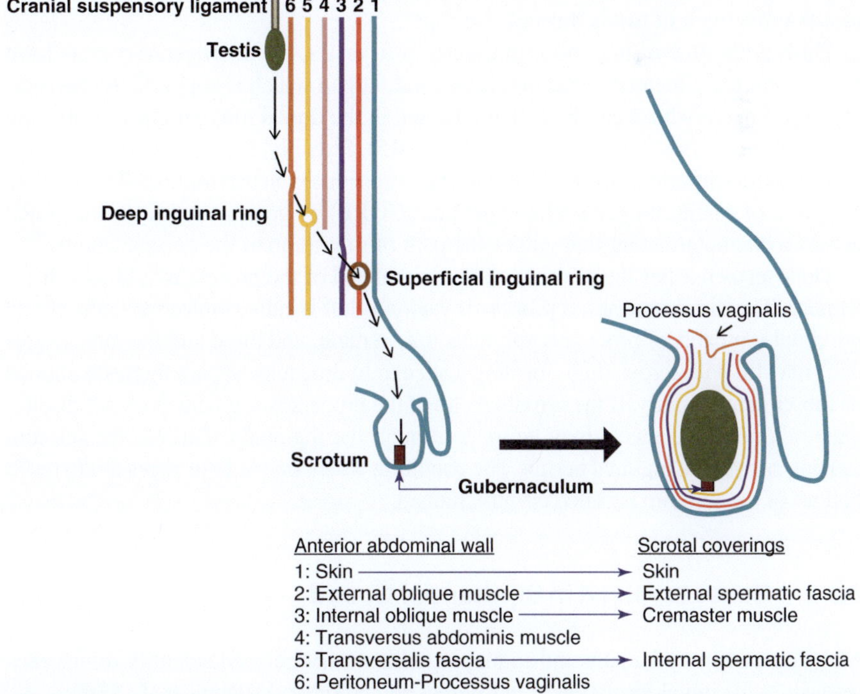

Fig. 6.2 Schematic diagram showing descent of testis

the developing external and internal oblique abdominal muscles in the region of labioscrotal swellings (Hunter 1762).

Descent of testes occurs in two phases – transabdominal phase followed by inguinoscrotal phase. After 8 weeks of gestation, the peritoneum just ventral to the gubernaculum evaginates forming processus vaginalis. Further evagination of processus vaginalis pushes out extensions of transversalis fascia, internal oblique muscle, and external oblique muscle which eventually form the internal spermatic fascia, cremaster muscles, and external spermatic fascia, respectively, all forming the coverings of the scrotum (see Fig. 6.2).

Due to the evagination of processus vaginalis, inguinal canal is formed which extends from deep inguinal ring to superficial inguinal ring. The deep ring is situated at the point where fascia transversalis is pushed inferiorly by the processus vaginalis. Similarly, superficial ring is formed where external oblique muscle is pushed inferiorly by the processus vaginalis.

The processus vaginalis evaginates up to scrotum after which gubernaculum shortens and pulls gonads through inguinal canal into scrotum. The gubernaculum expresses homeobox gene (HOX A10) and fibroblast growth factor-10 (FGF10) which control gubernacular enlargement by cell proliferation (Nightingale et al. 2008). Gubernacular swelling has an additional role of enlarging inguinal canal. The increased abdominal pressure created by the growth of abdominal viscera also assists movement of testes through the canal.

During the time of inguinoscrotal descent of testes, the androgen receptors have been found to be located in the inguinoscrotal fat pad which is supplied by the genitofemoral nerve whose cell bodies are located in the dorsal root ganglion at first and second lumbar spinal nerves (Hutson et al. 2015). Possibly, androgens bind to androgen receptors in inguinoscrotal fat pad which produces neurotrophins that regulate secretion of calcitonin gene-related peptide (CGRP) by genitofemoral nerve. CGRP acts as a chemoattractant that guides the path of migration of the gubernaculum.

Once testis reaches the scrotum, the proximal end of the processus vaginalis disintegrates. The distal remnant of processus vaginalis left is called tunica vaginalis. Some proximal remnants of processus vaginalis may remain, and these and the tunica vaginalis may fill with serous fluid, forming testicular hydrocoeles in pathologic conditions or subsequent to injury. If the proximal end of the processus vaginalis does not disintegrate, abdominal contents may herniate through the inguinal canal into the scrotum causing congenital inguinal hernia. The condition where one or both testes fail to reach bottom of the scrotum is called cryptorchidism.

6.3.2 Genes Involved in Bipotential Gonad

Mutation analysis in humans and animal experiments has helped to identify some genes relevant in the initial formation of indifferent genital ridge (Wilhelm et al. 2007).

Wt1 (Wilms' tumor suppressor 1) The sites of its location are the mesonephros, urogenital ridge, and developing gonads. Wt1 is expressed in the developing granulosa cells in females and Sertoli cells in males (Pritchard-Jones et al. 1990; Hanley et al. 1999; Hammes et al. 2001).

SF1 (Steroidogenic Factor 1) SF1 is present in the developing urogenital ridge, hypothalamus, and the anterior pituitary gland. After sexual differentiation, it is detected in the testes, in both Leydig and Sertoli cells (Hanley et al. 1999; El-Khairi and Achermann 2012; Suntharalingham et al. 2015).

Lhx9 (LIM homeobox gene 9) Lhx9 is present in the developing urogenital ridge and is involved in protein-protein interactions. The gene has a nucleic acid-binding homeobox domain (Birk et al. 2000).

Emx2 (Empty-spiracles homeobox gene 2) Emx2 is expressed in the developing dorsal telencephalon and in the epithelial parts of the urogenital system. In Emx2-/- mutants, the Mullerian duct never develops (Miyamoto et al. 1997).

6.4 Stages of Errors in Development

6.4.1 Chromosomal or Genetic Sex

The SRY is an important gene locus for testicular development (Swain and Lovelt-Badge 1999). Many transcription factors like SF-1, WT-1, DAX-1, and SOX-9 are also important for gonadal development (Migeon et al. 2002; Mendonca et al. 2002).

Two genes on the SRY region have a major role in male differentiation, SOX9 and FGF9, contributing to testicular cord formation. A novel mutation of the SRY gene has been reported in a XY sex reversal patient (Zhou et al. 2003). A homozygous SOX9 deletion with female external genitalia results in a XY karyotype (Chen et al. 2012). 46, XY sex-reversed patients are associated with mutations in known genes such as SRY, SOX9, WT1, and SF1, but the association is found in less number of patients. This indicates that other unknown genes may influence sex determination (Pannetier et al. 2004). Chromosome X duplications, chromosomes 9 and 10 deletions, and chromosome 17 translocations have been associated with abnormal testicular differentiation, in male-to-female sex reversal in 46, XY individuals (Flejter et al. 1998). SRY protein has been expressed in tubules of streak gonads and rete testis (Salas –Cortes et al. 2000). SRY protein expression is reported in both testicular and ovarian tissues in 46, XX true hermaphrodites. In this case, SRY protein does not inhibit ovary development. It may be suggested that other factors are required for complete testis development, especially downstream of the SRY protein (Salas –Cortes et al. 2000).

6.4.2 Gonadal Sex or Primary Sex Determination

Somatic mesenchyme forms the gonadal stroma. PGCs migrate and colonize in the bipotential gonad. In mosaic gonads, some XX cells may form Sertoli cells, while XY cells may form granulosa cells (Palmer and Burgoyne 1991a). Transgenic mice that express a reporter protein under the control of the SRY promoter shows expression in granulosa cells in an XX gonad, apart from expression in Sertoli cells in XY gonads (Albrecht and Eicher 2001). Steroid-secreting Leydig cells of testis and theca cells of ovary are probably derived from a single precursor.

6.4.2.1 Precursors for Internal and External Genitalia

The primordia for both male and female internal genital ducts are derived respectively from the Wolffian ducts in males and para-mesonephric or Mullerian duct in females.

The external genitalia in both sexes develop from a single bipotential precursor through the action of androgens. At 6 weeks gestation, failure of conversion of testosterone to DHT or nonaction of DHT leads to incomplete masculinization. There is associated rise in testosterone in response to luteinizing hormone (LH) surge. The rise in testosterone levels remains until the 14th week which promotes phenotypic differentiation during this period. After the 14th week, fetal testosterone levels decrease. But, continued action of testosterone contributes to the growth of the phallus.

In females, urethral folds and genital swellings remain separate forming labia minora and majora, respectively, while genital tubercle forms the clitoris. Androgens in males cause urethral folds to fuse forming urethra. The genital swellings fuse in the midline and form scrotum. The genital tubercle expands and forms glans penis.

6.4.2.2 Genes Involved in Duct Formation

Pax2 is important during development of urogenital structures, and it is expressed in the epithelial derivatives, condensing mesenchyme, Wolffian ducts, and mesonephric tubules (Torres et al. 1995).

Wnt 7a is expressed at the anterior mesonephros and throughout the Mullerian duct (Parr and McMahon 1998).

6.4.3 Phenotypic Sex or Secondary Sex Differentiation

The development of male phenotypic sex is governed by the testicular hormones – AMH and androgens (Jost 1972). The response of urogenital sinus and external genitalia to androgen requires conversion of testosterone into DHT by 5α-reductase enzyme. DHT binds directly to specific high affinity receptor proteins; the hormone-receptor complex induces an increase in transcription of specific mRNA inducing, as final effect, male phenotypic differentiation and in particular virilization of the external genitalia and male urethra. In the ovary, lack of differentiation persists.

6.5 Male Accessory Sex Glands

In males, two epithelial tissues-the epithelial mesodermal origin of the Wolffian duct and the epithelial endodermal origin of the urogenital sinus, are the source of origin of the accessory sex glands. Prostate gland and Cowper's gland are endodermal in origin. The seminal vesicle has a mesodermal origin. The details of development of genital ducts and male accessory sex glands have been described in the chapter, Genital Ducts and Other Accessory Sex Glands.

Key Questions

1. Describe the types of sexual development.
2. Describe the role of SRY gene in gonadal development.
3. What are the various genes that influence sexual development?
4. Discuss the two phases of testicular descent.

References

Albrecht K, Eicher E. Evidence that Sry is expressed in pre- Sertoli cells and Sertoli and granulosa cells have a common precursor. Dev Biol. 2001;240:92–107.

Birk OS, Casiano DE, Wassif CA, Cogliati T, Zhao L, Zhao Y, et al. The LIM homeobox gene Lhx9 is essential for mouse gonad formation. Nature. 2000;403:909–13.

Capel B, Albrecht KH, Washburn LL, Eicher EM. Migration of mesonephric cells into the mammalian gonad depends on Sry. Mech Dev. 1999;84:127–31.

Chen SY, Lin SJ, Tsai LP, Chou YY. Sex-reversed acampomelic campomelic dysplasia with a homozygous deletion mutation in SOX9 gene. Urology. 2012;79:908–11.

El-Khairi R, Achermann JC. Steroidogenic factor-1 and human disease. Semin Reprod Med. 2012;30:374–81.

Flejter WL, Fergestad J, Gorski J, Varvill T, Chandrasekharappa S. A gene involved in XY sex reversal is located on chromosome 9, distal to marker D9S1779. Am J Hum Genet. 1998;63:794–802.

Hammes A, Guo JK, Lutsch G, Leheste JR, Landrock D, Ziegler U, et al. Two splice variants of the Wilms' tumor 1 gene have distinct functions during sex determination and nephron formation. Cell. 2001;106:319–29.

Hanley NA, Ball SG, Clement-Jones M, Hagan DM, Strachan T, Lindsay S, et al. Expression of steroidogenic factor 1 and Wilms' tumour 1 during early human gonadal development and sex determination. Mech Dev. 1999;87:175–80.

Hilscher B, Hilscher W, Bulthoff-Ohnolz B, Kramer U, Birke A, Pelzer H, et al. Kinetics of gametogenesis. I. Comparative histological and autoradiographic studies of oocytes and transitional prospermatogonia during oogenesis and prespermatogenesis. Cell Tissue Res. 1974;154:443–70.

Hunter J. Observations on the state of the testis in the foetus and on the hernia congenita. In: Hunter W, editor. William Hunter's medical commentaries. London: Hamilton; 1762. p. 75–89.

Hutson JM, Li R, Southwell BR, Newgreen D, Cousinery M. Regulation of testicular descent. Pediatr Surg Int. 2015;31:317–25.

Jost A. A new look at the mechanism controlling sex differentiation in mammals. Johns Hopkins Med J. 1972;130(1):38–53.

Jost A. Basic sexual trends in the development of vertebrates. Sex, hormones and behavior. Amsterdam: Elsevier; 1979. CIBA Foundation Symposium no. 62.

Mendonca BB, Domenice S, Costa EMF, Arnhold IJP. Male pseudohermaphroditism. In: New M, editor. Paediatric Endocrinology. www.Endotext.org. 2002.

Migeon CJ, Wisniewski AB, Brown TR. Androgen insensitivity syndrome. In: Chrousos G, editor. Hormone resistance and hypersensitivity States. Baltimore/Lippincott: Williams and Wilkins; 2002. p. 409–39.

Miyamoto N, Yoshida M, Kuratani S, Matsuo I, Aizawa S. Defects of urogenital development in mice lacking Emx2. Development. 1997;124:1653–64.
Nightingale SS, Western P, Hutson JM. The migrating gubernaculum grows like a "limb bud". J Pediatr Surg. 2008;43(2):387–90.
Palmer SJ, Burgoyne PS. In situ analysis of fetal, prepuberal and adult XX_XY chimaeric mouse testes: Sertoli cells are predominantly, but not exclusively, XY. Development. 1991a;112:265–8.
Palmer SJ, Burgoyne PS. XY follicle cells in the ovaries of XO/XY and XO/XY/XXY mosaic mice. Development. 1991b;111:1017–9.
Pannetier M, Mandon-Pépin B, Copelli S, Fellous M. Molecular aspects of female and male gonadal development in mammals. Paediatr Endocrinol Rev. 2004;1:274–87.
Parr BA, McMahon AP. Sexually dimorphic development of the mammalian reproductive tract requires Wnt-7a. Nature. 1998;395:707–10.
Pritchard-Jones K, Fleming S, Davidson D, Bickmore W, Porteous D, Gosden C, et al. The candidate Wilms' tumour gene is involved in genitourinary development. Nature. 1990;346:194–7.
Salas-Cortés L, Jaubert F, Nihoul-Feketé C, Brauner R, Rosemblatt M, Fellous M. SRY protein is expressed in ovotestis and streak gonads from human sex-reversal. Cytogenet Cell Genet. 2000;91:212–6.
Siitera PK, Wilson JD. Testosterone formation and metabolism during male sexual differentiation in the human embryo. J Clin Endocrinol Metab. 1974;38(1):113–25.
Suntharalingham JP, Buonocore F, Duncan AJ, Achermann JC. DAX-1 (NR0B1) and steroidogenic factor-1 (SF-1, NR5A1) in human disease. Best Pract Res Clin Endocrinol Metab. 2015;29:607–19.
Swain A, Lovell-Badge R. Mammalian sex determination: a molecular drama. Genes Dev. 1999;13:735–67.
Torres M, Gómez-Pardo E, Dressler GR, Gruss P. Pax-2 controls multiple steps of urogenital development. Development. 1995;121:4057–65.
Wilhelm D, Palmer S, Koopman P. Sex determination and Gonadal development in mammals. Physiol Rev. 2007;87:1–28.
Zhou C, Li LY, Fu JJ, Mo YQ, Zhong CG, Lu GX. 46, XY female sex reversal patient with a novel point mutation in the coding sequence of the SRY gene. Zhonghua Yi Xue Yi Chuan Xue Za Zhi. 2003;20(5):369–72.

The Human Y Chromosome

7

Manish Jain, Amanpreet Kaur Kalsi, Prashant Kumar, and Ashutosh Halder

Learning Objectives

- Structure of the Y chromosome
- Evolution of Y chromosome from autosomes
- Genes on human Y chromosome
- SRY gene and sex determination
- Azoospermia factors
- Copy number variations
- Functions
- Y chromosome-linked disorders

7.1 Introduction

Y chromosome is an important chromosome that plays a crucial role in determination and differentiation of testes in male. Y chromosome has genes both testes specific and ubiquitous; thus, it also performs functions other than reproduction. Deletions, duplications, inversions, and other mutations are very common with Y chromosome due to

M. Jain, PhD • A.K. Kalsi, MSc • P. Kumar, MSc, MPhil
A. Halder, MD, DNB, DM, FAMS, MRCOG-I (✉)
Department of Reproductive Biology, All India Institute of Medical Sciences, New Delhi, India
e-mail: om_704@yahoo.co.in; kalsi.amanpreetkaur@gmail.com; prashant03micro@gmail.com; ashutoshhalder@gmail.com

A. Kumar, M. Sharma (eds.), *Basics of Human Andrology*,
DOI 10.1007/978-981-10-3695-8_7

absence of homologous pair and minimal crossing over in meiosis. Holandrically transmitted non-recombining region of Y (NRY) has made possible unraveling the mystery of human evolution. With only 1/6 the size of its partner X chromosome and merely 1/12 the number of genes that it carries, Y chromosome presents clear signs of a degenerated chromosome. We shall now describe the Y chromosome in detail.

7.2 Structural Components

Humans have two sex chromosomes – X chromosome and Y chromosome in male and two X chromosomes in female. The length of human Y chromosome is about 60 megabase (Mb) and represents about 2% of human genome (Murci and Fellous 2001). The Y chromosome is a small acrocentric chromosome and is placed in G group of human karyogram. Y chromosome has very short "p" arm (Yp) and relatively long "q" arm (Yq; see Fig. 7.1).

Y chromosome is one of the major contributors of sex determination in male. Based on the genetic deletion map, human Y chromosome is divided into 7 intervals – deletion interval 1 to 7 which are further subdivided into subintervals (see Fig. 7.2). Intervals 1, 2, and 3 are present on Yp; interval 4 lies at peri-centromere, while intervals 5, 6, and 7 are present on Yq (Vergnaud et al. 1986).

Cytogenetic banding study divides human Y chromosome into distinct regions – pseudoautosomal region (PAR) 1, 2, and 3, euchromatic regions, and heterochromatic region (see Fig. 7.2). PAR is the region which pairs with the respective region on X chromosome and undergoes exchange during meiosis. PAR 1 is 2771.47 kb (genomic coordinate Y:10,000–2,781,479) and is present at Yp terminal; PAR2 is 329.5 kb in size (genomic coordinate Y:56,887,902–57,217,415) and is present at Yq terminal. PAR 3 lies 700 kb away from PAR1 at Yp11.2 and is found in only 2% of general population. PAR3 shows unequal allelic recombination with its X homologue, i.e., Xq21.3 (Veerappa et al. 2013).

PAR1 and PAR2 comprise about 5% of the total Y chromosome (see Fig. 7.2). The rest 95% of Y chromosome comprises the NRY, also called male-specific Y (MSY) region, which includes both heterochromatic and euchromatic regions. The heterochromatic region lies at Yq12 locus on the distal end of Y chromosome. The heterochromatic region is genetically inert and comprises of mainly two extremely repetitive sequences – DYZ1 and DYZ2 – each having 5000 and 2000 copies, respectively. The length of the heterochromatic region varies in different populations due to

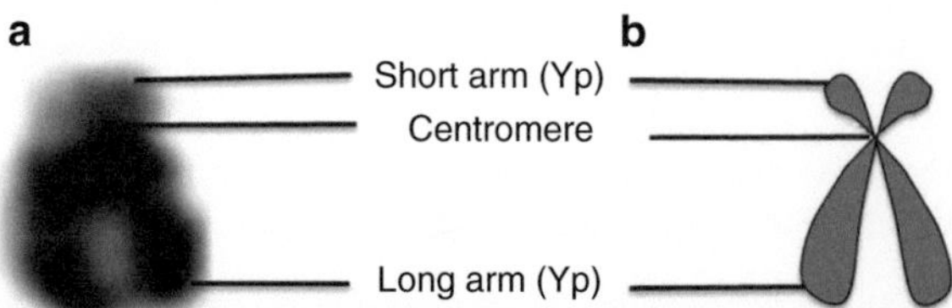

Fig. 7.1 (**a**) Unstained image of the Y chromosome taken from metaphase spread of male blood lymphocytes and (**b**) schematic diagram showing very short p arm and relatively long q arm of Y chromosome

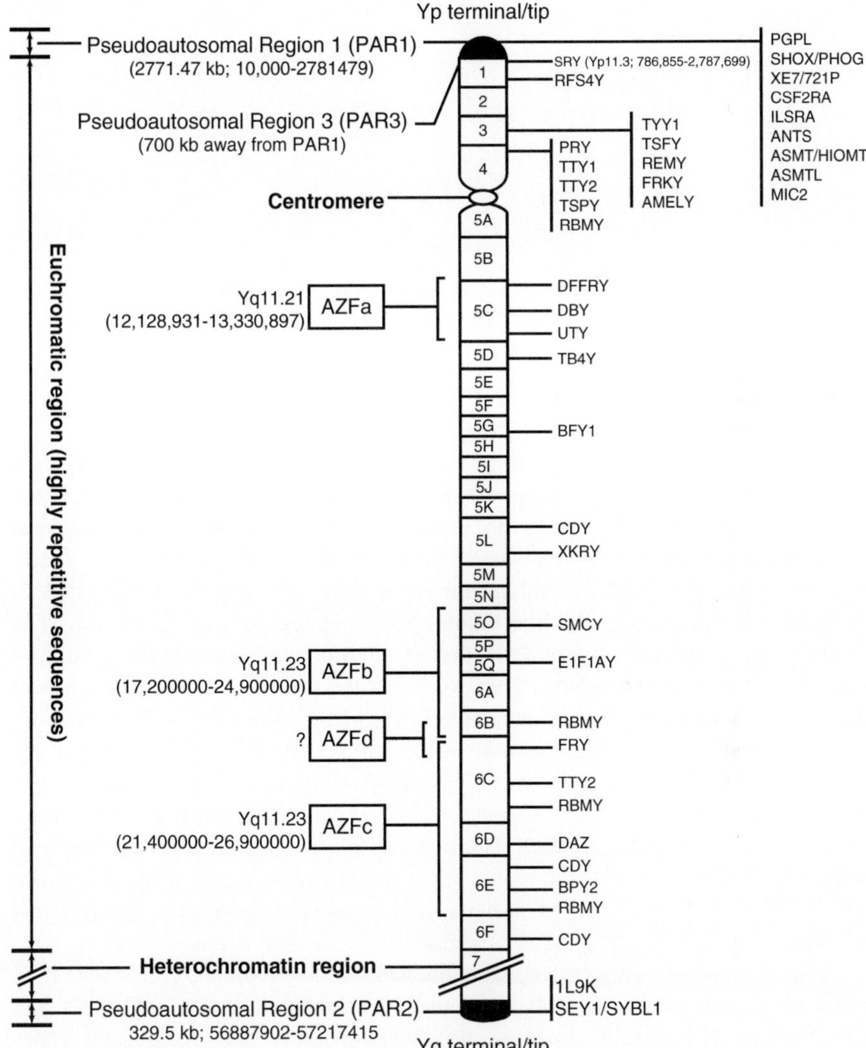

Fig. 7.2 Schematic diagram showing genetic deletion mapping dividing Y chromosome into 7 intervals. The candidate genes belonging to each region are shown on right; *AZFa–d* azoospermia factors (Adapted and modified from Murci and Fellous (2001))

copy number variations (CNVs). The euchromatic region of Y chromosome is present near the centromere of both arms. Many highly repetitive sequences are present in the euchromatic region in addition to several genes, having major biological roles like sex determination and spermatogenesis. The genes in the euchromatic region of Y chromosome belong to one of the three categories: X transposed, X degenerate, or ampliconic, depending upon the mode of evolution of the segment of Y to which they belong (Skaletsky et al. 2003). Different regions of Y chromosome as seen in fluorescent in situ hybridization (FISH) are shown in Fig. 7.3.

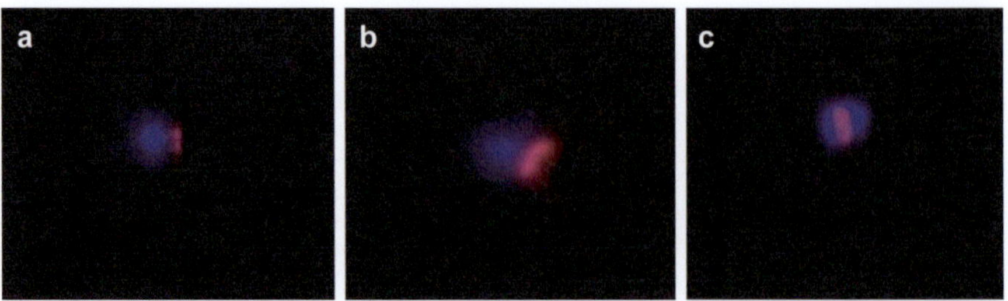

Fig. 7.3 Fluorescent in situ hybridization (FISH) images of chromosome Y (stained with DAPI: blue) is showing (**a**) SRY (*red*), (**b**) centromere (*red*), and heterochromatin region: (**c**) (*red*)

7.3 Evolution

The present form of human Y chromosome has evolved from autosomes and comprises only 3% of its ancestral genes (Skaletsky et al. 2003; Bellott et al. 2010). In the beginning, the changes leading to the formation of Y chromosome from autosomes were very fast, but have slowed down for past 25 million years (Hughes et al. 2012). When chromosome pairs recombine during meiosis in the germ cells, DNA is exchanged in such a way that the accumulated mutations get corrected via DNA exchange or recombination. The same process does not happen with the Y chromosome as it does not have counter pair except for the PAR regions which contribute only 5% of Y genome. Thus, degeneration might have occurred secondary to non-recombination leading to high mutation, inefficient selection, and genetic drift giving rise to the present form of Y (Graves 1995). The euchromatic region constitutes 156 transcription units known so far, together with 78 protein-coding genes which collectively encode 27 specific proteins. The transcription units belong to three types of sequences in the Y euchromatic region depending upon the source from which they are derived to become a part of Y, namely, X-transposed sequences, X-degenerate sequences, and ampliconic sequences (see Table 7.1 and Fig. 7.4).

X-transposed genes are derived from the genes on the X-linked region which, in turn, are derived from ancestral autosomes. X degenerate containing genes and pseudogenes are derived from autosomes (Watson et al. 1991). Ampliconic genes are the result of three converging processes:

(i) X-degenerate gene amplification (RBMY and VCY)
(ii) Autosomal gene transposition and amplification (DAZ)
(iii) Autosomal gene re-transposition and amplification (CDY)

The X-transposed sequences show 99% similarity to the X chromosome. The X-transposed region came to lie on human NRY/MSY around 3–4 million years ago after X to Y transition (see Fig. 7.4; Page et al. 1984; Mumm et al. 1997; Schwartz et al. 1998). Within the X-transposed segments of about 3.4 Mb, there are only two genes, both having homologues in Xq21.The nucleotide sequences of single copy MSY genes show 60–96% homology to their X-linked counterparts. Remarkably,

Table 7.1 Characteristics of three sequence classes in NRY/MSY euchromatic region

Sequence class	Evolutionary origin	Defining characteristic	Distribution	Total length (Mb)	No. of coding genes	No. of transcription units per Mb	No. of noncoding transcription units	Tissue expression
X transposed	Single transposition from X	99% identity to X	2 segments on Yp	3.4	2	0.6	0	TGIF2LX testis specific; PCDH11X in brain
X degenerate	Relics of ancient autosomes from which X and Y evolved	Genes or pseudogenes present as single copy and having X homologues	8 segments:2 on Yp; 6 on Yq	8.6	16, most expressed widely	2.2	4	Widely expressed
Ampliconic	Acquired from diverse sources, then amplified	Repetitive sequences with 99.9% similarity to MSY regions	7 segments: 2 on Yp; 7 on Yq	10.2	60, belonging to 9 families	13.3	74 (9 single-copy; 65 in 15 families)	Testis specific

Adapted and modified from Skaletsky et al. (2003)

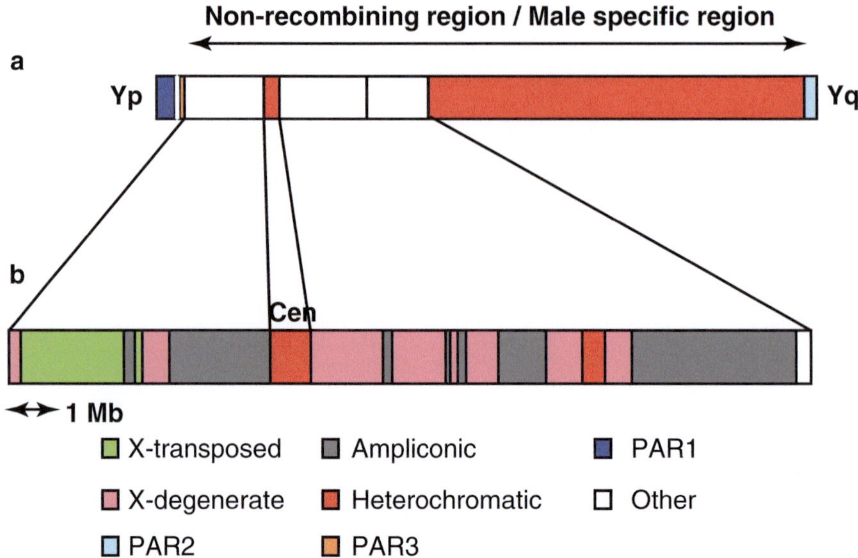

Fig. 7.4. Evolution of the Y chromosome: (**a**) whole Y chromosome, including the pseudoautosomal regions (*PAR 1, 2 and 3*) and the non-recombining region (NRY)/male-specific region (MSY) which includes euchromatic and heterochromatic regions. (**b**) Enlarged view of a 24-Mb euchromatic portion of the NRY, extending from the distal boundary of PAR1 on Yp to proximal boundary of the large heterochromatic region of Yq displaying the ways by which various parts of the Y euchromatic region are derived to finally become part of Y chromosome (Adapted and modified from Skaletsky et al. 2003)

all of the 12 universally expressed MSY genes are present in X-degenerate regions, and none of these genes have been recognized anywhere other than MSY. On the other side, out of the 11 MSY genes, which are expressed chiefly in testes, only one gene, the sex-determining region Y (SRY), is located in the X-degenerate region.

The third category of the euchromatic region is the ampliconic sequence which constitutes about 30% or 10.2 Mb of the MSY euchromatin. Ampliconic sequences present 99.9% similarity to other regions within the MSY. Such regions, extending over tens or hundreds of Kb and belonging to different families, are referred to as MSY-specific long repeat units. The amplicons are situated in seven segments scattered across 10.2 Mb length covering long arm and proximal short arm of the euchromatic region. Maximum genes are present in the ampliconic region, which includes both coding and noncoding genes (see Table 7.1). There are nine MSY-specific protein-coding gene families having different copy numbers ranging from two for VCY, XKRY, HSFY, and PRY; three for BPY2; four for CDY and DAZ; six for RBMY; around 35 CNVs for TSPY (Skaletsky et al. 2003). The copy numbers may vary in human populations. In contrast, all the nine gene families in the ampliconic region show highly restricted testis-specific expression (see Table 7.2).

Table 7.2 Genes of NRY/MSY euchromatic region

NRY/MSY sequence class	Gene name (symbol)	Number of copies	Homologous gene on X	Homologous gene on Autosome	Tissue expression
X transposed	TGF (beta)-induced transcription factor 2-like Y (TGIF2LY)	1	TGIF2LX	–	Testes
	Protocadherin 11 Y (PCDH11Y)	1	PCDH11X	–	Fetal brain, brain
X degenerate	Sex-determining region Y (SRY)	1	SOX3	–	Predominantly testes
	Ribosomal protein S4 Y isoform 1 (RPS4Y1)	1	RPS4X	–	Universal
	Zinc-finger Y (ZFY)	1	ZFX	–	Universal
	Amelogenin Y (AMELY)	1	AMELX	–	Teeth
	Transducin (beta)-like 1 protein Y (TBL1Y)	1	TBL1X	–	Prostate, fetal brain
	Protein kinase Y (PRKY)	1	PRKX	–	Universal
	Ubiquitin-specific protease 9 Y (USP9Y)	1	USP9X	–	Universal
	Dead box Y (DBY)	1	DBX	–	Universal
	Ubiquitous TPR motif Y (UTY)	1	UTX	–	Universal
	Thymosin (beta)-4 Y (TMSB4Y)	1	TMSB4X	–	Universal
	Neuroligin 4 isoform Y (NLGN4Y)	1	NLGN4X	–	Prostate, testes, fetal and adult brain
	Chromosome Y open reading frame 15A (CYorf15A)	1	CXorf15	–	Universal
	Chromosome Y open reading frame 15B (CYorf15B)	1	CXorf15	–	Universal
	SMC (mouse) homologue, Y (SMCY)	1	SMCX	–	Universal
	Translation-initiation factor 1A Y (EIF1AY)	1	EIF1AX	–	Universal
	Ribosomal protein S4 Y isoform 2 (RPS4Y2)	1	RPS4X	–	Universal

(continued)

Table 7.2 (continued)

NRY/MSY sequence class	Gene name (symbol)	Number of copies	Homologous gene on X	Homologous gene on Autosome	Tissue expression
Ampliconic	Testis-specific protein Y (TSPY)	~35	–	–	Testes
	Variable charge Y (VCY)	2	VCX		Testes
	XK-related Y (XKRY)	2	–	–	Testes
	Chromodomain Y (CDY)	4	–	CDYL	Testes
	Heat shock transcription factor Y (HSFY)	2	–	–	Testes
	RNA-binding motif Y (RBMY)	6	RBMX	–	Testes
	PTP-BL-related Y (PRY)	2	–	–	Testes
	Basic protein Y 2 (BPY2)	3	–	–	Testes
	Deleted in azoospermia (DAZ)	4	–	DAZL	Testes
Grand total		~78			

Adapted and modified from Skaletsky et al. (2003)

7.4 Genes on the Human Y Chromosome

Human Y chromosome has limited number of genes as compared to other chromosomes. The reduction in number of genes may have resulted from degeneration process during evolution (Graves 1995). The ancestral autosomal genes passed through X and Y chromosome to ultimately lie on NRY in present time (Bull 1983). Table 7.3 provides a summary of the genes present on NRY, PAR1, PAR2, and PAR3 of Y chromosome. The genes belonging to the NRY region can be further subdivided into two groups. The first group has genes which are expressed ubiquitously, exist as single copy having X homologues, and perform housekeeping functions. The second group includes genes that have testis-specific expression performing additional specialized functions and exist in multiple copies except the SRY gene which is present in single copy. The findings regarding X homologues of NRY genes are worth mentioning which provides an alternate answer for gene dose compensation. The NRY genes have escaped X inactivation and encode functionally interchangeable proteins (Lahn and Page 1997).

7.4.1 Sex-Determining Region Y (SRY) Gene

The SRY gene is present on Yp11.3 (genomic coordinate Y:786,855–2,787,699) and is responsible for testis determination (Sinclair et al. 1990). The SRY protein has 204 amino acid sequence encoded by a single exon spanning the SRY gene. The protein

Table 7.3 Genes on the human Y chromosome

Gene name (symbol)	Location (genomic coordinate)	Associate pathology/function	References
Phosphatidylinositol-specific phospholipase C, X domain containing 1 (PLCXD1)	PAR1 (Y:142,991–170,020)	Unknown	Kimura et al. (2006)
GTP-binding protein 6 (GTPBP6)	PAR1 (Y:171,426–180,887)	Unknown	Gianfrancesco et al. (1998)
Protein phosphatase 2, regulatory subunit B (PPP2R3B)	PAR1 (Y:333,933–386,955)	Regulates DNA replication. Overexpression causes G1 cell cycle arrest	Yan et al. (2000)
Short stature homeobox (SHOX)	PAR1 (Y:624,344–659,411)	Transcription factor associated to short stature syndromes	Rao et al. (1997)
Cytokine receptor-like factor 2 (CRLF2)	PAR1 (Y:1,264,869–1,281,529)	Receptors that bind TSLP cytokine; enhances dendritic cell maturation and $CD4^+$ T-cell proliferation	Al-Shami et al. (2005)
Colony-stimulating factor 2 receptor, alpha (CSFR2Rα)	(Y:1,337,701–1,378,827)	Receptor that binds granulocyte-macrophage colony-stimulating factor (GM-CSF); regulates eosinophil and macrophage development in the bone marrow and cell viability in human embryos	Sjoblom et al. (2002)
Interleukin 3 receptor, alpha (IL3RA)	(Y:1,405,509–1,451,582)	Receptors for interleukin 3	Kremer et al. (1993)
Solute carrier family 25 member A6 (SLC25A6) or ANT3, ANT3Y, MGC17525	PAR1 (Y:1,455,045–1,460,998)	A member of the ADP/ATP translocase family, which has a potential role in Th cell survival and immune cell homeostasis	Jang and Lee (2006)
Acetyl serotonin O-methyltransferase-like (ASMTL)	(Y:1,403,139–1,453,795)	Unknown	Ried et al. (1998)

(continued)

Table 7.3 (continued)

Gene name (symbol)	Location (genomic coordinate)	Associate pathology/function	References
Purinergic receptor P2Y, G-protein-coupled 8 (P2RY8)	(Y:1,462,572–1,537,506)	A member of the purine nucleotide G-protein coupled receptor gene family	Cantagrel et al. (2004)
CXYorf3 (XE7, XE7Y, DXYS155E, MGC39904, B lymphocyte surface antigen 721P, X escapee, CCDC133)	(Y:1,591,593–1,602,520)	Alternative splicing regulator	Mangs et al. (2006)
Acetyl serotonin O-methyltransferase (ASMT) or HIOMT, ASMTY, HIOMTY	(Y:1,595,455–1,643,081)	Catalyzes final reaction in melatonin synthesis	Slim et al. (1993)
Dehydrogenase/reductase (SDR family) X linked (DHRSXY)	PAR1 (Y:2,087,547–2,369,015)	Encodes an oxidoreductase of the short-chain dehydrogenase/reductase family	Gianfrancesco et al. (2001)
Zinc finger, BED-type containing 1 (ZBED1)	(Y:2,486,414–2,500,967)	Transposition of other transposable elements	Esposito et al. (1999)
CD99: CD99 molecule (MIC2)	PAR1 (Y:2,691,187–2,741,309)	Cell-surface molecule involved in T-cell adhesion and activation of death pathway in T cell	Pettersen et al. (2001)
XG blood group (XG)	PAR1 (Y:2,751,820–2,816,500)	Encodes cell-surface antigen	Ellis et al. (1994)
Sex Reversal/determining region Y (SRY[a])	NRY (Yp:1A1A;Y:786,855–2,787,699)	Sex reversal	Sinclair et al. (1990)
Ribosomal protein S4,Y(RPS4Y)	NRY (Yp:1A1B; Y:2,709,623–2,734,997)	Turner syndrome?	Fisher et al. (1990)
Zinc-finger Y (ZFY)	NRY (Yp:1A2; Y:2,934,416–2,982,508)	Turner syndrome?	Page et al. (1987)
Protein kinase, Y (PRKY)	NRY (Yp:3C-4A; Y:7,273,972–7,381,547)	Unknown	Shiebel (1997)
Testis transcript, Y1(TTY1[a])	NRY (Yp:4A; Y:9,753,156–9,774,289)	Unknown	Reijo et al. (1995)
Testis sepsis protein, Y (TSPY[a])	NRY(Yp:3C + 5; Y:410,723–413,797)	Gonadoblastoma	Arnemann et al. 1987
Amelogenin, Y (AMELY)	NRY (Yp:4A; Y:6,865,918–6,911,937)	Unknown	Nakahori et al. (1991)
Putative tyrosine phosphate protein-related Y (PRY[a])	NRY (Y:4A, 6E; Y:22,484,280–22,515,543)	Infertility?	Reijo et al. (1995)

Table 7.3 (continued)

Gene name (symbol)	Location (genomic coordinate)	Associate pathology/function	References
Testis transcript, Y2 (TTY[a])	NRY Y:4A-6C	Unknown	Reijo et al. (1995)
Ubiquitin-specific protease (or *Drosophila* fat-facets related, Y) (USP9Y) or DFFRY	NRY (Yq:5C; Y:14,813,160–14,972,768)	Azoospermia?	Lanh and Page (1997)
DEAD box, Y (DBY)	NRY (Yq:5C; Y:12,904,108–12,920,478)	Infertility?	Lanh and Page (1997)
Ubiquitous TRY motif, Y (UTY)	NRY (Yq:5C; Y:13,233,920–13,480,670)	Infertility	Lanh and Page (1997)
Thymosin?4, Y isoform (TB4Y)	NRY (Yq:5D; Y:13,703,567–13,706,024)	Infertility	Lanh and Page (1997)
Basic protein,Y1 (BPY1[a])	NRY (Yq:5G; Y:13,985,772–13,986,512)	Turner?	Reijo et al. (1995)
Chromodomain, Y (CDY)	NRY (Yq:5 L, 6F; Y:8,813,079–8,815,071)	Infertility?	Lanh and Page (1997)
XK-related, Y (XKRY[a])	NRY (Yq:5 L; Y:17,768,980–17,770,560)	Infertility?	Reijo et al. (1995)
RNA-binding motif, Y (RBM[a])	NRY (Yp + q; Y:21,534,902–21,561,014)	Infertility?	Ma et al. (1993)
Selected mouse cDNA, Y (SMCY)	NRY (Yq:5P; Y:19,705,415–19,744,939)	Unknown	Agulnik et al. (1994)
Translation-initiation factor 1A,Y (E1F1AY)	NRY (Yq:5Q; Y:22,737,611–22,755,040)	Infertility?	Lanh and Page (1997)
Deleted in azoospermia (DAZ[a])	NRY (Yq:6F; Y:25,365,622–25,437,497)	Infertility?	Reijo et al. (1995)
Variably charged protein, Y2 (VCY2)	NRY (Yq:6A; Y:22,984,263–23,005,465)	Infertility	Reijo et al. (1995)
Interleukin 9 receptor (IL9R)	PAR2 (Y:59,330,252–59,343,488)	Unknown	Vermeesh et al. (1997)
Synaptobrevin-like 1 (SYBL1)	PAR2 (Y:57,067,800–57,130,289)	Unknown	Kermouni et al. (1995)
Human-sprouty 3 (HSPRY3)	PAR2 (Y:59,100,457–59,115,123)	Unknown	Ciccododicola et al. (2000)
Chromosomes X And Y Open Reading Frame 1 (CXYORF1)	PAR2 (Y:156,020,826–156,025,710)	Unknown	Ciccododicola et al. (2000)
Protocadherin 11 Y linked (PCDH11Y)	PAR3 (Y:4,868,267–5,610,265)	Unknown	Veerappa et al. (2013)
TGFB-induced factor homeobox 2-like, Y linked (TGIF2LY)	PAR3 (Y:3,579,085–3,580,041)	Unknown	Veerappa et al. (2013)

Adapted and modified from Murci and Fellous (2001)
[a]Testis-specific genes or families

comprises of nucleic acid-binding domain and, thus, has a role in regulating gene expression. SRY gene initiates testis development as it differentiates the bipotential gonad to testicular pathway. Many other genes and loci interact with SRY protein, such as SOX-9, steroidogenic factor 1 (SF-1), and Wilm's tumor gene 1 (WT-1).

7.4.2 Azoospermia Factors

Azoospermia factors (AZFs) are the spermatogenesis loci present in the Yq11.2 region. The types of AZFs are AZFa, AZFb, and AZFc (see Fig. 7.2). AZFa (Yq11.21; genomic coordinate Y: 12,128,931–13,330,897) is located on the anterior side of deletion interval 5C.The size of the AZFa region is around 1 Mb. AZFb (Yq11.23; genomic coordinate Y:17,200,000–24,900,000) covers the region from deletion interval 5 to deletion interval 5O to 6B. AZFb covers 1–3 Mb DNA. AZFc (Yq11.23; genomic coordinate Y:21,400,000–26,900,000) is present on deletion interval 6C to 6E and covers 3.5 Mb DNA. Another region AZFd is described amid AZFb and AZFc, but existence of this region needs further verification (Kent-First et al. 1999). In all, 12 genes have been defined in AZF regions (Reijo et al. 1995).

7.4.2.1 AZFa

The AZFa region contains three genes (Lahn and Page 1997; Mazeyrat et al. 1998; Sargent et al. 1999). The first known gene in the AZFa region is DFFRY (*Drosophila* fat facet-related Y), also known as USP9Y (ubiquitin-specific protease 9; Mazeyrat et al. 1998; Brown et al. 1998). The DFFRY gene is located at Yq11.21 (genomic coordinate Y: 12,701,231–12,860,844). DFFRY gene is ubiquitously expressed and exists in single copy. It has an X homologue which escapes X inactivation and acts as a C-terminal ubiquitin hydrolase without encoding for RNA-binding protein (Lahn and Page 1997; Brown et al. 1998). USP9Y occupies less than half of the AZFa interval (Sargent et al. 1999). The other two X-Y homologous genes present in AZFa are DBY (dead box on the Y), located at Yq11.221 (genomic coordinate Y:12,904,108–12,920,478), and UTY (ubiquitous TPR motif on the Y), located at Yq11.221 (genomic coordinate Y:13,233,920–13,480,673; Lahn and Page 1997; Mazeyrat et al. 1998). AZFaT1 is another unique sequence described in AZFa (Sargent et al. 1999). The genes in AZFa may work singly or in combination with USP9Y. Absence of a single gene or combination of genes in cases of AZFa deletion has been associated with spermatogenic disruption, in particular with Sertoli cell-only syndrome or azoospermia leading to male infertility (Vogt et al. 1996; Sargent et al. 1999).

7.4.2.2 AZFb

AZFb has two genes – RBMY (RNA-binding motif on the Y) located on Yq11.223 (genomic coordinate Y:21,534,902–21,561,014) and EIF1AY (translation-initiation factor 1A, Y isoform) located on Yq11.223 (genomic coordinate Y:20,575,711–20,593,154). RBMY was identified and cloned from deletion interval 6 (Ma et al. 1993). Initially, it was proposed to have two motifs – YRRM1 and YRRM2 (Y-specific RNA recognition motif). Later on, it was found that there is actually a family of 20–50 genes and pseudogenes extending from Yp to Yq arm with a lot of genes concentrated in AZFb (Prosser et al. 1996; Elliott et al. 1997). So, the name changed from YRRM to "RBMY gene family" (Vogt et al. 1997). The genes belonging to RBMY gene family are divided into six subfamilies (Chai et al. 1997, 1998). The only actively transcribed gene is RBMY-I with majority of functional copies present in deletion interval 6B and so is the candidate gene in AZFb (Vogt et al. 1997). The protein encoded by RBMY gene has a single RRM (RNA recognition motif)-type RNA-binding domain at the N-terminal and four 37-amino acid repeats at the C-terminal. The C-terminal domain has a serine-arginine-glycine-tyrosine sequence and, so, is named SRGY box. All eutherians have multiple copies of RBMY gene (Delbridge et al. 1997). The active X homologue of the gene has retained widespread functions, while Y homologue has male-specific role in spermatogenesis (Delbridge et al. 1999; Mazeyrat et al. 1999). In accordance with its role in spermatogenesis, the RBMY genes are expressed only in the testicular germ cells (Elliott et al. 1997). Big AZFb deletions are known to cause azoospermia, while microdeletions are related to oligozoospermia and a variety of other infertile phenotypes. But, the exact role of RBMY in male germ cell development is still unclear because differential functions are associated with pre-mRNA splicing and variations in spatial localization in different testicular germ cells (Elliott et al. 1998). EIF1AY has ubiquitous expression and codes for Y isoform of eukaryotic translation-initiation factor, eIF-1A which also has a X homologue (Lahn and Page 1997). No specific deletion of this gene is known that could suggest its role in spermatogenesis. But, EIF1AY definitely plays some role in AZFb phenotype as it has many testis-specific transcripts along with the ubiquitously expressed transcripts (Lahn and Page 1997).

7.4.2.3 AZFc

The AZFc region has seven gene families performing different roles in spermatogenesis. The candidate gene of the AZFc region is DAZ (deleted in azoospermia) located on Yq11.223, deletion interval 6 (genomic coordinateY:23,129,355–23,199,123). DAZ belongs to a multigene family. Several copies of DAZ gene are present in the AZFc region of the Y chromosome (Saxena et al. 1996; Vogt et al. 1997; Glaser et al. 1998; Yen 1998). The DAZ gene has structural similarity with RBMY. DAZ gene is a 42 kb gene comprising 16 exons (Reijo et al. 1995; Saxena et al. 1996). There is individual variation in copy number and order of DAZ repeats (Yen et al. 1997). DAZ also has testicular germ cell specific expression (Menke et al. 1997; Habermann et al. 1998; Lee et al. 1998; Ferlin et al. 1999). Deletion of DAZ gene has been associated with spermatogenic defects (Reijo et al. 1995).

7.4.2.4 AZFd

So far, no gene has been described in the AZFd region. However, the DYS237 locus of AZFd may have some genes which play a role in spermatogenesis as predicted by deletion studies (Muslumanoglu et al. 2005). The length of deleted segment may vary. The most commonly observed deletion spans the region between the STS marker (SY153) in AZFd and the junction of euchromatic and heterochromatic regions. Microdeletions specific to AZFd have been shown to cause mild oligozoospermia or normal sperm count with defective sperm morphology (Kent-First et al. 1999).

7.5 Copy Number Variations (CNVs)

Chromosomal rearrangements can result in deletions, duplications, or unbalanced translocations which subsequently result in inappropriate gene dosage. Meiotic nondisjunction events in physiologically normal carriers of balanced translocations may also lead to a disturbance of gene dosage in offspring. Most of the rearrangements occur in specific regions of the genome, suggesting specific mechanisms causing these modifications. CNVs are submicroscopic DNA segments of 1 kb or more that are duplicated or deleted in the genome (Lee et al. 2007). CNVs can be pathogenic if they cause overt disease or benign if they predispose to disease or apparently have no effect. CNVs lie at the interface between microscopically visible rearrangements and point mutations; these are increasingly being investigated using microarray methods. Recent studies have shown that CNVs may affect around 20% of human genome. Currently, we are working on genotype-phenotype corelation in male infertility. Our initial findings indicate that CNVs of sex chromosomes, in particular of PARs, are more frequent with azoospermia (ongoing work; Halder 2016).

7.6 Functions

The Y chromosome has many functions to its credit that are restricted to the NRY. The characteristic roles of Y chromosome in humans include sex determination and development as well as maintenance of male gonads, including germ cells. Because Y chromosome is present only in males and is transmitted holandrically, it stores information on human evolution because the genes on the NRY region have never undergone recombination and are transmitted vertically as it is over generations (Murci and Fellous 2001).

7.7 Y-Linked Holandric Traits

Y-linked traits are controlled by alleles present on Y chromosome. Since only males have Y chromosome, Y-linked traits are passed from father to son. Approximately three dozen Y-linked traits have been discovered, and one such trait is hairy ears (Murci and Fellous 2001).

7.8 Y Chromosome-Linked Disorders

Mutations or deletions in Y chromosome result in Y-linked disorders, such as mutation in SRY, which results in defective testicular development, and deletion in the AZF region results in azoospermia. Protein encoded by AZF genes has a crucial role in germ cell development.

7.8.1 Defective Y Chromosome

Deleted SRY, incomplete Y, or fragmented Y may present as feminine phenotype, ambiguous genitalia, or infertility. As a result, there is defective testicular development along with defective internal and/or external male genitalia. In conditions of minimal or nonfunctioning Y fragment, the subject is usually a female having the features of mixed gonadal dysgenesis or Turner syndrome (Chang et al. 1990).

7.8.2 XX Male

The XX male syndrome is a rare genetic disorder. The phenotype is a variable that ranges from a severe impairment of external genitalia to a normal male phenotype with infertility. The genotype generally results from an unequal crossing over between the short arms of the sex chromosomes X and Y.

We observed a 38-year-old patient presented to us with concern of infertility. Physical examination and laboratory investigations showed normal external male genitalia with hypogonadism, gynecomastia, and glaucoma. The patient was azospermic with high gonadotropin levels, low anti-Mullerian hormone (AMH), and low inhibin B levels. Karyotyping revealed 46, XX with absence of Y chromosome. FISH and PCR showed presence of SRY gene (see Fig. 7.5). Testicular fine-needle aspiration cytology (FNAC) revealed Sertoli cell-only syndrome (Jain et al. 2013).

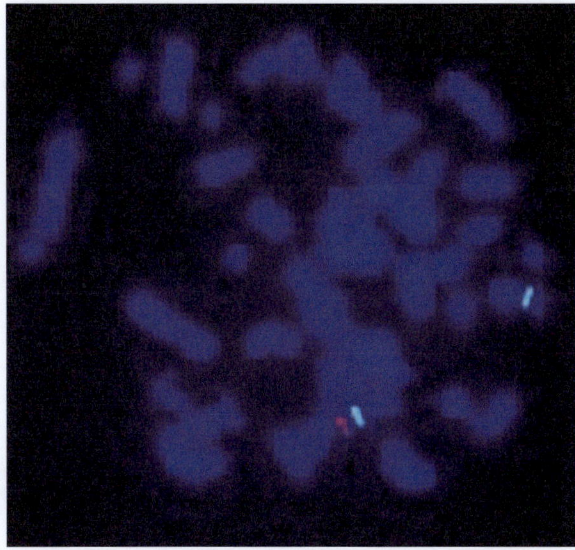

Fig. 7.5 Fluorescent in situ hybridization (FISH) image by using triple probe – *green* (X centromere), *red* (SRY), and yellow (Y centromere) signals. Two green signals indicate two centromeres of X chromosome. Absence of yellow signal indicates absence of Y centromere

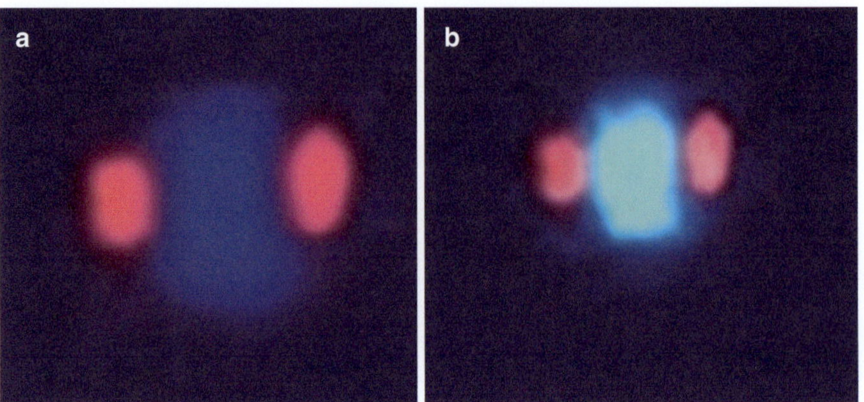

Fig. 7.6 Fluorescent in situ hybridization (FISH) image of Y dicentric chromosome: (**a**) CY3 (*red*) labeled Y centromere probe and DAPI stain showing two centromeres at the ends of rearranged Y chromosome; (**b**) FITC (*green*)-labeled Yqh probe and CY3 (*red*)-labeled Y centromere probe showing two centromeres (*red*) and in between heterochromatin region (*green*) of Y chromosome

7.8.3 Dicentric Y Chromosome

Dicentric Y chromosome is usually considered unstable since inappropriate position of two centromeres on metaphase spindle can cause bridge formation in anaphase. In some cases, if centromeres are close to each other, they can act as monocentric elements. In such situations, they replicate as normal chromosomes (Sears and Camera 1952; Darlington and Wiley 1953). Dicentric chromosomes are perhaps the most frequently encountered example of structural rearrangements of the Y chromosome. Several genes exist as multiple copies in the dicentric Y chromosome.

Dicentric Y chromosome (see Fig. 7.6) formation may occur in three ways: (1) by a break in Yp and reunion of the proximal ends of chromatid in meiosis I, resulting in a dicentric chromosome following meiosis II and the reduction of the centromere; (2) by a break in Yp or Yq followed by reunion of sister chromatid strands leading to duplication deficiency of short or long arm of Y chromosome; and (3) by breaks in either arm of the two Y chromosomes in an XYY individual, followed by exchange. The former two are the most likely mechanisms for the origin of the dicentric Y (Cohen et al. 1973).

7.8.4 Yqh Polymorphism and Inversions

The Y chromosome contains a major proportion of segmental duplication and shows cytogenetically observable structural polymorphisms such as length variation and inversions (Bobrow et al. 1971; Verma et al. 1978; Verma et al. 1982a; Bernstein et al. 1986; Skaletsky et al. 2003). The former comprises large Y (Yq+)

and small Y (Yq-) and the latter includes pericentric and paracentric inversions. Pericentric inversion is the rearrangement in which a segment, including the centromere, is rotated, while in paracentric inversion, the rotated segment does not include centromere. The Y chromosome is more polymorphic in the Asians (3.37%) and Hispanics (1.82%) compared to the Whites or Blacks (Hsu et al. 1987).

Y chromosome varies in length in different racial groups (Cohen et al. 1966). Variable length of Y chromosome has also been documented with different clinical conditions such as psychological disorders, Down syndrome, abnormal embryo development, birth complications, and bad obstetric history (Funderburk et al. 1978; Genest 1979; Verma et al. 1982a; Podugolnikova and Blumina 1983; Videbech et al. 1984; Minocherhomji et al. 2009).

The pericentric inversion of Y chromosome (invY) is familial (Solomon et al. 1964; Jacobs et al. 1964; Verma et al. 1982b). The incidence of invY ranges from 1 to 2 per 1000 male fetuses with an average of 1.15 per 1000 males (Shapiro et al. 1984). The incidence of invY in various populations has been studied worldwide (Bhasin 2005). 30.5% of invY has been reported in the immigrant Gujarati Muslim community which originated from Surat and settled in South Africa (Bernstein et al. 1986). Such type of polymorphic frequency of trait presents an example of random genetic drift occurring in reproductively isolated populations that follows rigid endogamy and without any reproductive abnormality. Other studies have shown association between pericentric inversion in different human chromosomes and congenital anomalies, repeated fetal loss, mental retardation, and infertility, possibly predisposing inter-chromosomal effect and nondisjunction (Krishna et al. 1992; Gardner and Sutherland 1996).

7.8.5 Gonadoblastoma

Gonadoblastoma is a rare benign tumor but can be potentially malignant (Scully 1953, 1970). Patients with disorders of sex development are at higher risk of developing gonadoblastoma. Two factors that predict development of gonadoblastoma are (1) presence of Y chromosome or a part of Y chromosome and (2) abdominal gonads. However, a limited number of cases of gonadoblastoma have been reported in patients with a 46,XX karyotype (Esin et al. 2012). The gene responsible for gonadoblastoma is gonadoblastoma locus on Y chromosome (GBY) which functions as an oncogene in conditions of dysgenetic gonads (Scully 1953). Studies from sex-reversed and gonadoblastoma patients reveal the position of GBY locus close to centromere of Y chromosome (Tsuchiya et al. 1995). The GBY locus has many genes. RNA-binding motif (RBM), protein tyrosine phosphatase (PTP) BL-related Y (PRY), amelogenin Y (AMELY), testis transcripts Y1 and Y2 (TTY1 and Y2), and protein kinase Y (PRKY) are some of known genes present in this locus.

Key Questions

Describe the euchromatin and heterochromatin regions of the Y chromosome.
What are the differences between X-linked and X-degenerate genes?
Explain briefly the evolution of Y chromosome from autosomes.
Describe the Azoospermic factor (AZF) region.
Describe the part of Y chromosome involved in Male fertility.
Name the Y-linked diseases.

References

Agulnik AI, Mitchell MJ, Mattei MG, Borsani G, Avner PA, Lerner JL, et al. A novel X gene with a widely transcribed Y-linked homologue escapes X-inactivation in mouse and human. Hum Mol Genet. 1994;3:879–84.

Al-Shami A, Spolski R, Kelly J, Keane-Myers A, Leonard WJ. A role for TSLP in the development of inflammation in an asthma model. J Exp Med. 2005;6:829–39.

Arnemann J, Epplen JT, Cooke HJ, Sauermann U, Engel W, Schmidtke J. A human Y chromosomal DNA sequence expressed in testicular tissue. Nucleic Acids Res. 1987;15:8713–24.

Bellott DW, Skaletsky H, Pyntikova T, Mardis ER, Graves T, Kremitzki C, et al. Convergent evolution of chicken Z and human X chromosomes by expansion and gene acquisition. Nature. 2010;466:612–6.

Bernstein R, Wadee A, Rosendorff J, Wessels A, Jenkins T. Inverted Y chromosome polymorphism in the Gujarati Muslim Indian population of South Africa. Hum Genet. 1986;74:223–9.

Bhasin MK. Human population cytogenetics: A review. Int J Hum Genet. 2005;5:83–152.

Bobrow M, Pearson PL, Pike MC, El-Alfi OS. Length variation in the quinacrine binding segment of human Y chromosomes of different sizes. Cytogenetics. 1971;10:190–8.

Brown GM, Furlong RA, Sargent CA, Erickson RP, Longepied G, Mitchell M, et al. Characterisation of the coding sequence and fine mapping of the human DFFRY gene and comparative expression analysis and mapping to the Sxrb interval of the mouse Y chromosome of the Dffry gene. Hum Mol Genet. 1998;7:97–107.

Bull JJ. Evolution of sex determining mechanisms. Menlo Park, California: Benjamin/Cummings Publ. Co.; 1983.

Cantagrel V, Lossi AM, Boulanger S, Depetris D, Mattei MG, Gecz J, et al. Disruption of a new X linked gene highly expressed in brain in a family with two mentally retarded males. J Med Genet. 2004;10:736–42.

Chai NN, Salido EC, Yen PH. Multiple functional copies of the RBM gene family, a spermatogenesis candidate on the human Y chromosome. Genomics. 1997;45:355–61.

Chai NN, Zhou H, Hernandez J, Najmabadi H, Bhasin S, Yen PH. Structure and organization of the RBMY genes on the human Y chromosome: transposition and amplification of an ancestral autosomal hnRNPG gene. Genomics. 1998;49:283–9.

Chang HJ, Clark RD, Bachman H. The phenotype of 45,X/46,XY mosaicism: an analysis of 92 prenatally diagnosed cases. Am J Hum Genet. 1990;46(1):156–67.

Ciccodicola A, D'Esposito M, Esposito T, Gianfrancesco F, Migliaccio C, Miano MG, et al. Differentially regulated and evolved genes in the fully sequenced Xp/Yq pseudoautosomal region. Hum Mol Genet. 2000;9:395–401.

Cohen MM, Shaw MW, Mac Clauer JW. Racial differences in the length of the human Y chromosome. Cytogenetics. 1966;5:34–52.

Cohen MM, MacGillivray MH, Capraro VJ, Aceto TA. Human Dicentric Y Chromosomes: case report and review of the literature. J Med Genet. 1973;10:74–9.

Darlington CD, Wylie AP. A dicentric cycle in narcissus. Heredity. 1953;6(supplement):197–213.

Delbridge ML, Harry JL, Toder R, O'Neill RJ, Ma K, Chandley AC, et al. A human candidate spermatogenesis gene, RBM1, is conserved and amplified on the marsupial Y chromosome. Nat Genet. 1997;15(2):131–6.

Delbridge ML, Lingenfelter PA, Disteche CM, Graves JA. The candidate spermatogenesis gene RBMY has a homologue on the human X chromosome. Nat Genet. 1999;22:223–4.

Elliott DJ, Millar MR, Oghene K, Ross A, Kiesewetter F, Pryor J, et al. Expression of RBM in the nuclei of human germ cells is dependent on a critical region of the Y chromosome long arm. Proc Natl Acad Sci U S A. 1997;94:3848–53.

Elliott DJ, Oghene K, Makarov G, Makarova O, Hargreave TB, Chandley AC, et al. Dynamic changes in the subnuclear organisation of pre-mRNA splicing proteins and RBM during human germ cell development. J Cell Sci. 1998;111:1255–65.

Ellis NA, Ye TZ, Patton S, German J, Goodfellow PN, Weller P. Cloning of PBDX, an MIC2-related gene that spans the pseudoautosomal boundary on chromosome Xp. Nat Genet. 1994;4:394–400.

Esin S, Baser E, Kucukozkan T, Magden HA. Ovarian gonadoblastoma with dysgerminoma in a 15-year-old girl with 46, XX karyotype: case report and review of the literature. Arch Gynecol Obstet. 2012;285(2):447–51.

Esposito T, Gianfrancesco F, Ciccodicola A, Montanini L, Mumm S, D'Urso M, et al. A novel pseudoautosomal human gene encodes a putative protein similar to Ac-like transposases. Hum Mol Genet. 1999;1:61–7.

Ferlin A, Moro E, Onisto M, Toscano E, Bettella A, Foresta C. Absence of testicular DAZ gene expression in idiopathic severe testiculopathies. Hum Reprod. 1999;14:2286–92.

Fisher EM, Beer-Romero P, Brown LG, Ridley A, McNeil JA, Lawrence JB, et al. Homologous ribosomal protein genes on the human X and Y chromosome: escape from X inactivation and possible implication for Turner Syndrome. Cell. 1990;63:1205–18.

Funderburk SJ, Guthrie D, Lind RC, Muller HM, Sparkes RS, Westlake JR. Minor chromosome variants in child psychiatric patients. Am J Med Genet. 1978;1:301–8.

Gardner RJ, Sutherland GR. Chromosomal abnormalities and genetic counseling. In: Gardner RJ, Sutherland GR, editors. Elements of Medical Cytogenetics. Oxford: Oxford University Press; 1996. p. 6–9.

Genest P. Chromosome variants and abnormalities detected in 51 married couples with repeated spontaneous abortions. Clin Genet. 1979;16:387–9.

Gianfrancesco F, Esposito T, Montanini L, Ciccodicola A, Mumm S, Mazzarella R, et al. A novel pseudoautosomal gene encoding a putative GTP-binding protein resides in the vicinity of the Xp/Yp telomere. Hum Mol Genet. 1998;3:407–14.

Gianfrancesco F, Sanges R, Esposito T, Tempesta S, Rao E, Rappold G, et al. Differential divergence of three human pseudoautosomal genes and their mouse homologs: implications for sex chromosome evolution. Genome Res. 2001;12:2095–100.

Glaser B, Yen PH, Schempp W. Fibre-fluorescence in situ hybridization unravels apparently seven DAZ genes or pseudogenes clustered within a Y-chromosome region frequently deleted in azoospermic males. Chromosome Res. 1998;6:481–6.

Graves JA. The origin and function of the mammalian Y chromosome and Y-borne genes an evolving understanding. Bioessays. 1995;17:311–20.

Habermann B, Mi HF, Edelmann A, Bohring C, Bäckert IT, Kiesewetter F, et al. DAZ (Deleted in Azoospermia) genes encode proteins located in human late spermatids and in sperm tails. Hum Reprod. 1998;13:363–9.

Halder A. An investigation to find out mechanism for phenotypic heterogenicity/variability in maturation arrest cases. 2016. Manuscript under preparation.

Hsu LYF, Benn PA, Tannenbaum HL, Perlis TE, Carlson AD. Chromosomal polymorphisms of 1, 9, 16, and Y in 4 major ethnic groups: a large prenatal study. Am J Med Genet. 1987;26:95–101.

Hughes JF, Skaletsky H, Brown LG, Pyntikova T, Graves T, Fulton RS, et al. Strict evolutionary conservation followed rapid gene loss on human and rhesus Y chromosomes. Nature. 2012;483:82–6.

Jacobs PA, Brunton M, Court Brown WM. Cytogenetic study in leucocytes on the general population: subjects of ages 65 years and more. Ann Hum Genet. 1964;27:353–65.

Jain M, Mohan V, Chaudhary I, Halder A. Sertoli cell only syndrome and glaucoma in a SRY positive XX infertile male. J Clin Diagn Res. 2013;7(7):1457–9.

Jang JY, Lee CE. IL-4-induced upregulation of adenine nucleotide translocase 3 and its role in Th cell survival from apoptosis. Cell Immunol. 2006;1:14–25.

Kent-First M, Muallem A, Shultz J, Pryor J, Roberts K, Nolten W, et al. Defining regions of the Y chromosome responsible for male infertility and identification of a fourth AZF region (AZFd) by Y chromosome microdeletion detection. Mol Reprod Dev. 1999;53(1):27–41.

Kermouni A, Vanroose E, Arden KC. The IL9 receptor gene (IL9R)genomic structure, chromosomal localization in the pseudoautosomal region of the long arm of the sex chromosomes, and identification of IL9R pseudogenes at 9qter, 10pter, 16qter and 18pter. Genomics. 1995;29:371–82.

Kimura K, Wakamatsu A, Suzuki Y, Ota T, Nishikawa T, Yamashita R, et al. Diversification of transcriptional modulation: large-scale identification and characterization of putative alternative promoters of human genes. Genome Res. 2006;1:55–65.

Kremer E, Baker E, D'Andrea RJ, Slim R, Phillips H, Moretti PA, et al. A cytokine receptor gene cluster in the X-Y pseudoautosomal region? Blood. 1993;1:22–8.

Krishna DS, SA AL-A, Farag TI. Pericentric inversion and recombinant aneusomy and other associatedchromosomalaberrations:randomornon-random.AmJHumGenet.1992;51(1146):A291.

Lahn BT, Page DC. Functional coherence of the human Y chromosome. Science. 1997;278(5338):675–80.

Lee JH, Lee DR, Yoon SJ, Chai YG, Roh SI, Yoon HS. Expression of DAZ (deleted in azoospermia), DAZL1 (DAZ-like) and protamine-2 in testis and its application for diagnosis of spermatogenesis in non-obstructive azoospermia. Mol Hum Reprod. 1998;4:827–34.

Lee C, Iafrate AJ, Brothman AR. Copy number variations and clinical cytogenetic diagnosis of constitutional disorders. Nat Genet. 2007;39:S48–54.

Ma K, Inglis JD, Sharkey A, Bickmore WA, Hill RE, Prosser EJ, et al. A Y chromosome gene familywithRNA-binding protein homology: candidates for the azoospermia factor AZF controlling spermatogenesis. Cell. 1993;75:1287–95.

Mangs AH, Speirs HJ, Goy C, Adams DJ, Markus MA, Morris BJ. XE7: a novel splicing factor that interacts with ASF/SF2 and ZNF265. Nucleic Acids Res. 2006;17:4976–86.

Mazeyrat S, Saut N, Sargent CA, Grimmond S, Longepied G, Ehrmann IE, et al. The mouse Y chromosome interval necessary for spermatogonial proliferation is gene dense with syntenic homology to the human AZFa region. Hum Mol Genet. 1998;7:1713–24.

Mazeyrat S, Saut N, Mattei MG, Mitchell MJ. RBMY evolved on the Y chromosome from a ubiquitously transcribed X-Y identical gene. Nat Genet. 1999;22:224–6.

Menke DB, Mutter GL, Page DC. Expression of DAZ, an azoospermia factor candidate, in human spermatogonia. Am J Hum Genet. 1997;60:237–41.

Minocherhomji S, Athalye AS, Madon PF, Kulkarni D, Uttamchandani SA, Parikh FR. A case-control study identifying chromosomal polymorphic variations as forms of epigenetic alterations associated with the infertility phenotype. Fertil Steril. 2009;92:88–95.

Mumm S, Molini B, Terrell J, Srivastava A, Schlessinger D. Evolutionary features of the 4-Mb Xq21.3 XY homology region revealed by a map at 60-kb resolution. Genome Res. 1997;7:307–4.

Müslümanoglu MH, Turgut M, Cilingir O, Can C, Ozyürek Y, Artan S. Role of the AZFd locus in spermatogenesis. Fertil Steril. 2005;84:519–22.

Murci LQ, Fellous M. The human Y chromosome: the biological role of a "functional wasteland". J Biomed Biotechnol. 2001;1:18–24.

Nakahaori Y, Takenaka O, Nakagome Y. A human X-Y homologous region encodes "amelogenin". Genomics. 1991;9:264–9.

Page DC. Hypothesis: a Y-chromosomal gene causes gonadoblastoma in dysgenetic gonads. Development. 1987;101(Suppl):151–5.

Page DC, Harper ME, Love J, Botstein D. Occurrence of a transposition from the Xchromosome long arm to the Y-chromosome short arm during human evolution. Nature. 1984;311:119–3.

Pettersen RD, Bernard G, Olafsen MK, Pourtein M, Lie SO. CD99 signals caspase-independent T cell death. J Immunol. 2001;8:4931–42.

Podugolnikova OA, Blumina MG. Heterochromatic regions on chromosomes 1, 9, 16, and Y in children with some disturbances occurring during embryo development. Hum Genet. 1983;63:183–8.

Prosser J, Inglis JD, Condie A, Ma K, Kerr S, Thakrar R, et al. Degeneracy in human multicopy RBM (YRRM), a candidate spermatogenesis gene. Mamm Genome. 1996;7:835–42.

Rao E, Weiss B, Fukami M, Rump A, Niesler B, Mertz A, et al. Pseudoautosomal deletions encompassing a novel gene cause growth failure in idiopathic short stature and Turner syndrome. Nat Genet. 1997;16:54–63.

Reijo R, Lee TY, Salo P, Alagappan R, Brown LG, Rosenberg M, et al. Diverse spermatogenic defects in humans caused by Y chromosome deletions encompassing a novel RNA-binding protein gene. Nat Genet. 1995;10(4):383–93.

Ried K, Rao E, Schiebel K, Rappold GA. Gene duplications as a recurrent theme in the evolution of the human pseudoautosomal region 1: isolation of the gene ASMTL. Hum Mol Genet. 1998;7:1771–8.

Sargent CA, Boucher CA, Kirsch S, Brown G, Weiss B, Trundley A, et al. The critical region of overlap defining the AZFa male infertility interval of proximal Yq contains three transcribed sequences. J Med Genet. 1999;36:670–7.

Saxena R, Brown LG, Hawkins T, Alagappan RK, Skaletsky H, Reeve MP, et al. The DAZ gene cluster on the human Y chromosome arose from an autosomal gene that was transposed, repeatedly amplified and pruned. Nat Genet. 1996;14:292–9.

Schiebel K, Winkelmann M, Mertz A, Xu X, Page DC, Weil D, et al. Abnormal XY interchange between a novel isolated protein kinase gene, PRKY, and its homologue, PRKX, accounts for one third of all (Y+)XX males and (Y-)XY females. Hum Mol Genet. 1997;6:1985–9.

Schwartz A, Chan DC, Brown LG, Alagappan R, Pettay D, Disteche C, et al. Reconstructing hominid Y evolution: X-homologous block, created by X-Y transposition, was disrupted by Yp inversion through LINE-LINE recombination. Hum Mol Genet. 1998;7:1–11.

Scully RE. Gonadoblastoma; a gonadal tumor related to the dysgerminoma (seminoma) and capable of sex-hormone production. Cancer. 1953;6(3):455–63.

Scully RE. Gonadoblastoma: a review of 74 cases. Cancer. 1970;25(6):1340–56.

Sears ER, Camara A. A transmissible dicentricchromosome. Genetics. 1952;37:125–35.

Shapiro LR, Pettersen RO, Wilmot PL, Warburton D, Benn PA, Hsu LYF. Pericentric inversion of the Y chromosome and prenatal diagnosis. Prenat Diagn. 1984;4:463–5.

Sinclair AH, Berta P, Palmer MS, Hawkins JR, Griffiths BL, Smith MJ, et al. A gene from the human sex-determining region encodes a protein with homology to a conserved DNA-binding motif. Nature. 1990;346:240–4.

Sjoblom C, Wikland M, Robertson SA. Granulocyte-macrophage colony-stimulating factor (GM-CSF) acts independently of the beta common subunit of the GM-CSF receptor to prevent inner cell mass apoptosis in human embryos. Biol Reprod. 2002;6:1817–23.

Skaletsky H, Kuroda-Kawaguchi T, Minx PJ, Cordum HS, Hillier L, Brown LG, et al. The male-specific region of the human Y chromosome is a mosaic of discrete sequence classes. Nature. 2003;423:825–7.

Slim R, Levilliers J, Lüdecke HJ, Claussen U, Nguyen VC, Gough NM, et al. A human pseudoautosomal gene encodes the ANT3 ADP/ATP translocase and escapes X-inactivation. Genomics. 1993;1:26–33.

Solomon IL, Hamm CW, Green OC. Chromosome studies on testicular tissue cultures and blood leukocytes of a male previously reported to have no Y chromosome. N Engl J Med. 1964;271:586–92.

Tsuchiya K, Reijo R, Page DC, Disteche CM. Gonadoblastoma: molecular definition of the susceptibility region on the Y chromosome. Am J Hum Genet. 1995;57(6):1400–7.

Veerappa AM, Ramachandra NB, Padakannaya P. Copy number variation-based polymorphism in a new pseudoautosomal region 3 (PAR3) of a human X-chromosome-transposed region (XTR) in the Y chromosome. Funct Integr Genomics. 2013;13(3):285–93.

Vergnaud G, Page DC, Simmler MC, Brown L, Rouyer F, Noel B, et al. Deletion map of the human Y chromosome based on DNA hybridization. Am J Hum Genet. 1986;38:109–24.

Verma RS, Dosik H, Scharf T, Lubs HA. Length heteromorphisms of fluorescent (f) and non-fluorescent (nf) segments of human Y chromosome: classification, frequencies, and incidence in normal Caucasians. J Med Genet. 1978;15:277–81.

Verma RS, Huq A, Madahar C, Qazi Q, Dosik H. Higher incidence of small Y chromosome in humans with trisomy 21 (Down syndrome). Pediatr Res. 1982a;16:769–70.

Verma RS, Rodriguez J, Dosik H. The clinical significance of pericentric inversion of the human Y chromosome: a rare "third" type of heteromorphism. J Hered. 1982b;73:236–8.

Vermeesch JR, Petit P, Kermouni A, Renauld JC, Van Den Berghe H, Marynen P. The IL-9 receptor gene, located in the Xq/Xp pseudoautosomal region, has an autosomal origin, escapes X inactivation and is expressed from the Y. Hum Mol Genet. 1997;1:1–8.

Videbech P, Nielsen J, Wohlert M, Eriksen G, Hansen KB, Hvidman L, et al. The impact of large Y chromosome on pregnancy, fetus and birth. Clin Genet. 1984;26:281–8.

Vogt PH, Edelmann A, Kirsch S, Henegariu O, Hirschmann P, Kiesewetter F, et al. Human Y chromosome azoospermia factors (AZF) mapped to different sub-regions in Yq 11. Hum Mol Genet. 1996;5:933–43.

Vogt PH, Affara N, Davey P, Hammer M, Jobling MA, Lau YF, et al. Report of the third international workshop on Y chromosome mapping. Cytogenet Cell Genet. 1997;79:1–20.

Watson JM, Spencer JA, Riggs AD, Graves JA. Sex chromosome evolution: platypus gene mapping suggests that part of the human X chromosome was originally autosomal. Proc Natl Acad Sci U S A. 1991;88:11256–60.

Yan Z, Fedorov SA, Mumby MC, Williams RS. PR48, a novel regulatory subunit of protein phosphatase 2A, interacts with Cdc6 and modulates DNA replication in human cells. Mol Cell Biol. 2000;3:1021–9.

Yen PH. A long-range restriction map of deletion interval 6 of the human Y chromosome: a region frequently deleted in azoospermic males. Genomics. 1998;54:5–12.

Yen PH, Chai NN, Salido EC. The human DAZ genes, a putative male infertility factor on the Y chromosome, are highly polymorphic in the DAZ repeat regions. Mamm Genome. 1997;8(10):756–9.

Part II

Neuroendocrine Regulation of Spermatogenesis

Testicular Hormones

8

Tarala D. Nandedkar

Learning Objectives

- Regulation of testicular functions
- *Testosterone*
  Structure and synthesis
  Mechanism of action
  Role in development and maturation
  Androgens in young and elderly men
- *Inhibins and activins*
  Function, regulation and mechanism of action
  Pathophysiology
- *Mullerian-inhibiting substance/anti-Mullerian hormone*
  Function in normal males
  Pathophysiology

8.1 Introduction

Testes are the male gonads located in the scrotal sacs. The testis secretes steroid as well as protein hormones. Androgens are the main steroids produced by the testis. Inhibins and activins are the proteins secreted by testes. Anti-Mullerian hormone (AMH) or Mullerian-inhibiting substance (MIS) is another protein of testicular origin. All these hormones play an important role in testicular development and maturation.

T.D. Nandedkar, PhD, FNASc
National Institute for Research in Reproductive Health, Mumbai, India
e-mail: tarla45@gmail.com

A. Kumar, M. Sharma (eds.), *Basics of Human Andrology*,
DOI 10.1007/978-981-10-3695-8_8

8.2 Regulation of Testicular Functions

The major role of the testis is production of sperm and testosterone. The hypothalamus secretes gonadotropin-releasing hormone (GnRH) which causes basophilic cells of the pituitary to secrete follicle-stimulating hormone (FSH) and luteinizing hormone (LH).

LH mainly acts on the Leydig cells of the testis. LH binds to receptors on the Leydig cells to increase cyclic adenosine monophosphate (cAMP) which in turn increases conversion of cholesterol to testosterone. FSH acts on the Sertoli cells. FSH binding induces elevation in cAMP which subsequently increases production of androgen-binding protein, inhibin, aromatase and many other proteins. FSH increases aromatization of testosterone to estradiol within the Sertoli cells. Thus, both the gonadotropins- FSH and LH, are essential in steroid production and thereby normal development of spermatogenesis. The details of regulation of testicular functions are described in the next chapter, Hypothalamic-Pituitary-Testicular Axis.

Testicular hormones – estradiol and inhibin – through their negative feedback action on pituitary and hypothalamus suppress FSH production (Meachem et al. 2001). Estradiol, through feedback action on hypothalamus, reduces GnRH pulse frequency, which in turn decreases LH pulse frequency. Testosterone secreted by the testis is converted into dihydrotestosterone (DHT) peripherally by 5α-reductase. Both testosterone and DHT exert feedback effect on pituitary, thereby decreasing LH release. Testosterone secretion shows daily variation. The circadian rhythm involves increased levels of testosterone in circulation during early morning and decreased levels in the late afternoon. Loss of circadian rhythm is observed in elderly men (Bremner et al. 1983). We shall now discuss the details of testicular hormones.

8.3 Testicular Hormones

8.3.1 Testosterone

Testosterone is a sex steroid hormone secreted primarily by the testis. Adrenal glands and ovaries in females also produce testosterone but in small quantities. In males, testosterone plays a major role in development of the testis and accessory glands and promotes secondary sex characteristics. In addition to this, testosterone also increases muscle mass, bone mass and growth of body hair. It is produced in mammals and in most of the vertebrates. Fish produces 11-ketotestosterone. Thus, testosterone is a ubiquitous steroid and has an ancient evolutionary history.

Testicular production of testosterone was first demonstrated in fowl (Berthold 1849). The mechanisms and effects of androgen actions in human beings have been reported (Brown-Sequard 1889).

8.3.1.1 Structure and Synthesis

Testosterone synthesis from cholesterol was reported in 1935 (Butenandt and Hanish 1935; Ruzicka and Wettstein 1935). The anabolic and androgenic effects of

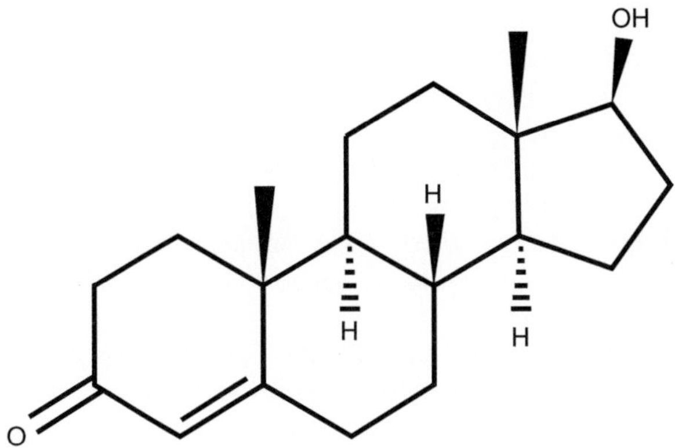

Fig. 8.1 Structure of testosterone (17β-hydroxyandrost-4-en-3-one)

testosterone-propionate have been shown in eunuchoidal men, boys as well as in women (Kenyon et al. 1940).

Testosterone is 17β-hydroxyandrost-4-en-3-one ($C_{19}H_{28}O_2$), a polycyclic alcohol (see Fig. 8.1).

Testosterone is synthesized from cholesterol in the Leydig cells. The LDL-cholesterol ester in the systemic circulation binds to low-density lipoprotein (LDL) receptors on Leydig cell surface and subsequently enters the cell by receptor-mediated endocytosis. A small fraction of cholesterol is also synthesized in Leydig cell endoplasmic reticulum. In the de novo synthesis of cholesterol, three molecules of acetyl-CoA condense to form mevalonate by the action of 3-hydroxy-3-methyl-glutaryl-coenzyme A reductase (HMG-CoA-reductase). Mevalonate is converted to squalene which undergoes cyclization and produces lanosterol. Lanosterol, after removal of three carbons, produces cholesterol. Cholesterol is esterified by cholesterol acyl transferase. Cholesteryl esters are then stored in Leydig cell cytoplasmic lipid droplets (Zmuda et al. 1997).

When Leydig cells are stimulated by LH, there is mobilization of cholesterol from lipid droplets. Cholesterol is then transported to the outer membrane and then to the inner membrane of mitochondria. LH and its second messenger cAMP activate protein kinase A (PK-A) which subsequently phosphorylates and activates cholesteryl ester hydrolase which causes hydrolysis of cholesteryl esters. The cholesterol liberated from lipid droplets is then transported to the outer mitochondrial membrane either by binding to proteins which make it soluble or by vesicle formation. Sterol carrier protein 2 (SCP2), a nonspecific lipid-transfer protein and a subfamily of lipid-binding proteins, is involved in cholesterol transport (Hu et al. 2010).

Soluble N-ethylmaleimide-sensitive factor attachment protein receptor (SNARE) proteins are responsible for the fusion between transport vesicles and target membranes during protein transport. These proteins facilitate the transport of cholesterol possibly due to the functional interaction between lipid droplets and mitochondria.

Delivery of cholesterol to the inner mitochondrial membrane is an important step in steroidogenesis. The cholesterol side chain cleavage P450scc (Cyt $P450_{scc}$) enzyme is located at the inner mitochondrial membrane. Cholesterol is hydrophobic and, therefore, does not diffuse easily through the aqueous intermembrane area of the mitochondria. The inter-mitochondrial transport requires the participation of steroidogenic acute regulatory protein (StAR).

StAR is a short-lived cytoplasmic 37-KDa protein which is cleaved after entering the mitochondria to yield a more stable intramitochondrial 30-kDa form (Hu et al. 2010).

Cholesterol to pregnenolone is the first step of conversion in testosterone biosynthesis; this step is catalyzed by cytochrome $P450_{scc}$ enzyme using nicotinamide adenine dinucleotide phosphate (NADPH) as a cofactor. Pregnenolone then diffuses out to cytoplasmic endoplasmic reticulum where remaining steps of testosterone biosynthesis are carried out. The conversion of pregnenolone to testosterone occurs via two distinct pathways- Δ4 and Δ5 pathway, as shown in Fig. 8.2.

Conversion of testosterone to DHT is by 5α-reductase enzyme, while aromatization to estradiol is by aromatase enzyme. About 7% of testosterone is reduced to DHT in male accessory sex organs and hair follicles. Approximately 0.3% of testosterone is aromatized to estradiol in the gonads, brain, liver and adipose tissues (Mooradian et al. 1987).

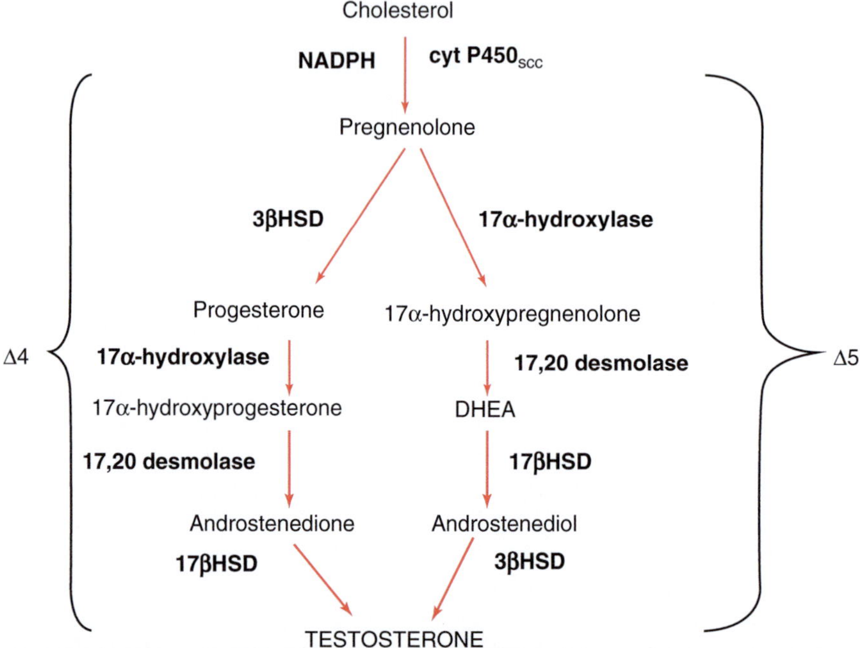

Fig. 8.2 Flow chart showing steroid biosynthesis in Leydig cells. *HSD*, hydroxysteroid dehydrogenase; *DHEA*, dehydroepiandrosterone; *NADPH*, nicotinamide adenine dinucleotide phosphate

8.3.1.2 Mechanism of Action

Testosterone is the major circulating androgen in men. Testosterone is adequate for the development of internal genitalia, skeletal muscles and activation of sexual behavior. During puberty, testosterone stimulates the development of secondary sexual characteristics and acquisition of fertility. Testosterone is metabolized either to DHT or estradiol. DHT is mainly responsible for the development of external genitalia, accessary glands and secondary sexual hair. Estradiol is important for sexual differentiation of the brain, bone mass accretion and fusion of epiphyses during puberty.

In the morning, serum levels of total testosterone less than 300 ng/dL, are considered low (Holt and Zieve 2008). Total testosterone is the sum of free and bound testosterone. About 40% of testosterone is bound to sex hormone-binding globulin (SHBG), and only 2% testosterone is in free form. The remaining testosterone is weakly bound to albumin and can be easily available to tissue when required. Therefore, along with free testosterone, albumin-bound testosterone is also considered to be bioactive (Basaria and Dobs 2001). Free testosterone can reach the cytoplasm of target tissue where it can bind to androgen receptor in the cells, and the complex undergoes structural changes. Free testosterone then moves to the nucleus to bind to hormone-responsive element of the chromosomal DNA, influencing transcriptional activity of specific genes causing anabolic and androgenic effects. Anabolic effects of testosterone are growth and muscle strength, increased bone density and bone maturation. Testosterone has androgenic effects after birth on maturation of sex organs and development of secondary sex characteristics.

8.3.1.3 Androgens in Development and Maturation

Prenatal differentiation

Between 4 and 6 weeks of gestation, androgens cause genital virilization. DHT is more active than testosterone. Virilization includes midline fusion, phallic urethra, phallic enlargement, scrotal thinning and rugation. In the second trimester, high androgen levels observed in male foetuses cause masculinization.

Postnatal development

Androgen levels are high in first week and remain high until 2–3 months of age, although its function is not understood. Possibly, these high levels cause brain masculinization. Interestingly, testosterone is aromatized to estrogen in males prior to brain masculinization. In females, alpha-fetoprotein binds to estrogen to block its action (Weinbauer et al. 1997).

Prepubertal development

A rise in androgen levels is observed prior to puberty, also referred as pubarche. Pubic hair growth as well as axillary hair growth, hair on upper lip and side buns are the changes observed. Skin becomes oily and hairy. Due to accelerated bone maturation and growth spurt, increase in height is seen in boys.

Puberty
In puberty, physiological transition from childhood to adulthood occurs. There is growth of hair over the pubic area, chest and face. Increase in activity and size of sebaceous glands is also observed. Increased libido and penile erections are also the pubertal changes. Voice changes and bone maturation are the other changes occurring at puberty. The details of puberty changes are described in the chapter Male Puberty.

Adulthood

- Regulation of hypothalamo-pituitary-gonadal axis
- Differentiation of spermatogonia to sperm
- Ejaculation of semen

Most of these changes are regulated by steroid hormones along with other protein hormones such as growth factors, prostaglandins, etc.

8.3.1.4 Androgen Replacement in Young Men

Androgen replacement has been used in patients with disorders of hypothalamic-pituitary-testicular axis such as hypogonadism. Androgen therapy can be in the form of injectables, mostly intramuscular, transdermal patches or gels, subcutaneous pellets or oral therapy. Minor side effects are acne and oily skin, while major side effects are increased haematocrit, sleep apnoea, hair loss and acceleration of pre-existing prostate cancer. Monitoring haematocrit and prostate-specific antigen (PSA) levels during therapy is, therefore, beneficial. The treatment of testosterone in young boys/men is mainly for management of delayed puberty, hypogonadism or as male hormonal contraception (Wu 2000).

Delayed puberty In delayed puberty, boys do not show signs of onset of puberty by 13.5–14 years of age. Although this occurs in 3% of boys, majority of them have constitutional delayed puberty (CDP), i.e. it is a temporary extension of hypogonadotropic hypogonadism (HH), some of them having family history of delayed puberty. It is at times difficult to distinguish between CDP and permanent isolated HH (IHH). The only distinction is that CDP shows spontaneous onset of puberty while IHH remains unchanged with time. Morning samples of plasma testosterone and clinical monitoring can be the useful parameters (Wu 2000). The use of testosterone in delayed puberty was aimed to achieve efficacy and long-term safety by varying testosterone formulation and its dose and duration. Low doses of parenteral testosterone due to safety and efficacy have been preferred over other formulations such as testosterone-enanthate, oral testosterone-undecanoate, oral oxandrolone and mesterolone tried by various investigators. Transdermal patches of testosterone are popular due to painless administration.

Hypogonadotropic hypogonadism (HH) Androgen deficiency is either due to testicular failure or HH. The gonadal failure causes decrease in Leydig cell number and volume. Secondary hypogonadism is the result of tumors or trauma or radio or chemotherapy (AACE 2002). Androgen deficiency is also observed in rheumatic arthritis and HIV infection. Loss of androgen production causes physical and behavioral alterations, and testosterone treatment improves these symptoms.

Testosterone for male contraception Synergistic action of testosterone and FSH is required for normal spermatogenesis (Weinbauer et al. 1997; Sharpe 1994). Testosterone can suppress pituitary gonadotropins, thereby blocking spermatogenesis without affecting spermatogonial stem cells ensuring reversibility (El Shannawy et al. 1998). However, high levels of testosterone are required to suppress spermatogenesis. A weekly intramuscular injection of testosterone could achieve oligoazoospermia. However, marked fluctuations in testosterone levels suggested that androgen-only hormonal contraception in men may not have clinical application. Selective androgen receptor modulators (SARM) have been investigated for androgen replacement as well as for male contraception. A potent progesterone 19-nortestosterone having higher affinity for androgen receptor than testosterone could achieve azoospermia only in 50% of the volunteers suggesting testosterone in combination with another hormone, or antihormone may be more useful (Knuth et al. 1985).

Progestogens are potent inhibitors of gonadotropins. Therefore, depot medroxyprogesterone acetate (DMPA) or desogestrel along with testosterone-enanthate has been evaluated and found to induce azoospermia. Due to heterogeneity in population worldwide, androgen-only regimen is recommended for Asian men, while androgen/progesterone combination is suited for Caucasians (Wu 2000).

8.3.1.5 Androgens in Elderly Men

In two longitudinal studies, linear decline in total and free testosterone has been reported along with the corresponding increase in SHBG levels in aeging men above the age of 60 (Zmuda et al. 1997; Morley et al. 1997). In another report, decline in both testosterone and DHEA is shown after the age of 40 (Feldman et al. 2002). There is also a rise in SHBG. Therefore, it is suggested that a change in hypothalamo-hypophyseal-testicular axis may have a role in overall health of elderly men. The biochemical and physiological state is referred to as androgen decline in aeging males (ADAM), the decline being gradual and never absolutely absent. Further, a decrease in biologically available androgen or insensitivity of target organs to androgens may be observed (Chatterji and Roy 1990). Androgen deficiency can be symptomatic or asymptomatic. The symptoms can be not feeling well, reduced energy and enthusiasm; and diminished sexual function including loss of libido (Tan and Philip 1999). Hypogonadism can cause osteoporosis, muscle atrophy and cognitive changes for a long duration.

Effect of testosterone therapy can be through the testosterone receptors in the brain and in the penis, thereby improving libido and erectile function of the penis. Testosterone treatment may not affect cardiovascular system in aeging men although it is controversial (Tenover 2000). Testosterone therapy can prevent fractures in elderly men by increasing bone and muscle strength. Testosterone levels are low in old men who developed prostate cancer. Also it has been reported that testosterone administration does not cause prostate cancer (Bhasin et al. 2001). Gynaecomastia may be caused due to aromatization of exogenous testosterone to estrogen but can be treated by aromatase inhibitors. Monitoring PSA and haematocrit levels is useful for long-term treatment.

8.3.2 Inhibins and Activins

McCullagh in 1932 coined the term inhibin describing a nonsteroidal substance released from the testis regulating pituitary function. Inhibin was later purified from follicular fluid and characterized (Ling et al. 1985; Robertson et al. 1985). Inhibin is produced mainly by the gonads. It is a glycoprotein hormone and consists of disulphide-linked dimers of two subunits α and β. The α-subunit of inhibin A and inhibin B is common, while β-subunit is βA or βB, respectively. The 32-kDa dimers of inhibin A and inhibin B are bioactive (O'Conner and de Kretser 2004). Dimers of the β-subunit form activins: activin A is βA-βB, activin B is βB-βB, while activin AB is βA-βB (Ling et al. 1986). Inhibin and activin are structurally related to transforming growth factor beta (TGF β) superfamily. These glycoproteins either inhibit or stimulate FSH production and are thereby named as inhibin or activin, respectively. Precursors of inhibin and activin subunits form homodimers and heterodimers in endoplasmic reticulum in the cytoplasm of the cell during peptide folding and processing (Walton et al. 2011). Follistatin is also a protein which neutralizes action of activin by binding to it (Philips 2003).

Orchidectomy reduces circulating levels of inhibin but not activin indicating the testis to be the main source of inhibin (Robertson et al. 1988). In addition to LH, Leydig cell steroidogenesis is controlled by autocrine mechanisms and paracrine regulation by Sertoli cells, namely, by activins and inhibins. Although the intratesticular concentrations of activin and inhibin are high, these glycoproteins may not be involved in regulation of steroid hormone synthesis, but alteration in their production may cause pathological changes (Hedger and Winnall 2012). Transgenic mice, with overexpression of activin βA subunit, increased production of activin A which may lead to damage to spermatogenesis (Tanimoto et al. 1999). Expression of activin receptor 2B is predominant than 2A in normal human testis. However, 2A is selectively up-regulated in seminomas and non-seminomas and in the immune cells in testicular carcinomas (Dias et al. 2008).

In most of the species, inhibin B is the major form of inhibin produced by the Sertoli cells of testes, while spermatogonial cells also express inhibin B (Anderson et al. 1998). Inhibin can be a useful clinical marker for testicular function as inhibin levels show positive correlation with Sertoli cell number, sperm concentration and spermatogonial status (Sharpe et al. 1999; Ramaswamy et al. 1999).

8.3.2.1 Functions of Inhibin

The presence of activin A and inhibin B has been reported in the testis at 13 weeks of gestation in humans (Anderson et al. 2002). After birth, inhibin B levels rise in infants coinciding with increase in gonadotropins and testosterone levels. A peak of inhibin B is noted at 3–6 months and remains elevated until 15 months of age. Testosterone and gonadotropin levels decline by 6–9 months and remain at basal levels during prepubertal age and increase markedly at the onset of puberty. Basal levels of inhibin observed during prepuberty increase gradually from midpuberty and reach adult level by puberty. Thus, a positive correlation exists between inhibin B and both testosterone and gonadotropin. However, in adulthood a negative correlation is noted between inhibin and FSH (Crofton et al. 1997; Andersson et al. 1997). Inhibin B levels are highest at 20–29 years of age in men and then decrease gradually with a corresponding rise in FSH (Byrd et al. 1998).

8.3.2.2 Regulation of Inhibin B

Sertoli cell proliferation is in two phases: first 12 months after birth and at puberty in humans which coincides with the increase in inhibin B (Cortes et al. 1987). This may be due to FSH-stimulated Sertoli cell proliferation. The Sertoli cell proliferation occurs prior to puberty, in the absence of germ cells and positive correlation in FSH, and inhibin B is observed. While at puberty, with the onset of spermatogenesis, a negative correlation is seen (Andersson et al. 1998). Thus, feedback control between pituitary FSH and inhibin B from Sertoli cells depends on the presence or absence of germ cells. Inhibin B levels are undetectable in boys with Sertoli cell-only syndrome (SCOS) suggesting that the Sertoli cell function declines in the absence of germ cells. The details of SCOS are given in the chapter Primary Testicular Failure. Similarly, inhibin levels in seminal plasma correlate with the sperm number. Inhibin B levels in seminal plasma are undetectable after vasectomy (Anderson et al. 2002).

8.3.2.3 Mechanism of Action

Activin and inhibin are the members of TGF-β family. Activin acts through the pathway of serine/threonine kinase receptor complex (Massague 1998). Inhibin coreceptors betaglycan and inhibin-binding protein p120 are involved in mediating inhibin signalling pathway (Lewis et al. 2000). The expression of stage-specific betaglycan-mRNA has been reported in the testis (Chong et al. 2000).

8.3.2.4 Inhibin in Infertile Men

Inhibin B levels have been shown to be low in men with SCOS (Meachem et al. 2001). Decline in serum inhibin B and increase in FSH have been observed in men with non-azoospermia (Bohring and Krause 1999; von Eckardstein et al. 1999; Brugo-Olmedo et al. 2001). However, in three cases with no spermatogenesis, inhibin levels are normal. Thus, inhibin possibly is not a good marker for spermatogenesis when compared to FSH levels, semen analysis and clinical evaluation of male fertility (Halder et al. 2005).

8.3.2.5 Inhibin-Activin-Associated Cancers in Males

The presence of inhibin in seminal plasma has been reported by a number of investigators. In addition to production of inhibin from the testis, prostatic inhibin-like peptide (PIP) having FSH-suppressing property has been demonstrated. PIP has been suggested to be as sensitive marker as PSA and prostatic acid phosphatase (PAP) in prostatic adenocarcinoma (Garde et al. 1994). Interestingly, serum PIP levels are higher in benign prostatic hyperplasia (BPH) than in prostate cancer patients and lowest in control men. Further, urinary PIP can possibly be a biomarker for BPH as the levels of inhibin B are strikingly higher than those in prostate cancer and controls (Teni et al. 1988).

Prostate cancer is associated with the loss of inhibin α-expression (Mellor et al. 1998). Possibly, some of the gene variants are located within the 3′ untranslated region of a gene which is in association with abnormal miRNA binding (Nicoloso et al. 2010).

Thus, normally, inhibins prevent activin formation, thereby decreasing its signalling activity. However, loss of inhibin function may be responsible for increase in action of activin leading to cancer (Shelling 2012).

Inhibin-α has been suggested as a tumor suppressor gene (Stenvers and Findley 2009). Cancer phenotype of inhibin-α knockout mice has been reported (Matzuk et al. 1992). In mice, functional deletion of alpha subunit leads to development of gonadal and adrenal tumors. However, elevated expression of inhibin-α was found in high-grade prostatic tumors, thus questioning the role of inhibin-α in cancer (Risbridger et al. 2004).

It is possible that cancer may be caused because of a defect in the activin signalling pathway which in turn elevates inhibin levels due to resistance during development (Fuller and Chu 2004).

8.3.3 Anti-Mullerian Hormone (AMH)/Mullerian-Inhibiting Substance (MIS)

8.3.3.1 In Normal Males

There is a factor other than testosterone which is involved in normal male sexual differentiation; it was named in 1986 as AMH (Jost 1947; Josso and Picard 1986). AMH is a 140-kDa gonadal glycoprotein of TGF-β family, and its primary role is to promote involution of the Mullerian ducts during normal male sexual differentiation. However, postnatally, it is secreted by Sertoli cells of the testis, and its receptors are seen on the Leydig cells, suggesting its importance in the development of the testis. AMH possibly affects androgen production by downregulating steroidogenic enzymes.

MIS/AMH- mRNA and protein levels decrease considerably after birth but gradually increase at puberty and are detectable in adulthood (Kuroda et al. 1990; Lee et al. 1992). Expression of receptors is downregulated at birth, appeared by 2 weeks of age and then, remained abundant in prepubertal and adult testis (Baarends et al. 1995). During infancy, immature Sertoli cells secrete both inhibin B and AMH abundantly. However, during prepuberty, when the gonadotropin levels are low, inhibin levels are also low, but AMH is elevated which declines during early puberty, probably correlating with Sertoli cell maturation (Andersson et al. 1998; Rajpert De Meyts et al. 1999).

In MIS/AMH transgenic mice, serum testosterone levels are lower than those in control mice. In male mice with deletions of MIS gene or its receptor, Mullerian ducts are retained, but Leydig cell hyperplasia and neoplasia are observed (Lyet et al. 1995; Behringer et al. 1990). These data demonstrate that MIS expression can alter Leydig cell number and function, thereby regulating Leydig cell proliferation and steroidogenic capacity of developing testis. Thus, along with LH, MIS/AMH also appears to regulate Leydig cell maturation and differentiation (Lee 2000).

8.3.3.2 In Male Disorders

MIS/AMH may serve as an index of gonadal integrity, differentiating between CDP and IHH delayed puberty, and may diagnose bilateral cryptorchidism (Donahoe et al. 1987; Misra et al. 2002; Coutant et al. 2010). Low levels of AMH in patients with CDP and prepubertal testis volume while very low levels of AMH in boys with congenital HH may be of clinical value in early diagnosis (Hero et al. 2012).

Testicular hormones play a major role in development and maturation of the testis. In elderly men, testosterone declines, and hormone replacement therapy may be beneficial. Alterations in testicular hormones may lead to disorders in men suggesting their importance in pathophysiology.

Acknowledgement Author is grateful to Dr. SD Kholkute, Director, NIRRH, Mumbai, for the support. The help from Ms. Shanti Ganeshan, Documentation and Information Officer, NIRRH, and the staff of the NIRRH library is acknowledged.

Key Questions

- Describe the pathways of testosterone synthesis.
- Discuss the mechanism of action of androgens.
- Discuss the role of androgens in prenatal differentiation and puberty changes.
- Write briefly the role of testosterone replacement in young and elderly males.
- Explain the mechanism of actions of inhibins and activins.
- Discuss the role of Mullerian-inhibiting hormone in male reproductive disorders.

References

AACE. Medical guidelines for clinical practice for the evaluation and treatment of hypogonadism in adult male patients. Endocr Pract. 2002;8:439–56.

Anderson RA, Irvine DS, Balfour C, Groome NP, Riley SC. Inhibin B in seminal plasma: testicular origin and relation to spermatogenesis. Hum Reprod. 1998;13:920–6.

Anderson RA, Cambray N, Hartley PS, McNeilly AS. Expression and localization of inhibin alpha, inhibin/activin beta and beta B and the activin type II and inhibin betaglycan receptors in the developing human testis. Reproduction. 2002;123:779–88.

Andersson AM, Juul A, Petersen JH, Muller J, Groome NP, Skakkebaek NE. Serum inhibin B in healthy pubertal and adolescent boys: relation to age, stage of puberty, and follicular-stimulating hormone, luteinizing hormone, testosterone and estradiol levels. J Clin Endocrinol Metab. 1997;82:3976–81.

Andersson AM, Muller J, Skakkebaek NE. Different roles of prepubertal and postpubertal germ cells and Sertoli cells in the regulation of serum inhibin B levels. J Clin Endocrinol Metab. 1998;83:4451–8.

Baarends WM, Hoogerbrugge JW, Post M, Visser JA, De Rooij D, Parvinen M, et al. Anti-Mullerian hormone and anti-Mullerian hormone type II receptor mRNA expression during postnatal testis development and in the adult testis of the rat. Endocrinology. 1995;136:5614–22.

Basaria S, Dobs AS. Hypogonadism and androgen replacement therapy in elderly men. Am J Med. 2001;110:563–72.

Behringer RR, Cate RL, Froelick GA, Palmiter RD, Brinster RL. Abnormal sexual development in transgenic mice chronically expressing Mullerian inhibiting substance. Nature. 1990;345(6271):167–70.

Berthold AA. Transplantation der Hoden [Transplantation of testis] (in German). Arch Anat Physiol Wissensch. 1849;16:42–6.

Bhasin S, Woodhouse L, Casaburi R, Singh AB, Bhasin D, Berman N, et al. Testosterone dose-response relationships in healthy young men. Am J Physiol Endocrinol Metab. 2001;281:E1172–E81.

Bohring C, Krause W. Serum levels of inhibin B in men with different causes of spermatogenic failure. Andrologia. 1999;31:137–41.

Bremner WJ, Vitiello MV, Prinz PN. Loss of circadian rhythmicity in blood testosterone levels with aging men. J Clin Endocrinol Metabol. 1983;56:1278–81.

Brown-Sequard CE. The effects produced on man by subcutaneous injections of liquid obtained from the testicles of animals. Lancet. 1889;3438:105–7.

Brugo-Olmedo S, De Vincentiis S, Calamera JC, Urrutia F, Nodar F, Acosta AA. Serum inhibin may be a reliable marker of the presence of testicular spermatozoa in patients with nonobstructive azoospermia. Fertil Steril. 2001;76:1124–9.

Butenandt A, Hanisch G. Uber die Umwandlung des Dehydroandrosterons in Androstenol-(17)-one-(3) (Testosterone); um Weg zur Darstellung des Testosterons auf Cholesterin (Vorlauf Mitteilung). [The conversion of dehydroandrosterone into androstenol-(17)-one-3 (testosterone); a method for the production of testosterone from cholesterol (preliminary communication)] (in German). Chem Ber. 1935;68:1859–62.

Byrd W, Bennett MJ, Carr BR, Dong Y, Wians F, Rainey W. Regulation of biologically active dimeric inhibin A and B from infancy to adulthood in the male. J Clin Endocrinol Metab. 1998;83:2849–54.

Chatterjee B, Roy AK. Changes in hepatic androgen sensitivity and gene expression during aging. J Steroid Biochem Mol Biol. 1990;37:437–45.

Chong H, Pangas SL, Bernard DJ, Wang E, Gitch J, Chen W, et al. Structure and expression of a membrane component of the inhibin receptor system. Endocrinology. 2000;141:2600–7.

Cortes D, Muller J, Skakkebaek NE. Proliferation of Sertoli cells during development of the human testis assessed by stereological methods. Int J Androl. 1987;10:589–96.

Coutant R, Biette-Demeneix E, Bouvattier C, Bouhours-Nouet N, Gatelais F, Dufresne S, et al. Baseline inhibin B and anti-Mullerian hormone measurements for diagnosis of hypogonadotropic hypogonadism (HH) in boys with delayed puberty. J Clin Endocrinol Metab. 2010;95:5225–32.

Crofton PM, Illingworth PJ, Groome NP, Stirling HF, Swanston I, Gow S, et al. Changes in dimeric inhibin A and B during normal early puberty in boys and girls. Clin Endocrinol (Oxf). 1997;46:109–14.

Dias V, Meachem S, Rajpert-De Meytes E, MacLachlan R, Manuelpillai U, Loveland KL. Activin receptor subunits in normal and dysfunctional adult human testis. Hum Reprod. 2008;23:412–20.

Donahoe PK, Cate RL, MacLaughlin DT, Epstein J, Fuller AF, Takahashi M, et al. Mullerian inhibiting substance: gene structure and mechanism of action of a fetal regressor. Recent Prog Horm Res. 1987;43:431–67.

El Shannawy A, Gates R, Russel L. Hormonal regulation of spermatogenesis in the hypophysectomized rat: cell viability after hormonal replacement.in adults after intermediate periods of hypophysectomy. J Androl. 1998;19:320–34.

Feldman HA, Longcope C, Derby CA, Johannes CB, Araujo AB, Coviello AD, et al. Age trends in level of serum testosterone and other hormones in middle-aged men: longitudinal results from the Massachusetts Male Aging Study. J Clin Endocrinol Metabol. 2002;87:589–98.

Fuller PJ, Chu S. Signalling pathways in the molecular pathogenesis of ovarian granulosa cell tumours. Trends Endocrinol Metab. 2004;15:122–8.

Garde SV, Sheth AR, Venkatesan VM, Panchal CJ, Porter AT, Grignon DJ. Prostate inhibin peptide (PIP) in prostate cancer: a comparative immunohistochemical study with prostate-specific antigen (PSA) and prostatic acid phosphatase (PAP). Cancer Lett. 1994;78:11–7.

Halder A, Fauzdar A, Kumar A. Serum inhibin B and follicle-stimulating hormone levels as markers in the evaluation of azoospermic men: a comparison. Andrologia. 2005;37:173–9.

Hedger MP, Winnall WR. Regulation of activin and inhibin in the adult testis and the evidence for functional roles in spermatogenesis and immunoregulation. Mol Cell Endocrinol. 2012;359:30–42.

Hero M, Tomiska J, Vaaralahti K, Laitinen E-M, Sipila I, Puhakka L, et al. Circulating anti-Mullerian hormone levels in boys decline during early puberty and correlate with inhibin B. Fertil Steril. 2012;97:1242–7.

Holt EH, Zieve D. Testosterone. MedlinePlus Medical Encyclopedia. U.S. National Library of Medicine. 2008.

Hu J, Zhang Z, Shen WJ, Azhar S. Cellular cholesterol delivery, intracellular processing and utilization for biosynthesis of steroid hormones. Nutr Metab (Lond). 2010;7(47):1–25.

Josso N, Picard JY. Anti-Mullerian hormone. Physiol Rev. 1986;66:1038–90.

Jost A. Recherches sur la differentiayion sexuelle de l'embryon de lapin. Arch Anat Microsc Morphol Exp. 1947;36:271–315.

Kenyon AT, Knowlton K, Sandiford I, Koch FC, Lotwin G. A comparative study of the metabolic effects of testosterone propionate in normal men and women and in eunuchoidism. Endocrinology. 1940;26:26–45.

Knuth UA, Behre H, Belkien L, Bents H, Nieschlag E. Clinical trial of 19-nortestosterone-hexoxylphenyl-propionate (Anador) for male fertility regulation. Fertil Steril. 1985;44(6):814–21.

Kuroda T, Lee MM, Haqq CM, Powell DM, Manganaro TF, Donahoe PK. Mullerian inhibiting substance ontogeny and its modulation by follicle-stimulating hormone in the rat testis. Endocrinology. 1990;127:1825–32.

Lee MM-C. MIS actions in the developing testis. In: Goldberg E, editor. The testis: from stem cell to sperm function. New York: Springer-Verlag, NY Inc.; 2000. p. 89–105.

Lee MM, Cate RL, Donahoe PK, Waneck GL. Developmentally regulated polyadenylation of two discrete mRNA for Mullerian inhibiting substance. Endocrinology. 1992;130:847–53.

Lewis KA, Gray PC, Blount AL, MacConell LA, Wiater E, Bilezikjian LM, et al. Betaglycan binds inhibin and can mediate functional antagonism of activin signaling. Nature. 2000;404:411–4.

Ling N, Ying SY, Ueno N, Esch F, Denoroy L, Guillemin R. Isolation and partial characterization of a Mr 32,000 protein with inhibin activity from porcine follicular fluid. Proc Natl Acad Sci U S A. 1985;82:7217–21.

Ling N, Ying SY, Ueno N, Shimasaki S, Esch F, Hotta M, et al. Pituitary FSH is released by a heterodimer of the b-subunit from the two forms of inhibin. Nature. 1986;321:779–82.

Lyet L, Louis F, Forest MG, Josso N, Behringer RR, Vigier B. Ontogeny of reproductive abnormalities induced by deregulation of anti-Mullerian hormone expression in transgenic mice. Biol Reprod. 1995;52:444–54.

Massague J. TGF-beta signal transduction. Annu Rev Biochem. 1998;67:753–91.

Matzuk MM, Finegold MJ, Su JG, Hsueh AJ, Bradley A. Alpha-inhibin is a tumour-suppressor gene with gonad specificity in mice. Nature. 1992;360:313–9.

McCullagh DR. Dual endocrine activity of the testes. Science. 1932;76:19–20.

Meachem SJ, Nieschlag E, Simoni M. Inhibin B in male reproduction: pathophysiology and clinical relevance. Eur J Endocrinol. 2001;145:561–71.

Mellor SL, Richards MG, Pedersen JS, Robertson DM, Risbridger GP. Loss of the expression and localization of inhibin alpha-subunit in high grade prostate cancer. J Clin Endocrinol Metab. 1998;83:969–75.

Misra M, MacLaughlin DT, Donahoe PK, Lee MM. Measurement of Mullerian inhibiting substance facilitates management of boys with microphallus and cryptorchidism. J Clin Endocrinol Metab. 2002;87:3598–602.

Mooradian AD, Morley JE, Korenman SG. Biological actions of androgens. Endocr Rev. 1987;8:1–28.

Morley JE, Kaiser FE, Perry 3rd HM, Patrick P, Morley PM, Stauber PM, et al. Longitudinal changes in testosterone, luteinizing hormone, and follicle-stimulating hormone in healthy older men. Metabolism. 1997;46:410–3.

Nicoloso MS, Sun H, Spizzo R, Kim H, Wickramasinghe P, Shimizu M, et al. Single-nucleotide polymorphisms inside miRNA target sites influence tumour susceptibility. Cancer Res. 2010;70:2789–98.

O'Connor AE, de Kretser DM. Inhibins in normal male physiology. Semin Reprod Med. 2004;22:177–85.

Philips DJ. The activin/inhibin family. In: Thomson AW, Lotze MT, editors. The cytokine handbook. London: Elsevier Science; 2003. p. 1153–77.

Rajpert De Meyts E, Jorgensen N, Graem N, Muller J, Cate RL, Skakkebaek NE. Expression of anti-Mullerian hormone during normal and pathological gonadal development associated with differentiation of Sertoli and granulosa cells. J Clin Endocrinol Metab. 1999;84:38–46.

Ramaswamy S, Marshall GR, McNeilly AS, Plant TM. Evidence that in a physiological setting Sertoli cell number is the major determinant of circulating concentration of inhibin B in the adult male rhesus monkey (Macaca mulatta). J Androl. 1999;20:430–4.

Risbridger GP, Shibata A, Ferguson KL, Stamey TA, McNeil JE, Peehl DM. Elevated expression of inhibin alpha in prostate cancer. J Urol. 2004;171:192–6.

Robertson DM, Foulds LM, Leversha L, Morgan FJ, Hearn MT, Burger HG, et al. Isolation of inhibin from bovine follicular fluid. Biochem Biophys Res Commun. 1985;126:220–6.

Robertson DM, Hayward S, Irby D, Jacobsen J, Clarke L, MacLachlan RI, et al. Radioimmunoassay of rat inhibin: changes after PMSG stimulation and gonadectomy. Mol Cell Endocrinol. 1988;58:1–8.

Ruzicka L, Wettstein A. Uber die kristallinische Herstellung des Testikelhormons, Testosteron (Androsten-3-ol-17-ol) [The crystalline production of the testicle hormone, testosterone (Androsten-3-ol-17-ol] (in German). Helvetica Chimica Acta. 1935;18:1264–75.

Sharpe RM. Regulation of spermatogenesis. In: Knobil E, Neill JD, editors. The Physiology of Reproduction. New York: Raven Press; 1994. p. 1363–434.

Sharpe RM, Turner KJ, McKinnell C, Groome MP, Atanassova N, Millar MR, et al. Inhibin levels in plasma of the male rat from birth to adulthood: effect of experimental manipulation of Sertoli cell number. J Androl. 1999;20:94–101.

Shelling AN. Mutations in inhibin and activin genes associated with human disease. Mol Cell Endocrinol. 2012;359:113–20.

Stenvers KL, Findley JK. Inhibins: from reproductive hormones to tumour suppressors. Trends Endocrinol Metab. 2009;21:174–80.

Tan RS, Philip PS. Perceptions of and risk factors for andropause. Arch Androl. 1999;43:97–103.

Tanimoto Y, Tanimoto K, Sugiyama F, Horiguchi H, Murakami K, Yagami K, et al. Male sterility in transgenic mice expressing activin bA subunit gene in testis. Biochem Biophys Res Commun. 1999;259:699–705.

Teni TR, Sheth AR, Kamath MR, Sheth NA. Serum and urinary prostatic inhibin-like peptide in benign prostatic hyperplasia and carcinoma of prostate. Cancer Lett. 1988;43:9–14.

Tenover JL. HRT for the aging male: a therapy that has come of age. In: Goldberg E, editor. The testis: from stem cell to sperm function. New York: Springer-Verlag, NY Inc.; 2000. p. 82–8.

Von Eckardstein S, Simoni M, Bergmann M, Weinbauer GF, Gassner P, Schepers AG, et al. Serum inhibin B in combination with follicle-stimulating hormone (FSH) is a more sensitive marker than serum FSH alone for impaired spermatogenesis in men, but cannot predict the presence of sperm in testicular tissue samples. J Clin Endocrinol Metab. 1999;84:2496–501.

Walton KL, Makanji Y, Robrtson DM, Harrison CA. The synthesis and secretion of inhibins. Vitam Horm. 2011;85:149–84.

Weinbauer GF, Gromoll J, Simoni M, Nieschlag E. Physiology of testicular function. In: Nieschlag E, Behre HM, editors. Andrology, Male Reproductive Health and Dysfunction. Springer, Berlin: Springer; 1997. p. 25–57.

Wu CW. Androgen replacement in young men. In: Goldberg E, editor. The testis: from stem cell to sperm function. New York: Springer-Verlag, NY Inc.; 2000. p. 89–105.

Zmuda JM, Cavely JA, Kriska A, Glynn NW, Gutai JP, Kuller LH. Longitudinal relation between endogenous testosterone and cardiovascular risk factors in middle aged men, a 13-year follow-up of former multiple risk factor intervention trial participants. Am J Epidemiol. 1997;146:609–17.

Hypothalamic-Pituitary-Testicular Axis

9

Bodhana Dhole and Anand Kumar

Learning Objectives

- Structure of gonadotropin-releasing hormone (GnRH)
- Structure of GnRH receptor
- Synthesis and regulation of GnRH receptor
- GnRH-GnRH receptor interactions
- GnRH pulses
- Regulation of GnRH secretion
- Structure of gonadotropins
- Regulation of gonadotropin secretion

9.1 Introduction

Male reproduction, development, and maintenance of male sexual characteristics are principally governed by the hypothalamic-pituitary-testicular (HPT) axis. The axis is composed of the hypothalamic gonadotropin-releasing hormone (GnRH), luteinizing hormone (LH), follicle-stimulating hormone (FSH), and the gonadal steroids. GnRH is the central regulator of the male reproductive hormonal cascade. The HPT axis maintains a dynamic equilibrium of serum levels of reproductive hormones through a closed-loop feedback mechanism. A cause leads to an effect and that may be the end of a process. The cause and effect continue to modify each other. The effect may suppress the cause for attenuation of the effect. At other times, the effect may promote the cause for a still larger effect. The former is the negative

B. Dhole, PhD • A. Kumar, MD, FAMS (✉)
Department of Reproductive Biology, All India Institute of Medical Sciences, New Delhi, India
e-mail: bodhana.dhole@gmail.com; anandkumarrepbiol@hotmail.com

A. Kumar, M. Sharma (eds.), *Basics of Human Andrology*,
DOI 10.1007/978-981-10-3695-8_9

feedback from the effect to the cause, and the latter is the positive feedback from the effect to the cause. Numerous hypothalamic, pituitary, and testicular factors tightly regulate the HPT axis and ensure its proper functioning.

9.2 Structure of Gonadotropin-Releasing Hormone (GnRH)

Localized within the hypothalamus is a group of 1,000 neurons which synthesize and secrete GnRH (Rance et al. 1994). GnRH neurons are fusiform in shape and generally bipolar. The axons project to the median eminence and infundibular stalk.

During the development of the embryo, GnRH-synthesizing neurons migrate from the olfactory area to the preoptic and arcuate nuclei of the hypothalamus. The migration of GnRH neurons is facilitated by a glycoprotein adhesion factor encoded by X-linked Kal-1 gene. Deficiency in the glycoprotein adhesion factor results in a condition referred to as Kallmann syndrome which is characterized by anosmia, GnRH deficiency, and hypogonadism (Rance et al. 1994).

In the mammalian hypothalamus, GnRH was first isolated as a decapeptide having pyro-Glu_1, His_2-Trp_3-Ser_4-$Tyrs_5$-$Glyg_6$-$Leur_7$-Arg_8-Pro_9-Gly_{10}-NH_2 structure (Millar et al. 2004). The gene for pre-pro-GnRH is located on the short arm of chromosome 8. GnRH is synthesized as a much larger 92-amino acid precursor peptide. The precursor peptide has four segments from N-terminal to the C-terminal: (1) signal peptide containing 23 amino acids, (2) GnRH decapeptide, (3) proteolytic site of 3 amino acids and (4) GnRH-associated peptide (GAP) containing 56 amino acids (Nikolics et al. 1985). The pre-pro-GnRH mRNA is encoded in four exons. Exon 1 contains the 5′ untranslated region of the gene; exon 2 contains the signal peptide GnRH and the N-terminus of GAP; exon 3 contains the central portion of GAP; and exon 4 contains the C-terminus of GAP and the 3′ untranslated region (see Fig. 9.1). The nucleotide sequences which

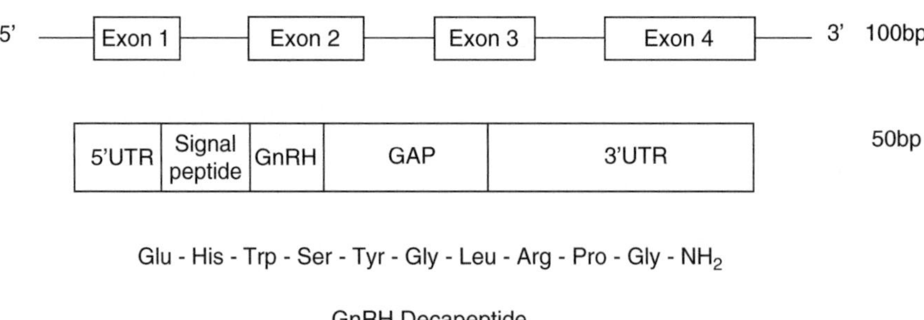

Fig. 9.1 Schematic diagram of the human gene for gonadotropin-releasing hormone (GnRH), the complimentary DNA (cDNA), and the gonadotropin-releasing hormone peptide; *bp* base pair, *UTR* untranslated region, *GAP* GnRH-associated peptide

encode the GnRH decapeptide are highly homologous. The precursor peptide is cleaved into decapeptide. The decapeptide is packed in storage granules. The storage granules are then transported to the median eminence (Milton et al. 1986).

9.2.1 Structure of GnRH Receptor

The specific receptor for GnRH is a typical seven-transmembrane domain G-protein receptor that belongs to type I rhodopsin-like receptor family. The seven hydrophobic transmembrane domains are connected via three extracellular loops (EL) and three intracellular loops (IL; see Fig. 9.2). The EL binds to GnRH, while the IL couples with G-protein and is involved in intracellular signal transmission. In humans, three GnRH receptors or receptor-like sequences have been identified: GnRH I, GnRH II, and GnRH II receptor-like homologue. The genes encoding the three receptors are located on chromosome 4, 1, and 14, respectively (Neill 2002). The gene encoding GnRH I and GnRH II is 40% identical to each other. The open reading frame of GnRH I and GnRH II gene consists of three exons spanning 18.9 kb. The three exons encode amino acids 1–174, 175–248, and 249–328, respectively. Between amino acids 174–175 in the transmembrane domain 4, intron 1 is located. Intron 2 is located between amino acids 248–249. The GnRH II receptor-like homologue gene is 40% and 100% identical to exon 2 and 3 of GnRH I and GnRH II gene, respectively. The GnRH II receptor-like homologue gene is encoded on the antisense DNA strand of the RM8 gene (Neill 2002).

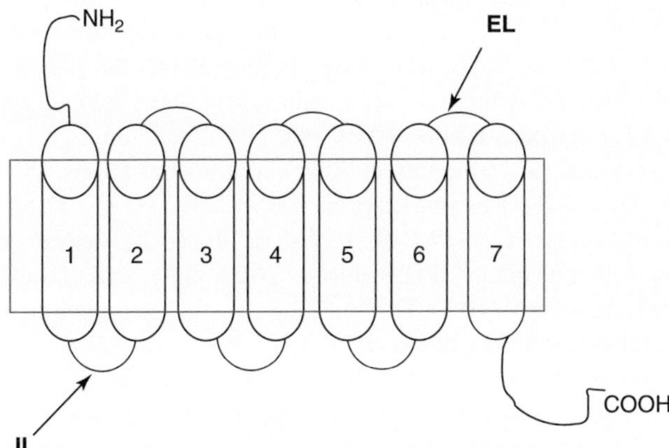

Fig. 9.2 The gonadotropin hormone (GnRH) receptor (*EL* extracellular loop, *IL* intracellular loop)

9.2.1.1 Synthesis and Regulation of GnRH Receptor

In rat pituitary cells, GnRH and protein kinase C (PKC) stimulate synthesis of GnRH receptor. However, GnRH and PKC utilize different mechanisms to stimulate GnRH receptor synthesis (Braden et al. 1991). Pulsatile secretion of GnRH every 30 min results in a maximum synthesis of GnRH receptor (Kaiser et al. 1997).

Estrogen stimulates GnRH receptor mRNA levels in rat pituitary tissue. Estrogen also increases the number of pituitary cells expressing GnRH receptor (Quiñones-Jenab et al. 1996). However, in adult rat pituitary cells, short-term administration of estradiol decreases GnRH receptor concentrations, while long-term administration increases the concentration of GnRH receptors (Emons et al. 1988). Progesterone inhibits the expression of GnRH receptor in the pituitary gland (Wu et al. 1994). Inhibin, a protein hormone, has no effect on the synthesis of GnRH receptors under basal conditions. However, inhibin blocks GnRH-stimulated synthesis of GnRH receptor (Braden et al. 1990). Activin A, another protein hormone, normally functions opposite to inhibin and stimulates synthesis of GnRH receptor in rat pituitary cell cultures. Inhibin does not block the stimulatory effect of activin A on GnRH receptor synthesis (Braden and Conn 1992).

9.2.2 GnRH-GnRH Receptor Interactions

When GnRH couples to its G-protein receptor, it can activate three G-proteins – stimulatory G-proteins: G_s and G_q, and inhibitory G-protein: G_i. When low concentrations of GnRH bind to its receptor, it activates Gs, which in turn activates adenyl cyclase (AC) that catalyzes the conversion of adenosine triphosphate (ATP) to 3′,5′-cyclic adenosine monophosphate (cAMP) and pyrophosphate (Krsmanovick et al. 2009). cAMP interacts with plasma membrane cyclic nucleotide-gated K^+ channels (CNG). The flow of ions across the plasma membrane causes a conformational change in the plasma membrane voltage-gated calcium channels (VGCC); it opens up the VGCC and allows calcium (Ca^{+2}) inflow from extracellular sites and raises the intracellular Ca^{+2} concentrations. When low concentrations of GnRH bind to its receptor, it also activates $G_{q.}$ The activated G_q stimulates phospholipase C-β (PLC-β). PLC-β catalyzes the cleavage of phosphatidylnositol-4, 5-bisphosphate (PIP_2) into 1, 2-diacylglycerol (DAG) and inositol-1, 4, 5-trisphosphate (IP_3). The binding of IP_3 to calcium channels in endoplasmic reticulum (ER) turns on a downstream signaling cascade which results in the release of Ca^{2+} into the cytoplasm. The increase in Ca^{+2} by the above two pathways causes phosphorylation of certain downstream Ca^{+2}-dependent kinases that facilitates cytoplasmic GnRH release from the secretory organelles. At high GnRH concentrations, GnRH binding to its receptor activates G_i that inactivates cAMP and also interacts with G-protein-activated inwardly rectifying potassium channels (GIRK) that blocks plasma membrane VGCC, decreases intracellular Ca^{+2} concentrations, and therefore reduces GnRH production to baseline levels (see Fig. 9.3; Krsmanovick et al. 2009).

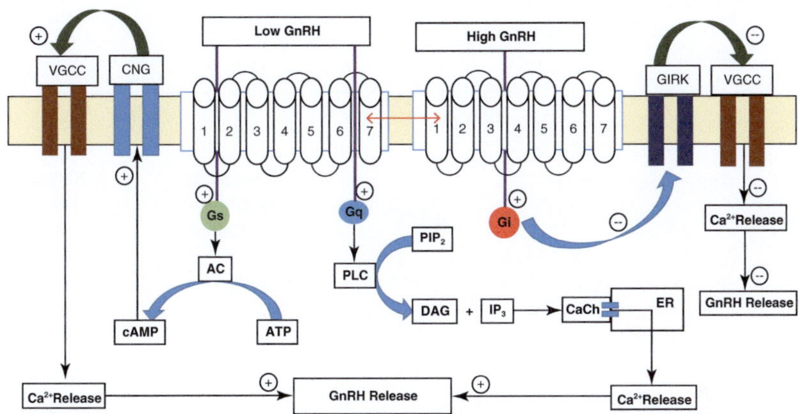

Fig. 9.3 Schematic diagram showing GnRH-GnRH receptor interactions (*GnRH* gonadotropin-releasing hormone, *Ca2+* calcium, *CaCh* calcium channel, *ER* endoplasmic reticulum, *DAG* 1, 2-diacylglycerol, *IP3* inositol-1, 4, 5-trisphosphate, *PIP2* phosphatidylnositol-4, 5-bisphosphate, *cAMP* cyclic adenosine monophosphate, *ATP* adenosine triphosphate, *AC* adenyl cyclase, *VGCC* voltage-gated calcium channels, *GIRK* G-protein-activated inwardly rectifying potassium channels, *PLC* phospholipase C, *CNG* cyclic nucleotide-gated K^+ channels, *Gs* and *Gq* stimulatory G-proteins, *Gi* inhibitory G-protein

9.2.3 GnRH Pulses

GnRH is secreted from the hypothalamic neurons in a pulsatile manner. The pulse-generating capacity of the GnRH neurons is intrinsic, i.e., even in the absence of any external influences, GnRH is released in pulses (Low 2011). The frequency of the GnRH pulses regulates the FSH to LH ratio from the pituitary. One pulse of GnRH every hour leads to a low ratio of FSH to LH secretion from the pituitary. One pulse of GnRH every 3 hr has an opposite effect on the FSH to LH ratio (Low 2011).

9.2.4 Regulation of GnRH Secretion

GnRH neurons are adjacent to other neuronal systems; thus, GnRH secretion can be modified by central nervous system inputs and neurotransmitters. Figure 9.4 shows different factors that regulate the HPT axis.

9.2.4.1 Gamma-Aminobutyric Acid

Gamma-aminobutyric acid (GABA) is synthesized from glutamate catalyzed by the enzyme glutamate decarboxylase (Soghomonian and Martin 1998). GABAergic neurons located in the anteroventral periventricular area (AVPV), lateral preoptic area, and suprachiasmatic nucleus show a direct action on GnRH neurons. All GnRH neurons express $GABA_A$ receptor, and majority of them express $GABA_B$ receptor (Herbison 2015). GABA is known to inhibit GnRH secretion. In female rats, GABA infusion into the preoptic area or intraperitoneal injection of the $GABA_A$ receptor agonist, muscimol, blocks the LH surge (Adler and Crowley 1986; Herbison

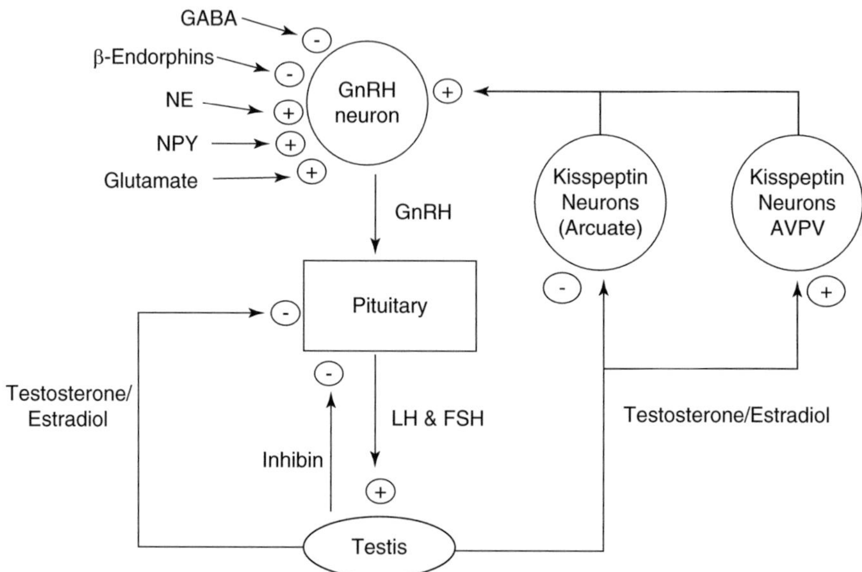

Fig. 9.4 Regulation of the hypothalamic-pituitary-testicular axis (*NE* norepinephrine, *NPY* neuropeptide Y, *GABA* gamma-aminobutyric acid, *GnRH* gonadotropin-releasing hormone, *LH* luteinizing hormone, *FSH* follicle-stimulating hormone, *AVPV* anteroventral periventricular nucleus

and Dyer 1991). In female rats, intracerebroventricular injection of muscimol suppresses GnRH mRNA levels in the preoptic area (Bergen et al. 1991). Testosterone stimulates the expression of $GABA_A$ receptors (Zhang et al. 1999). Therefore, testosterone mediates its negative feedback effect on GnRH through GABA.

9.2.4.2 Glutamate

Excitatory amino acid, such as glutamate, is found in large concentrations in a variety of important hypothalamic nuclei, including the arcuate nucleus, supraoptic nucleus, suprachiasmatic nucleus, paraventricular nucleus, and preoptic area. Glutamate exerts its action through ionotropic and metabotropic receptors located on the GnRH neurons. Glutamate stimulates LH secretion by stimulating the release of GnRH (Brann 1995).

9.2.4.3 Norepinephrine

Norepinephrine (NE) is known to stimulate GnRH secretion. The stimulation is mediated by the β-1-adrenergic receptors (Segovia et al. 1996). Estrogen stimulates the release of NE from the mediobasal hypothalamus of female rats (Ohtsuka et al. 1989).

9.2.4.4 Neuropeptide Y

Neuropeptide Y (NPY) is a 36-amino acid peptide. NPY neurons are located in the arcuate nucleus and brainstem. NPY receptors are located on the GnRH neurons

(Herbison 2015). In male rats, NPY upregulates GnRH synthesis; the upregulation is mediated via Y1–NPY receptor (Li et al. 1994). NPY also acts at the level of pituitary, together with GnRH, to increase the LH secretion (Kalra et al. 1988).

9.2.4.5 Kisspeptin

Kisspeptin is a neuropeptide synthesized and secreted by the neurons in the hypothalamic arcuate, periventricular, and the AVPV nuclei (Dungan et al. 2006). Kisspeptin and GnRH-secreting neurons are located in a few common hypothalamic areas. The KISS1 gene encodes a 145-amino acid prepro-kisspeptin protein which is enzymatically cleaved into a 54-amino acid active protein known as kisspeptin-54. The kisspeptin protein is encoded by the KISS1 gene located on chromosome 1q32.1 in humans (see Fig. 9.5).

Kisspeptin binds to GPR54, a G-protein-coupled receptor. When kisspeptin binds to GPR54, it activates G-protein, Gq/G_{11}. Gq protein activates PLC. PLC in turn hydrolyzes PIP_2 into DAG and IP_3. IP_3 is released as a soluble structure into the cytosol. IP_3 then diffuses through the cytosol to bind to IP_3 receptors on Ca^{2+} channels in the ER. These channels open and only allow the passage of Ca^{2+}into the cytoplasm. DAG remains bound to the membrane. DAG activates the mitogen-activated protein kinase (MAPK) pathway (see Fig. 9.6; Pinilla et al. 2012).

Kisspeptin acts directly on the hypothalamic GnRH neurons where GPR54 mRNA has been identified (Irwig et al. 2004). Kisspeptin stimulates GnRH secretion and consequently LH and FSH secretion. The activation of GnRH is GPR54 receptor mediated as it is evident from mice lacking a functional GPR54 gene (Dungan et al. 2006).

The kisspeptin neurons in the arcuate and AVPV nuclei express androgen receptors and estrogen receptor-α (Smith et al. 2005). Kisspeptin neurons in the AVPV nucleus are different from those in the arcuate nucleus. In the AVPV, castration reduces both kisspeptin cell number and intracellular kisspeptin mRNA per cell (Smith et al. 2005). The stimulatory effect of estrogen on kisspeptin in the AVPV nucleus is responsible for the LH surge at mid-cycle in females. In female rats, the concentration of kisspeptin neurons in the AVPV nucleus is highest during the evening of proestrus (Smith et al. 2006). Most of the kisspeptin neurons in the AVPV nucleus co-express the Fos gene at the time of the LH surge. The product of

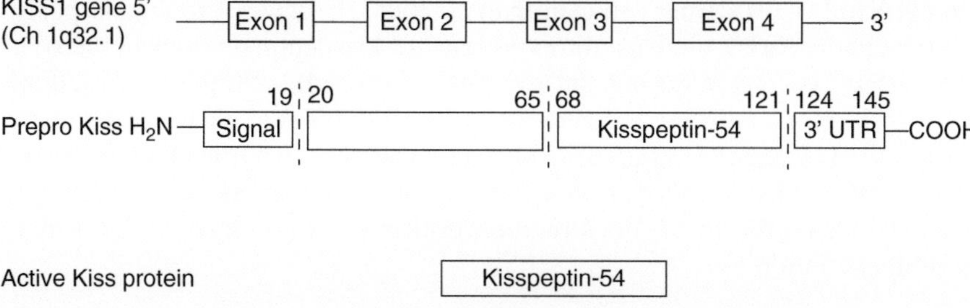

Fig. 9.5 Schematic diagram of kisspeptin gene and the corresponding protein

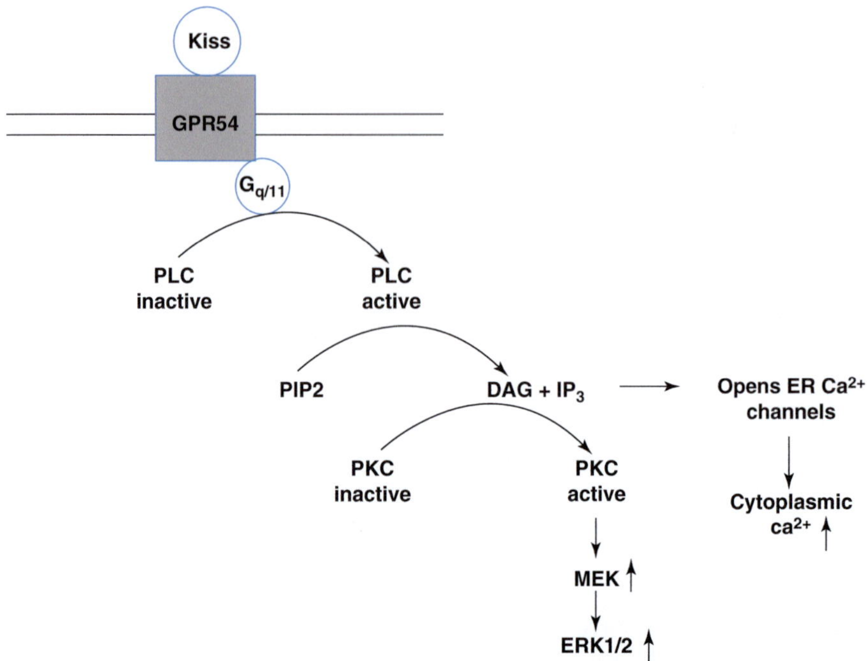

Fig. 9.6 The PLC-IP_3-DAG pathway activated by kisspeptin-GPR54 interaction (*KISS* kisspeptin, *GPR54* kisspeptin receptor, *PLC* phospholipase C, *PIP_2* phosphatidylinositol-4, 5-bisphosphate, *DAG* diacylglycerol, *IP_3* inositol-1, 4, 5-triphosphate, *ER* endoplasmic reticulum, *MEK* mitogen-activated protein kinase, *ERK* extracellular signal-regulated kinase

the Fos gene, a transcription factor, stimulates the proliferation of GnRH neurons and secretion of GnRH (Smith et al. 2006). In mice, histone of the kisspeptin gene locus is highly acetylated by estrogen which promotes the interactions among RNA polymerase and other transcriptional factors and thereby induce gene transcription.

Both estradiol and testosterone inhibit KISS1 gene expression in arcuate nucleus (see Fig. 9.7). Reduction of serum gonadal steroid levels by gonadectomy increases the number of kisspeptin neurons and also enhances the expression of KISS1–mRNA per cell (Smith et al. 2005). In contrast to the kisspeptin neurons in the AVPV nucleus, the expression of kisspeptin neurons in the arcuate nucleus is lowest during the evening of proestrus (Smith et al. 2006). Kisspeptin neurons in the arcuate nucleus do not co-express Fos gene at the time of the LH surge (Smith et al. 2006). In mice, histone of the kisspeptin gene locus in the arcuate nucleus show histone deacetylation in response to estrogen which condense the chromatin structure resulting in a subsequent inhibition in gene transcription.

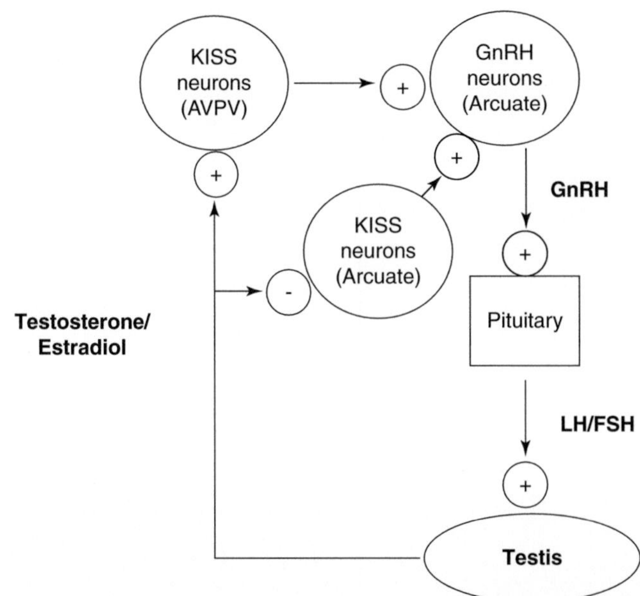

Fig. 9.7 Flow chart showing kisspeptin-mediated regulation of GnRH secretion (*AVPV* anteroventral periventricular, *GnRH* gonadotropin-releasing hormone, *LH* luteinizing hormone, *FSH* follicle-stimulating hormone)

9.2.4.6 Endogenous Opioids

Endogenous opioid peptides are considered as important neuroendocrine regulators of GnRH secretion. Endogenous opioid peptides are located in cell bodies within the hypothalamus in the areas rich in GnRH neurons. Opioid receptors are located on the GnRH neurons (Low 2011). β-Endorphin from opioidergic system can decrease the secretion of GnRH, whereas NE neurons can stimulate its release (see Fig. 9.8; Ferin et al. 1984). Additionally, males addicted to heroin have been associated with central hypogonadism and decreased levels of gonadotropins and gonadal steroids (Katz 2005). In rats, treatment with morphine, an opiate, significantly decreases basal and human chorionic gonadotropin (hCG)-stimulated testosterone levels and testicular interstitial fluid volume (Adams et al. 1993). Intratesticular opiates inhibit Sertoli cell growth and androgen-binding protein secretion (Gerendai et al. 1986).

9.2.4.7 Gonadal Steroids

Gonadal steroids – testosterone and estradiol – exert a negative feedback effect on the HPT axis (see Fig. 9.9). The GnRH neurons lack androgen or estrogen receptors (Low 2011). Therefore, the gonadal steroids regulate GnRH by acting on the afferent neurons synapsing with GnRH neurons. At the hypothalamic level, all the

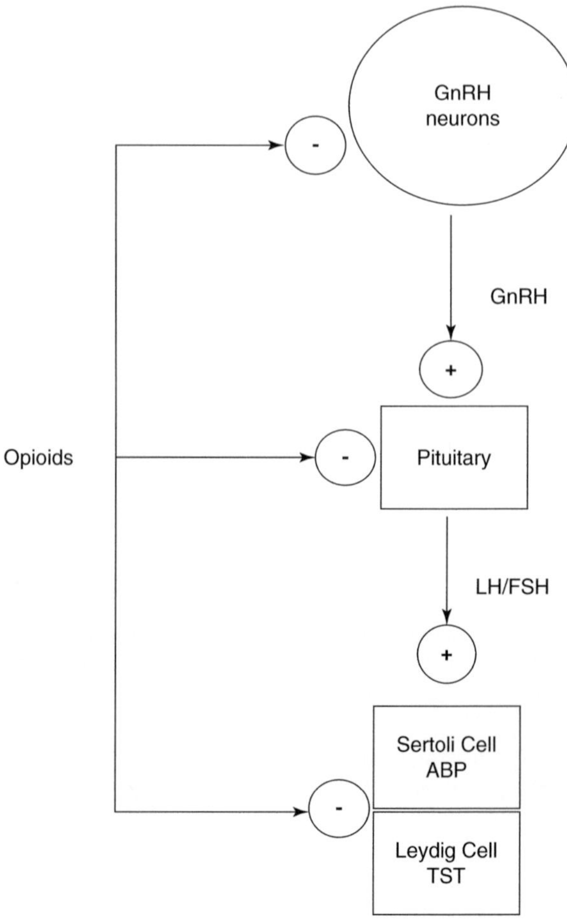

Fig. 9.8 Flow chart showing interaction between opioids and the hypothalamic-pituitary-testicular axis. *GnRH* gonadotropin-releasing hormone, *LH* luteinizing hormone, *FSH* follicle-stimulating hormone, *ABP* androgen-binding protein, *TST* testosterone

gonadal steroids reduce the frequency of GnRH pulse probably by altering neurotransmitter inputs from the afferent neurons impinging on GnRH neurons. Gonadal steroids also regulate KISS1 mRNA expression which further regulates GnRH secretion as discussed in the previous section.

9.3 Structure of Gonadotropins

The pituitary gland or hypophysis is found at the base of the brain below the hypothalamus. The infundibulum or pituitary stalk which carries both axons and blood vessels, serves as a link between the hypothalamus and the pituitary gland. The pituitary gland can be divided into two functionally and embryologically distinct parts/lobes – the anterior pituitary or adenohypophysis and the posterior pituitary or

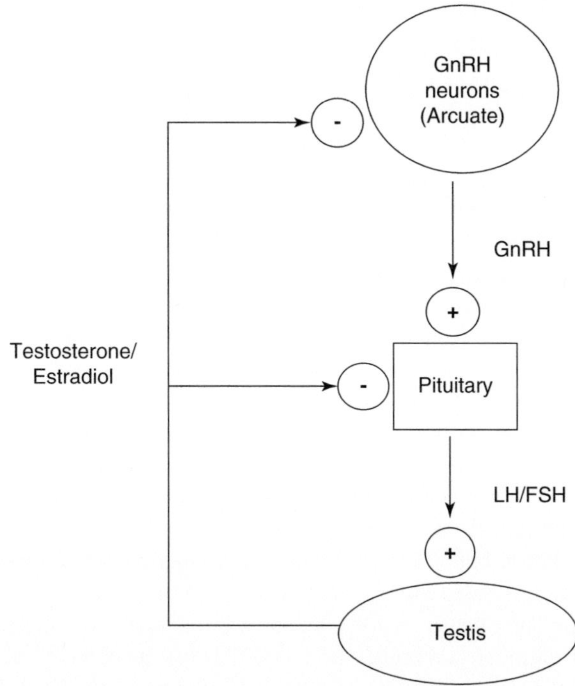

Fig. 9.9 Flow chart showing regulation of hypothalamic-pituitary-testicular axis by gonadal steroids; *GnRH* gonadotropin-releasing hormone, *LH* luteinizing hormone, *FSH* follicle-stimulating hormone

neurohypophysis. The anterior pituitary cells which synthesize and secrete gonadotropins are called gonadotrophs. Gonadotrophs secreting LH and FSH constitute 7–14% of the cells in the pituitary (Manjithaya and Dighe 2006). Pituitary cells also synthesize adrenocorticotropic hormone, growth hormone, thyroid-stimulating hormone (TSH), and prolactin. A specialized capillary network, the pituitary portal system, located between the median eminence of the hypothalamus and the pituitary, transports neuropeptides between these two structures. The portal circulatory system delivers neuropeptides from the hypothalamus to pituitary. Through retrograde blood flow, the circulatory system allows communication between anterior pituitary and the hypothalamus. The communication is also possible between anterior and posterior parts of pituitary gland (Lechan and Toni 2000).

GnRH is transported directly to the pituitary gland via the portal circulation where it binds with its specific cell surface receptors and stimulates the anterior pituitary gland to release two glycoproteins – LH and FSH in a pulsatile fashion. LH and FSH are protein dimers consisting of two glycosylated polypeptide chains-alpha (α) and beta (β), which are non-covalently linked together. The α-subunit of the glycoprotein hormones LH, FSH, TSH, and hCG is identical. However, the β-subunit structure is unique for each hormone and determines specificity for their different receptors. In humans, the genes encoding the common α-subunit are located on chromosome 6, while the β-subunits of FSH and LH are located on chromosome 11 and 19, respectively. Circulating free-α and free-β subunits are not recognized by gonadotropin receptors and are, therefore, are not biologically active (Bousfield et al. 2006).

Posttranslational processing of the gonadotropins involves glycosylation or addition of carbohydrate side chains by microsomal enzymes in the ER (Catt and Pierce 1986). The common α-subunit has two glycosylation sites at amino acid 52 and 78, while the glycosylation sites on their β-subunits are 7, 24, and 30, respectively. The oligosaccharide structures in both LH and FSH have a central mannose core, bound to an asparagine (Asn) residue. In LH, the Asn residue is bound through two residues of N-acetylglucosamine and a terminal sulfate while in FSH through a sialic acid residue. The terminal N-acetylglucosamine sulfate in LH is quickly removed from circulation by liver receptors that recognize the terminal sulfate. In contrast, the terminal sialic acid residue in FSH protects it from being captured by liver receptors. As a result the half-lives of LH and FSH in peripheral circulation vary widely, i.e., approximately 20 and 180 min, respectively (Bousfield et al. 2006).

The net surface charge on LH and FSH also determines their biological activity and half-lives. The more basic form of LH and FSH results in a greater in vitro potency but a shorter half-life in circulation, while reverse effects are observed for the less basic forms (Stanton et al. 1996).

The release of gonadotropins in response to GnRH has two unique properties. Continuous administration of GnRH results in an initial rise in the levels of LH and FSH. However, after several days of exposure to GnRH, secretion of LH and FSH gradually diminishes. Continuous exposure to GnRH for a prolonged period results in diminished GnRH binding or downregulation of its receptors. The downregulation is a reversible process and is accompanied with decoupling of receptor from G-protein and internalization of the receptor from the surface and its degradation in lysosomes (Clayton and Catt 1981). Downregulation and diminished secretion of LH and FSH can also be induced by the administration of long-acting synthetic agonist analogues of GnRH. The GnRH agonists can reduce LH and FSH levels and abolish gonadal function. Pulsatile administration of GnRH at a physiological frequency of 60 min to 4 h increases GnRH receptor binding on the gonadotroph and consequently increases LH and FSH levels by a process called upregulation (Bousfield et al. 2006).

The differential release of FSH and LH is also related to GnRH pulse frequency (Belchetz et al. 1978). One pulse of GnRH every hour may lead to greater release of LH. A slower frequency, i.e., one pulse of GnRH every 3 h, may stimulate FSH secretion with low secretion of LH. It has been observed that at the onset of puberty when GnRH pulse generator begins to operate, the ratio of FSH to LH is greater than the adults (Burr et al. 1970).

When FSH binds to its receptors located on the plasma membrane of Sertoli cells, it activates the cAMP-dependent protein kinase A (PKA) pathway. PKA phosphorylates cAMP-dependent response element-binding proteins. Through this pathway, specific Sertoli cell genes for various proteins such as inhibin, activins, anti-Mullerian hormone, estrogens, androgen-binding proteins, transferrin, ceruloplasmin, and a variety of proteases including plasminogen activator, cathepsin L, and several metalloproteases are activated. Leydig cells have specific membrane receptors for LH. Binding of LH to its receptors on Leydig cells activates a cAMP-dependent PKA pathway which activates enzymes required for testicular steroidogenesis (Lo and Lamb 2004).

9.3.1 Regulation of Gonadotropin Secretion

The secretion of gonadotropin is regulated by GnRH and testicular steroids. The regulation of gonadotropin secretion by GnRH has been discussed in the previous section. We shall now describe the role of testicular steroids in regulating gonadotropin secretion.

9.3.1.1 Role of Androgens

The production of gonadotropins is negatively controlled through an inhibitory action of androgens both at the central nervous system and the pituitary level. Androgens exert a negative feedback effect on gonadotropin release from the pituitary. The negative feedback effect of androgen decreases responsiveness of the pituitary to GnRH, thereby resulting in decrease in LH pulse amplitude and fall in plasma LH levels (Sheckter et al. 1989). Testosterone also exerts its negative feedback effect on GnRH release by stimulating release of opioid peptides from hypothalamic neurons. Opioids inhibit pulsatile GnRH secretion from the hypothalamus and consequently reduce LH levels (Veldhuis et al. 1984). Testosterone is reduced to non-aromatizable dihydrotestosterone (DHT) in the brain and peripheral tissues. DHT does not have any effect at the pituitary level but exerts a negative feedback effect on LH secretion at the hypothalamic level (Cailleux-Bounacer et al. 2009). Administration of DHT to young men has been shown to significantly decrease immunoactive FSH levels without any change in bioactive FSH levels (Urban et al. 1991).

9.3.1.2 Role of Estrogens

Testosterone aromatizes to estradiol in the brain and peripheral tissues. Estrogens inhibit pituitary LH secretion by decreasing hypothalamic GnRH pulse frequency and also the pituitary responsiveness to GnRH. All estrogen-mediated effects can be reversed by administration of anti-estrogens or aromatase inhibitors (Hayes et al. 2000). It is the level of circulating estrogen, but not of testosterone, that regulates the LH levels in males. In hypothyroid men, the testosterone levels decrease, but estrogen and LH levels do not change suggesting that estradiol is of primary importance in the regulation of LH secretion in males (Kumar et al. 2007). In hyperthyroid men, a hyperestrogenemic condition exerts a positive feedback effect on LH secretion instead of a negative feedback (Kumar et al. 2012). Injection of 17β-estradiol to normal adult males results in an initial fall in FSH and LH followed by an increase in LH alone (Kulin and Reiter 1976). Maintenance of serum estradiol concentration similar to that present in women at mid-cycle, for a period of 96–122 h, leads to a surge of LH in both intact and castrated men (Barbarino et al. 1983). However, the magnitude of LH surge was not as huge as found in females which is probably because of high testosterone levels seen in men (Barbarino et al. 1983, Kumar et al. 2012). The exact cause for the switch from a negative feedback effect to a positive feedback effect of estrogen on LH levels is not known in males. However, the mechanism of positive feedback effect of estradiol in males can be the same as that seen in mid-cycle women. Estradiol levels of 200–400 pg/ml persisting for at least 36 h are adequate to generate LH surge in women at mid-cycle (Karsch et al. 1973). The role of kisspeptin in the AVPV nucleus during the LH surge has already been discussed previously in this chapter. Within the pituitary, estradiol increases GnRH receptor expression, resulting in increased pituitary sensitivity to GnRH pulses (Nakai et al. 1978). Hypothalamic GnRH secretion is also increased at the time of LH surge, as directly measured in sheep and rats; however, this increase is not essential for production of LH surge (Ehlers and Halvorson 2013). In lower animals such as rodents, the behavior of HPT axis is rigidly dimorphic and cannot be altered by changing serum hormonal levels. It can be concluded that in primates and humans, steroid-mediated feedback mechanism is epigenetically and endocrinologically determined.

Estrogen administration has been shown to delay the clearance of FSH from circulation by decreasing the half-life of FSH, but it does not change FSH pulse frequency.

9.3.1.3 Role of Inhibin and Activin

FSH secretion is also partly regulated by inhibin and activin, which belong to the tumor growth factor-β (TGFβ) protein superfamily. Inhibin, a nonsteroidal protein hormone, is secreted by Sertoli cells under the stimulation of FSH. The protein is composed of two dissimilar subunits, an α-subunit with molecular weight 18,000 Da and a β-subunit with molecular weight 14,000 Da. Both subunits are cross-linked by disulfide bridges. Two forms of inhibin have been identified – inhibin A and inhibin B. Inhibin B is the predominant form in primates including humans (Childs 2006). Both forms share the same α-subunit but have differences in the peptide sequence of the β-subunit. Inhibin has been shown to suppress FSH secretion preferentially (Mason et al. 1985). When castrated male ram is infused with physiological concentration of inhibin, it results in suppression of FSH levels to that in normal male ram. Moreover, when inhibin is administered in these castrated males treated with GnRH, it still suppresses FSH levels. This suggests that inhibin acts exclusively on the pituitary gland (Tilbrook et al. 1993). Inhibin also exerts its negative effect on pituitary gland by antagonizing the effect of activin. Activin is composed of two β-subunits of inhibin and stimulates FSH synthesis and secretion. Activins are produced by anterior pituitary gonadotrophs and testicular Sertoli cells. Pituitary activins behave as autocrine modulator of basal FSH secretion as suggested in experiments using anti-activin antibody. Administration of anti-activin antibody inhibits pituitary FSH β-subunit levels without affecting α-subunit levels or LH levels (Corrigan et al. 1991).

Key Questions

- Describe the segments of the precursor peptide of GnRH.
- Describe the GnRH receptor-mediated signaling cascade downstream of GnRH receptor binding.
- How do kisspeptin and opioids mediate regulation of GnRH secretion?
- Describe the structure of LH and FSH.
- Discuss the role of testicular steroids and protein hormones in regulation of gonadotropin secretion.

References

Adams ML, Sewing B, Forman JB, Meyer ER, Cicero TJ. Opioid-induced suppression of rat testicular function. J Pharmacol Exp Ther. 1993;266:323–8.

Adler BA, Crowley WR. Evidence for gamma-aminobutyric acid modulation of ovarian hormonal effects on luteinizing hormone secretion and hypothalamic catecholamine activity in the female rat. Endocrinology. 1986;118:91–7.

Barbarino A, Marinis LD, Mancini A. Estradiol modulation of basal and gonadotropin-releasing-hormone-induced gonadotropin release in intact and castrated men. Neuroendocrinology 1983;36:105–11.

Belchetz PE, Plant TM, Nakai Y, Keogh EJ, Knobil E. Hypophysial responses to continuous and intermittent delivery of hypopthalamic gonadotropin-releasing hormone. Science. 1978;202:631–3.

Bergen H, Hejtmancik J, Pfaff D. Effects of gamma-aminobutyric acid receptor agonists and antagonist on LHRH-synthesizing neurons as detected by immunocytochemistry and in situ hybridization. Exp Brain Res. 1991;87:46–56.

Bousfield GR, Jia L, Ward DN. Gondotropins: Chemistry and Biosynthesis. In: Neill JD, editor. Knobil and Neill's physiology of reproduction. Amsterdam: Elsevier Academic Pres; 2006. p. 1581–634.

Braden TD, Bervig T, Conn PM. Protein kinase-C activation stimulates synthesis of gonadotropin-releasing hormone (GnRH) receptors, but does not mediate GnRH-stimulated receptor synthesis. Endocrinology. 1991;129:2486–90.

Braden TD, Conn PM. Activin-A stimulates the synthesis of gonadotropin-releasing hormone receptors. Endocrinology. 1992;130:2101–5.

Braden TD, Farnworth PG, Burger HG, Conn PM. Regulation of the synthetic rate of gonadotropin-releasing hormone receptors in rat pituitary cell cultures by inhibin. Endocrinology. 1990;127:2387–92.

Brann DW. Glutamate: a major excitatory transmitter in neuroendocrine regulation. Neuroendocrinology. 1995;61:213–25.

Burr IM, Sizonenko PC, Kaplan SL, Grumbach MM. Hormonal changes in puberty. Pediatr Res. 1970;4:25–35.

Cailleux-Bounacer A, Rohmer V, Lahlou N, Lefebvre H, Roger M, Kuhn JM. Impact level of dihydrotestosterone on the hypothalamic-pituitary-Leydig cell axis in men. Int J Androl. 2009;32:57–65.

Catt KJ, Pierce JG. Gonadotropic hormones of the adenohypophysis. In: Yen SC, Jaffe RB, editors. Reproductive endocrinology. Philadelphia: Saunders; 1986. p. 75–114.
Childs GV. Gonadotropes and Lactotropes. In: Neill JD, editor. Knobil and Neill's physiology of reproduction. Amsterdam: Elsevier Academic Press; 2006. p. 1483–580.
Clayton RN, Catt KJ. Gonadotropin releasing hormone receptors: characterization, physiological regulation and relationship to reproductive function. Endocr Rev. 1981;2:186–209.
Corrigan AZ, Bilezikjian LM, Carroll RS, Bald LN, Schmelzer CH, Fendly BM, et al. Evidence for an autocrine role of activin B within rat anterior pituitary cultures. Endocrinology. 1991;128:1682–4.
Dungan HM, Clifton DK, Steiner RA. Minireview: kisspeptin neurons as central processors in the regulation of gonadotropin-releasing hormone secretion. Endocrinology. 2006;147:1154–8.
Ehlers K, Halvorson LM. Gonadotropin-releasing Hormone (GnRH) and the GnRH Receptor (GnRHR). Glob. libr. women's med. (ISSN: 1756-2228); 2013. doi:10.3843/GLOWM.10285.
Emons G, Hoffmann HG, Brack C, Ortmann O, Sturm R, Ball P, et al. Modulation of gonadotropin-releasing hormone receptor concentration in cultured female rat pituitary cells by estradiol treatment. J Steroid Biochem. 1988;31:751–6.
Ferin M, Van Vugt D, Wardlaw S. The hypothalamic control of the menstrual cycle and the role of the endogenous opioid peptides. Recent Prog Horm Res. 1984;40:441–85.
Gerendai I, Shaha C, Gunsalus GL, Bardin CW. The effects of opioid receptor antagonists suggest that testicular opiates regulate Sertoli and Leydig cell function in the neonatal rat. Endocrinology. 1986;118:2039–44.
Hayes FJ, Seminara SB, DeCruz S, Boepple PA, Crowley WF. Aromatase inhibition in the human male reveals a hypothalamic site of estrogen feedback. J Clin Endocrinol Metab. 2000;85:3027–35.
Herbison AE, Dyer RG. Effect on luteinizing hormone secretion of GABA receptor modulation in the medial preoptic area at the time of proestrous luteinizing hormone surge. Neuroendocrinology. 1991;53:317–20.
Herbison AE. Physiology of the adult gonadotropin-releasing hormone-neuronal network. In: Plant TM, Zeleznik AJ, editors. Knobil and Neill's physiology of reproduction. 4th ed. London: Academic Pres; 2015. p. 399–468.
Irwig MS, Fralay GS, Smith JT, Acohido BV, Popa SM, Cunningham MJ, et al. Kisspeptin activation of gonadotropin releasing hormone neurons and regulation of KISS-1 mRNA in the male rat. Neuroendocrinology. 2004;80:264–72.
Kaiser UB, Jakubowiak A, Steinberger A, Chin WW. Differential effects of gonadotropin-releasing hormone (GnRH) pulse frequency on gonadotropin subunit and GnRH receptor messenger ribonucleic acid levels in vitro. Endocrinology. 1997;138:1224–31.
Kalra SP, Allen LG, Sahu A, Kalra PS, Crowley WR. Gonadal steroids and neuropeptide Y-opioid-LHRH axis: interactions and diversities. J Steroid Biochem. 1988;30:185–93.
Karsch FJ, Weick RF, Butler WR, Dierschke DJ, Krey LC, Weiss G, et al. Induced LH surges in the rhesus monkey: strength-duration characteristics of the estrogen stimulus. Endocrinology. 1973;92:1740–7.
Katz N. The impact of opioids on the endocrine system. Pain Management. 2005;1:1–6.
Krsmanovick LZ, Hu L, Leung PK, Feng H, Catt KJ. The hypothalamic GnRH pulse generator: multiple regulatory mechanisms. Trends Endocrinol Metab. 2009;20:402–8.
Kumar A, Dewan R, Suri J, Kohli S, Shekhar S, Dhole B, et al. Abolition of endocrine dimorphism in hyperthyroid males? An argument for the positive feedback effect of hyperestrogenaemia on LH secretion. Andrologia. 2012;44:217–25.
Kumar A, Mohanty BP, Rani L. Secretion of testicular steroids and gonadotrophins in hypothyroidism. Andrologia. 2007;39:253–60.
Kulin HE, Reiter EO. Gonadotropin and testosterone measurements after estrogen administration to adult men, prepubertal and pubertal boys, and men with hypogonadotropism: Evidence for maturation of positive feedback in the male. Pediat. Res. 1976;10:46–51.
Lechan RM, Toni R. Functional anatomy of the hypothalamus and pituitary. In: De Groot LJ, editor. Endotext. 2000. South Dartmouth (MA): MDText.com, Inc. www.endotext.org.

Li S, Hong M, Fournier A, St-Pierre S, Pelletier G. Role of neuropeptide Y in the regulation of gonadotropin-releasing hormone gene expression in the rat preoptic area. Brain Res Mol Brain Res. 1994;26:69–73.

Lo KC, Lamb DJ. The testis and male accessory organs. In: Yen SC, Jaffe RB, editors. Reproductive endocrinology. Philadelphia: Saunders; 2004. p. 367–88.

Low MJ. Neuroendocrinology. In: Melmed S, Polonsky KS, Larsen PR, Kronenberg HM, editors. Williams Textbook of Endocrinology. 12th ed. Philadelphia: Saunders; 2011. p. 103–74.

Manjithaya RR, Dighe RR. Regulation of gonadotropins synthesis. PNAS Acad. 2006;72:1–12.

Mason AJ, Hayflick JS, Esch F, Ueno H, Ying SY, Guillemin R, et al. Complementary DNA sequences of ovarian follicular fluid inhibin show precursor structure and homology with transforming growth factor 3. Nature (Lond). 1985;318:659–63.

Millar RP, Lu Z, Pawson AJ, Flanagan CA, Morgan K, Maudsley SR. Gonadotropin-releasing hormone receptors. Endocr Rev. 2004;25:235–75.

Milton RC, Wormald PJ, Brandt W, Millar RP. The delineation of a decapeptide gonadotropin-releasing sequence in the carboxyl-terminal extension of the human gonadotropin-releasing hormone precursor. JBC. 1986;261:16990–7.

Nakai Y, Plant TM, Hess DL, Keogh EJ, Knobil E. On the sites of the negative and positive feedback actions of estradiol in the control of gonadotropin secretion in the rhesus monkey. Endocrinology. 1978;102:1008–14.

Neill JD. GnRH and GnRH receptor genes in the human genome. Endocrinology. 2002;143:737–43.

Nikolics K, Mason AJ, Szönyi E, Ramachandran J, Seeburg PH. A prolactin-inhibiting factor within the precursor for human gonadotropin-releasing hormone. Nature. 1985;316:511–7.

Ohtsuka S, Nishizaki T, Tasaka K, Miyake A, Tanizawa O, Yamatodani A, et al. Estrogen stimulates gonadotropin-releasing hormone release from rat hypothalamus independently through catecholamine and histamine in vitro. Acta Endocrinol. 1989;120:644–8.

Pinilla L, Aguilar E, Dieguez C, Millar RP, Tena-Sempere M. Kisspeptins and reproduction: physiological roles and regulatory mechanisms. Am Physiol Rev. 2012;92:1235–316.

Quiñones-Jenab V, Jenab S, Ogawa S, Funabashi T, Weesner GD, Pfaff DW. Estrogen regulation of gonadotropin-releasing hormone receptor messenger RNA in female rat pituitary tissue. Brain Res Mol Brain Res. 1996;38:243–50.

Rance NE, Young 3rd WS, McMullen NT. Topography of neurons expressing luteinizing hormone-releasing hormone gene transcripts in the human hypothalamus and basal forebrain. J Comp Neurol. 1994;339:573–86.

Segovia S, del Cerro MC, Ortega E, Pérez-Laso C, Rodriguez-Zafra C, Izquierdo MA, et al. Role of GABAA receptors in the organization of brain and behavioural sex differences. Neuroreport. 1996;7:2553–7.

Sheckter CB, Matsumoto AM, Bremmer WJ. Testosterone administration inhibits gonadotropin secretion by an effect directly on the human pituitary. J Clin Endocrinol Metab. 1989;68:397–401.

Smith JT, Dungan HM, Stoll EA, Gottsch ML, Braun RE, Eacker SM, et al. Differential regulation of KISS-1 mRNA expression by sex steroids in the brain of the male mouse. Endocrinology. 2005;146:3686–92.

Smith JT, Popa SM, Clifton DK, Hoffman GE, Steiner RA. Kiss1 neurons in the forebrain as central processors for generating the preovulatory luteinizing hormone surge. J Neurosci. 2006;26:6687–94.

Soghomonian JJ, Martin DL. Two isoforms of glutamate decarboxylase: why? Trends Pharmacol Sci. 1998;19:500–5.

Stanton PG, Burgon PG, Hearn MTW, Robertson DM. Structural and functional characteristics of hFSH and hLH isoforms. Mol Cell Endocrinol. 1996;125:133–41.

Tilbrook AJ, de Kretser DM, Clarke IJ. Human recombinant inhibin A suppresses plasma follicle stimulating hormone to intact levels but has no effect on luteinizing hormone in castrated rams. Biol Reprod. 1993;49:779–88.

Urban RJ, Padmanabhan V, Beitins I, Veldhuis JD. Metabolic clearance of human follicle-stimulating hormone assessed by radioimmunoassay, immunoradiometric assay, and in vitro Sertoli cell bioassay. J Clin Endocrinol Metab. 1991;73:818–23.

Veldhuis JD, Rogol AD, Samojlik E, Ertel NH. Role of endogenous opiates in the expression of negative feedback actions of androgen and estrogen on pulsatile properties of luteinizing hormone secretion in man. J Clin Invest. 1984;74:47–55.

Wu JC, Sealfon SC, Miller WL. Gonadal hormones and gonadotropin-releasing hormone (GnRH) alter messenger ribonucleic acid levels for GnRH receptors in sheep. Endocrinology. 1994;134:1846–50.

Zhang L, Chang YH, Feldman AN, Ma W, Lahjouji F, Barker JL, et al. The expression of GABA(A) receptor alpha2 subunit is upregulated by testosterone in rat cerebral cortex. Neurosci Lett. 1999;265:25–8.

Spermatogenesis

10

Bodhana Dhole and Anand Kumar

Learning Objectives

- Stages of spermatogenesis
- Spermatogenic wave and cycle
- Male germ cell apoptosis
- Factors regulating spermatogenesis
- Effects of endocrine disruptors on spermatogenesis
- Genetic disorders of spermatogenesis

10.1 Introduction

Spermatogenesis is a complex cellular event which is divided into four distinct steps, namely, (1) mitotic proliferation of spermatogonial cells; (2) meiotic division to yield four haploid, round spermatids; (3) spermiogenesis process which leads to transformation of round spermatids to elongated spermatids; and (4) spermiation process that causes release of mature sperm into tubular lumen. Various types of spermatogenic cells are shown in Fig. 10.1. In humans, it takes an average of 74 days for a primordial germ cell to transform into a mature sperm (Gilbert 2000).

B. Dhole, PhD • A. Kumar, MD, FAMS (✉)
Department of Reproductive Biology, All India Institute of Medical Sciences,
New Delhi, India
e-mail: anandkumarrepbiol@hotmail.com

A. Kumar, M. Sharma (eds.), *Basics of Human Andrology*,
DOI 10.1007/978-981-10-3695-8_10

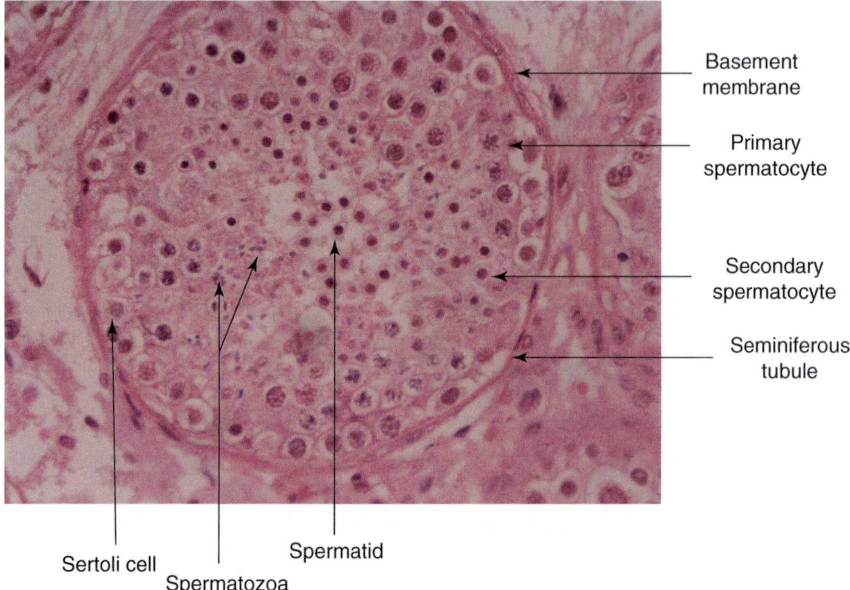

Fig. 10.1 Hematoxylin and eosin-stained section of testis showing cells in the seminiferous tubule (Courtesy: Dr. Venkateswaran Iyer, Professor, Department of Pathology, All India Institute of Medical Sciences, New Delhi)

10.2 Stages of Spermatogenesis

10.2.1 Mitotic Proliferation of Spermatogonial Cells

The spermatogonial stem cells are present on the basement membrane of seminiferous tubule. Primates, including humans, have two distinct types of spermatogonial population based on their state of differentiation – type A and type B. Spermatogonia are large cells with an approximate diameter of 6–7 μm (Kim et al.2013). Type A spermatogonia can be further distinguished as dark (Ad) and pale (Ap) subtypes on the basis of their nuclear staining in histological sections. Ad is the non-proliferative, reserve spermatogonial population in the testis. The type Ad spermatogonia have ovoid nuclei with intensely basophilic, finely granular chromatin. Under normal physiological conditions, it exhibits no mitotic activity. However, the Ad spermatogonia transform into Ap only after certain conditions such as radiation, long-term estrogen therapy, and cryptorchidism to replenish the testicular spermatogonial population. Ap spermatogonia are larger cells with more spherical nuclei and better developed nucleoli. The Ap nuclei contain lightly staining, finely granular chromatin (Ross and Pawlina 2006). Ap spermatogonia are considered true stem cells of the testis which renew their own population and also differentiate to produce type B spermatogonia (McLachlan et al. 2002a). Type B spermatogonia have spherical nuclei with chromatin that is condensed into large clumps along the nuclear

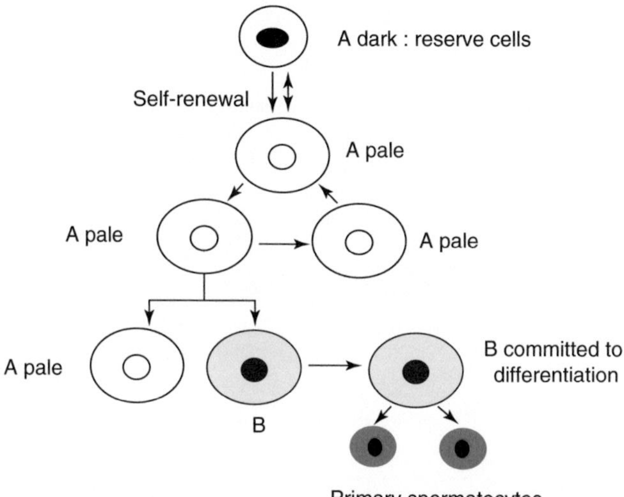

Fig. 10.2 Schematic diagram showing differentiation of spermatogonial cells

envelope and around a central nucleolus (Ross and Pawlina 2006). In humans, type B spermatogonia divide mitotically once to ultimately form two primary spermatocytes (see Fig. 10.2; Waheeb and Hofmann 2011). The nuclei of primary spermatocytes have more abundant and more darkly stained chromatin. At this time the nucleolus is no longer prominent (Kim et al. 2013).

10.2.1.1 Movement of Primary Spermatocyte Across the Blood-Testis Barrier

Primary spermatocytes move from the basal compartment of seminiferous tubule to its adluminal compartment where meiosis and further differentiation occur. Movement of primary spermatocytes through the blood-testis barrier (BTB) requires timely disassembly of cell-cell junctions followed by their reassembly. Before the movement of primary spermatocyte, the cell-cell junctions cleave, and new junctions are formed after their migration (see Fig. 10.3). The structural components of BTB have been discussed in detail in chapter "The Testis."

Cytokines facilitate cleavage of tight junctions (TJs) and ectoplasmic specialization (ES) in the region of BTB by activating proteases and downregulating expression of junctional proteins. Tumor necrosis factor-alpha (TNF-α) inhibits production of tight junction protein occludin by decreasing occludin promoter activity. Interleukin (IL)-4 and IL-13 disrupt BTB by inhibiting expression of TJ protein-occludin and ZO-1. Fibroblast growth factor dissociates the cadherin-catenin complex and translocates surface catenin proteins to the nucleus (Mruk and Cheng 2004).

Vascular endothelial growth factor (VEGF) possibly plays a role in Sertoli-Sertoli cell junction disassembly although no supportive study is available at present. VEGF has been shown to stimulate tyrosine phosphorylation of VE-cadherin,

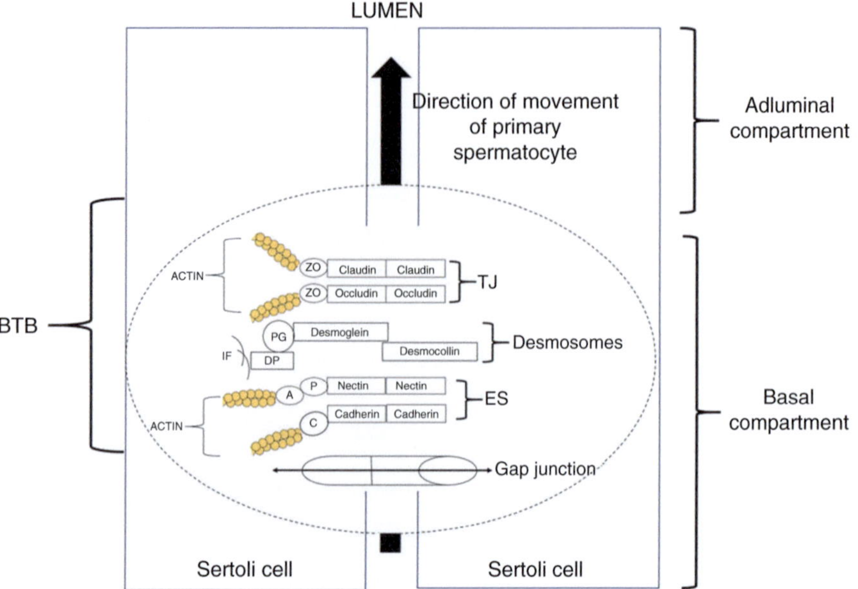

Fig. 10.3 Movement of primary spermatocyte through the blood-testis barrier (BTB). *ZO* zonula occludens, *PG* plakoglobin, *DP* desmoplakin, *IF* intermediate filaments, *P* ponsin, *A* afadin, *C* catenin, *TJs* tight junctions, *ES* ectoplasmic specialization

β-catenin, plakoglobin, and p120 in human umbilical vein endothelial cells (HUVECs) resulting in destabilization of cell-cell junctions between adjacent endothelial cells (Esser et al. 1998).

The reassembly of disrupted junctions is facilitated by cytokine-activated protease inhibitors. Plasminogen activator inhibitor 1 (PAI-1), a specific serine protease inhibitor, is synthesized by the Sertoli cells and peritubular myoid cells. Basic fibroblast growth factor, tumor growth factor-alpha (TGF-α) and TNF-α, is known to increase both PAI-1-mRNA and protein levels (Le Magueresse-Battistoni et al. 1996, 1997). Tissue inhibitors of metalloproteinases 1 and 2 (TMP-1, TMP-2) are synthesized by Sertoli cells. mRNA and protein levels of TMP-1 are significantly increased by IL-1α in Sertoli cell cultures, demonstrating the role of cytokines in resealing TJs and ES (Mruk and Cheng 2004; Gronning et al. 2000).

Testosterone also plays an important role in junction assembly and disassembly. Testosterone accelerates internalization of TJ and ES proteins in clathrin-coated pits and also stimulates recycling of these proteins on cell surface (Yan et al. 2008).

10.2.2 Meiosis

Primary spermatocyte undergoes two rounds of meiotic division. Meiosis ensures the reduction of chromosome number from diploid to haploid and generation of genomic diversity by reshuffling information between homologous pair of

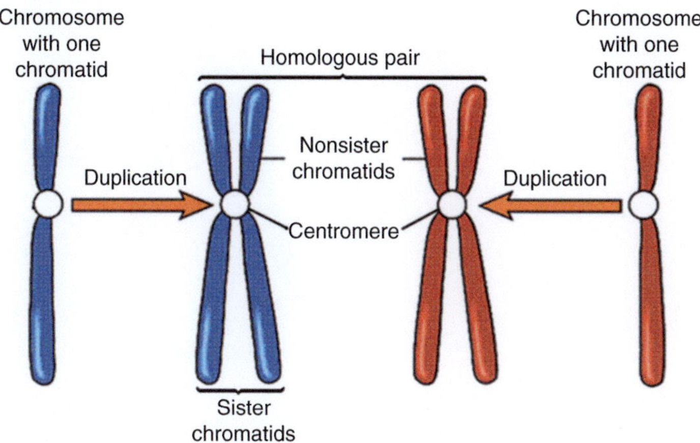

Fig. 10.4 Diagrammatic representation of chromosomal events in prophase I

chromosomes. In *meiosis I, prophase* is the longest step and consists of several substages – (1) leptotene, (2) zygotene, (3) pachytene, (4) diplotene, and (5) diakinesis. Prophase I is followed by metaphase I, anaphase I, and telophase I (Gilbert 2000).

Homologous chromosomes usually refer to a pair of chromosomes with the same length and centromere position and with genes for the same characteristics at corresponding loci. During the reduction division, the DNA content in each chromosome duplicates itself. Each replicated chromosome contains two genetically identical sister chromatids connected at the centromere. During later stages of meiotic prophase I, homologous chromosomes come close to each other and pair up (see Fig. 10.4). Pairing of homologous chromosomes is essential for exchange of genetic information between them (Gilbert 2000).

In the *diploid pre-leptotene or interphase* spermatocyte, the cell grows in size, and DNA content in each chromosome duplicates. The chromosomes at the pre-leptotene stage are long, filamentous, and decondensed. The centriole also duplicates at this stage (Hermo et al. 2010; Handel and Schimenti 2010).

Pre-leptotene is followed by the *leptotene* stage at which chromosomes condense and become visible under light microscope as long threads. Homologous chromosomes start moving toward each other and begin to align longitudinally but are not yet paired. Next, the synaptonemal complex (SC), a zipper-like, proteinaceous scaffolding structure, begins to assemble (see Fig. 10.5). The SC consists of two lateral elements and a central element which are held together by transverse filaments. Lateral elements first appear as axial elements before the pairing of homologous chromosomes. The SC connects homologous chromosomes along their entire length and is essential for their pairing or synapsis. Each of the axial element associates with a pair of sister chromatids (Hermo et al. 2010; Handel and Schimenti 2010).

Each axial element of SC consists of cohesion proteins and SC-specific proteins, SYCP2 and SYCP3. SYCP2, a 173 kDa protein, is uniquely expressed in meiotic

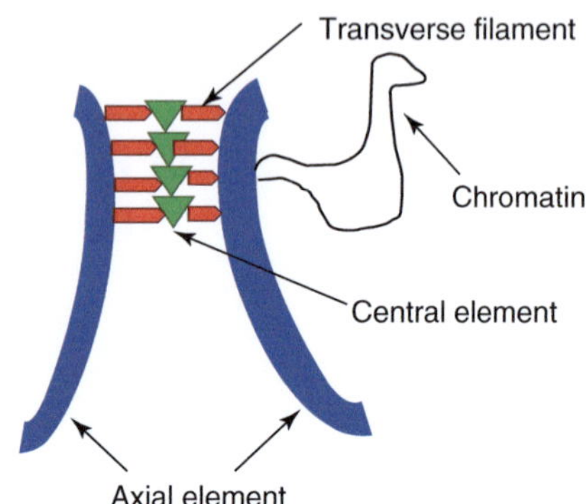

Fig. 10.5 Structural components of a synaptonemal complex

germ cells. SYCP1 is found in transverse filaments. The central element contains SYCE1 and SYCE2 proteins which interact with the transverse filament protein SYCP1. Additionally, central element also contains TEX12 which interacts with SYCE2 (Costa and Cooke 2007).

In the *zygotene* stage, the chromosomes further shorten and condense. Homologous chromosome pairing or synapsis is facilitated by complete assembly of the SC structure. Transverse filaments connect both the chromosome-bound axial elements to the central element, thereby associating the homologous chromosomes. The paired homologous chromosomes are called tetrad or bivalent chromosome. The paired chromosomes are attached to the nuclear envelope at their telomeres resulting in a bouquet arrangement (Hermo et al. 2010; Handel and Schimenti 2010).

In the *pachytene* stage, exchange of genomic information between non-sister chromatids takes place. The exchange at the pachytene stage is termed as DNA recombination or crossing over which occurs at specific sites known as chiasmata. DNA recombination is initiated when a double-stranded break (DSB) is introduced in one non-sister chromatid. The DSB is catalyzed by a highly conserved SPO11 trans-esterase protein, yielding a covalent Spo11-DNA complex. An endonuclease cleaves and releases the SPO11 bound to a short oligonucleotide. The endonuclease also degrades the 5′ end of each strand leaving two 3′ tails of single-stranded DNA. The single-stranded tails are stabilized by binding to several proteins such as Rad51 and Dmc1. One of the single-stranded tails inserts itself between the double strands of the non-sister chromatid, separating the two strands. When the invading single strand finds a complementary sequence in the non-sister chromatid, it pairs up and DNA synthesis is initiated from the invading 3′ end using the non-sister chromatid single strand as the template. The displaced single-stranded DNA of the non-sister chromatid pairs up with the other 3′ tail. As a result, there is formation of two Holliday junctions. Finally, all the DNA strands are cut, and the cut ends of one chromatid strand are exchanged and ligated covalently to the cut end of the other chromatid resulting in completion of DNA recombination (Hermo et al. 2010; Handel and Schimenti 2010; Neale and Keeney 2006).

In the *diplotene* stage, the SC degrades and the homologous chromosomes partially separate from each other, but they remain attached at chiasmata. The primary spermatocytes in the diplotene stage are the largest male germ cells and have thickened chromatin threads (Johnson et al. 2001). During *diakinesis*, the last step of prophase I, chromosomes become fully condensed. At the end of prophase I, nucleolus disappears and the nuclear membrane breaks down. During prophase I, duplicated centrioles move to the opposite pole of the cell. The centrioles are the regions from where meiotic microtubules (MTs) originate and grow in the nuclear region after the nuclear envelope breaks down. The MTs are connected to the chromosome via kinetochore protein complexes which assemble on the centromere. The kinetochore contains two regions – 1) an inner kinetochore, which is tightly associated with the centromere DNA, and 2) an outer kinetochore, which interacts with MTs (see Fig. 10.6).

During *metaphase I*, the homologous chromosome pair moves along the metaphase plate and aligns on the equatorial plane, a region which divides the meiotic spindle into two halves. During *anaphase I*, the MTs shorten; centromeres cleave and pull away the homologous chromosomes. The two homologous chromosomes move to the opposite poles of the cell. During *telophase I*, the haploid chromosomes reach two opposite poles of the cell, chromosomes decondense, MT spindle disappears, and nuclear membrane starts forming around the haploid chromosomes. Next event is incomplete cytokinesis where two haploid secondary spermatocytes are formed from each primary spermatocyte (Gilbert 2000).

Meiosis II follows meiosis I and is similar to mitosis. Prophase II is very short where chromosomes condense, nucleoli disappear, nuclear membrane disintegrates, centrioles move to opposite poles, and meiotic spindle formation is initiated. Prophase II is followed by anaphase II and telophase II which are similar to the corresponding stages in meiosis I. Meiosis II ends with the formation of four round haploid spermatids (Gilbert 2000).

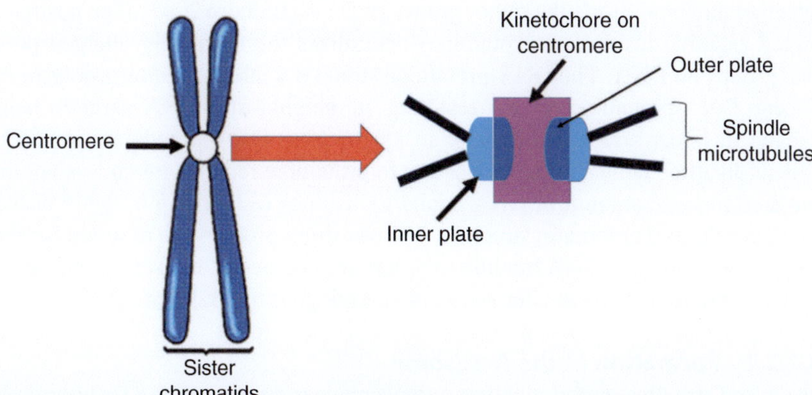

Fig. 10.6 Structural components of kinetochore

10.2.3 Spermiogenesis

Spermiogenesis involves transformation of round spermatids into mature sperm. The events associated with spermiogenesis include nuclear modifications, acrosome formation, and assembly of tail structures. In the early stage of spermiogenesis, spermatids have spherical nucleus and occupy the center of the cell. The cytoplasm contains a number of mitochondria, Golgi complex, granules which are formed by the Golgi complex and vacuoles (Kim et al. 2013).

10.2.3.1 Nuclear Modifications

The condensation of sperm nuclear chromatin material is a highly ordered and complex process. Two DNA strands of each chromosome attach to the nuclear annulus (NA). NA is a 2 μm bent ring-shaped structure unique to sperm. Specific nucleotide sequences on each DNA, alternatively called NA-DNA, bind to NA resulting in a unique sperm-specific DNA organization, termed as chromosomal anchoring (Ward and Coffey 1989). The chromosomes anchored on the NA organize into DNA loop domains. The DNA attaches to parts of nuclear matrix at specific regions termed as matrix attachment regions (MARs) which are located every 30–50 kb resulting in a series of loop domains (Ward and Coffey 1991).

In somatic cells and early round spermatids, DNA is organized into nucleosomes. Each nucleosome consists of a histone octamer containing core histone proteins – H2A, H2B, H3, and H4 in doublets. The histone octamer binds and wraps 146 base pairs or about 1.7 turns of DNA. Adjacent nucleosomes are joined by the linker DNA containing an average of about 20 base pairs. Another histone, H1, is located at the linker region. The nucleosomes are further coiled into a 30 nm solenoid fiber with the help of H1 (Ward and Coffey 1991).

In maturing elongated spermatids, DNA supercoiling is eliminated by transient DNA strand breaks catalyzed by topoisomerases to relieve torsional stress. During spermiogenesis, histones are replaced by transition proteins – TP1 and TP2 and subsequently by highly basic arginine-rich protamines. Protamines align parallel to each other and bind inside the minor groove of DNA (Balhorn 2007). The positively charged arginine residue on protamines neutralizes the negatively charged phosphate groups on DNA. The DNA-protamine strand fits into the major groove of the adjacent DNA-protamine strand resulting in neighboring DNA strands being arranged in linear arrays. Protamines in neighboring DNA strands are joined by covalent disulfide bonds. The DNA bound to protamine requires much less volume compared to same amount of DNA bound to histones resulting in a highly packed doughnut or toroid chromatin structure. The transition proteins are required for processing of protamines. DNA binding to protamines ceases all transcriptional activities in the sperm nucleus in later stages of spermiogenesis (Balhorn 2007).

10.2.3.2 Formation of the Acrosome

Acrosome formation and elongation follows nuclear condensation (Toshimori and Ito 2003). Acrosome biogenesis includes four distinct phases – (1) Golgi phase, (2) cap phase, (3) acrosome phase, and (4) maturation phase. The Golgi phase is characterized by the presence of periodic acid-Schiff (PAS)-positive granules known as proacrosomal granules (PAG; Ross and Pawlina 2006). Golgi apparatus is the principal supplier of proteins required for acrosome biogenesis. Many acrosome-specific proteins, such as proacrosin and acrogranin, are packed in electron-dense vesicles or PAGs in the early meiotic stage. The PAGs remain in the boundaries of the Golgi apparatus until the completion of meiosis and subsequent formation of round spermatids. Initially, PAGs are close to the concave face of Golgi apparatus, facing the nucleus. In the Golgi phase, PAGs start to translocate toward the nuclear surface, where they fuse together to form a single acrosomal vesicle (Moreno et al. 2000; Ramalho-Santos et al. 2002). The transport of PAGs requires MTs and actin-based motor proteins such as kinesin, dynein, and myosin Va (Kierszenbaum and Tres 2004). A small number of 40–50 nm-diameter vesicles, similar to coatomer-coated vesicle proteins (COPs) in somatic cells, are also involved in acrosome formation. In somatic cells, COPs are coated vesicular carriers involved in transport of secretary proteins between endoplasmic reticulum and Golgi apparatus. β-COP, a component of coated vesicles, has been identified in Golgi apparatus and acrosome membrane in rat spermatids (Martinez-Menarguez et al. 1996).

Acroplaxome, a cytoskeletal plate, is present in the subacrosomal space in mammalian spermatids and links the inner acrosomal membrane to the nuclear envelope. The acroplaxome contains F-actin and keratin and may be responsible for holding the primary acrosomal vesicle during the early stages of acrosomal biogenesis. Acroplaxome has a marginal ring consisting of intermediate filaments attached to the inner acrosomal membrane (Kierszenbaum et al. 2003). In the next cap and acrosome phase, the acrosomal vesicle gradually flattens and spreads around the nucleus from the luminal end to occupy two-third of it laterally like a cap. The Golgi apparatus is located close to the acrosome during its entire development phase. When acrosome formation is complete, the Golgi assumes a spherical shape and migrates away from the acrosome to the opposite side (see Fig. 10.7; Toshimori and Ito 2003; Moreno et al. 2000).

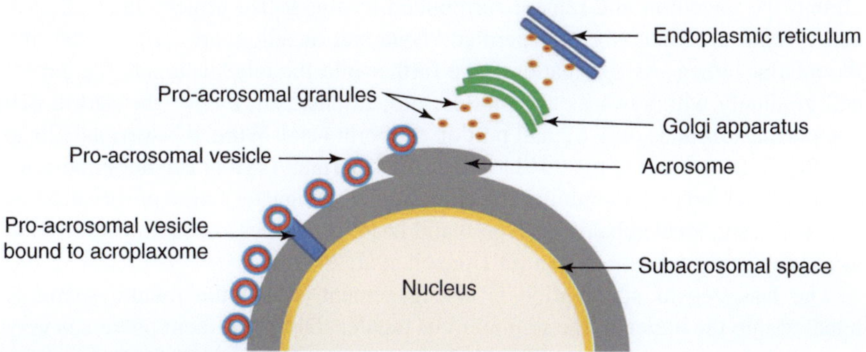

Fig. 10.7 Schematic diagram showing formation of acrosome

10.2.3.3 Formation and Assembly of the Tail Structures

The sperm tail or flagellum is a highly conserved cytoskeleton structure which is divided into four segments – (1) the connecting piece or neck, (2) the midpiece, (3) the principal piece, and (4) the endpiece. The tail is anchored to the sperm head via the connecting piece. During acrosome development, the two centrioles move on the side of the nucleus opposite to that of the acrosome. The pair of centrioles, proximal and distal from the nucleus, together with the surrounding accessory proteins, is called centrosome. The axoneme consists of a core MT doublet surrounded by nine outer MT doublets. Axoneme is the first structure to appear from the distal centriole during early spermiogenesis. The proximal centriole lies at an angle of 75–90° to the longitudinal axis of the sperm flagellum (Hermo et al. 2010). The axoneme elongates in a direction away from the sperm nucleus. During mid-spermiogenesis, outer dense fibers (ODFs) lie outside each of the nine axonemal outer MT doublets. ODFs originate in the most proximal part of the midpiece toward the nucleus. ODFs, which arise as thin electron-dense filaments, gradually increase in thickness and elongate in a direction toward the nucleus. In the principal piece of the tail, ODFs 3 and 8 terminate and are replaced by two longitudinal columns of fibrous sheath (FS). Remaining seven ODFs run the variable lengths of the principal piece of the tail. The FS tapers in a direction away from the nucleus and terminates at the end of the principal piece of tail. The endpiece of the tail contains only the axoneme. Mitochondrial sheath is formed by the division and elongation of sperm mitochondria which assemble in an end-to-end fashion resulting in two helices that wind around the ODFs in the midpiece of tail (Turner 2003). The detailed structure of the sperm tail will be discussed in chapter "The Sperm."

10.2.4 Spermiation

The process of spermiation is characterized by release of mature spermatids from the connecting Sertoli cells into the lumen of the seminiferous tubule. During the early stages of spermiation, elongated spermatids have large amount of cytoplasm around the flagellum and remain surrounded by finger-like projections of Sertoli cell cytoplasm. Gradually, the spermatid head and flagellum are pulled away into the tubular lumen. As spermatids move further into the tubular lumen, the Sertoli cell gradually withdraws its cytoplasmic projections, and, finally, the Sertoli cell cytoplasm surrounds only a small portion of sperm head. While the spermatid head and flagellum move into the tubular lumen, the spermatid cytoplasm does not move and is left behind in the seminiferous epithelium. During later stages of spermiation, the cytoplasm, located below the spermatid head, condenses and is now referred to as a cytoplasmic lobe or droplet (O'Donnell et al. 2011).

The last step of spermiation is disengagement, where the mature sperm is released into the lumen of the seminiferous tubules. Disengagement process is very fast, and the released sperm are washed away by movement of tubular fluid. Spermatids are subject to sufficient amount of shear force by the tubular fluid which helps to overcome the spermatid links to the Sertoli cells. The released spermatids move toward the rete testis and then to the epididymis. As the ES is removed several hours before disengagement, therefore, non-ES junctions are responsible for the spermatid-Sertoli cell interaction before and during disengagement. During disengagement, a small amount of sperm cytoplasm is attached to its midpiece which is later lost during their transit through the epididymis. The remaining cytoplasm along with several unwanted cytoplasmic organelles is left behind in the epithelium and discarded as residual body. The residual body moves toward the basement membrane and fuses with Sertoli cell lysosome and eventually gets phagocytosed (O'Donnell et al. 2011).

10.3 Spermatogenic Wave and Cycle

One of the characteristic features of spermatogenesis in mammals is continuous production of sperm. During spermatogenesis, sperm are produced in waves, sequentially along the length of the seminiferous tubules to ensure that sperm are produced continuously. In a cross section of a seminiferous tubule, sperm are found in different stages of development. The groups containing sperm in a particular stage of spermatogenesis are arranged in spherical circles in the seminiferous epithelium. A spermatogenic wave is the distance between two identical stages in the seminiferous tubule (see Fig. 10.8). The stages are found in ascending order from the rete testis to the center of the seminiferous tubule from where the stages are reversed. Each seminiferous tubule contains multiple complete waves. The duration of each stage of spermatogenesis to reappear within a given segment of the tubule is known as the spermatogenic cycle. The time required for each stage of spermatogenesis and total time of the cycle is constant for each species (Yoshida et al. 2006). A cross section through an individual seminiferous tubule of the human and primate testis shows a degree of spatial organization that is more limited to wedges. In these species the activator

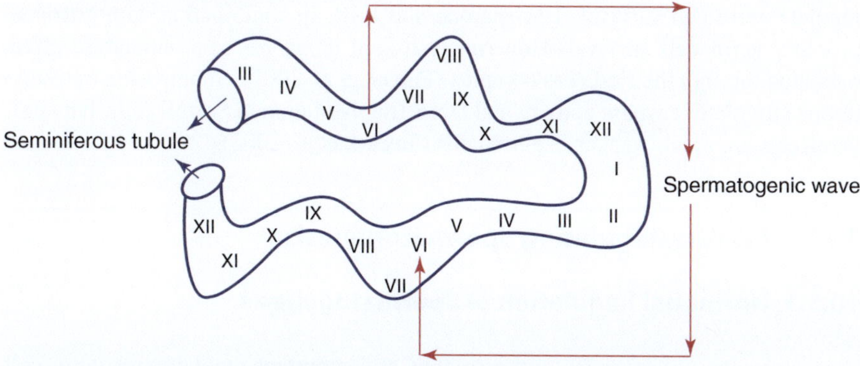

Fig. 10.8 Diagrammatic representation of the spermatogenic wave

message does not get all the way round a cross section of a seminiferous tubule, and so the coordinated development of different spermatogonial cells is initiated over a smaller area (Johnson 2013).

10.4 Male Germ Cell Apoptosis

The number of germ cells in the seminiferous tubule is maintained by a balance between new germ cell proliferation and death of germ cells by apoptosis. Apoptosis or programmed cell death is essential for maintaining homeostasis and eliminating defective germ cells. In testis, about 75% of germ cells are removed by apoptosis. Apoptosis of germ cells can occur via both the extrinsic and the intrinsic pathways (Shaha et al. 2010).

Germ cell apoptosis is evident in the fetus in very early stages of testicular development. During migration of primordial germ cells to the genital ridge, a large number of germ cells, particularly those showing aberrant migration, are removed by apoptosis. Apoptosis of mouse germ cells is dependent on Bax, a proapoptotic protein. Bcl-x, an antiapoptotic protein, plays a crucial role in mouse germ cell survival during testicular development (Rucker et al. 2000). In the postnatal life, when primordial germ cell differentiates into spermatogonia, germ cell population is maintained by a fine balance between Bax and Bcl-2 or Bcl-x and their downstream targets, the caspase 2, 3, 8, and 9 (Furuchi et al. 1996; Russell et al. 2002; Zheng et al. 2006; Moreno et al. 2006).

The extrinsic apoptotic pathway is also initiated when Fas ligand (FasL) binds to its receptor Fas. Fas is a transmembrane receptor containing 281 amino acids. The activated FasL/Fas complex turns on a proapoptotic signal in cells expressing Fas. In humans, aberrant meiotic and postmeiotic germ cells show upregulated expression of Fas gene suggesting that the FasL/Fas may be involved in removal of defective germ cells (Francavilla et al. 2002). The testicular levels of FasL are upregulated in patients suffering from maturation arrest and Sertoli cell-only syndrome. Therefore, it can be suggested that the FasL/Fas complex and its downstream target caspase 3 may be involved in removing defective germ cells.

Follicle-stimulating hormone (FSH), testosterone, and estrogen are known to regulate germ cell survival. Testosterone and FSH, in combination with estrogen, promote germ cell survival. However, estrogen alone has a proapoptotic effect mediated through the FasL/Fas complex (Pareek et al. 2007). Suppression of testosterone stimulates caspase activity and DNA fragmentation in Sertoli cells, but similar effects are much weaker in germ cells (Tesarik et al. 2002).

10.5 Factors Regulating Spermatogenesis

10.5.1 Hormonal Regulation of Spermatogenesis

Gonadotropins, testosterone, and estrogens are important regulators of germ cell development. Testosterone binds to androgen receptors located on Sertoli cells, Leydig cells, and peritubular cells, while FSH receptors are present exclusively on

Sertoli cells. Both FSH and testosterone act alone or in concert during different stages of spermatogenesis (McLachlan et al. 2002a).

The effect of gonadotropin suppression on different germ cell number has been studied (Zhengwei et al. 1998). Testosterone-enanthate treatment in normal fertile men results in a significant increase in serum testosterone levels, while the luteinizing hormone (LH) and FSH levels become undetectable. Gonadotropin suppression induces azoospermia in 80% and oligospermia in 20% of testosterone-enanthate-treated men. In addition, there is a significant decrease in testicular volume and seminiferous tubule diameter after the treatment. In response to gonadotropin suppression, there is no change in the number of type A spermatogonia, whereas significant reduction is seen in the case of type B spermatogonia, meiotic germ cells, round spermatids, elongating spermatids, and mature spermatocytes. The results suggest that transformation of spermatogonia A to spermatogonia B is inhibited in absence of gonadotropins (Zhengwei et al. 1998).

Testosterone implants given to adult macaque monkeys induce azoospermia in 44% and oligospermia in 55% of the monkeys. In the same study, testosterone implants cause significant decrease in serum bioactive LH and FSH levels. Short-term treatment of testosterone implants for 2 weeks results in a marked decrease in Ap spermatogonia, increase in Ad spermatogonia whereas no change is evident in type B spermatogonia number. Longer exposure of 14 weeks with exogenous testosterone restores Ad spermatogonia to control levels, but Ap spermatogonia still remain suppressed. Longer exposure also results in a significant reduction in type B spermatogonia. Thus, gonadotropin suppression inhibits type Ap spermatogonia proliferating to type B spermatogonia and promotes their differentiation to type Ad spermatogonia (O'Donnell et al. 2001).

Testosterone withdrawal suppresses the conversion of round to elongated spermatids suggesting a role of testosterone in facilitating the maturation of round spermatids to elongated spermatids in adult Sprague-Dawley rats. The failure of maturation is partly due to the premature detachment of round spermatids from the seminiferous epithelium. The immature cells are found in the epididymis, where the cells subsequently degenerate (O'Donnell et al. 1996). Androgens regulate adhesion between Sertoli cells and round spermatids, either through their effects on the cell adhesion molecules located between the two cell types or on the intracellular junctional complexes located in the Sertoli cell. The role of testosterone in maintaining cell-cell junctions has been discussed in the previous section of this chapter. Spermiogenesis is not majorly affected in men following gonadotropin suppression (O'Donnell et al. 2001). Spermiation is mediated by Sertoli cells under the influence of both FSH and testosterone. Within a few days of hormone suppression, a significant failure in spermiation occurs. It is suggested that suppressed hormone levels either result in a loss of spermiation signals or initiate a process required for spermatid retention. In primates including men, loss of spermiation is more efficient in causing azoospermia (McLachlan et al. 2002b).

Administration of testosterone leads to azoospermia in only 70% of normal men and variable degree of oligospermia in the remaining 30% men. The results can be explained based on the role of dihydrotestosterone (DHT). Abolition of androgen action in the testis by the administration of the androgen receptor antagonist, flutamide, increases the reduction of testosterone to its 5α-reduced metabolite DHT.

DHT is responsible for low-level sperm production in oligospermic individuals even when the gonadotropin and testosterone levels are suppressed (O'Donnell et al. 1996).

Medroxyprogesterone acetate, together with testosterone, has been trialed as potential male contraceptive. Depot medroxyprogesterone acetate (DMPA) and testosterone-undecanoate injected to healthy men at 8-week intervals result in azoospermia or severe oligospermia in majority of men (Gu et al. 2004). Medroxyprogesterone acetate and testosterone injections increase the number of apoptotic germ cells at various stages of differentiation including spermatogonial cell, spermatocytes, and round and elongated spermatids. There is also an increase in proapoptotic protein, caspase 9, in the apoptotic germ cells (Ruwanpura et al. 2008).

Estrogen receptors-ERα and ERβ, have been identified in the developing germ cells from the zygotene and early pachytene stages to elongating spermatid stage. 17β-estradiol is able to inhibit germ cell apoptosis in human testicular germ cell cultures suggesting that estrogens are directly involved in the regulation of germ cell survival. Even at lower concentrations, between 10^{-9} and 10^{-10} mol/L, 17β-estradiol successfully inhibits germ cell apoptosis. DHT is also capable of inhibiting male germ cell apoptosis but at a higher concentration of 10^{-7} mol/L than estrogen (Pentikainen et al. 2000). Estrogen, too, influences spermatogenesis by reabsorbing fluid in rete testis and epididymis. Deficiency in estrogen levels or aromatase activity or estrogen receptor functions promotes fluid retention in rete testis. Fluid accumulation in testes builds up pressure leading to damage to seminiferous epithelium and impairment of germ cell development (O'Donnell et al. 2001).

Phytoestrogens are plant-derived products that are structurally similar to estrogens and behave as estrogen agonist or antagonist. Phytoestrogens have significant effects on spermatogenesis. Feeding adult rats on phytoestrogen-rich diet results in significant decrease in sperm count, reduction in round and elongated spermatid volumes, and increased diameter of seminiferous tubule lumen. Apoptosis of spermatocytes and round spermatids is higher in the diet-treated group (Assinder et al. 2007). However, feeding of aromatase knockout mice on diet containing soy, a source of estrogenic isoflavones, partly prevents the decline in germ cell numbers and maintains Sertoli cell function. Therefore, a role of phytoestrogens in maintaining testicular function can be suggested (Robertson et al. 2002). Phytoestrogenic sesame treatment increases the motility and number of morphologically normal sperm in adult rats (Ukwenya et al. 2007).

10.5.2 Effect of Temperature on Spermatogenesis

Testis is maintained at temperatures 2–4 °C lower than the core body temperature. Testicular temperatures are maintained by its location in the scrotum, exchange of heat between testicular arteries and veins, and loss of heat from the scrotal skin. The testis originates within the abdominal cavity and in most mammals, it descends into scrotum during the fetal and neonatal life. Testicular descent has been discussed in chapter "Development of Male Gonad." Testicular descent provides a hypothermic environment away from the core body temperature. The testis is supplied by a pair of testicular arteries that arise from the abdominal aorta. The arteries pass through the inguinal canal to enter the scrotum. Testicular arteries run a straight course in the inguinal canal and become tortuous in the scrotum. The testicular veins arising from the testis form interconnecting network of fine channels among the arteries forming the pampiniform plexus of veins that runs in spermatic cord. Spermatic cord suspends the testis within the scrotum, contains vessels and nerves of the testis, and extends from the inguinal canal downward into the scrotum. Arterial and venous blood flows in close apposition but in opposite directions in the spermatic cord. The vascular arrangement in the spermatic cord facilitates the exchange of heat between spermatic vein and arteries. The warm spermatic arterial blood supplying the testis is cooled by exchange of heat with the venous blood. Hypothermia in testis is further controlled by two muscles – the dartos muscle in the scrotum that regulates scrotal surface area and the cremaster muscle that controls the position of the scrotum (Setchell and Breed 2006).

A mild increase in testicular temperature disrupts spermatogenesis and subsequently leads to impaired fertility. Pachytene, diplotene, and early round spermatids are most susceptible to heat stress. Exposure of adult monkey scrotum to 43 °C for 30 min once daily for 6 consecutive days induces azoospermia in 75% and oligospermia in 25% of monkeys by 6–8 weeks after exposure. An increase in germ cell apoptosis is found 3 and 7 days after the heat exposure. Accumulation of the pro-apoptotic Bax protein is found in the heat-induced apoptotic germ cells (Lue et al. 2002). However, in rat testicular cells, following heat exposure, Bax mRNA and protein levels remain unchanged (Yamamoto et al. 2000). Transient heating of monkey testis amplifies and hastens the suppression of spermatogenesis in monkeys in response to testosterone implant (Lue et al. 2006). Elevated temperature results in a complex stress response, which includes induction of genes associated with oxidative stress and hypoxia. Scrotal heating in mice at 42 °C results in a significant increase in hypoxia-inducible factor-1alpha (HIF-1α)-mRNA in germ cell nucleus within 6 h. Moreover, immunostaining data shows that HIF-1α is mainly localized in the nucleus of heated testicular germ cells where HIF-1α is able to activate downstream genes. Heat stress also leads to increased expression of two antioxidant enzymes – glutathione peroxidase 1 (GPX1) and glutathione S-transferase alpha (GSTA)-mRNA and protein levels suggesting a strong oxidative stress response (Paul et al. 2009).

The BTB is also affected by a rise in testicular temperature. When scrotum is heated to 43 °C for 30 min, it results in reduced expression of TJ components such as occludin, claudin-3, and ZO-1. Immunostaining studies show that occludin and ZO-1 are lost from BTB site, whereas claudin-11 becomes diffuse and cytoplasmic within 2 days following heat exposure. Heat stress expands the gap between adjacent Sertoli cells and increases permeability of TJs (Cai et al. 2011).

Rise in testicular temperature has been associated with two common pathologies-varicocele and cryptorchidism.

Varicocele is characterized by an abnormal dilation of the spermatic veins within the scrotum. Varicocele develops when there is a reversal of blood flow down the spermatic veins into pampiniform plexus. The retrograde venous flow occurs due to absent or incompetent venous valves. Impaired blood circulation created by a varicocele results in improper cooling of the warm arterial blood. The rise in blood temperature results in higher testicular temperature. Varicocele causes an average increase of 2.6 °C in the scrotal temperature. The rise in testicular temperature affects testicular function resulting in infertility in about 40% of men (Marmar 2001). Most men with varicocele are not azoospermic but exhibit varying degree of oligospermia. The mean volume of the testis of varicocele patients is significantly smaller. Semen parameters such as pH, liquefaction time, and sperm concentration are normal, whereas sperm motility and viability are significantly lower in varicocele patients (Xue et al. 2012).

Varicocele is associated with an increase in oxidative stress due to increase in reactive oxygen species (ROS) and decrease in the total antioxidant capacity of the semen in varicocele patients (Pasqualotto et al. 2008). The rise in ROS impairs sperm motility. The sperm plasma membrane has a high amount of polyunsaturated fatty acids such as docosahexaenoic acid which are rapidly oxidized by ROS, thereby decreasing the flexibility and motility of the sperm tail (Tremellen 2008). ROS also decrease the adenosine triphosphate (ATP) generated by sperm mitochondrial sheath which provides energy for sperm motility (De Lamirande and Gagnon 1992). Teratozoospermic sperm, in most cases, retain excess residual body which is normally lost during sperm maturation. The residual bodies have large amount of glucose-6-phosphate dehydrogenase which generates NADPH. NADPH produces ROS catalyzed by the NADPH oxidase enzyme present in sperm plasma membrane (Gomez et al. 1996).

In *cryptorchidism*, one or both testes are retained within the abdominal cavity or within the inguinal canal instead of descending normally into the scrotum. The prevalence of cryptorchidism in full-term newborns ranges between 1% and 3%, while in premature boys, it is reported to be as high as 30% (Chung and Brock 2011). The undescended testes are in a hyperthermic environment and have temperatures similar to core body temperature. Long-term consequences of cryptorchidism may include disturbed spermatogenesis. Adult men with persistent bilateral cryptorchidism usually show azoospermia. The patients with cryptorchidism have severely reduced number of testicular germ cells. There is also an increase in testicular ROS levels resulting in significant sperm DNA damage (Kurpisz et al. 2010).

10.5.3 Effect of Oxygen Tension on Spermatogenesis

Testes are unique organs in the body. The testicular germ cells, like tumor cells, are always actively proliferating. There ought to be a mechanism to limit proliferation and actions of gonadotropins.

Testes function in a hypoxic environment. Testicular oxygen tension has been shown to be at least twice lower than core body tissues in rat, rabbit, ram, and dog (Cross and Silver 1962; Massie et al.1969; Klotz et al. 1996; Carlsson et al. 2001; Yu and Cringle 2006).

Lower testicular temperatures lead to a significant drop in oxygen tension (Klotz et al. 1996). During hypothermia, the decrease in blood flow can be the principal cause for the drop of testicular oxygen tension. Local heating of the scrotum of rabbits causes a 10–40% increase in testicular oxygen tension (Cross and Silver 1962). The oxygen tension in the cryptorchid testis is twice higher than scrotal testis in rats (Massie et al. 1969). However, multiple other causes of low oxygen tension in the testis can be possible besides lower temperature.

The testis has evolved different adaptive responses to sustain in a hypoxic environment. Immunolocalization studies have shown the localization of HIF-1α, an intracellular factor that regulates the adaptive measures in hypoxic condition in the interstitial Leydig cells in rat adult testis (Lysiak et al. 2009). Low-oxygen environment is as important as gonadotropins in regulating testicular functions (Kumar et al. 2012, 2014). Both induced hypoxia and LH/cyclic adenosine monophosphate (cAMP) stimulate VEGF production via HIF-1α in Leydig cell-derived MA-10 cells (Kumar et al. 2012). VEGF is a heparin-binding disulfide-linked dimeric glycoprotein which increases capillary and venous permeability. VEGF plays an important role in the formation of new capillaries and blood vessels in normal physiological processes such as in growth, in development, and in the progression of pathological conditions such as tumors (Folkman 2002). Null mutations of VEGF are lethal and the loss of even single allele results in embryonic death (Carmeliet et al. 1996; Ferrara et al. 1996). Homozygous knockouts of VEGF receptor are also lethal (Millauer et al. 1994; Hiratsuka et al. 1998). Overexpression of VEGF in the testis and epididymis of transgenic mice causes infertility. The testes of the transgenic mice exhibit spermatogenic arrest and increase capillary density (Korpelainen et al. 1998). Transgenic mice expressing VEGF under the control of the polyepithelial mucin-1 (muc-1) promoter shows reduced male fertility due to impaired spermiogenesis and aberrant placentation (Huminiecki et al. 2001).

LH/cAMP and induced hypoxia have divergent actions on steroidogenesis. LH/cAMP stimulates progesterone production, whereas induction of hypoxia inhibits progesterone production in MA-10 cells (Kumar et al. 2014). The decrease in progesterone production could be attributed to different causes – (1) an increase in ROS production, (2) an increase in HIF-1α activity, and (3) a decrease in the mRNA level of cytP450scc. In primary Leydig cells, hydrogen peroxide (H_2O_2) results in a decrease in steroidogenic acute regulatory (StAR) and 3β-hydroxysteroid dehydrogenase (3β-HSD) protein levels (Allen et al. 2004). In Leydig cell-derived MA-10 cells, H_2O_2 decreases StAR protein levels and subsequently progesterone production (Diemer et al. 2003). The inhibitory action of hypoxia on steroid production can benefit the testis. Testosterone acts as a major factor in the regulation of spermatogenesis and hypoxia by restricting steroid production which probably prevents overconsumption of testicular resources and, thus, limits spermatogenesis within the physiological range.

Exposure of adult male rats to a hypoxic environment results in a significant drop in tetraploid cell population in rat testis. The ratio of haploid to diploid cells in testis is also reduced significantly. There is an increase in germ cell apoptosis in the seminiferous epithelium, primarily in the spermatogonial and spermatocyte cell population (Liao et al. 2010). There is also a significant increase in the proapoptotic Bax protein, and the ratio of Bax to Bcl-2 is also higher (Liao et al. 2007).

10.5.4 Effect of Radiation on Spermatogenesis

Exposure of testes to radiation can have profound effects on male reproduction. Sperm production in the irradiated men decreases, but sperm count recovers to normal values after several weeks of exposure. There is marked reduction in type A spermatogonia suggesting the ability of high-energy irradiation to eliminate the dividing pool of spermatogonial cells (Clifton and Bremner 1983). Pretreatment of adult monkeys with FSH followed by high-energy radiation helps to protect against reduction in spermatogonial cell number. The protective effects of FSH treatment are caused by the increase in the proliferative activity of the type A spermatogonia (Van Alphen et al. 1989). Pretreatment with medroxyprogesterone acetate even for a short period of time is successful in reducing the toxic effects of irradiation (Jegou et al. 1991). Radiation sensitivity of the testis decreases from late fetal life to the end of first week after birth. Exposure of mice at 17.5th day after conception or at 1st day after birth to gamma rays decreases sperm count, while exposure of 8-day-old mice has no effect, suggesting that irradiation of late fetal or early neonatal testes has a direct impact on the generation of the neonatal spermatogonial stem cell pool (Forand et al. 2009). Chemotherapy used for treatment of testicular cancer causes significant decrease in sperm parameters, which is most significant 3 months after the end of chemotherapy. However, in the radiotherapy-treated group, the decrease is most relevant 6 months after the end of radiotherapy (Gandini et al. 2006).

There are several reports suggesting an adverse role of mobile phone radiation on spermatogenesis. When male albino rats are exposed to mobile phone radiation for a short period, no testicular abnormalities are observed. However, long-term exposure causes hypospermatogenesis and maturation arrest in a small proportion of irradiated rats (Meo et al. 2011). Morphometrical analysis of human sperm, after exposure to mobile phone radiation, shows a significant reduction in sperm head area and acrosomal area (Falzone et al. 2011). Human sperm count, motility, viability, and normal morphology significantly decrease on exposure to mobile phone radiation. The severity of the reduction correlates with the time of exposure to mobile phone radiation (Agarwal et al. 2008).

10.6 Endocrine Disruptors and Spermatogenesis

Endocrine disruptors interfere with synthesis, metabolism, and action of endogenous hormones resulting in various disorders including cancers, birth defects, and sexual and developmental disorders. The endocrine disruptors can be a natural plant-derived compound, also known as phytochemical or some synthetic chemical known as xenobiotics. Endocrine disruptors usually have different routes of exposure such as food, drinks, or air. The chemicals are structurally similar to endogenous hormones and usually bind to their receptors. Chemically, they behave as hormone agonists or antagonists. Bisphenol A, dioxins, dichlorodiphenyltrichloroethane (DDT), and polychlorinated biphenyl are some major endocrine disruptors which are reported to have adverse effects on adult spermatogenesis (Yeung et al. 2011).

Bisphenol A, a xenobiotic estrogenic compound, is widely used as a plastic monomer. Subcutaneous administration of bisphenol A to adult mice and rats results in morphologically abnormal spermatids with deformed acrosomal vesicles, acrosomal caps, acrosomes, and nuclei. The ES between the spermatids and the Sertoli cells is also severely hampered, but these specializations between adjoining Sertoli cells are not hampered (Toyama et al. 2004). Bisphenol A, even at very low doses, significantly reduces testicular weight and daily sperm production (Ashby et al. 2003). Bisphenol A increases Leydig cells and germ cell apoptosis in mice. There is accumulation of proapoptotic markers such as Fas, FasL, and active caspase-3 in the apoptotic cells suggesting that bisphenol A induces apoptosis in these cells through the Fas signaling pathway (Li et al. 2009). A rise in urinary bisphenol A in infertile men is associated with a decline in sperm count, motility, and morphology (Meeker et al. 2010).

Dioxins, lipophilic by-products of industrial processes such as chlorine bleaching of pulp and paper, are resistant to biological and environmental degradation. Dioxins bind and activate aryl hydrocarbon receptor (AHR). AHR is a cytosolic transcription factor that is normally inactive, bound to several chaperon proteins. When 2,3,7,8-tetrachlorodibenzo-p-dioxin (TCDD) binds to AHR, chaperons dissociate resulting in translocation of these receptors into the nucleus, where it regulates the transcription of other genes. In utero exposure to TCDD significantly reduces sperm count in rats (Mably et al. 1992; Bjerke and Peterson 1994). The adverse effects of TCDD are AHR mediated because any alteration in the AHR reduces the severity of TCDD's effects on sperm count (Simanainen et al. 2004).

Male rats fed with DDT have a decrease in germ cell and Leydig cell number. When the DDT-treated rats are mated with normal rats, it results in a smaller litter size (Krause et al. 1975).

Polychlorinated biphenyls (PCBs), a class of organochlorines, are widely used as dielectric and coolant fluids in transformers, capacitors, and electric motors. PCBs are ubiquitous environmental contaminants which decrease serum thyroxine (T_4) concentrations. There is a fall in serum T_4 levels when neonatal rats are treated with PCB. There is a significant rise in testicular weight, Sertoli cell proliferation, and daily sperm production in the treated rats. T_4 replacement reduces testicular weights in the treated rats suggesting that PCBs produce these effects primarily by inducing hypothyroidism, which subsequently leads to increased Sertoli cell proliferation, testis weight, and daily sperm production (Cooke et al. 1996).

10.7 Genetic Disorders of Spermatogenesis

Infertility or the inability to conceive after at least 1 year of regular and unprotected intercourse affects about 10–15% of couples. In recent years, single-gene mutations or gene defects have emerged as a principal cause of spermatogenesis-related disorders.

Spermatogenesis-associated protein 16 or SPATA16 gene encodes a highly conserved, testis-specific protein belonging to the tetratricopeptide repeat-like superfamily. SPATA16, a 65 kDa protein, is composed of 11 exons which contain a tetratricopeptide repeat (TPR) domain. The encoded protein is located in sperm Golgi apparatus and the proacrosomic vesicles that are transported to the acrosome in round and elongated spermatids during spermiogenesis. Mutation in SPATA16 gene results in globozoospermia or round-headed sperm where the acrosome is completely or partially absent (Dam et al. 2007). Single mutation in the SPATA gene has been reported in three brothers in a Jewish family. Homozygous single-gene substitution in exon 4 of SPATA16 gene in the three brothers resulted in globozoospermia which caused infertility. The single-nucleotide substitution causes a change in a highly conserved amino acid residue located at the C-terminal end of the TPR domain. Additionally, it also disrupts the 5′ splice site of intron 4 (Dam et al. 2007).

Protein interacting with C kinase 1 or PICK1 gene encodes a 415 amino acid cytosolic protein that is involved in vesicle trafficking from the Golgi apparatus to the acrosome. PICK1 protein interacts with many membrane proteins via its conserved PDZ domain. The interactions of PICK1 protein facilitate the subcellular targeting and surface expression of the membrane proteins. PICK1 gene knockout mice show fragmented acrosome leading to round-headed sperm, reduced sperm count, and severely impaired sperm motility (Xiao et al. 2009). A homozygous missense mutation in the exon 13 of the PICK1 gene has been observed in the globozoospermia patients. The mutation results in a change in amino acid in the C-terminal domain of the protein suggesting a role of the PICK1 gene in the formation of the acrosome (Liu et al. 2010).

Aurora kinase C or AURKC gene encodes a protein belonging to the serine/threonine kinase protein subfamily. The protein is involved in organization of MTs and spindle formation and ensures correct chromosome alignment and segregation during mitosis and meiosis. In infertile men, sperm with large head and multiple flagella are seen along with a single-nucleotide mutation in the coding region of AURKC gene. The single-nucleotide mutation results in premature termination of translation and formation of a truncated protein completely lacking the kinase domain. Thus, the disruption of AURKC gene results in infertility due to production of morphologically abnormal sperm (Dieterich et al. 2007).

As mentioned earlier in this chapter, SYCP3 gene encodes a 27 kDa nuclear protein that forms an essential component of the SC formed during meiosis. Deletion in the SYCP3 gene in azoospermic patients with maturation arrest causes premature termination of translation and formation of a truncated protein resulting in disruption in SC formation and subsequently meiosis (Miyamoto et al. 2003). Mutational analysis of the SYCP3 gene in 58 patients with maturation arrest shows that SYCP3 mutations are uncommon in patients with azoospermia (Stouffs et al. 2005).

Meiosis defective 1 or MEI 1 gene encodes an 894-amino acid protein which is responsible for the formation of meiotic double-stranded breaks during homologous recombination. In MEI 1 gene knockout mouse models, meiotic arrest occurs due to impaired chromosomal synapsis (Libby et al. 2003). The single-nucleotide

Table 10.1 Genes involved in spermatogenesis and their associated diseases

Gene	Chromosome location	Function	Disease	Reference
SPATA16	3	Formation of acrosome	Globozoospermia	Dam et al. (2007)
PICK1	22	Formation of acrosome	Globozoospermia	Liu et al. (2010)
AURKC	19	Formation of spindle fibers during meiosis	Large-headed, multi-flagellar sperm	Dieterich et al. (2007)
SYCP3	12	Formation of synaptonemal complex	Disrupted meiosis	Miyamoto et al. (2003)
MEI1	22	Homologous DNA recombination	Disrupted spermatogenesis	Sato et al. (2006)

polymorphism (SNP) in the MEI 1 gene leads to defect in spermatogenesis (Sato et al. 2006). See Table 10.1 for genes causing spermatogenesis disorders.

Key Questions

- Describe briefly the different populations of spermatogonial cells.
- Explain briefly the migration of primary spermatocytes through blood-testis barrier.
- Describe the key events occurring in meiosis I in male germ cell.
- What are the morphological modifications occurring during spermiogenesis?
- How are sperm released from Sertoli cells into the lumen of the seminiferous tubule?
- What is the importance of germ cell apoptosis?
- What are the effects of endocrine disruptors on spermatogenesis?

References

Agarwal A, Deepinder F, Sharma RK, Ranga G, Li J. Effect of cell phone usage on semen analysis in men attending infertility clinic: an observational study. Fertil Steril. 2008;89:124–8.

Allen JA, Diemer T, Janus P, Hales KH, Hales DB. Bacterial endotoxin lipopolysaccharide and reactive oxygen species inhibit Leydig cell steroidogenesis via perturbation of mitochondria. Endocrine. 2004;25:265–75.

Ashby J, Tinwell H, Lefevre PA, Joiner R, Haseman J. The effect on sperm production in adult Sprague-Dawley rats exposed by gavage to bisphenol A between postnatal days 91–97. Toxicol Sci. 2003;74:129–38.

Assinder S, Davis R, Fenwick M, Glover A. Adult-only exposure of male rats to a diet of high phytoestrogen content increases apoptosis of meiotic and post-meiotic germ cells. Reproduction. 2007;133:11–9.

Balhorn R. The protamine family of sperm nuclear proteins. Genome Biol. 2007;8(9):227.

Bjerke DL, Peterson RE. Reproductive toxicity of 2, 3, 7, 8-tetrachlorodibenzo-p-dioxin in male rats: different effects of in utero versus lactational exposure. Toxicol Appl Pharmacol. 1994;127:241–9.

Cai H, Ren Y, Li XX, Yang JL, Zhang CP, Chen M, et al. Scrotal heat stress causes a transient alteration in tight junctions and induction of TGF-beta expression. Int J Androl. 2011;34:352–62.
Carlsson PO, Palm F, Andersson A, Liss P. Markedly decreased oxygen tension in transplanted rat pancreatic islets irrespective of the implantation site. Diabetes. 2001;50:489–95.
Carmeliet P, Ferreira V, Breier G, Pollefeyt S, Kieckens L, Gertsenstein M, et al. Abnormal blood vessel development and lethality in embryos lacking a single VEGF allele. Nature. 1996;380:435–9.
Chung E, Brock GB. Cryptorchidism and its impact on male fertility: a state of art review of current literature. Can Urol Assoc J. 2011;5:210–4.
Clifton DK, Bremner WJ. The effect of testicular x-irradiation on spermatogenesis in man. A comparison with the mouse. J Androl. 1983;4:387–92.
Cooke PS, Zhao YD, Hansen LG. Neonatal polychlorinated biphenyl treatment increases adult testis size and sperm production in the rat. Toxicol Appl Pharmacol. 1996;136:112–7.
Costa Y, Cooke HJ. Dissecting the mammalian synaptonemal complex using targeted mutations. Chromosome Res. 2007;15:579–89.
Cross BA, Silver IA. Neurovascular control of oxygen tension in the testis and epididymis. J Reprod Fertil. 1962;3:377–95.
Dam AH, Koscinski I, Kremer JA, Moutou C, Jaeger AS, Oudakker AR, et al. Homozygous mutation in SPATA16 is associated with male infertility in human globozoospermia. Am J Hum Genet. 2007;81:813–20.
De Lamirande E, Gagnon C. Reactive oxygen species and human spermatozoa. II. Depletion of adenosine triphosphate plays an important role in the inhibition of sperm motility. J Androl. 1992;13:379–86.
Diemer T, Allen JA, Hales KH, Hales DB. Reactive oxygen disrupts mitochondria in MA-10 tumour Leydig cells and inhibits steroidogenic acute regulatory (StAR) protein and steroidogenesis. Endocrinology. 2003;144:2882–91.
Dieterich K, Soto Rifo R, Faure AK, Hennebicq S, Ben Amar B, Zahi M, et al. Homozygous mutation of AURKC yields large-headed polyploid spermatozoa and causes male infertility. Nat Genet. 2007;39:661–5.
Esser S, Lampugnani MG, Corada M, Dejana E, Risau W. Vascular endothelial growth factor induces VE-cadherin tyrosine phosphorylation in endothelial cells. J Cell Sci. 1998;111:1853–65.
Falzone N, Huyser C, Becker P, Leszczynski D, Franken DR. The effect of pulsed 900-MHz GSM mobile phone radiation on the acrosome reaction, head morphometry and zona binding of human spermatozoa. Int J Androl. 2011;34:20–6.
Ferrara N, Carver-Moore K, Chen H, Dowd M, Lu L, O'Shea KS, et al. Heterozygous embryonic lethality induced by targeted inactivation of the VEGF gene. Nature. 1996;380:439–42.
Folkman J. Role of angiogenesis in tumor growth and metastasis. Semin Oncol. 2002;29(Suppl 16):15–8.
Forand A, Messiaen S, Habert R, Bernardino-Sgherri J. Exposure of the mouse perinatal testis to radiation leads to hypospermia at sexual maturity. Reproduction. 2009;137:487–95.
Francavilla S, D'Abrizio P, Cordeschi G, Pelliccione F, Necozione S, Ulisse S, et al. Fas expression correlates with human germ cell degeneration in meiotic and post-meiotic arrest of spermatogenesis. Mol Hum Reprod. 2002;8:213–20.
Furuchi T, Masuko K, Nishimune Y, Obinata M, Matsui Y. Inhibition of testicular germ cell apoptosis and differentiation in mice misexpressing Bcl-2 in spermatogonia. Development. 1996;122:1703–9.
Gandini L, Sgrò P, Lombardo F, Paoli D, Culasso F, Toselli L, et al. Effect of chemo- or radiotherapy on sperm parameters of testicular cancer patients. Hum Reprod. 2006;21:2882–9.
Gilbert SF. The saga of the germ line. In: Developmental biology. 6th ed. Sunderland: Sinauer Associates; 2000.
Gomez E, Buckingham DW, Brindle J, Lanzafame F, Irvine DS, Aitken RJ. Development of an image analysis system to monitor the retention of residual cytoplasm by human spermatozoa:

correlation with biochemical markers of the cytoplasmic space, oxidative stress, and sperm function. J Androl. 1996;17:276–87.

Gronning LM, Wang JE, Ree AH, Haugen TB, Tasken K, Tasken KA. Regulation of tissue inhibitor of metalloproteinases-1 in rat Sertoli cells: induction by germ cell residual bodies, interleukin-1alpha, and second messengers. Biol Reprod. 2000;62:1040–6.

Gu YQ, Tong JS, Ma DZ, Wang XH, Yuan D, Tang WH, et al. Male hormonal contraception: effects of injections of testosterone undecanoate and depot medroxyprogesterone acetate at eight-week intervals in Chinese men. J Clin Endocrinol Metab. 2004;89:2254–62.

Handel MA, Schimenti JC. Genetics of mammalian meiosis: regulation, dynamics and impact on fertility. Nat Rev Genet. 2010;11:124–36.

Hermo L, Pelletier RM, Cyr DG, Smith CE. Surfing the wave, cycle, life history, and genes/proteins expressed by testicular germ cells. Part 1: background to spermatogenesis, spermatogonia, and spermatocytes. Microsc Res Tech. 2010;73:241–78.

Hiratsuka S, Minowa O, Kuno J, Noda T, Shibuya M. Flt-1 lacking the tyrosine kinase domain is sufficient for normal development and angiogenesis in mice. Proc Natl Acad Sci U S A. 1998;95:9349–54.

Huminiecki L, Chan HY, Lui S, Poulsom R, Stamp G, Harris AL, et al. Vascular endothelial growth factor transgenic mice exhibit reduced male fertility and placental rejection. Mol Hum Reprod. 2001;7:255–64.

Jegou B, De La Calle JFV, Bauche F. Protective effect of medroxyprogesterone acetate plus testosterone against radiation-induced damage to the reproductive function of male rats and their offspring. Proc Natl Acad Sci U S A. 1991;88:8710–4.

Johnson MH. Making sperm. In: Johnson MH, editor. Essential reproduction. 7th ed. Chichester/West Sussex: Wiley-Blackwell; 2013. p. 105–21.

Johnson L, Staub C, Neaves WB, Yanagimachi R. Live human germ cells in the context of their spermatogenic stages. Hum Reprod. 2001;16:1575–82.

Kierszenbaum AL, Tres LL. The acrosome-acroplaxome-manchette complex and the shaping of the spermatid head. Arch Histol Cytol. 2004;67:271–84.

Kierszenbaum AL, Rivkin E, Tres LL. Acroplaxome, an F-actin-keratin-containing plate, anchors the acrosome to the nucleus during shaping of the spermatid head. Mol Biol Cell. 2003;14:4628–40.

Kim JH, Chung JS, Lee KY. Ultrastructural characteristics of the testis, spermatogenesis and taxonomic values of sperm morphology in male ruditapes philippinarum in Western Korea. Dev Reprod. 2013;17:121–32.

Klotz T, Vorreuther R, Heidenreich A, Zumbe J, Engelmann U. Testicular tissue oxygen pressure. J Urol. 1996;155:1488–91.

Korpelainen EI, Karkkainen MJ, Tenhunen A, Lakso M, Rauvala H, Vierula M, et al. Overexpression of VEGF in testis and epididymis causes infertility in transgenic mice: evidence for non-endothelial targets for VEGF. J Cell Biol. 1998;143:1705–12.

Krause W, Hamm K, Weissmuller J. The effect of DDT on spermatogenesis of the juvenile rat. Bull Environ Contam Toxicol. 1975;14:171–9.

Kumar A, Rani L, Dhole B, Chaturvedi PK. Oxygen as a regulator of MA-10 cell functions: effect of cobalt chloride on vascular endothelial growth factor production. Andrologia. 2012;44(Suppl 1):615–20.

Kumar A, Rani L, Dhole B. Role of oxygen in the regulation of Leydig tumour derived MA-10 cell steroid production: the effect of cobalt chloride. Syst Biol Reprod Med. 2014;60:112–8.

Kurpisz M, Havryluk A, Nakonechnyj A, Chopyak V, Kamieniczna M. Cryptorchidism and long-term consequences. Reprod Biol. 2010;10:19–35.

Le Magueresse-Battistoni B, Pernod G, Kolodie L, Benahmed M. Plasminogen activator inhibitor-1 regulation in cultured rat peritubular cells by basic fibroblast growth factor and transforming growth factor-alpha. Endocrinology. 1996;137:4243–9.

Le Magueresse-Battistoni B, Pernod G, Kolodie L, Morera AM, Benahmed M. Tumor necrosis factor-alpha regulates plasminogen activator inhibitor-1 in rat testicular peritubular cells. Endocrinology. 1997;138:1097–105.

Li YJ, Song TB, Cai YY, Zhou JS, Song X, Zhao X, et al. Bisphenol A exposure induces apoptosis and upregulation of Fas/FasL and caspase-3 expression in the testes of mice. Toxicol Sci. 2009;108:427–36.

Liao WG, Gao YQ, Cai MC, Wu Y, Huang J, Fan YM. Hypoxia promotes apoptosis of germ cells in rat testes. Zhonghua Nan Ke Xue. 2007;13:487–91.

Liao W, Cai M, Chen J, Huang J, Liu F, Jiang C, et al. Hypobaric hypoxia causes deleterious effects on spermatogenesis in rats. Reproduction. 2010;139:1031–8.

Libby BJ, Reinholdt LG, Schimenti JC. Positional cloning and characterization of Mei1, a vertebrate-specific gene required for normal meiotic chromosome synapsis in mice. Proc Natl Acad Sci U S A. 2003;100:15706–11.

Liu G, Shi QW, Lu GX. A newly discovered mutation in PICK1 in a human with globozoospermia. Asian J Androl. 2010;12:556–60.

Lue YH, Lasley BL, Laughlin LS, Swerdloff RS, Hikim AP, Leung A, et al. Mild testicular hyperthermia induces profound transitional spermatogenic suppression through increased germ cell apoptosis in adult cynomolgus monkeys (*Macaca fascicularis*). J Androl. 2002;23:799–805.

Lue YH, Wang C, Liu YX, Hikim AP, Zhang XS, Ng CM, et al. Transient testicular warming enhances the suppressive effect of testosterone on spermatogenesis in adult cynomolgus monkeys (*Macaca fascicularis*). J Clin Endocrinol Metab. 2006;91:539–45.

Lysiak JJ, Kirby JL, Tremblay JJ, Woodson RI, Reardon MA, Palmer LA, et al. Hypoxia-inducible factor-1alpha is constitutively expressed in murine Leydig cells and regulates 3beta-hydroxysteroid dehydrogenase type 1 promoter activity. J Androl. 2009;30:146–56.

Mably TA, Bjerke DL, Moore RW, Gendron-Fitzpatrick A, Peterson RE. In utero and lactational exposure of male rats to 2, 3, 7, 8-tetrachlorodibenzo-p-dioxin. 3. Effects on spermatogenesis and reproductive capability. Toxicol Appl Pharmacol. 1992;114:118–26.

Marmar JL. The pathophysiology of varicoceles in the light of current molecular and genetic information. Hum Reprod Update. 2001;7:461–72.

Martinez-Menarguez JA, Geuze HJ, Ballesta J. Identification of two types of beta-COP vesicles in the Golgi complex of rat spermatids. Eur J Cell Biol. 1996;71:137–43.

Massie ED, Gomes WR, VanDemark NL. Oxygen tension in testes of normal and cryptorchid rats. J Reprod Fertil. 1969;19:559–61.

McLachlan RI, O'Donnell L, Meachem SJ, Stanton PG, de Kretser DM, Pratis K, et al. Identification of specific sites of hormonal regulation in spermatogenesis in rats, monkeys, and man. Recent Prog Horm Res. 2002a;57:149–79.

McLachlan RI, O'Donnell L, Meachem SJ, Stanton PG, de K, Pratis K, et al. Hormonal regulation of spermatogenesis in primates and man: insights for development of the male hormonal contraceptive. J Androl. 2002b;23:149–62.

Meeker JD, Ehrlich S, Toth TL, Wright DL, Calafat AM, Trisini AT, et al. Semen quality and sperm DNA damage in relation to urinary bisphenol A among men from an infertility clinic. Reprod Toxicol. 2010;30:532–9.

Meo SA, Arif M, Rashied S, Khan MM, Vohra MS, Usmani AM, et al. Hypospermatogenesis and spermatozoa maturation arrest in rats induced by mobile phone radiation. J Coll Physicians Surg Pak. 2011;21:262–5.

Millauer B, Shawver LK, Plate KH, Risau W, Ullrich A. Glioblastoma growth inhibited in vivo by a dominant-negative Flk-1 mutant. Nature. 1994;367:576–9.

Miyamoto T, Hasuike S, Yogev L, Maduro MR, Ishikawa M, Westphal H, et al. Azoospermia in patients heterozygous for a mutation in SYCP3. Lancet. 2003;362:1714–9.

Moreno RD, Ramalho-Santos J, Sutovsky P, Chan EK, Schatten G. Vesicular traffic and golgi apparatus dynamics during mammalian spermatogenesis: implications for acrosome architecture. Biol Reprod. 2000;63:89–98.

Moreno RD, Lizama C, Urzua N, Vergara SP, Reyes JG. Caspase activation throughout the first wave of spermatogenesis in the rat. Cell Tissue Res. 2006;325:533–40.

Mruk DD, Cheng CY. Sertoli-Sertoli and Sertoli-germ cell interactions and their significance in germ cell movement in the seminiferous epithelium during spermatogenesis. Endocr Rev. 2004;25:747–806.

Neale MJ, Keeney S. Clarifying the mechanics of DNA strand exchange in meiotic recombination. Nature. 2006;442:153–8.

O'Donnell L, McLachlan RI, Wreford NG, de Kretser DM, Robertson DM. Testosterone withdrawal promotes stage-specific detachment of round spermatids from the rat seminiferous epithelium. Biol Reprod. 1996;55:895–901.

O'Donnell L, Narula A, Balourdos G, Gu YQ, Wreford NG, Robertson DM, et al. Impairment of spermatogonial development and spermiation after testosterone-induced gonadotropin suppression in adult monkeys (*Macaca fascicularis*). J Clin Endocrinol Metab. 2001;86:1814–22.

O'Donnell L, Nicholls PK, O'Bryan MK, McLachlan RI, Stanton PG. Spermiation: the process of sperm release. Spermatogenesis. 2011;1:14–35.

Pareek TK, Joshi AR, Sanyal A, Dighe RR. Insights into male germ cell apoptosis due to depletion of gonadotropins caused by GnRH antagonists. Apoptosis. 2007;12:1085–100.

Pasqualotto FF, Sundaram A, Sharma RK, Borges Jr E, Pasqualotto EB, Agarwal A. Semen quality and oxidative stress scores in fertile and infertile patients with varicocele. Fertil Steril. 2008;89:602–7.

Paul C, Teng S, Saunders PT. A single, mild, transient scrotal heat stress causes hypoxia and oxidative stress in mouse testes, which induces germ cell death. Biol Reprod. 2009;80:913–9.

Pentikainen V, Erkkila K, Suomalainen L, Parvinen M, Dunkel L. Estradiol acts as a germ cell survival factor in the human testis in vitro. J Clin Endocrinol Metab. 2000;85:2057–67.

Ramalho-Santos J, Schatten G, Moreno RD. Control of membrane fusion during spermiogenesis and the acrosome reaction. Biol Reprod. 2002;67:1043–51.

Robertson KM, O'Donnell L, Simpson ER, Jones ME. The phenotype of the aromatase knockout mouse reveals dietary phytoestrogens impact significantly on testis function. Endocrinology. 2002;143:2913–21.

Ross MH, Pawlina W. Testis. In: Ross MH, Pawlina W, editors. Histology: a text and atlas with correlated cell and molecular biology. 5th ed. Princeton: Recording for the Blind & Dyslexic; 2006. p. 728–71.

Rucker 3rd EB, Dierisseau P, Wagner KU, Garrett L, Wynshaw-Boris A, Flaws JA, et al. Bcl-x and Bax regulate mouse primordial germ cell survival and apoptosis during embryogenesis. Mol Endocrinol. 2000;14:1038–52.

Russell LD, Chiarini-Garcia H, Korsmeyer SJ, Knudson CM. Bax-dependent spermatogonia apoptosis is required for testicular development and spermatogenesis. Biol Reprod. 2002;66:950–8.

Ruwanpura SM, McLachlan RI, Stanton PG, Loveland KL, Meachem SJ. Pathways involved in testicular germ cell apoptosis in immature rats after FSH suppression. J Endocrinol. 2008;197:35–43.

Sato H, Miyamoto T, Yogev L, Namiki M, Koh E, Hayashi H, et al. Polymorphic alleles of the human MEI1 gene are associated with human azoospermia by meiotic arrest. J Hum Genet. 2006;51:533–40.

Setchell BP, Breed WG. Anatomy, vasculature and innervation of the male reproductive tract. In: Jimmy D, editor. Neill Knobil and Neill's physiology of reproduction. 3rd ed. New York: Elsevier; 2006. p. 771–826.

Shaha C, Tripathi R, Mishra DP. Male germ cell apoptosis: regulation and biology. Philos Trans R Soc Lond B Biol Sci. 2010;365:1501–15.

Simanainen U, Haavisto T, Tuomisto JT, Paranko J, Toppari J, Tuomisto J, et al. Pattern of male reproductive system effects after in utero and lactational 2,3,7,8-tetrachlorodibenzo-p-dioxin (TCDD) exposure in three differentially TCDD-sensitive rat lines. Toxicol Sci. 2004;80:101–8.

Stouffs K, Lissens W, Tournaye H, Van Steirteghem A, Liebaers I. SYCP3 mutations are uncommon in patients with azoospermia. Fertil Steril. 2005;84:1019–20.

Tesarik J, Martinez F, Rienzi L, Iacobelli M, Ubaldi F, Mendoza C, et al. In-vitro effects of FSH and testosterone withdrawal on caspase activation and DNA fragmentation in different cell types of human seminiferous epithelium. Hum Reprod. 2002;17:1811–9.

Toshimori K, Ito C. Formation and organization of the mammalian sperm head. Arch Histol Cytol. 2003;66:383–96.

Toyama Y, Suzuki-Toyota F, Maekawa M, Ito C, Toshimori K. Adverse effects of bisphenol A to spermiogenesis in mice and rats. Arch Histol Cytol. 2004;67:373–81.

Tremellen K. Oxidative stress and male infertility--a clinical perspective. Hum Reprod Update. 2008;14:243–58.

Turner RM. Tales from the tail: what do we really know about sperm motility? J Androl. 2003;24:790–803.

Ukwenya V, Oluyemi K, Ashamu E, Saalu C, Oyewo O, Makanjuola V. Profertility effects of alcoholic extract of sesame in male Sprague-Dawley rats. ISPUB. 2007;5:1–5.

Van Alphen MM, van de Kant HJ, de Rooij DG. Protection from radiation-induced damage of spermatogenesis in the rhesus monkey (*Macaca mulatta*) by follicle-stimulating hormone. Cancer Res. 1989;49:533–6.

Waheeb R, Hofmann MC. Human spermatogonial stem cells: a possible origin for spermatocytic seminoma. Int J Androl. 2011;34:e296–305.

Ward WS, Coffey DS. Identification of a sperm nuclear annulus: a sperm DNA anchor. Biol Reprod. 1989;41:361–70.

Ward WS, Coffey DS. DNA packaging and organization in mammalian spermatozoa: comparison with somatic cells. Biol Reprod. 1991;44:569–74.

Xiao N, Kam C, Shen C, Jin W, Wang J, Lee KM, et al. PICK1 deficiency causes male infertility in mice by disrupting acrosome formation. J Clin Invest. 2009;119:802–12.

Xue J, Yang J, Yan J, Jiang X, He LY, Wu T, et al. Abnormalities of the testes and semen parameters in clinical varicocele. Nan Fang Yi Ke Da Xue Xue Bao. 2012;32:439–42.

Yamamoto CM, Sinha Hikim AP, Huynh PN, Shapiro B, Lue Y, Salameh WA, et al. Redistribution of Bax is an early step in an apoptotic pathway leading to germ cell death in rats, triggered by mild testicular hyperthermia. Biol Reprod. 2000;63:1683–90.

Yan HH, Mruk DD, Lee WM, Cheng CY. Blood-testis barrier dynamics are regulated by testosterone and cytokines via their differential effects on the kinetics of protein endocytosis and recycling in Sertoli cells. FASEB J. 2008;22:1945–59.

Yeung BH, Wan HT, Law AY, Wong CK. Endocrine disrupting chemicals: multiple effects on testicular signalling and spermatogenesis. Spermatogenesis. 2011;1:231–9.

Yoshida S, Sukeno M, Nakagawa T, Ohbo K, Nagamatsu G, Suda T, et al. The first round of mouse spermatogenesis is a distinctive program that lacks the self-renewing spermatogonia stage. Development. 2006;133:1495–505.

Yu DY, Cringle SJ. Oxygen distribution in the mouse retina. Invest Ophthalmol Vis Sci. 2006;47:1109–12.

Zheng S, Turner TT, Lysiak JJ. Caspase 2 activity contributes to the initial wave of germ cell apoptosis during the first round of spermatogenesis. Biol Reprod. 2006;74:1026–33.

Zhengwei Y, Wreford NG, Royce P, de Kretser DM, McLachlan RI. Stereological evaluation of human spermatogenesis after suppression by testosterone treatment: heterogeneous pattern of spermatogenic impairment. J Clin Endocrinol Metab. 1998;83:1284–91.

Part III

The Semen

The Human Semen

11

Surabhi Gupta and Anand Kumar

Learning Objectives

- Physical properties
- Composition
- Fructolysis
- Coagulation and liquefaction

11.1 Introduction

Human semen is a protein-rich body fluid produced by the male reproductive organs. It is a complex cell suspension in a fluid containing an array of heterogeneous substances produced by different male reproductive glands like the testis, epididymis, seminal vesicles, prostate, Cowper's gland (bulbourethral) and glands of Littre (periurethral glands). Its main function is to act as a buffered, nutrient-rich medium which transports the sperm through the male reproductive tract into the female reproductive tract.

During coitus, a heterogeneous ejaculate is deposited in the female tract. This is because the accessory sex glands discharge their secretions by contracting in an organ-specific sequence during emission/ejaculation. This ensures that the various components of semen are delivered in sequential order. The order of

S. Gupta, PhD (✉) • A. Kumar, MD, FAMS
Department of Reproductive Biology, All India Institute of Medical Sciences,
New Delhi, India
e-mail: surabhi72@rediffmail.com

A. Kumar, M. Sharma (eds.), *Basics of Human Andrology*,
DOI 10.1007/978-981-10-3695-8_11

Table 11.1 Sequence of secretions during ejaculation

Order of secretion	Contributing gland	% of total ejaculate volume
Ist	Cowper's gland (bulbourethral) and glands of Littre (periurethral)	1–5
IInd – A	Testis/Epididymis	5–10
IInd – B	Prostate	20–30
IIIrd	Seminal vesicles	65–75

secretion and the relative contribution of each gland are listed in Table 11.1. The initial secretion, known as pre-ejaculate, comprises of secretions from the Cowper's and Littre glands. This mucinous secretion lubricates the urethra and neutralizes any traces of residual acidic urine. The next fraction results from the simultaneous contractions of the epididymis and prostate. It contains the maximum concentration of sperm along with epididymal and prostatic secretions. The final and largest fraction of the ejaculate is contributed by the seminal vesicles.

Two fates exist for the semen components that remain in the male tract after ejaculation: (i) passive resorption by surrounding tissue or (ii) expulsion during urination (Prins and Lindgren 2015).

11.2 Physical Properties

The volume of a typical human ejaculate is about 3 ml although it can range from 2 to 5 ml (Owen and Katz 2005). Normal semen is a greyish, opalescent fluid with a density between 1.043 and 1.102 g/ml. The colour may appear whitish due to the presence of high number of sperm or leukocytes. If red blood cells are present (hemospermia), the colour may appear reddish brown (WHO 2010).

Semen is slightly alkaline which helps in neutralizing the acidic environment of vagina. The pH measured can vary from 7.2 to 7.8 depending on the time elapsed since ejaculation. Decrease in pH of whole semen over time is attributed to fructolysis and production of lactic acid. However, semen has a buffering capacity much higher than other body fluids. This buffering capacity is contributed by bicarbonate/carbon dioxide (HCO_3/CO_2), high protein content and low molecular weight compounds like citrate, pyruvate and phosphate (Wolters-Everhardt et al. 1987). Another peculiar property of semen is its high osmolarity which is due to the presence of high concentration of organic components rather than inorganic ions.

11.3 Composition

The components of semen can be divided into 'cellular' and 'acellular' components (see Fig. 11.1). The acellular component, obtained after removal of the cells by centrifugation, is termed as seminal fluid and comprises >90% of the semen volume.

11.3.1 Cellular Components

An average human ejaculate has about 100 million sperm/ml though they contribute less than 1% of the ejaculate volume (Prins and Lindgren 2015). The total number of sperm per ejaculate correlates with the length of abstinence as well as the testicular volume (Schwartz et al. 1979; Cooper 2010). Detailed description of sperm is given in the subsequent chapters, The Sperm and Sperm Function Tests.

The other cellular components of semen are epithelial cells of the urogenital tract, leukocytes and even spermatogenic cells. The presence of immature germ

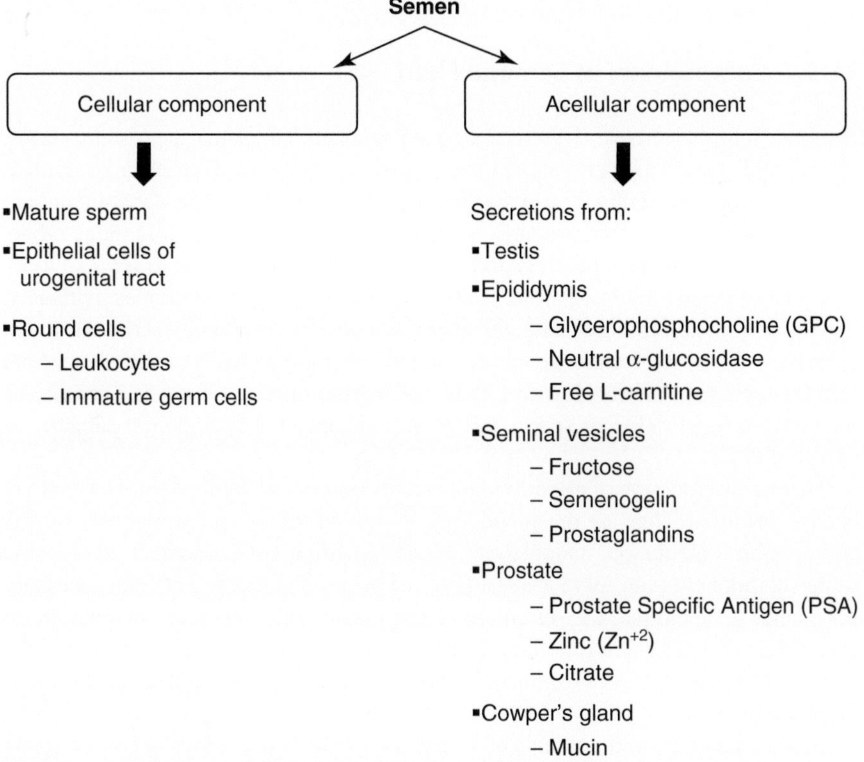

Fig. 11.1 Composition of human semen

cells in the semen may indicate testicular damage or defective spermatogenesis, while the presence of leukocytes may be suggestive of inflammation in the accessory glands (WHO 2010).

A simple microscopic examination of semen is not able to differentiate between leukocytes and spermatogenic cells, which are collectively labelled as 'round cells' (Johanisson et al. 2000). Most human ejaculates show the presence of leukocytes with a predominance of granulocytes. The leukocytes present in the semen are predominantly peroxidase-positive granulocytes (polymorphonuclear leukocytes). They can be easily distinguished from the peroxidase-free multinucleated spermatids by histochemical staining for peroxidase. However, activated granulocytes, which have lost their granules, and other leukocytes, e.g. lymphocytes, monocytes and macrophages, are peroxidase negative. They can be differentiated by immunostaining for CD45 which is the common leukocyte antigen (WHO 2010). If the number of leukocytes present in the ejaculate is greater than the threshold value (1.0×10^6 peroxidase-positive cells/ml), it is termed as leukospermia. An increase in the total number of leukocytes present in the ejaculate correlates with the severity of the inflammatory condition (Wolff 1995).

11.3.2 Components of Seminal Fluid

Seminal fluid comprises of secretions from the seminal vesicles, prostate, testes, epididymides and Cowper's and Littre glands with the greatest molecular content being provided by seminal vesicles (see Fig. 11.1). Some of the constituents are found in the serum and are probably exudates from the circulation, but many others are produced exclusively by the reproductive organs and are unique to seminal fluid. Individual seminal fluid constituents are not essential for fertilization as evidenced by the fact that epididymal/testicular sperm, obtained by testicular sperm extraction (TESE), can be used for assisted reproductive technology (ART) to achieve normal fertilization rates in vitro. However, they may be important under normal conditions for transport/maturation of sperm and greatly enhance the fertilization capacity of sperm in vivo.

The precise function of all individual constituents of the seminal fluid has not yet been determined. They are presumed to be important for the sperm function, during and/or after ejaculation. Qualitative and/or quantitative assessment of specific semen components can serve as marker of the proper functioning of each accessory sex gland, for example, measurement of citric acid, zinc and acid phosphatase to assess the prostatic gland function; fructose and prostaglandins for seminal vesicle; free-L-carnitine, glycerophosphocholine (GPC); and neutral α-glucosidase for epididymal function.

Human seminal fluid contains a diverse set of molecules ranging from organic constituents like proteins, peptides, sugars and lipids to inorganic ions like zinc (see Table 11.2). Average protein concentration of human seminal fluid is 25–55 g/L with albumin making up about one third of the total protein present (Owen and Katz 2005; Rodriguez-Martinez et al. 2011). Albumin in semen is mainly of prostatic

Table 11.2 Important constituents of human seminal fluid

Name of the constituent	Concentration (mg/ml)	Major source
Phosphorylcholine	10.0	Epididymis
Prostate-specific antigen (PSA)	0.5–5.0	Prostate
Citric acid	3.76	Prostate
Spermine	0.5–3.5	Prostate
Prostatic acid phosphatase (PAP)	0.3–1.0	Prostate
Zinc	0.14	Prostate
Fructose	2.0	Seminal vesicles
Total lipids (cholesterol + phospholipids)	1.85	Seminal vesicles
Prostaglandins	0.1–0.3	Seminal vesicles

origin, but the majority of other proteins present are contributed by seminal vesicles (Hirsch et al. 1991). Some of the most important components of the human seminal plasma are listed in Table 11.2 and discussed in subsequent section of the chapter.

11.3.2.1 Originating from Seminal Vesicles

The most important constituents in the seminal vesicle secretions include fructose, semenogelin and prostaglandins. Fructose serves as the primary energy source for sperm in semen. It is produced exclusively by the seminal vesicles, and, hence, its absence in semen is a sign of potential ejaculatory duct obstruction. Semenogelin is a 52-kDa protein which is involved in coagulation of semen. The cleavage products of semenogelin formed following liquefaction have biological functions, such as inhibition of sperm motility and antibacterial activity. The seminal fluid contains about 15 different prostaglandins, predominantly prostaglandin E. The prostaglandins induce smooth muscle contractions in the female genital tract, thereby helping in rapid sperm transport independent of sperm motility. Seminal vesicles are also the major contributor of phospholipids present in semen. The ratio of cholesterol to phospholipids in the semen is proposed to help stabilize the sperm against temperature and environmental shock (White et al. 1976). The other proteins secreted by seminal vesicles include fibronectin, lactoferrin, protein C inhibitor and prolactin-inducible protein (Rodriguez-Martinez et al. 2011; Drabovich et al. 2014).

11.3.2.2 Originating from Prostate

The major proteins secreted by the prostate include prostate-specific antigen (PSA), prostatic acid phosphatase (PAP) and cysteine-rich prostate-specific protein-94 (PSP94). PSA is a zinc-binding serine protease of the Kallikrein family, which hydrolyzes semenogelin leading to liquefaction of the coagulum. PAP is a 102-kDa glycoprotein dimer with enzymatic activity. The main substrate for PAP in seminal fluid is phosphorylcholine phosphate. Prostate also produces spermine which gives semen its unique odour. Spermine has four positive charges and can bind to acidic or negatively charged molecules like phosphate ions, phospholipids or nucleic acids. Enzymatic oxidation of spermine by diamine oxidase, which is present in the seminal fluid, yields aldehyde products which are toxic to both sperm and bacteria. Hence, prolonged exposure of sperm to seminal fluid reduces their fertilization capability (Folk et al. 1980; Prins and Lindgren 2015).

Concentration of zinc in the normal human seminal fluid is more than 100 times compared to concentration in serum. It is involved in regulating liquefaction by binding to semenogelin. It also has an antibacterial activity. Similar to zinc, the concentration of citrate in the semen is 500–1000 times higher than that in blood. It is a potent binder of metal ions, and its concentration (20 mM) compares to the combined concentration of divalent metals (calcium, 7 mM; magnesium, 4.5 mM; and zinc, 2.1 mM).

11.4 Fructolysis

Due to the high motility of sperm, their energy requirement is very high. The major energy source for sperm in the semen is fructose which is produced by the seminal vesicle. Typical concentration of fructose in human semen is 200 mg/dl. To maintain a high adenosine triphosphate/adenosine diphosphate (ATP/ADP) ratio, the sperm utilize anaerobic glycolysis of fructose termed as fructolysis. The process of fructolysis has been described in the chapter, Seminal Vesicles. Each fructose molecule yields 2 lactate − ions and 2 hydrogen ions (H^+).

A positive correlation exists between the degree of sperm motility and the rate of fructolysis in human semen (Peterson and Freund 1976). However, immobilization by spermicidal agent (lipid peroxidase) leads to irreversible loss of fructolytic ability of the sperm (Mann et al. 1980).

11.5 Coagulation and Liquefaction

Human semen coagulates spontaneously after ejaculation and subsequently liquefies within 15–60 min at room temperature. Although the exact mechanism underlying the process of semen coagulation/liquefaction is not clearly understood, it is believed to be regulated through a series of enzymes, mainly proteases, inhibitory factors and metal ions (Emami et al. 2008). Components of the semen are stored in separate glands and get mixed only upon ejaculation. The prostatic secretion containing Zn^{+2} and zinc-inhibited PSA are mixed with the seminal vesicle-produced semenogelin proteins and protein C inhibitor (PCI). Since zinc has a higher affinity for semenogelins in comparison to PSA, it preferentially binds to semenogelins after ejaculation. This induces a conformational change of semenogelin leading to formation of an insoluble, fibrous coagulum. Sperm are immobilized in this coagulum. Chelation of zinc ions diminishes the concentration of free Zn^{+2}, thus activating PSA. Activated PSA cleaves the semenogelins resulting in liquefaction of the gel and release of motile sperm (Malm et al. 2007). Zinc and PCI are also released into solution, and these in turn bind to PSA, preventing further undesirable proteolysis. The details of the coagulation and liquefaction are shown schematically in

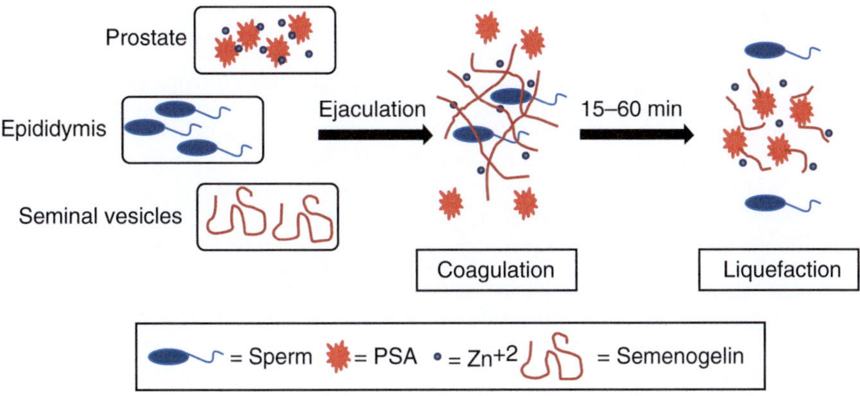

Fig. 11.2 Schematic diagram showing the coagulation and liquefaction process of human semen

Fig. 11.2. The coagulation/liquefaction process allows suitable exposure of sperm to seminal fluid that stimulates motility, increases fertilizing ability and also permits orderly entry of sperm into the female genital tract (Hafez 1976). Absent or incomplete liquefaction process correlates with reduced fertilizing capability (Prins and Lindgren 2015).

11.6 Future Directions

It is important to understand the salient physical and chemical properties of normal human semen in order to formulate a standardized semen simulant. This simulant fluid would be helpful in research related to intravaginal drug delivery for contraceptive and prophylactic drugs (Owen and Katz 2005).

Another area with untapped potential is the use of seminal fluid as a non-invasive clinical sample to identify biomarkers for infertility as well as reproductive tract diseases like prostatitis, cancer, etc. Seminal fluid contains many molecules which are produced by specific male reproductive organs/glands, and, hence, any pathological condition of these organs would influence the molecular composition of semen. Discovery of PSA as a marker of prostatic diseases, both benign prostatic hyperplasia and prostate cancer, is the best example to illustrate this point.

Key Questions

- In what sequence are the various fractions of semen secreted during ejaculation?
- Name the constituents of seminal fluid and discuss their function.
- Describe the key steps in the process of coagulation and liquefaction of human semen.

References

Cooper TG. Semen analysis. In: Nieschlag E, Behre HM, Nieschlag S, editors. Andrology: male reproductive health and dysfunction. Heidelberg/New York: Springer; 2010. p. 125–38.

Drabovich AP, Saraon P, Jarvi K, Diamandis EP. Seminal plasma as a diagnostic fluid for male reproductive system disorders. Nat Rev Urol. 2014;11:278–88.

Emami N, Deperthes D, Malm J, Diamandis EP. Major role of human KLK14 in seminal clot liquefaction. J Biol Chem. 2008;283:19561–9.

Folk JE, Park MH, Chung SI, Schrode J, Lester EP, Cooper HL. Polyamines as physiological substrates for transglutaminases. J Biol Chem. 1980;255:3695–700.

Hafez ESE. Human semen and fertility regulation in men. Saint Louis: Mosby; 1976.

Hirsch IH, Jeyendran RS, Sedor J, Rosecrans RR, Staas WE. Biochemical analysis of electroejaculates in spinal cord injured men: comparison to normal ejaculates. J Urol. 1991;145:73–6.

Johanisson E, Campana A, Luthi R, de Agostini A. Evaluation of 'round cells' in semen analysis: a comparative study. Hum Reprod Update. 2000;6:404–12.

Malm J, Jonsson M, Frohm B, Linse S. Structural properties of semenogelin I. FEBS J. 2007;274:4503–10.

Mann T, Jones R, Sherins RJ. Fructolysis in human spermatozoa under normal and pathological conditions. J Androl. 1980;1:229–33.

Owen DH, Katz DF. A review of the physical and chemical properties of human semen and the formulation of a semen simulant. J Androl. 2005;26:459–69.

Peterson RN, Freund M. Metabolism of human spermatozoa. In: Hafez ESE, editor. Human semen and fertility regulation in men. Saint Louis: Mosby; 1976. p. 176–86.

Prins GS, Lindgren M. Accessory sex glands in the male. In: Plant TM, Zeleznik AJ, editors. Knobil and neill's physiology of reproduction. 4th ed. Amsterdam: Academic Press; 2015. p. 773–804.

Rodriguez-Martinez H, Kvist U, Ernerudh J, Sanz L, Calvete JJ. Seminal plasma proteins: what role do they play? Am J Reprod Immunol. 2011;66(Suppl 1):11–22.

Schwartz D, Laplanche A, Jouannet P, David G. Within-subject variability of human semen in regard to sperm count, volume, total number of spermatozoa and length of abstinence. J Reprod Fertil. 1979;57:391–5.

White IG, Darin-Bennett A, Poulos A. Lipids of human semen. In: Hafez ESE, editor. Human semen and fertility regulation in men. Saint Louis: Mosby; 1976. p. 144–52.

WHO. WHO laboratory manual for the examination and processing of human semen. 5th ed. Geneva: WHO Press; 2010.

Wolff H. The biologic significance of white blood cells in semen. Fertil Steril. 1995;63:1143–57.

Wolters-Everhardt E, Dony JM, Peters WH, De Pont JJ. Buffering substances of human semen. Fertil Steril. 1987;48:159–61.

The Sperm

12

Mona Sharma and Anand Kumar

Learning Objectives
Structure of the sperm
- Strict criteria of normal sperm morphology
- Post-testicular maturation
  - Epididymal maturation
  - Capacitation
- Functional attributes of the sperm
  - Motility
  - Fertilization
  - Egg activation

12.1 Introduction

Sperm or spermatozoon is a specialized cell with potential function of transferring paternal genome into the egg. Sperm is formed by process of cell division and maturation starting from spermatogonial stage, progressing to primary spermatocyte, secondary spermatocyte, spermatid, and mature sperm.

The structural components of sperm correspond with its highly specialized functions. The first ever detailed description of sperm was given by Antoni van Leeuwenhoek in 1677. Sperm consists of the head and flagellum joined by a connecting piece or neck. The head consists of nucleus and acrosome (see Fig.

M. Sharma, MD, DNB, MAMS (✉) • A. Kumar, MD, FAMS
Department of Reproductive Biology, All India Institute of Medical Sciences, New Delhi, India
e-mail: dr.mona18sharma@gmail.com

A. Kumar, M. Sharma (eds.), *Basics of Human Andrology*,
DOI 10.1007/978-981-10-3695-8_12

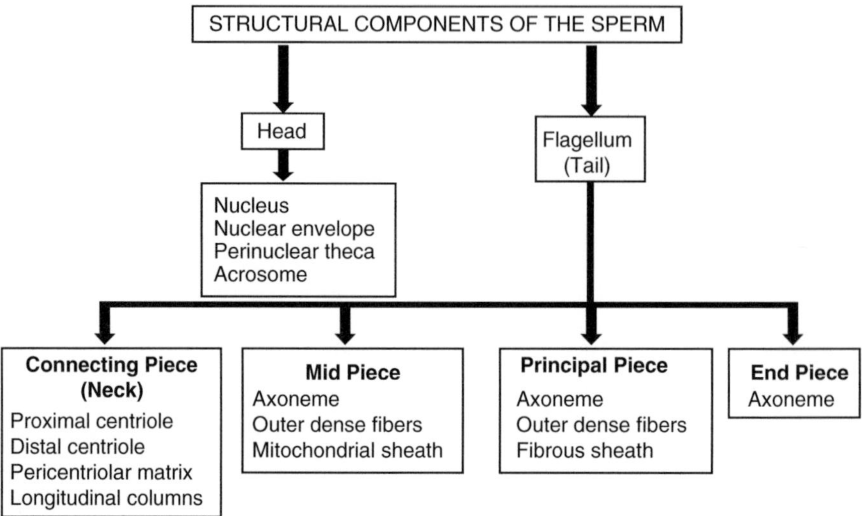

Fig. 12.1 Structural components of the sperm

12.1). Nucleus contains paternal genome. Acrosome is enzyme-filled cap covering the anterior part of the sperm head. Acrosomal enzymes facilitate fertilization. The flagellum is the storehouse of energy needed for sperm motility. The structural uniqueness of sperm is due to the presence of unique proteins such as nuclear packaging proteins, cytoskeleton proteins in the head, acrosomal proteins, and flagellar proteins. Sperm undergoes various structural modifications in the testis and epididymis as well as during its journey to the egg/oocyte through female reproductive tract. The structural modifications make sperm capable of surviving in female reproductive tract, reaching the egg vestments, and penetrating and fertilizing the egg.

We shall now describe the detailed structure and functional attributes of the sperm. The structural and functional details are based on various animal studies and few studies done on human sperm.

12.2 Structure

The main structural components of sperm are the head and the tail or flagellum (Fawcett 1975). Sperm is covered by plasma membrane that differs in various regions of the sperm. Human sperm is about 50–60 μm long with about 50–55 μm long flagellum. The diameter of flagellum is more than 1 μm in connecting piece and tapers toward the end piece (Baccetti 1984). Formation and assembly of the head and flagellar components have been described in the chapter, Spermatogenesis.

12.2.1 Head

The sperm head is about 4 μm long and consists of nucleus, acrosome, and cytoskeleton. The structural arrangement of these components from inside out consists of nucleus, nuclear envelope, perinuclear theca (PNt), acrosome, and plasma membrane (see Fig. 12.2).

Nucleus is highly condensed. Around 15% of human sperm DNA is complexed with histone proteins – H2A, H2B, and H3 (Gatewood et al. 1990). Protamines replace histones during nuclear condensation. Protamines are small and basic proteins rich in arginine and cysteine and are produced during spermatogenesis (Grimes 1986; Hecht 1989). Protamines stabilize chromatin by forming disulfide linkages with adjacent DNA strands, thereby making sperm nucleus hypercondensed. Various other proteins also undergo degradation during sperm maturation. The degraded proteins are transported into neck cytoplasm through nuclear pores present in nuclear envelope below the level of posterior ring (Haraguchi et al. 2007). These proteins get incorporated in the cytoplasmic droplets or residual bodies which finally get shed during epididymal maturation.

Nucleus is covered with *nuclear envelope* having inner and outer nuclear membranes. During spermiogenesis, nuclear condensation causes reduction in size of nuclear envelope and reorganization of excess nuclear envelope into redundant

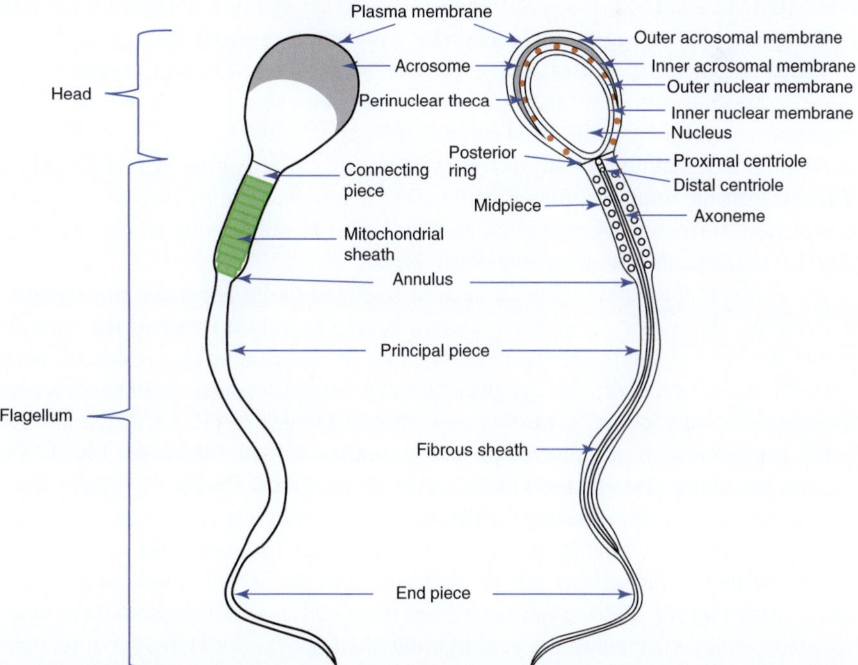

Fig. 12.2 (**a**) Schematic diagram showing different parts of the sperm; (**b**) longitudinal section of sperm showing detailed internal structural components

nuclear envelope (Toshimori et al. 1985; Kerr 1991). The redundant nuclear envelope in mature sperm facilitates hyperactivated motility by providing intracellular calcium (Ca^{2+}) stores (Ho and Suarez 2003).

Posterior ring or nuclear ring is located at the transition of the head and connecting piece (see Fig. 12.2). It represents the line of fusion of plasma membrane with outer and inner membranes of nuclear envelope (Fawcett 1970). Posterior ring is present in all mammalian sperm. Inner surface of the inner membrane of the nuclear envelope shows the presence of cytoskeleton protein meshwork made up of intermediate filament-type lamins. This protein meshwork is known as nuclear lamina (Gerace et al. 1984; Krohne and Benavente 1986).

PNt is present outside the nuclear envelope (see Fig. 12.2). It is formed by cytoskeleton proteins such as actin and actin-binding proteins, e.g., spectrin, tubulins, etc. It maintains the shape of the head and is removed at the time of fertilization. PNt is divided into three segments – subacrosomal, equatorial, and postacrosomal.

Subacrosomal segment is present underneath the acrosome and is fused with inner acrosomal membrane (IAM). Fusion of this segment with IAM strengthens the receptors present on IAM for binding of the sperm head to zona pellucida (ZP). The subacrosomal PNt contains proteins such as T-actin 2 (Tanaka et al. 2003). During spermiogenesis, filamentous (F) actin is depolymerized into globular (G) actin (Camatini et al. 1992). Actin polymerization is seen in ZP-induced acrosome reaction.

Equatorial segment of PNt is folded and consists of PNt, outer acrosomal membrane (OAM), and IAM. It contains receptors such as equatorin or MN9 for binding of the sperm head to egg plasma membrane (Toshimori et al. 1992).

Postacrosomal segment of PNt is present between nucleus and plasma membrane posterior to acrosome and extends up to posterior ring. Proteins present in this segment are cyclin 1, cyclin 2 and calicin (Hess et al. 1993; Lecuyer et al. 2000). It also contains signaling proteins or sperm-borne oocyte-activating factors (SOAFs). These factors are important for activating egg, thereby facilitating meiotic cell cycle completion, formation of pronuclei, and inhibition of polyspermy (Sutovsky et al. 2003). These events initiate zygote development.

Acrosome is a unique organelle derived from the Golgi apparatus during spermiogenesis. Acrosome is a double membranous cap situated around the anterior part of the head of sperm and contains enzymes for penetrating egg vestments such as cumulus oophorus, ZP, and egg plasma membrane. Acrosome contains receptors for sperm-ZP interaction. Acrosome consists of IAM and OAM. IAM overlies the outer membrane of nuclear envelope and is continuous with OAM (see Fig. 12.2). Plasma membrane covers the OAM. OAM disintegrates during acrosome reaction. IAM remains patent during fertilization. Acrosomal matrix is present between OAM and IAM that contains proteases (Yoshinaga and Toshimeri 2003). The major acrosomal matrix protease is acrosin or serine protease which is unique to sperm and is important for ZP penetration (Adham et al. 1997). Other acrosomal enzymes are acrogranin for ZP adhesion, acid hydrolases, and hyaluronidases for zona penetration (Zaneveld et al. 1973; Anakwe and Gerton 1990). The nonenzymatic contents which are important for ZP binding are MC41, SP56, and SP-10 (Foster et al. 1994; Kim et al. 2001; Tanii et al. 2001).

Plasma membrane of the sperm head is organized into acrosomal region and postacrosomal region. Plasma membrane covering postacrosomal region is present between the posterior end of acrosome and connecting piece.

12.2.2 Flagellum

The base of a flagellum is near the nucleus at the junction of connecting piece and the head (Zamboni and Stefanini 1971). Basic structure of flagellum is axoneme which is present throughout the length of flagellum (Turner 2003). In animals with internal fertilization, such as mammals, axoneme is surrounded by outer dense fibers (ODFs) and fibrous sheath (FS). ODFs are present in midpiece and principal piece. FS surrounds ODFs in principal piece (see Fig. 12.5). The other important component of flagellum is mitochondrial sheath that surrounds ODFs in midpiece.

Axoneme is the innermost component of flagellum and consists of microtubules (MTs) arranged in 9 + 2 orientation. This structural arrangement refers to axonemal motility complex of the sperm. It represents nine outer MT doublets (MT-1 and MT-2) and a single central MT doublet which is attached to outer MT doublets by radial spokes (see Fig. 12.3a, b). Outer MT doublets are numbered I to IX clockwise from a plane perpendicular to line bisecting central pair of MTs.

MTs form the basic structural unit of axoneme. MTs are made up of heterodimers of alpha (α) and beta (β) tubulin proteins. α and β tubulins associate and form protofilaments (PFs). Thirteen PFs associate and form round MT-1 and 11 PFs form C-shaped MT-2. Tubulins undergo posttranslational modifications such as acetylations and phosphorylations which are essential for sperm motility (Huitorel et al. 2002).

Dynein extends from MT-1 to MT-2 of adjacent outer doublets (see Fig. 12.3). Dyneins are the actin-based motor proteins as they have adenosine triphosphatase (ATPase) activity which generates sliding force and mechanical energy needed for motility (Gibbons and Fronk 1972). Dynein is made up of three types of protein chains – heavy, intermediate, and light. Dynein has two set of arms – outer and inner. Outer arm is attached to tubulins by set of three polypeptides that form dynein docking unit (Takada et al. 2002). Outer arm is implicated for regulating velocity of sliding. Inner arm is responsible for generating forward propulsive force and forming flagellar wave form. Dynein regulatory polypeptide unit is attached to the outer MT doublets and it is the site where dynein activity is regulated. Dynein is also required during acrosome formation for translocation of Golgi-derived proacrosomal granules toward nuclear surface. These proacrosomal granules fuse near the nuclear surface and form acrosomal vesicle.

Adjacent outer MT doublets are joined by *nexin* proteins such as calcium-regulated nucleoside diphosphate kinases (Bozkurt and Wooley 1993; Patel-King et al. 2002).

Radial spokes extend from outer MT doublets to the central MT doublet (see Fig. 12.3; Smith and Yang 2004). Radial spokes provide structural connection between central MT doublet, dynein arms, and dynein regulatory unit. It has been shown that

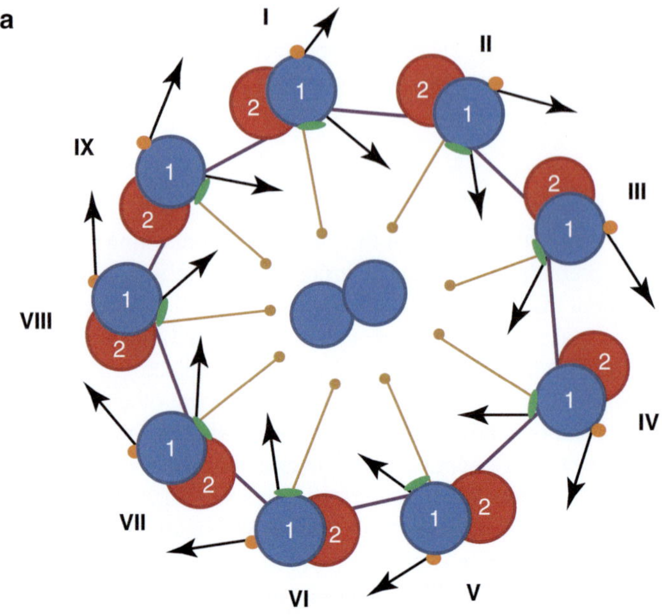

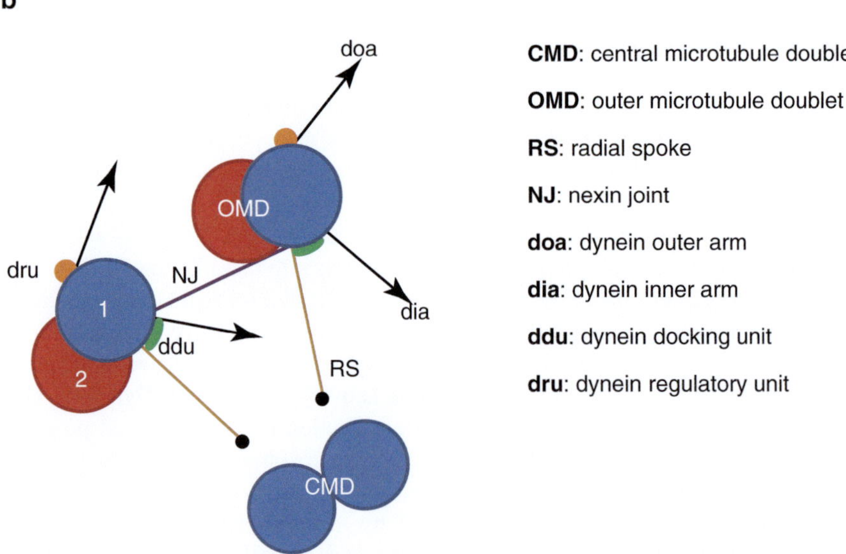

Fig. 12.3 Schematic diagram showing (**a**) structure of axonemal motility complex; (**b**) components of the axoneme

interaction between radial spokes and central MT doublet is a signal for regulating sliding velocity of dynein arms and creating flagellar wave form (Porter and Sale 2000; Smith and Yang 2004).

The sperm flagellum has four parts based on external structures surrounding the centrally placed axoneme – connecting piece, midpiece, principal piece, and end piece.

12.2.2.1 Connecting Piece

Connecting piece or neck is about 0.5 μm long. It consists of a basal plate which is adherent to outer membrane of nuclear envelope and a capitulum which is a dense fibrous plate. One of the proteins present in connecting piece is SPA-TC1 which has been reported in mouse and human sperm (Zahler and Doak 1975).

Connecting piece contains nine segmented columns which are direct continuations of ODFs. It anchors ODFs to base of flagellum. Inner to columns is the proximal centriole. The distal centriole becomes continuous with axoneme (see Fig. 12.2). Proximal and distal centrioles together with pericentriolar matrix form centrosome. Centrosome is a zone from where MTs arise and is commonly known as MT-organizing center (MTOC; Pedersen 1972). Sperm centrosome plays a critical role after sperm-egg fusion. Egg centrosomal proteins assemble on the sperm centrosomal components and form array of MTs known as sperm aster (Sathananthan et al. 1997). Sperm aster is needed for guiding movements and union of male and female pronuclei. Defective sperm centrosome is considered a factor for sperm-derived infertility.

12.2.2.2 Midpiece

Midpiece or middle piece is about 7 μm long. It consists of mitochondrial sheath that forms structural helix of about 75–100 condensed and elongated mitochondria. The presence of mitochondrial sheath in midpiece makes it the storehouse of energy required for sperm flagellar motility (see Fig. 12.2). Sperm mitochondria are special as these are resistant to hypotonic environments and can use lactate as energy substrate (Oko and Clermont 1990). Sperm mitochondria are also the sites containing paternal genome (Sutovsky et al. 2004).

Mitochondrial sheath is wrapped around the ODFs and is adhered to the underlying submitochondrial reticulum (Olson and Winfrey 1990). The submitochondrial reticulum is made up of filamentous meshwork fused with the ring or annulus at the junction of midpiece and principal piece. Mitochondrial sheath contains a protein – spergen 1; a function of this protein is not completely known but is suggested to be involved in the formation of mitochondrial sheath (Doiguchi et al. 2002).

Inner to mitochondrial sheath are the ODFs which extend from connecting piece to the anterior part of principal piece. ODFs consist of cysteine-, serine-, or proline-rich filaments with disulfide crossbridges. Each peripheral MT doublet is associated with an ODF, thereby making the complex having 9 + (9 + 2) arrangement (see Fig. 12.4). ODFs 1, 5, and 6 are larger fibrils as compared to the remaining ODFs. ODF 3 and 8 terminate at the annulus or junction of midpiece and principal piece, from where these are replaced by longitudinal columns of FS.

The shape of the ODFs appears as teardrop in cross section (see Fig. 12.4). ODFs are thickest in the middle of midpiece and tapers toward the proximal half of PP in human sperm (Fawcett 1975). ODFs cover about 60% length of principal piece in human sperm flagellum and contain 30% of sperm proteins (Serres et al. 1984a; Perez-Ramirez and Castells 1991). ODFs are protective for sperm during its passage through the epididymis (Baltz et al. 1990). ODFs are formed during spermiogenesis at late spermatid phase. During this process, ODFs incorporate zinc (Zn) that is required for accurate topology of ODF (Baccetti et al. 1976). The amount of Zn in ejaculate is 100 times higher than in blood serum and is majorly contributed by prostate (Janick et al. 1971; Iyengar 1987). More than 93% of sperm intracellular Zn is localized in the ODFs (Calvin et al. 1973).

Axoneme produces kinetic energy which is transferred to the base of flagellum by stiff disulfide bridges in ODFs (Lindemann and Kanous 1995). Zn binds to sulfhydryl group of cysteine and forms Zn-mercaptide complex (Calvin et al. 1973). This prevents ODFs from premature oxidation (Henkel et al. 2001). During post-testicular sperm maturation in the epididymis, more than 60% Zn is removed from ODFs that leads to oxidation of sulfhydryl group to disulfide bridges which makes ODFs stiff and rigid (Calvin 1981; Cornwall et al. 1988). ODF stiffness is responsible for whiplike progressive movements of the sperm. Therefore, Zn elimination is an essential step to acquire progressive motility. Flagellar Zn has been inversely correlated with sperm motility and fertility (Henkel et al. 1999).

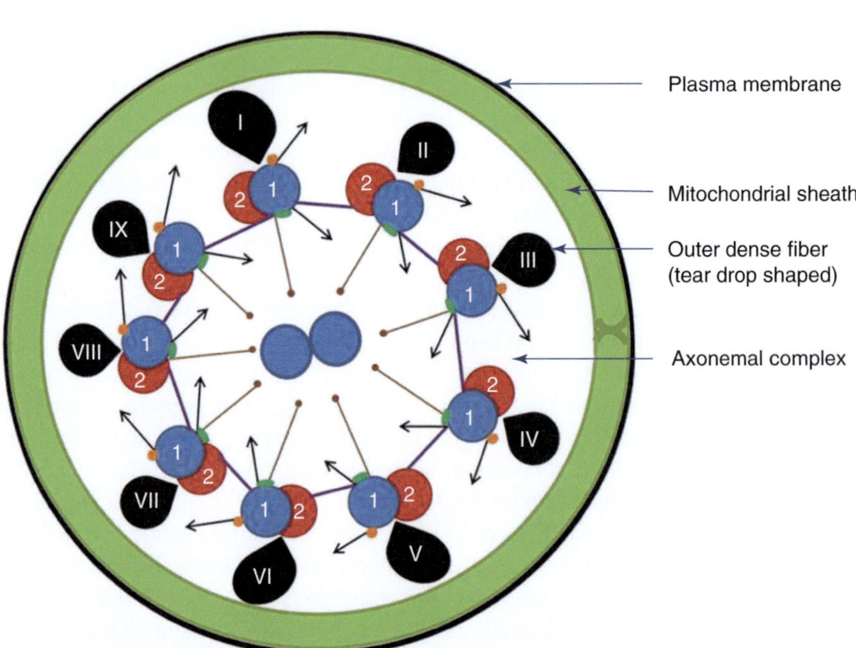

Fig. 12.4 Schematic diagram showing transverse section of sperm flagellum at midpiece

12.2.2.3 Principal Piece

Principal piece is situated distal to midpiece and is the longest segment of about 40 μm length (Fawcett 1970). The proximal extent of principal piece is identified by the presence of a transverse ring or Jansen's ring or annulus at the junction of midpiece and principal piece (see Fig. 12.2). Annulus contains septin group of proteins and maintains shape of the flagellum (Garty et al. 1988).

Principal piece consists of two filamentous longitudinal columns of FS, made up of proteins with disulfide cross-linkages. FS runs along the entire length of principal piece and surrounds ODFs (Oko and Clermont 1990). ODF 3 and 8 end at annulus. At the same position, principal piece is supported by longitudinal columns which send inward extensions that get attached to the projections from outer MT doublet 3 and 8. The columns are connected by circumferential ribs (Fawcett 1970; see Fig. 12.5). Ribs are wider where they merge with the longitudinal columns. In human sperm, ribs in FS are 10–20 nm apart and 50 nm wide (Baccetti 1984).

The most abundant protein in FS is AKAP4 (A kinase anchoring protein 4; Carrera et al. 1994). AKAP4 anchors protein kinase A (PKA) which is the target of cyclic adenosine monophosphate (cAMP)-mediated signaling cascade. PKA further stimulates tyrosine phosphorylation of target proteins involved in capacitation, acrosome reaction, and sperm motility (Eddy et al. 2003). Protein present in principal piece is tektin-2 (TEKT2) which is involved in sperm motility (Wolkowicz et al. 2002).

FS is proposed to provide flagellar flexibility and motility (Eddy et al. 2003). Structural integrity between ODFs and FS is maintained by binding proteins such as rhophilin present at ODFs and ropporin located at FS (Fujita et al. 2000). FS also

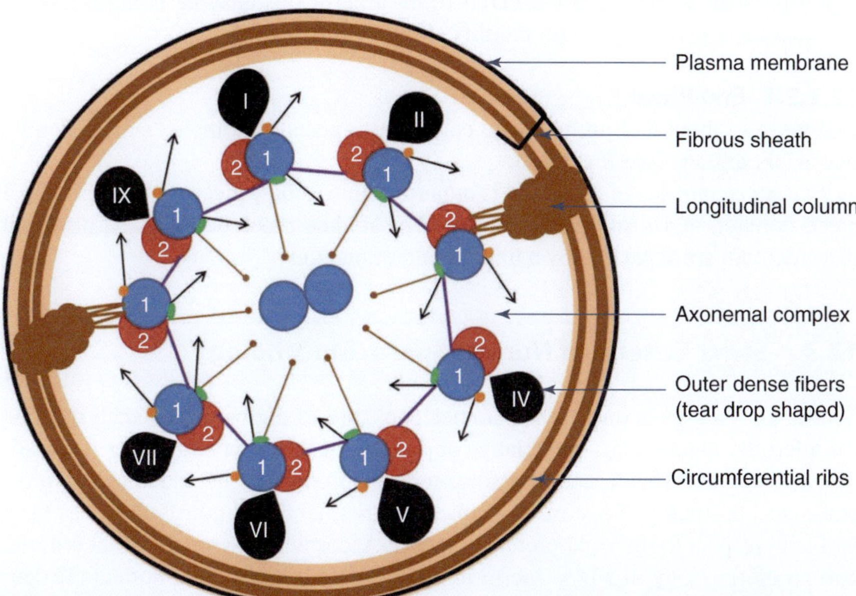

Fig. 12.5 Schematic diagram showing cross section of sperm flagellum at principal piece

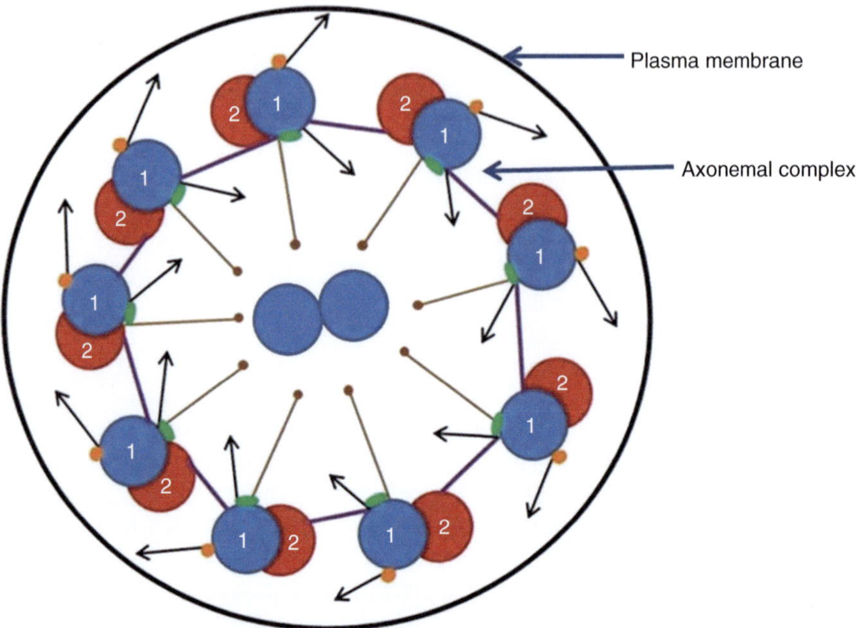

Fig. 12.6 Schematic diagram showing cross section of sperm flagellum at end piece

contains glycolytic enzymes such as sperm-specific glyceraldehyde 3-phosphodehydrogenase-S (G3PDH-S); the role of this enzyme is implicated in providing energy during sperm motility (Miki et al. 2004).

12.2.2.4 End Piece

End piece is about 5–7 μm long and consists of axonemal complex surrounded by plasma membrane (see Fig. 12.6).

Plasma membrane of flagellum is organized into middle piece domain, principal piece domain, and distal piece domain. The middle piece domain and principal piece domain are separated by a fibrous ring or annulus.

12.3 Strict Criteria of Normal Sperm Morphology

Sperm morphology is one of the parameters examined during semen analysis done for infertility assessment. The relative objectivity of the observers makes morphology assessment difficult, and therefore, certain criteria have been defined for morphological normality of sperm. The concept of normal sperm morphology has been laid with respect to the zona-preferred sperm. According to Kruger's strict criteria, normal morphology in >14% sperm in an ejaculate is considered normal (Kruger et al. 1987). Sperm with borderline morphology are considered abnormal.

Assessment of sperm morphology is important as morphologically abnormal sperm are associated with increased DNA fragmentation, immature chromatin, and poor fertilization potential.

Sperm selection for intracytoplasmic sperm injection (ICSI) needs time lapse assessment under higher magnification >X6000, by using motile sperm organelle morphology examination (MSOME). It has been shown in various studies that morphology assessment by MSOME correlates with fertilization rate, embryo quality, and live births (Bartoov et al. 2001, 2002; Cassuto et al. 2009). MSOME includes assessment of the head, acrosome, nucleus, cytoplasmic vacuoles (CVs), cytoplasmic droplets, and position of tail (Cassuto et al. 2009). Under higher magnification, some fine morphological details such as CVs or sperm vacuoles can be observed. CVs appear as surface concavities under light microscopy due to different optical densities of chromatin and CVs. The origin of CVs, whether from nucleus, acrosome, or any membranous structure, is still uncertain. CVs are present in normal as well as abnormal sperm, both in acrosomal and postacrosomal parts of the head. Atomic force microscopy shows slightly depressed plasma membrane in the areas of vacuole (Boitrelle et al. 2011). Association between size of CVs, DNA fragmentation, and chromosomal abnormalities has been controversial (Franco Jr et al. 2008; Watanabe et al. 2011; Boitrelle et al. 2011; Perdrix et al. 2011). Under transmission electron microscopy, CVs appear as bulges at the surface and contain membranous lamellae or whorls covered by acrosome and plasma membrane. The membranous whorls, at places, are continuous with nuclear envelope (Fekonja et al. 2014). Nuclear envelope reorganization during nuclear condensation is a known phenomenon during spermiogenesis. The association between nuclear condensation defects and CVs has been observed (Boitrelle et al. 2013). But, the exact origin of CVs, whether from nucleus, is still to be elucidated.

Criteria of classifying sperm morphology are given by World Health Organization (WHO) which has changed the normality percentage of morphologically normal sperm to 4% (WHO 2010; see Table 12.1 and Fig. 12.7).

12.4 Sperm Maturation

During passage through reproductive tracts, both male and female, sperm is exposed to varying extracellular milieu that stimulates sperm to undergo structural and biochemical modifications. Complete structural and functional maturation of sperm after its production occurs in testis by spermiogenesis and post-testicular maturation in the epididymis and female reproductive tract.

The underlying mechanisms of post-testicular sperm maturation in male reproductive tract are still not very clear, but the role of extracellular vesicles has been proposed. The extracellular vesicles are secreted by different types of cells and are classified into ectosomes, exosomes, and microvesicles (Bobrie et al. 2011). The epididymis and prostate also secrete extracellular vesicles which are called epididymosomes and prostasomes, respectively.

Table 12.1 Classification of sperm morphology

Component	Normal	Abnormal
Head	Oval shaped, smooth contour Length 3–5 μm; width 2–3 μm Acrosome covering 40–70% of head area (Menkveld et al. 2001) Up to two small vacuoles in acrosomal area No vacuole in postacrosomal area	Tapered, round, or pyriform shaped Larger or smaller size Acrosome covering <40% or >70% of head area Vacuoles >2 or covering >20% of head area Vacuole in postacrosomal area in any number Double head or no head
Midpiece	Same length as that of head Slender with no enlargement Residual cytoplasm or cytoplasmic droplets: membrane-bound vesicles <1/3rd of head size, not well stained, located at head-neck junction, may extend up to midpiece (Mortimer and Menkveld 2001; Abraham-Peskir et al. 2002; Fetic et al. 2006)	Sharp bend between neck and midpiece Thick midpiece Excess residual cytoplasm ≥1/3rd head size (Mortimer and Menkveld 2001)
Principal piece	Length ten times as that of head (around 45 μm) Can be curved but not making sharp angle	Short length Multiple in number Hairpin bends Sharply angulated bends Coiled

12.4.1 Post-testicular Maturation in the Epididymis

Soon after release from testis, sperm is not able to fertilize egg. Sperm must pass through the length of the epididymis for complete maturation which takes 10 days to 2 weeks. The epididymis secretes proteins which interact with sperm. This leads to modification of sperm surface proteins, changes in sperm plasma membrane, and incorporation of epididymal proteins into sperm cellular components.

The epididymis secretes proteins through merocrine and apocrine pathways (Rejraji et al. 2006). Epididymal proteins undergo posttranslational modifications in the Golgi apparatus which produces secretory vesicles. These secretory vesicles liberate proteins by exocytosis. The presence of epididymosomes has been reported in humans (Thimon et al. 2008). Sperm acquire epididymal proteins by glycosylphosphatidylinositol (GPI) anchor which is formed after post-translational modification in the Golgi apparatus. Epididymal proteins with anchor are contained within secretory vesicles which fuse with sperm plasma membrane.

Epididymosomes are rich in proteins involved in sperm remodeling, motility, and sperm-egg interactions. One of the constituents of epididymosomes is

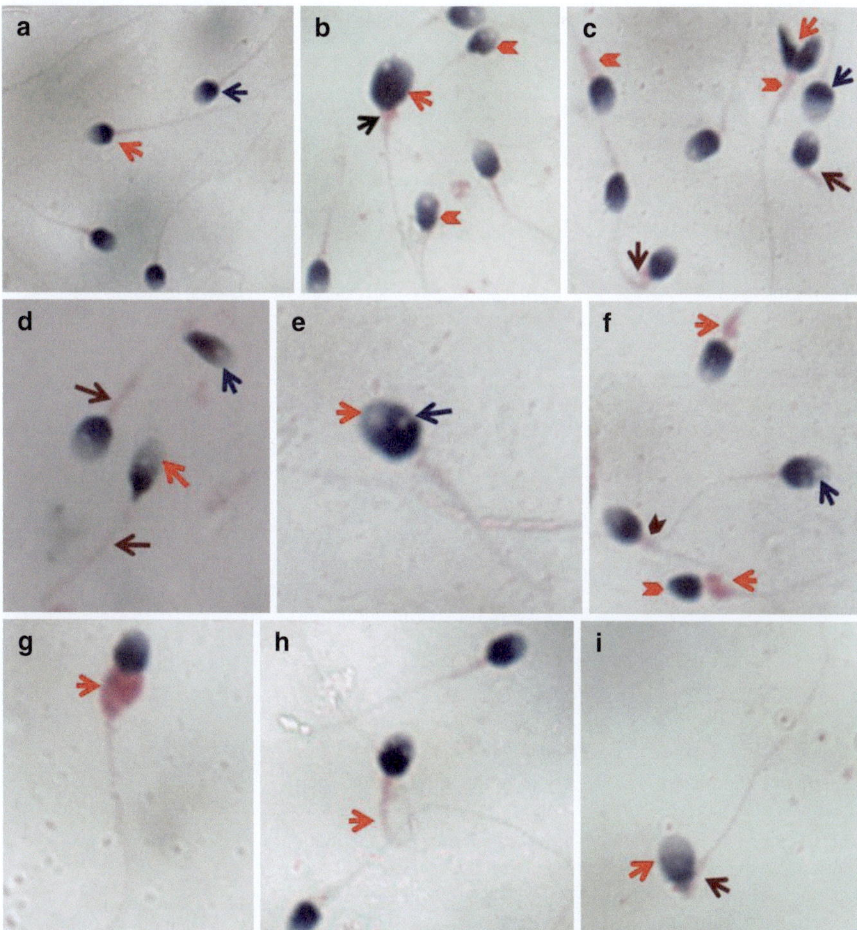

Fig. 12.7 Sperm morphological analysis by Papanicolaou staining. (**a**) Normal morphology (*red* and *blue* As). (**b**) Large head with acrosome < 40% (*red* A); RC (*blue* A); abnormal shape (*red* AH). (**c**) Double head (*red* A); acrosome >70% (*blue* A); bent in tail (*brown* As); thick tail (*red* AHs). (**d**) Acrosome >70% (*red* A); abnormal head shape (*blue* A); thick tail (*brown* As). (**e**) CV in acrosomal area (*red* A) and in postacrosomal area (*blue* A). (**f**) CVs in acrosomal area (*blue* A); RC (*red* As); thick tail (*brown* AH); no acrosome (*red* AH). (**g**) ERC (*red* A). (**h**) Coiled tail (*red* A). (**i**) Acrosome >70% (*red* A); bent in tail (*brown* A) (Courtesy: Mr. Pradeep, Medical Lab Technologist, Andrology Research Unit, Department of Reproductive Biology, All India Institute of Medical Sciences, New Delhi). *A* arrow, *AH* arrow head, *RC* residual cytoplasm, *ERC* excess residual cytoplasm, *CV* cytoplasmic vacuole

macrophage migration inhibiting factor (MIF) which gets incorporated with sperm ODFs (Eickhoff et al. 2004, 2006). At ODFs, MIF is involved in Zn elimination which is an essential step to facilitate sperm motility (Frenette et al. 2002). By this, epididymosomes regulate motility during epididymal maturation. Other components of epididymosomes involved in acrosome reaction are liprin α3 and kinase

Table 12.2 Sperm maturation changes in the epididymis

Maturation changes	Functional consequences	Reference
90% water reabsorption in head of epididymis, increase in osmolality by 22%	Sperm concentration increases in the body of epididymis	Hinton et al. (1981), Clulow et al. (1994)
Decrease in cholesterol in plasma membrane, increase in unsaturated and saturated fatty acid ratio	Plasma membrane becomes fluidic and ready for acrosome reaction and fertilization	Nikolopoulou et al. (1985), Parks and Hammerstedt (1985)
Sperm interaction with epididymal fluid which is rich in glycosides and glycosyl transferases	Changes in glycosidic nature of sperm plasma membrane facilitate sperm-zona pellucida (ZP) binding	Brown et al. (1983)
Epididymal protein transfer to sperm plasma membrane	Proteins important for sperm-ZP interaction – P34H, 3B2F, HE5, SPAM-1 Protein for oocyte fusion – SOB2 Proteins for acrosome reaction and capacitation – liprin α3 and kinase cSrc Protein for sperm-egg fusion – ADAM 7 Epididymal sterol-binding proteins – HE1 – decrease in sperm membrane rigidity that facilitate acrosome reaction	Boue et al. (1994), Kirchhoff et al. (1996), Legare et al. (1999)
Sperm nuclear content becomes compact	Reduction in size of the sperm head	
Appearance of cytoplasmic droplets at the neck of the sperm, later get shed off	Indicative of sperm maturation; abnormal if ≥1/3rd head size	Anberg (1957)
Zn content decreases by 60% that increases stability of outer dense fibers (ODFs) by sulfhydryl oxidation into disulfide bridges	Tail becomes rigid and show bending and motility	Cornwall et al. (1988)

cSrc (Joshi et al. 2012; Krapf et al. 2012). Epididymal transit also causes sperm to acquire epididymal proteins such as human epididymal protein 5 (HE5; Kirchhoff 1996). Zn regulates the amount of proteins that transfer from epididymosomes to sperm (Frenette et al. 2002). Sperm adhesion molecule 1 (SPAM 1or PH 20) also gets transferred from epididymosomes to sperm (Griffiths et al. 2008). Ubiquitin proteins in epididymososmes are responsible for elimination of defective sperm. ADAM 7 or alpha disintegrin metalloproteases is a member of transmembrane superfamily and is exclusively expressed in epididymosomes (Oh et al. 2005, 2009). This protein is involved in fertilization process. Table 12.2 highlights the sperm maturation changes occurring in the epididymis.

12.4.2 Post-testicular Maturation by Prostatic Secretions

Sperm are mixed with secretions of accessory glands at ejaculation. Secretary vesicles arising from prostatic acinar or glandular cells are called prostasomes. Prostasomes are present within large vesicles in most of the glandular cells. When larger vesicular bodies fuse with plasma membrane of acinar cells, prostasomes are secreted. Prostasomes have high cholesterol: phospholipid ratio of 2:1 (Arienti et al. 1999). High cholesterol inhibits capacitation as sperm membrane cholesterol efflux is stimulatory for capacitation (Pons-Rejraji et al. 2011). Prostasomes contain Ca^{2+} signaling proteins which are important for sperm motility. Prostasomes are also protective to sperm against immune reaction in female reproductive tract by inhibiting complement pathway, neutrophil phagocytosis, and lymphocyte proliferation.

12.4.3 Capacitation

The post-testicular structural and functional modifications in sperm physiology during its passage in female reproductive tract are called capacitation or physiological priming. Sperm capacitation is a prerequisite for hyperactivated motility and acrosome reaction.

12.4.3.1 Molecular Events of Capacitation

Initiation of capacitation involves three important events: (1) lipid changes in the sperm plasma membrane; (2) ionic movements across plasma membrane, thereby changing its potential; and (3) tyrosine phosphorylation of proteins involved in hyperactivation and acrosome reaction.

Sperm plasma membrane changes The common phospholipid components of plasma membrane are sphingomyelin (SM), phosphatidylcholine (PC), phosphatidylserine (PS), and phosphatidylethanolamine (PE). Phospholipids are present asymmetrically across leaflets of plasma membrane. PS and PE are present in inner leaflet; SM and PC are present in outer leaflet. Plasma membrane asymmetry is maintained by three transferases or carrier proteins. Flippase facilitates movement from outer leaflet to inner leaflet; floppase allows movement from inner leaflet to outer leaflet and scramblase facilitates movement of phospholipids in both directions (Kol et al. 2002). Sperm plasma membrane also shows such asymmetry, as observed in bull and ram sperm (Pomorski et al. 1995; Nolan et al. 1995).

Bicarbonate (HCO_3^-) is an important ion of capacitation and is present in secretions of female reproductive tract (Gadella and Harrison 2000; Shi and Roldan 1995). Increase in intracellular HCO_3^- in sperm leads to increased intracellular pH during capacitation (Zeng et al. 1996). HCO_3^- also regulates cAMP turnover by directly stimulating adenylyl cyclase (AC; Garbers et al. 1982; Harrison and Miller 2000; Visconti et al. 1990). Increase in cAMP further stimulates PKA that causes phosphorylation of various head and flagellar proteins (see Fig. 12.8; Harrison and Miller 2000; Harrison 2004). PKA also stimulates enzymes for synthesis and degradation of cAMP (Hanoune and Defer 2001).

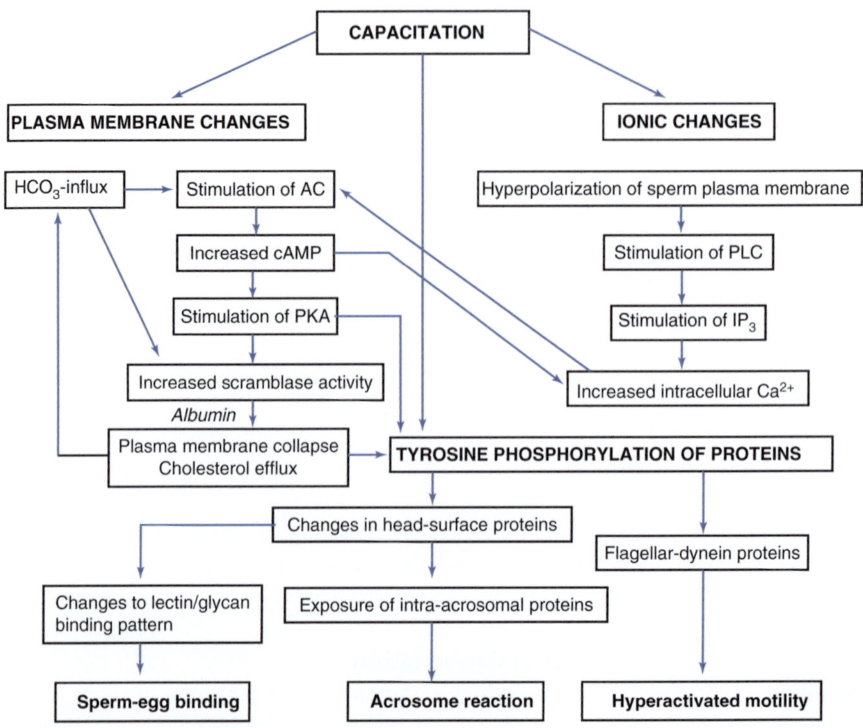

Fig. 12.8 Flow chart showing molecular events of capacitation. *HCO3−* bicarbonate, *AC* adenylyl cyclase, *cAMP* cyclic adenosine monophosphate, *PKA* protein kinase A, *PLC* phospholipase C, *IP3* inositol triphosphate, *Ca2+* calcium

HCO_3^- induces collapse of plasma membrane asymmetry by increasing PKA and scramblase activities. This leads to changes in phospholipid orientation and ultimately membrane collapse (Gadella and Harrison 2000). HCO_3^--induced collapse in plasma membrane asymmetry is also seen in human sperm (de Vries et al. 2003). Plasma membrane collapse is an early event of capacitation (see Fig. 12.8).

Another plasma membrane change occurring during capacitation is cholesterol removal by albumin present in female reproductive tract secretions. Cholesterol present in plasma membrane forms complexes with sphingolipids that are called lipid rafts (LRs; Simon and Vaz 2004). LRs associate with cytoskeleton proteins (Kusumi et al. 2004). Cholesterol removal or efflux causes changes in LRs that leads to stimulation of tyrosine kinase-mediated signaling pathways followed by tyrosine phosphorylation of sperm proteins (Brown and London 2000; Edidin 2001).

Plasma membrane collapse also facilitates albumin-mediated cholesterol removal (Gadella and Harrison 2000). It has been shown that only after HCO_3^- addition, albumin stimulates cholesterol removal (Flesch et al. 2001). Therefore, it can be suggested that HCO_3^-/ cAMP pathway is followed by PKA-dependent collapse of phospholipid asymmetry that leads to cholesterol loss. Loss of cholesterol from sperm plasma membrane also stimulates HCO_3^- transport that further upregulates cAMP pathway as a result of feed forward mechanism (Visconti et al. 1999). There is also removal of sterols from plasma membrane that leads to decreased cholesterol/phospholipid ratio (Hoshi et al. 1990). As a result, sperm plasma membrane becomes fluidic that further facilitates acrosome reaction (Wolf and Cardullo 1991).

Ionic changes Ion transport across sperm plasma membrane facilitates motility, intracellular messenger regulation, and sperm capacitation (Gatti et al. 1990; Visconti et al. 1995; Arnoult et al. 1999). Sperm is exposed to changing extracellular ionic environment. In the epididymis, there is increased potassium ion (K^+), decreased sodium (Na^{2+}), and HCO_3^- concentration; whereas after ejaculation, in seminal fluid and in female reproductive tract, concentration of K^+ decreases and of Na^{2+} and HCO_3^- increases (Setchell et al. 1994). Altered extracellular ion concentrations induce changes in intracellular ion concentration, and hence, there is change in sperm membrane potential (Demarco et al. 2003). During capacitation, there is increase in intracellular negative ion concentration which causes hyperpolarization of sperm plasma membrane in mammals (Zeng et al. 1995). This leads to increase in intracellular Ca^{2+} ion concentration by stimulation of phospholipase C (PLC) or inositol triphosphate receptors (IP_3; Arnoult et al. 1999). IP_3-sensitive Ca^{2+} channels or receptors bind to IP_3 that leads to increase in intracellular Ca^{2+} release and stimulation of calcium-dependent AC. This is followed by PKA-mediated tyrosine phosphorylation (see Fig. 12.8). Ca^{2+} also binds to calmodulin (CaM) that stimulates CaM kinases (CaMK). This leads to CaMK-mediated phosphorylation of head and flagellar proteins.

Tyrosine phosphorylation by PKA as posttranslational modification is one of the important signaling pathways stimulated during capacitation (Visconti et al. 1995; Manning et al. 2002). Protein phosphorylation is an essential process to initiate or stop biological events. Sperm lacks transcription and translation machinery. Thus, sperm maturational events depend solely on posttranslational modifications such as protein phosphorylation. As a result, there is either loss of sperm surface proteins or exposure of surface receptors. These changes later facilitate head plasma membrane to fuse with OAM during acrosome reaction and to oocyte plasma membrane at the time of sperm-egg fusion. Tyrosine phosphorylation of flagellar proteins leads to hyperactivated motility (see Fig. 12.8).

12.5 Functional Attributes of the Sperm

12.5.1 Motility

Sperm motility is an energy-consuming process. Sperm flagellum moves in plane perpendicular to central pair of MTs that passes through MT doublet 1 and between MT doublets 5 and 6 (Woolley 1977). The flagellum also twists while bending forward (Yeung and Woolley 1984; Wooley and Osborn 1984). The amplitude of curvature increases during flagellar wave propagation. ODFs taper toward the end, thereby decreasing resistance to bending (Serres et al. 1984a, b). Sperm exhibits forward propulsive force due to sliding of dynein arms progressively backward.

12.5.1.1 Motility Patterns

Changes in the motility pattern are observed during sperm maturation through its journey from the epididymis to the egg (see Table 12.3). Rapid and progressive motility exhibited by sperm extracted from cauda epididymis and in ejaculate is due to increased rigidity provided by disulfide bonds in ODFs and FS (Calvin and Bedford 1971). Sperm motility within semen coagulum is inhibited by a protein semenogelin (SG) which is produced by seminal vesicles and constitutes 20% of seminal proteins (Robert and Gagnon 1999). Prostate-specific antigen cleaves SG and facilitates motility. SG also prevents premature capacitation. Progressive motility is seen only when sperm are ejaculated. Till then, within the epididymis, some inhibitors such as immobilin, a highly viscoelastic glycoprotein secreted by epididymal cells, inhibits sperm motility (Turner and Giles 1982).

Hyperactivated motility is needed to penetrate highly viscous fluid in female reproductive tract and also to penetrate ZP (Stauss et al. 1995; Ho and Suarez 2001a). Capacitated sperm exhibit hyperactivated motility with increased lateral head movements and asymmetrical flagellar bends. Hyperactivated motility is maintained by propagation of Ca^{2+} influx and stimulation of Ca^{2+}-dependent ATPases along the length of the tail (Yanagimachi 1970; Ho and Suarez 2003).

12.5.1.2 Mechanism of Motility

Plane of tail bending is perpendicular to line joining the central MT doublets. Movement in this plane is facilitated by changing space between the ribs on either side of FS. Phosphorylation of dynein protein initiates motility (Tash 1989). This causes activation of dynein ATPases that leads to hydrolysis of adenosine triphosphates (ATPs) and energy production followed by dynein sliding. Further, there is MT sliding and the generation of sliding force that is transformed into a bend as axoneme is anchored to the base of the head (see Fig. 12.9; Brokaw 1972, 1989).

12.5.1.3 Regulators of Motility

Energy source Seminal plasma sugars such as glucose and fructose are sources of ATP generation. The propagation of bends during sperm motility is continuous along the length of flagellum and energy required is derived from ATPs. The site of ATP generated by oxidative phosphorylation is mitochondria which is located at the extreme anterior end of midpiece.

Table 12.3 Motility patterns of sperm

Caput epididymis	Cauda epididymis/ejaculate	Female reproductive tract
Basal motility with slow and circular movements (Bedford 1975)	Rapid motility which is progressively forward with linear and symmetrical bends having low amplitude and small arc of curvature (Bedford 1975)	Hyperactivated motility with increased velocity, decreased forward progressive movement, increased lateral displacement of head and asymmetrical flagellar bends (de Lamirande et al. 1997)

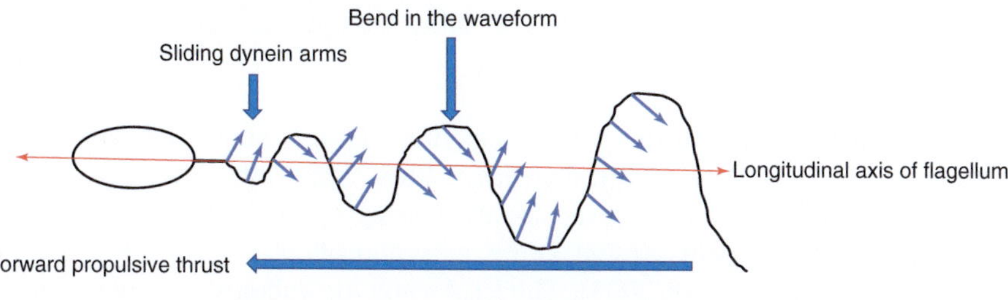

Fig. 12.9 Schematic diagram showing mechanism of sperm motility

ATP diffusion from midpiece to distal end of flagellum is essential to support active sliding of MTs during motility (Tombes and Shapiro 1985). The process of energy transfer from midpiece to principal piece is through enzymatic shuttles such as creatine kinase, adenylate kinase, and phosphokinase (Dzeja and Terzic 2003). ATP is hydrolyzed into adenosine diphosphate (ADP) and provides energy needed for sliding of MTs and propagation of bending along the length of the flagellum. ADP is rephosphorylated by phosphate transfer from creatine phosphate. In some animal species such as sea urchin sperm, phosphocreatine shuttle is present to transfer energy by facilitating transfer of phosphate groups. But, this kind of shuttle is absent in mammals (Kaldis et al. 1996).Creatine phosphokinase shuttles are present in human sperm. Isoforms of creatine phosphokinase (CK) –B and M are proposed to be the indices of male fertility (Huszar et al. 1988, 1992). But, the role of these shuttles as means of energy transfer has not been confirmed in mammalian and human sperm.

The extramitochondrial sources of energy have been suggested. The presence of hexokinase 1 in the sperm head and flagellum and G3PDH-S in FS has been identified (Visconti et al. 1996; Bunch et al. 1998). The presence of these enzymes suggests that glycolysis could be the source of energy for sperm motility. The glycolytic pathway for ATP generation is more efficient for sperm motility (Mukai and Okuno 2004). Glycolysis in principal piece is important way to produce ATP locally for sperm motility (Turner 2003).

Zinc (Zn) is an inhibitor of sperm motility (Henkel et al. 1999). Zn binds to sulfhydryl group of ODFs. Once Zn is removed during epididymal maturation, there is formation of disulfide bridges that further facilitates motility.

Calcium ions Intracellular Ca^{2+} levels are regulated by voltage-gated calcium channels (VLCC) which are of three types – (a) CatSper 1 and 2, (b) CaV 2.3, and (c) PMCA4 (Sakata et al. 2002; Carlson et al. 2003; Quill et al. 2003; Okunade et al. 2004). Release of intracellular Ca^{2+} stores via IP_3 signaling has been observed (Ho and Suarez 2001b). IP_3 receptors are localized to redundant nuclear envelope on the sides of flagellar base. The flagellar bends initiate at the base and propagate distally (Ho and Suarez 2003). The intracellular Ca^{2+}stores at the base of flagellum provide initial Ca^{2+} signal that further stimulate Ca^{2+} influx through CatSper channels in principal piece.

Role of cAMP ATP with the help of enzyme AC gets converted into cAMP which further stimulates cAMP-dependent PKA. PKA inhibitor H89 inhibits sperm motility (Holt and Harrison 2002). During epididymal maturation, the precursor AKAP protein present at FS gets phosphorylated to form mature AKAP (Yeung and Cooper 2003). The PKA catalytic subunits are found on FS in human sperm (Moos et al. 1998). PKA stimulation causes tyrosine phosphorylation of flagellar proteins (Leclerc et al. 1996; Si and Okuno 1999; Urner and Sakkas 2003).

Bicarbonate HCO_3^- present in seminal plasma and in female reproductive tract secretions is important for motility initiation. The role of HCO_3^- in stimulating motility has been proved by animal experiments (Si and Okuno 1999). HCO_3^- ions induce hyperactivation. HCO_3^- increases intracellular pH which further stimulates opening of VLCC and activation of AC (Wennemuth et al. 2003). AC increases cAMP production (Okamura et al. 1985). cAMP further stimulates Ca^{2+}channels; as a result there is increase in Ca^{2+} that leads to PKA-mediated hyperactivation (see Fig. 12.8; Ho and Suarez 2001b).

12.5.2 Fertilization

Fertilization is the process of fusion of male and female gametes, and it involves multiple proteins of cell-cell signaling, adhesion, and fusion. Fertilization occurs in the ampullary part of the fallopian tube. During fertilization, sperm has to pass through three barriers or egg vestments which are cumulus oophorus cells, ZP, and egg plasma membrane. Mammalian fertilization involves following sequential steps: binding of acrosome-intact sperm to ZP, sperm signaling to initiate acrosome reaction, acrosome exocytosis and penetration of ZP, binding of egg plasma membrane, fusion with egg plasma membrane, and egg activation (see Fig. 12.10).

Egg/oocyte is encased by ZP which is surrounded by cumulus oophorus cells. Cumulus is made up of granulosa cells surrounded by extracellular matrix composed of hyaluronan, heparin sulfate, etc. (Rodgers et al. 2003). Cumulus cells may act as selective barrier preventing uncapacitated sperm from penetration and fusion (Corselli and Talbot 1987). Cumulus cells also secrete sperm chemoattractant such as progesterone which is released at the time of ovulation (Yamashita et al. 2003; Sun et al. 2005). The role of progesterone is also implicated in stimulating sperm hyperactivated motility and acrosome reaction (Roldan et al. 1994; Jaiswal et al. 1999). The released progesterone acts on the sperm surface receptors and stimulate phospholipase C (PLC) that leads to increase in intracellular Ca^{2+} (Meizel and Turner 1991; Blackmore et al. 1991). In human sperm, the process of chemotaxis is regulated by G protein-coupled odorant receptors (OR) such as hOR17–4 (Ikawa et al. 2008).

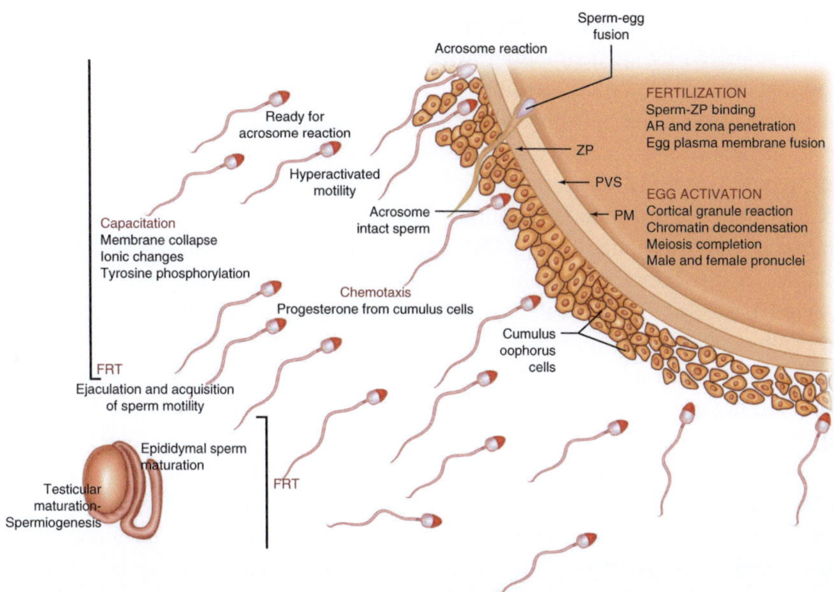

Fig. 12.10 Schematic diagram showing sperm functional events from testis to egg. *FRT* female reproductive tract, *ZP* zona pellucida, *PVS* perivitelline space, *PM* plasma membrane, *AR* acrosome reaction

Sperm acrosomal hyaluronidase enzyme helps in cumulus matrix dispersal (McClean and Rowlands 1942). Sperm that have completed acrosome reaction cannot penetrate cumulus which suggests that acrosome reaction begins while passing through cumulus matrix (Cummins and Yanagimachi 1986). Sperm plasma membrane also contains PH20, a glycosylphosphatidylinositol-linked protein which has hyaluronidase-like activity and is important for cumulus penetration (Lin et al. 1994). It has also been shown that acrosome-intact sperm can associate with ZP and ZP provides acrosome reaction inducing substrates (Wassarman et al. 2001). Sperm passing through cumulus matrix show enlargement in the anterior part of the head before acrosome reaction that may be related to transient exposure of acrosomal enzyme (Cummins and Yanagimachi 1986; Kim and Gerton 2003). Therefore, it can be suggested that acrosome-intact sperm passes through cumulus matrix with the help of PH 20 (Myles and Primakoff 1997). After passing through cumulus, sperm binds to ZP.

Binding of sperm to ZP is species specific and specificity is due to ZP (Gwatkin 1977). Zona-free hamster egg can be bound by human sperm, and this binding is not seen with intact ZP hamster egg (Yanagimachi 1984). ZP is a glycoprotein layer synthesized by the growing oocytes. The components of ZP are ZP1-4. In mice, it consists of ZP1-3. All components undergo posttranslational modifications by glycosylation (Bleil and Wassarman 1980; Wassarman 1988; Lefievre et al. 2004). ZP proteins are bound to ZP transmembranous domain made up of around 260 amino acids forming extracellular matrix proteins (Bork and Sander 1992).

The first step in binding of sperm to ZP is initial attachment (Gwatkin 1977). This is followed by sperm binding to zona adhesion molecules; ZP3 is the most important molecule which specifically binds to the sperm head (Mortillo and Wassarman 1991). Serine-/threonine-linked oligosaccharide chains in ZP3 bind to sperm surface b1,4-galactosyltransferase-I (GalT) that stimulates phosphatidylinositol signaling and PLC cascade which leads to Ca^{2+} influx.

Ca^{2+} influx creates ionic oscillations which further stimulate acrosome reaction (Shur et al. 2006; Florman et al. 2008). These events involve stimulation of SNARE (soluble N-ethylamaleimide-sensitive attachment protein receptors) proteins involved in exocytosis, Ca^{2+}-binding proteins, and Ca^{2+} sensor synaptogranin (Ungar and Hughson 2003; Sudhof and Rizo 1996). SNARE protein homologs in mammalian sperm are Rab 3A and N-ethylmaleimide-sensitive factor (NSF; Yunes et al. 2000). All these molecules stimulate acrosome reaction. Acrosomal proteins which help sperm to get adhered to ZP during acrosome reaction are zona adherens, SP56, P95, and proacrosin (Jones 1990; Foster et al. 1997; Olson et al. 2004). Human sperm without acrosome do not undergo ZP penetration (Bourne et al. 1995).

Actin cytoskeleton present in PNt takes part in acrosome reaction. Actin during physiological cell cycle undergoes polymerization-depolymerization. Monomeric G-actin polymerizes to form F-actin and then F-actin depolymerizes to regenerate G-actin (Aderem 1992). This process is regulated by depolymerizing factor cofilin and actin-binding proteins such as profilin, beta actinin, fragmin, etc. (Aderem 1992; Didry et al. 1998). Actin polymerization to F-actin is an event occurring during capacitation, acrosome reaction, and zona penetration. Polymerized actin is lost with acrosome reaction.

During acrosome reaction, OAM fuses with PM followed by exposure of acrosomal contents such as acrosin, proacrosin, hyaluronidase, glycohydrolases, acrosomal sperm protein SP-10, and aryl sulfatase (Yang et al. 1974; Tesarik et al. 1990; Coonard et al. 1996; Meyer and Rosenberger 1999). Exposure of these molecules leads to digestion of ZP and egg membranes (Abou-haila and Tulsiani 2009). After sperm-ZP binding, zona undergoes modifications to prevent polyspermy. Polyspermy is inhibited by modification of glycolytic activity of ZP3 just after fusion.

Once sperm and egg proteins interact, fusion occurs at equatorial segment. Sperm-egg fusion needs the presence of ligands on each surface and is characterized by transition of two lipid bilayers into single lipid bilayer (see Table 12.4).

12.5.2.1 Post-fusion Events

Fusion is followed by decrease in sperm tail motility by signaling to axonemal complex to decrease dynein-mediated sliding of MTs (Sato and Blandau 1979). The sperm head is then engulfed by egg plasma membrane extensions. The role of actin filaments has been proposed in incorporating sperm nucleus and other organelles into egg. Incorporation of sperm chromatin into egg is important for normal development of embryo; sperm centrosome facilitates pronuclear migration (Schatten 1994). Sperm plasma membrane as well as tail disperses in the egg membrane and mitochondria degenerates (Longo 1989; Sutovsky et al. 2000).

Table 12.4 Sperm-egg fusion proteins

Sperm ligands	References
Fertilin α and β: member of ADAM (alpha disintegrin and metalloprotease)	Primakoff et al. (1987), Talbot et al. (2003)
Testis-specific protein – IZUMO	Rubinstein et al. (2006)
Human epididymal protein ARP CRISP protein (cystein rich secretory protein)	Ellerman et al. (2002), Rubinstein et al. (2006)
Egg ligands	
Integrin group of proteins – α5β1, α6β1, αvβ3	Evans et al. (1995)
Tetraspanin protein CD9	Talbot et al. (2003), Lefevre et al. (2010)
Cell adhesion molecular protein	Sutovsky (2009)
Glycosylphosphatidylinositol (GPI)-anchored protein	Lefevre et al. (2010)

12.5.3 Egg Activation

Another function of the sperm is egg activation following gamete fusion. Sperm mediates egg activation by increasing intracellular Ca^{2+} through three proposed mechanisms: sperm may introduce Ca^{2+} directly into the egg following fusion. Ca^{2+} entry activates the egg (Parrington et al. 2007); sperm binding and fusion stimulates egg PLC-mediated signaling (Parrington et al. 2007); sperm may release SOAFs into egg during fusion which may cause IP_3-mediated Ca^{2+} release and oscillations (Swann 1990). One of the factors proposed is postacrosomal sheath WW domain-binding protein (PAWP; Aarabi et al. 2010). Another factor is sperm-derived PLC that is also responsible for triggering Ca^{2+} release and oscillations in the egg (Saunders et al. 2002). Apart from these factors, egg maturation is a prerequisite for egg activation (Cheung et al. 2000).

The process of egg activation involves activation of RNAs and proteins for transcription. Egg activation is important for embryogenesis (Horner and Wolfner 2008). Egg activation involves egg membrane physical and chemical changes to inhibit polyspermy after sperm-ZP binding and to protect embryo. Within first hour postfertilization, there is an egg cortical granule reaction and paternal chromatin decondensation. Following this, there is completion of female meiosis (Whitaker 2006).

Egg undergoes *cortical reaction* after sperm binding, in which there is release of cortical granules (Horner and Wolfner 2008). Cortical granules present beneath the egg plasma membrane are released following fertilization and diffuse in perivitelline space. The granules block sperm receptors and modify egg vestments (Hoodbhoy and Talbot 1994).

Before fertilization, oocyte remains in meiotic arrest. Meiotic arrest is maintained by M-phase promoting factor (MPF). MPF consists of cyclin B and cdc2 kinase subunits. After fertilization, there is increase in oocyte Ca^{2+} influx that stimulates CaM which further stimulates CaMK II. This enzyme inhibits cyclin subunit. As a result, there is loss of MPF and release of meiotic arrest. Cdc2 contributes in releasing meiotic arrest by phosphorylating certain substrates that leads to breakdown of nuclear envelop, spindle formation, and secondary polar body generation. Sperm nucleus undergoes remodeling by DNA replication followed by removal of sperm nuclear envelope and replacement of sperm-specific histones or protamines by maternal variants (Horner and Wolfner 2008). There is completion of meiosis and formation of male and female pronuclei followed by zygote formation.

Key Questions

- Describe the different parts of perinuclear theca with their protein contents.
- Describe the structure of axonemal motility complex.
- Discuss the criteria of classifying sperm morphology.
- How do epididymal proteins regulate sperm maturation?
- Discuss the role of HCO_3^- in capacitation.
- How does zinc regulate sperm motility?
- What are the sperm-egg fusion proteins?
- Describe the molecular events following gamete fusion.

References

Aarabi M, Qin Z, Xu W, Mewburn J, Oko R. Sperm borne protein PAWP, initiates zygotic development in *Xenopus laevis* by eliciting intracellular calcium release. Mol Reprod Dev. 2010;77:249–56.

Abou-haila A, Tulsiani DR. Signal transduction pathways that regulate sperm capacitation and th acrosome reaction. Arch Biochem Biophys. 2009;485:72–81.

Abraham-Peskir JV, Chantler E, Uggerhoj E, Fedder J. Response of midpiece vesicles on human sperm to osmotic stress. Hum Reprod. 2002;17:375–82.

Aderem A. Signal transduction and actin cytoskeleton: the roles of MARCKS and profiling. Trends Biochem Sci. 1992;17:438–43.

Adham IM, Nayernia K, Engel W. Spermatozoa lacking acrosin protein show delayed fertilization. Mol Reprod Dev. 1997;46:370–6.

Anakwe OO, Gerton GL. Acrosome biogenesis during meiosis: evidence from the synthesis and distribution of an acrosomal glycoprotein, acrogranin, during guinea pig spermatogenesis. Biol Reprod. 1990;42:317–28.

Ånberg Å. The ultrastructure of the spermatozoon. An electronmicroscopic study of the spermatozoa from sperm samples and the epididymis including some observations of the spermatid. Acta Obstet Gynecol Scand. 1957;36(2):1–133.

Arienti G, Saccardi C, Carlini E, Verdacchi R, Palmerini CA. Distribution of lipid and protein in human semen fractions. Clin Chim Acta. 1999;289:111–20.

Arnoult C, Kazam IG, Visconti PE, Kopf GS, Villaz M, Florman HM. Control of the low voltage-activated calcium channel of mouse sperm by egg ZP3 and by membrane hyperpolarization during capacitation. Proc Natl Acad Sci U S A. 1999;96:6757–62.

Baccetti B. The human spermatozoon. In: Van Blerkom J, Motta PM, editors. Ultrastructure of reproduction. The Hague: Martinus Nijhoff; 1984. p. 110–26.

Baccetti B, Pallini V, Burrini AG. The accessory fibers of the sperm tail. II: their role in binding zinc in mammals and cephalopods. J Ultrastruct Res. 1976;54:261–75.

Baltz JM, Williams PO, Cone RA. Dense fibers protect mammalian sperm against damage. Biol Reprod. 1990;43:485–91.

Bartoov B, Berkovitz A, Eltes F. Selection of spermatozoa with normal nuclei to improve the pregnancy rate with intracytoplasmic sperm injection. N Engl J Med. 2001;345:1067–8.

Bartoov B, Berkovitz A, Eltes F, Kogosowski A, Menezo Y, Barak Y. Real-time fine morphology of motile human sperm cells is associated with IVF-ICSI outcome. J Androl. 2002;23:1–8.

Bedford JM. Maturation, transport and fate of spermatozoa in the epididymis. In: Greep RO, editor. Endocrinology, male reproductive system, Handbook of Physiology, Section 7, Vol. 5. Washington, DC: American Physiological Society; 1975. p. 303–17.

Blackmore PF, Neulen J, Lattanzio F, Beebe SJ. Cell surface-binding sites for progesterone mediate calcium uptake in human sperm. J Bio Chem. 1991;266:18655–9.

Bleil JD, Wassarman PM. Structure and function of the zona pellucida: identification and characterization of the proteins of the mouse oocyte's zona pellucida. Dev Biol. 1980;76:185–202.

Bobrie A, Colombo M, Raposo G, Thery C. Exosome secretion: molecular mechanisms and roles in immune responses. Traffic. 2011;12:1659–68.

Boitrelle F, Ferfouri F, Petit JM, Selva J. Large human sperm vacuoles observed in motile spermatozoa under high magnification: nuclear thumbprints linked to failure of chromatin condensation. Hum Reprod. 2011;26(7):1650–8.

Boitrelle F, Albert M, Petit JM, Ferfouri F, Wainer R, Bergere M, et al. Small human sperm vacuoles observed under high magnification are pocket-like nuclear concavities linked to chromatin condensation failure. Reprod Biomed Online. 2013;27(2):201–11.

Bork P, Sander C. A large domain common to sperm receptors (Zp2 and Zp3) and TGF-b type III receptor. FEBS Lett. 1992;300:237–40.

Boué F, Berubé B, De Lamirande E, Gagnon C, Sullivan R. Human sperm-zona pellucid interaction in inhibited by an antiserum against a hamster sperm protein. Biol Reprod. 1994;51:577–87.

Bourne H, Liu DY, Clarke GN, Gordon Baker HW. Normal fertilization and embryo development by intracytoplasmic sperm injection of round-headed acrosomeless sperm. Fertil Steril. 1995;63:1329–32.

Bozkurt HH, Wooley DM. Morphology of nexin links in relation to interdoublet sliding in the sperm flagellum. Cell Motil Cytoskeleton. 1993;24:109–18.

Brokaw CJ. Flagellar movement: a sliding filament model. Science. 1972;178:455–62.

Brokaw CJ. Direct measurements of sliding between outer doublet microtubules in swimming sperm flagella. Science. 1989;243:1593–6.

Brown DA, London E. Structure and function of sphingolipid- and cholesterol-rich membrane rafts. J Biol Chem. 2000;275:17221–4.

Brown CR, Von GKI, Jones R. Changes in plasma membrane glycoproteins of rat spermatozoa during maturation in the epididymis. J Cell Biol. 1983;96:256–64.

Bunch DO, Welch JE, Magyar PL, Eddy EM, O'Brien DA. Glyceraldehyde 3-phosphate dehydrogenase-S protein distribution during mouse spermatogenesis. Biol Reprod. 1998;58:834–41.

Calvin HI. Comparative labelling of rat epididymal spermatozoa by intratesticularly administered $^{65}ZnCl_2$ and [^{35}S]cysteine. J Reprod Fert. 1981;61:65–73.

Calvin HI, Bedford JM. Formation of disulfide bonds in the nucleus and accessory structures of mammalian spermatozoa during maturation in the epididymis. J Reprod Fertil. 1971;13:65–75.

Calvin HI, Yu CC, Bedford JM. Effects of epididymal maturation, zinc (II) and copper (II) on the reactive sulfhydryl content of structural elements in rat spermatozoa. Exp Cell Res. 1973;81:333–41.

Camatini M, Colombo A, Bonfanti P. Cytoskeletal elements in mammalian spermiogenesis. Microsc Res Tech. 1992;20:232–50.

Carlson AE, Westenbroek RE, Quill T, Ren D, Clapham DE, Hille B, et al. CatSper1 required for evoked Ca^{2+}entry and control of flagellar function in sperm. Proc Natl Acad Sci U S A. 2003;100:14864–8.

Carrera A, Gerton GL, Moss SB. The major fibrous sheath polypeptide of mouse sperm: structural and functional similarities to the A-kinase anchoring proteins. Dev Biol. 1994;165:272–84.

Cassuto NG, Bouret D, Plouchart JM, Jellad S, Vanderzwalmen P, Balet R, et al. A new real-time morphology classification for human spermatozoa: a link for fertilization and improved embryo quality. Fertil Steril. 2009;92:1616–25.

Cheung A, Swann K, Carroll J. The ability to generate normal Ca2+ transients in response to spermatozoa develops during the final stages of oocyte growth and maturation. Hum Reprod. 2000;15:389–95.

Clulow J, Jones RC, Hansen LA. Micropuncture and cannulation studies of fluid composition and transport in the ductuli efferentes testis of the rat: comparisons with the homologous metanephric proximal tubule. Exp Physiol. 1994;79:915–28.

Coonard SA, Herr JC, Westhusin ME. Inhibition of bovine fertilization in vitro by antibodies to SP-10. J Reprod Fertil. 1996;107:287–97.

Cornwall GA, Vindivich D, Tillman S, Chang TS. The effect of sulfhydryl oxidation on the morphology of immature hamster epididymal spermatozoa induced to acquire motility in vitro. Biol Reprod. 1988;39:141–55.

Corselli J, Talbot P. In vitro penetration of hamster oocyte-cumulus complexes using physiological numbers of sperm. Dev Biol. 1987;122:227–42.

Cummins JM, Yanagimachi R. Development of ability to penetrate the cumulus oophorus by hamster spermatozoa capacitated in vitro, in relation to the timing of the acrosome reaction. Gamete Res. 1986;15:187–212.

de Lamirande E, Leclerc P, Gagnon C. Capacitation as a regulatory event that primes spermatozoa for the acrosome reaction and fertilization. Mol Human Reprod. 1997;3:175–94.

de Vries KJ, Wiedmer T, Sims PJ, Gadella BM. Caspase-independent exposure of aminophospholipids and tyrosine phosphorylation in bicarbonate responsive human sperm cells. Biol Reprod. 2003;68:2122–34.

Demarco IA, Espinosa F, Edwards J, Sosnik J, De La Vega-Beltran JL, Hockensmith JW, et al. Involvement of a Na+/HCO-3 cotransporter in mouse sperm capacitation. J Biol Chem. 2003;278:7001–9.

Didry D, Carlier MF, Pantaloni D. Synergy between actin depolymerizing factor/cofolin and profiling in increasing actin filament turnover. J Biol Chem. 1998;273:25602–11.

Doiguchi M, Mori T, Toshimori K, Shibata Y, Iida H. Spergen-1 might be an adhesive molecule associated with mitochondria in the middle piece of spermatozoa. Dev Biol. 2002;252:127–37.

Dzeja PP, Terzic A. Phosphotransfer networks and cellular energetics. J Exp Biol. 2003;206:2039–47.

Eddy EM, Toshimori K, O'Brien DA. Fibrous sheath of mammalian spermatozoa. Microsc Res Tech. 2003;61:103–15.

Edidin M. Membrane cholesterol, protein phosphorylation, and lipid rafts. Sci STKE. 2001;2001(67):pe1.

Eickhoff R, Baldauf C, Koyro HW, Wennemuth G, Suga Y, Seitz J, et al. Influence of macrophage migration inhibitory factor (MIF) on the zinc content and redox state of protein-bound sulphydryl groups in rat sperm: indications for a new role of MIF in sperm maturation. Mol Hum Reprod. 2004;10:605–11.

Eickhoff R, Jennemann G, Hoffbauer G, Schuring MP, Kaltner H, Sinowatz F, et al. Immunohistochemical detection of macrophage migration inhibitory factor in fetal and adult bovine epididymis: release by the apocrine secretion mode? Cells Tissues Organs. 2006;182:22–31.

Ellerman DA, Da RV, Cohen DJ, Busso D, Morgenfeld MM, Cuasnicu PS. Expressionand structure-function analysis of DE, a sperm cysteine-richsecretory protein that mediates gamete fusion. Biol Reprod. 2002;67:1225–31.

Evans JP, Schultz RM, Kopf GS. Identification and localization of integrin subunits in oocytesand eggs of the mouse. Mol Reprod Dev. 1995;40:211–20.

Fawcett DW. A comparative view of sperm ultrastructure. Biol Reprod. 1970;Suppl 2:90–127.

Fawcett D. The mammalian spermatozoon. Dev Biol. 1975;44:394–436.

Fekonja N, Strus J, Zniderič MT, Knez K, Bokal EV, Verdenik I, et al. Clinical and structural features of sperm head vacuoles in men included in the in vitro fertilization program. Biomed Res Int. 2014;2014:1–12.

Fetic S, Yeung CH, Sonntag B, Nieschlag E, Cooper TG. Relationship of cytoplasmic droplets to motility, migration in mucus, and volume regulation of human spermatozoa. J Androl. 2006;27:294–301.

Flesch FM, Brouwers JF, Nievelstein PF, Verkleij AJ, van Golde LM, Colenbrander B, et al. Bicarbonate stimulated phospholipid scrambling induces cholesterol redistribution and enables cholesterol depletion in the sperm plasma membrane. J Cell Sci. 2001;114:3543–55.

Florman HM, Jungnickel MK, Sutton KA. Regulating the acrosome reaction. Int J Dev Biol. 2008;52:503–10.

Foster JA, Klotz KL, Flickinger CL, Thomas TX, Wright RM, Castillo JR, et al. Human SP-10: acrosomal distribution, processing and fate after the acrosomal reaction. Biol Reprod. 1994;51:1222–31.

Foster JA, Friday BB, Maulit MT, Blobel C, Winfrey VP, Olson GE, et al. AM67, a secretory component of the guinea pig spermacrosomal matrix, is related to mouse sperm protein sp56and the complement component 4-binding proteins. J Biol Chem. 1997;272:12714–22.

Franco Jr JG, Baruffi RLR, Mauri AL, Petersen CG, Oliveira JBA, Vagnini L. Significance of large nuclear vacuoles in human spermatozoa: implications for ICSI. Reprod Biomed Online. 2008;17(1):42–5.

Frenette G, Lessard C, Sullivan R. Selected proteins of "prostasome like particles" from epididymal cauda fluid are transferred to epididymal caput spermatozoa in bull. Biol Reprod. 2002;67:308–13.

Fujita A, Nakamura K, Kato T, Watanabe N, Ishizaki T, Kimura K, et al. Ropporin, a sperm-specific binding protein of rophilin, that is localized in the fibrous sheath of sperm flagella. J Cell Sci. 2000;113:103–12.

Gadella BM, Harrison RA. The capacitating agent bicarbonate induces protein kinase A-dependent changes in phospholipid transbilayer behavior in the sperm plasma membrane. Development. 2000;127:2407–20.

Garbers DL, Tubb DJ, Hyne RV. A requirement of bicarbonate for Ca2 + –induced elevations of cyclic AMP in guinea pig spermatozoa. J Biol Chem. 1982;257:8980–4.

Garty NB, Galiani D, Aharonheim A, Ho YK, Phillips DM, Dekel N, et al. G-proteins in mammalian gametes: an immunocytochemical study. J Cell Sci. 1988;91:21–31.

Gatewood JM, Cook GR, Balhorn R, Schmid CW, Bradbury EM. Isolation of four core histones from human sperm chromatin representing a minor subset of somatic histones. J Biol Chem. 1990;265:20662–6.

Gatti JL, Billard R, Christen R. Ionic regulation of the plasma membrane potential of rainbow trout (Salmo gairdneri) spermatozoa: role in the initiation of sperm motility. J Cell Physiol. 1990;143:546–54.

Gerace L, Comeau C, Benson M. Organization and modulation of nuclear lamina structure. J Cell Sci. 1984;1:137–60.

Gibbons IR, Fronk E. Some properties of bound and soluble dynein from sea urchin flagella. J Cell Biol. 1972;54:365–81.

Griffiths GS, Galileo DS, Reese K, Martin-Deleon PA. Investigating the role of murine epididymosomes and uterosomes in GPI-linked protein transfer to sperm using SPAM1 as a model. Mol Reprod Dev. 2008;75:1627–36.

Grimes Jr SR. Nuclear proteins in spermatogenesis. Comp Biochem Physiol. 1986;83B:495–500.

Gwatkin RBL. Fertilization mechanisms in man and mammals. New York: Plenum; 1977.

Hanoune J, Defer N. Regulation and role of adenylyl cyclase isoforms. Annu Rev Pharmacol Toxicol. 2001;41:145–74.

Haraguchi CM, Mabuchi T, Hirata S, Shoda T, Tokumoto T, Hoshi K, et al. Possible function of caudal nuclear pocket: degradation of nucleoproteins by ubiquitin-proteasome system in rat-spermatids and human sperm. J Histochem Cytochem. 2007;55:585–95.

Harrison RA. Rapid PKA-catalysed phosphorylation of boar sperm proteins induced by the capacitating agent bicarbonate. Mol Reprod Dev. 2004;67:337–52.

Harrison RA, Miller NG. cAMP-dependent protein kinase control of plasma membrane lipid architecture in boar sperm. Mol Reprod Dev. 2000;55:220–8.

Hecht NB. Mammalian protamines and their expression. In: Hnilica L, Stein G, Stein J, editors. Histones and other basic nuclear proteins. Boca Raton: CRC Press; 1989. p. 347–73.

Henkel R, Bittner J, Weber R, Hther F, Miska W. Relevance of zinc in human sperm flagella and its relation to motility. Fertil Steril. 1999;6:1138–43.

Henkel R, Baldauf C, Bittner J, Weidner W, Miska W. Elimination of zinc from the flagella of spermatozoa during epididymal transit is important for motility. Reprod Technol. 2001;10:280–5.

Hess H, Heid H, Franke WW. Molecular characterization of mammalian cylicin, a basic protein of the sperm head cytoskeleton. J Cell Biol. 1993;122:1043–52.

Hinton BT, Pryor JP, Hirsch AV, Setchell BP. The concentration of some inorganic ions and organic compounds in the luminal fluid of the human ductus deferens. Int J Androl. 1981;4:457–61.

Ho HC, Suarez SS. Hyperactivation of mammalian spermatozoa: function and regulation. Reproduction. 2001a;122:519–26.

Ho HC, Suarez SS. An inositol 1,4,5-trisphosphate receptor-gated intracellular Ca2+ store is involved in regulating sperm hyperactivated motility. Biol Reprod. 2001b;65:1606–15.

Ho HC, Suarez SS. Characterization of the intracellular calcium store at the base of the sperm flagellum that regulates hyperactivated motility. Biol Reprod. 2003;68:1590–6.

Holt WV, Harrison RAP. Bicarbonate stimulation of boar sperm motility via a protein kinase A-dependent pathway: between-cell and between-ejaculate differences are not due to deficiencies in protein kinase A activation. J Androl. 2002;23:557–65.

Hoodbhoy T, Talbot P. Mammalian corticalgranules: contents, fate, and function. Mol Reprod Dev. 1994;39:439–48.

Horner VL, Wolfner MF. Transitioning from egg to embryo: triggers and mechanisms of egg activation. Dev Dyn. 2008;237:527–44.

Hoshi K, Aita T, Yanagida K, Yoshimatsu N, Sato A. Variation in the cholesterol/ phospholipid ratio in human spermatozoa and its relationship with capacitation. Hum Reprod. 1990;5:71–4.

Huitorel P, White D, Fouquet JP, Kann ML, Cosson J, Gagnon C. Differential distribution of glutamylated tubulin isoforms along the sea urchin sperm axoneme. Mol Reprod Dev. 2002;62:139–48.

Huszar G, Vigue L, Corrales M. Sperm creatine phosphokinase activity as a measure of sperm quality in normospermic, variablespermic, and oligospermic men. Biol Reprod. 1988;38:1061–6.

Huszar G, Vigue L, Morshedi M. Sperm creatine phosphokinase Misoform ratios and fertilizing potential of men: a blinded study of 84 couples treated with in vitro fertilization. Fertil Steril. 1992;57:882–8.

Ikawa M, Inoune N, Okabe M. Mechanisms of sperm-egg interactions emerging from gene manipulated animals. Int J Dev Biol. 2008;52:657–64.

Iyengar GV. Reference values for the concentration of As, Cd, Co, Cr, Cu, Fe, I, Hg, Mn, Mo, Ni, Pb, Se and Zn in selected human tissues and body fluids. Biol Trace Elem Res. 1987;12:263–95.

Jaiswal BS, Tur-Kaspa I, Dor J, Mashiach S, Eisenbach M. Human sperm chemotaxis: is progesterone a chemoattractant? Biol Reprod. 1999;60:1314–9.

Janick J, Zeitz L, Whitmore WF. Seminal fluid and spermatozoon zinc levels and their relationship to human spermatozoon motility. Fertil Steril. 1971;22:573–80.

Jones R. Identification and functions of mammalian sperm–egg recognition molecules during fertilization. J Reprod Fertil Suppl. 1990;42:89–105.

Joshi CS, Suryawanshi AR, Khan SA, Balasinor NH, Khole VV. Liprin a3: a putative estrogen regulated acrosomal protein. Histochem Cell Biol. 2012;139:535–48.

Kaldis P, Stolz M, Wyss M, Zanolla E, Rothen-Rutishauser B, Vorherr T, et al. Identification of two distinctly localized mitochondrial creatine kinase isoenzymes in spermatozoa. J Cell Sci. 1996;109:2079–88.

Kerr JB. Ultrastructure of the seminiferous epithelium and intertubular tissue of the human testis. J Electron Microsc Tech. 1991;19:215–40.

Kim KS, Gerton GL. Differential release of soluble and matrix components: evidence for intermediate states of secretion during spontaneous acrosomal exocytosis in mouse sperm. Dev Biol. 2003;264:141–52.

Kim KS, Cha MC, Gerton GL. Mouse sperm protein sp56 is a component of the acrosomal matrix. Biol Reprod. 2001;64:36–43.

Kirchhoff C. CD52 is the 'major maturation-associated' sperm membrane antigen. Mol Hum Reprod. 1996;2:9–17.

Kirchhoff C, Osterhoff C, Young L. Molecular cloning and characterization of HE1, a major secretory protein of the human epididymis. Biol Reprod. 1996;54:847–56.

Kol MA, de Kruijff B, de Kroon AI. Phospholipid flip-flop in biogenic membranes: what is needed to connect opposite sides. Semin Cell Dev Biol. 2002;13:163–70.

Krapf D, Ruan YC, Wertheimer EV, Battistone MA, Pawlak JB, Sanjay A, et al. Src is necessary for epididymal development and is incorporated into sperm during epididymal transit. Dev Biol. 2012;369:43–53.

Krohne G, Benavente R. The nuclear lamins. A multigene family of proteins in evolution and differentiation. Exp Cell Res. 1986;162:1–10.

Kruger TF, Acosta AA, Simmons KF, Swanson RJ, Matta JF, Veeck LL, et al. New method of evaluating sperm morphology with predictive value for human in vitro fertilization. Urology. 1987;30:248–51.

Kusumi A, Koyama-Honda I, Suzuki K. Molecular dynamics and interactions for creation of stimulation-induced stabilized rafts from small unstable steady-state rafts. Traffic. 2004;5:213–20.

Leclerc P, De Lamirande E, Gagnon C. Cyclic adenosine 3′,5′monophosphate-dependent regulation of protein tyrosine phosphorylation in relation to human sperm capacitation and motility. Biol Reprod. 1996;55:684–92.

Lecuyer C, Dacheux JL, Hermand E, Mazeman E, Rousseaux J, Rousseaux-Prevost R. Actin binding properties and colocalization with actin during spermiogenesis of mammalian sperm calicin. Biol Reprod. 2000;63:1801–10.

Lefevre B, Wolf JP, Ziyyat A. Sperm egg interaction: is there a link between tetraspanin (s) and GPI- anchored protein (s)? Bioessays. 2010;32:143–52.

Lefievre L, Conner SJ, Salpekar A, Olufowobi O, Ashton P, Pavlovic B, et al. Four zona pellucida glycoproteins are expressed in the human. Hum Reprod. 2004;19:1580–6.

Légaré C, Gaudreault C, St-Jacques S, Sullivan R. P34H sperm protein is preferentially expressed by the human corpus epididymidis. Endocrinology. 1999;140:3318–27.

Lin Y, Mahan K, Lathrop WF, Myles DG, Primakoff P. A hyaluronidase activity of the sperm plasma membrane protein PH-20 enables sperm to penetrate the cumulus cell layer surrounding the egg. J Cell Biol. 1994;125:1157–63.

Lindemann CB, Kanous KS. "Geometric clutch" hypothesis of axonemal function: key functions and testable predictions. Cell Motil Cytoskelet. 1995;31:1–8.

Longo FJ. Incorporation and dispersal of spermsurface antigens in plasma membranes of inseminated seaurchin (*Arbacia punctulata*) eggs and oocytes. Dev Biol. 1989;131:37–43.

Manning G, Plowman GD, Hunter T, Sudarsanam S. Evolution of protein kinase signaling from yeast to man. Trend Biochem Sci. 2002;27:514–20.

McClean D, Rowlands IW. The role of hyaluronidase in fertilization. Nature. 1942;150:627–8.

Meizel S, Turner KO. Progesterone acts at the plasma membrane of human sperm. Mol Cell Endocrinol. 1991;11:R1–5.

Menkveld R, Wong WY, Lombard CJ, Wetzels AMM, Thomas CMG, Merkus HMWM, et al. Semen parameters, including WHO and strict criteria morphology, in a fertile and subfertile population: an effort towards standardization of in-vivo thresholds. Hum Reprod. 2001;16:1165–71.

Meyer SA, Rosenberger AE. A plasma membrane associated hyaluronidase is localized to the posterior region of stallion sperm and is associated with sperm function. Biol Reprod. 1999;61:444–51.

Miki K, Qu W, Goulding EH, Willis WD, Bunch DO, Strader LF, et al. Glyceraldehyde 3-Phosphate dehydrogenase-S, a sperm-specific glycolytic enzyme, is required for sperm motility and male fertility. Proc Natl Acad Sci U S A. 2004;101(47):16501–6.

Moos J, Peknicova J, Geussova G, Philimonenko V, Hozac P. Association of protein kinase A type I with detergent-resistant structures of mammalian sperm cells. Mol Reprod Dev. 1998;50:79–85.

Mortillo S, Wassarman PM. Differential bindingof gold-labeled zona pellucida glycoproteins mZP2 and mZP3 to mouse sperm membrane compartments. Development. 1991;113:141–9.

Mortimer D, Menkveld R. Sperm morphology assessment—historical perspectives and current opinions. J Androl. 2001;22:192–205.

Mukai C, Okuno M. Glycolysis plays a major role for adenosine triphosphate supplementation in mouse sperm flagellar movement. Biol Reprod. 2004;71:540–7.

Myles DG, Primakoff P. Why did the sperm cross the cumulus—to get to the oocyte: functions of the sperm surface proteins PH-20 and fertilin in arriving at and fusing with the egg. Biol Reprod. 1997;56:320–7.

Nikolopoulou M, Soucek M, Vary JC. Changes in the lipid content of boar sperm plasma membranes during epididymal maturation. Biochem Biophys Acta. 1985;815:486–98.

Nolan JP, Magargee SF, Posner RG, Hammerstedt RH. Flow cytometric analysis of transmembrane phospholipid movement in bull sperm. Biochemistry. 1995;34:3907–15.

Oh J, Woo JM, Choi E, Kim T, Cho BN, Park ZY, et al. Molecular, biochemical, and cellular characterization of epididymal ADAMs, ADAM7 and ADAM28. Biochem Biophys Res Commun. 2005;331:1374–83.

Oh JS, Han C, Cho C. ADAM7 is associated with epididymosomes and integrated into sperm plasma membrane. Mol Cells. 2009;28:441–6.

Okamura N, Tajima Y, Soejima A, Masuda H, Sugita Y. Sodium bicarbonate in seminal plasma stimulates the motility of mammalian spermatozoa through direct activation of adenylate cyclase. J Biol Chem. 1985;260:9699–705.

Oko R, Clermont Y. Mammalian spermatozoa: structure and assembly of the tail. In: Gagnon C, editor. Controls of sperm motility: biological and clinical aspects. Boca Raton: CRC Press; 1990. p. 3–28.

Okunade GW, Miller ML, Pyne GJ, Sutliff RL, O'Connor KT, Neumann JC, et al. Targeted ablation of plasma membrane Ca^{2+}-ATPase (PMCA) 1 and 4 indicates a major housekeeping function of PMCA1 and a critical role in hyperactivated sperm motility and male fertility for PMCA4. J Biol Chem. 2004;279:33742–50.

Olson GE, Winfrey VP. Mitochondria– cytoskeletal interactions in the sperm midpiece. J Struct Biol. 1990;103:13–22.

Olson GE, Winfrey VP, Bi M, Hardy DM, Nagdas SK. Zonadhesin assembly into the hamster-sperm acrosomal matrix occurs by distinct targeting strategies during spermiogenesis and maturation in the epididymis. Biol Reprod. 2004;71:1128–34.

Parks JE, Hammerstedt RH. Developmental changes occurring in the lipids of ram epididymal spermatozoa plasma membrane. Biol Reprod. 1985;32:653–68.

Parrington J, Davis LC, Galione A, Wessel G. Flipping the switch: how a sperm activates the egg at fertilization. Dev Dyn. 2007;236:2027–38.

Patel-King RS, Benashski SE, King SM. A bipartite Ca2 + -regulated nucleosidediphosphate kinase system within the Chlamydomonas flagellum. The regulatory subunit p72. J Biol Chem. 2002;277:34271–9.

Pedersen H. Further observations on the fine structure of the human spermatozoon. Z Zellforsch. 1972;123:305–15.

Perdrix A, Travers A, Chelli MH, Escalier D, Do Rego JL, Milazzo JP, et al. Assessment of acrosome and nuclear abnormalities in human spermatozoa with large vacuoles. Hum Reprod. 2011;26(1):47–58.

Perez-Ramirez B, Castells M. In vitro biosynthesis of rat sperm outer dense fiber components. Life Sci. 1991;49:1549–54.

Pomorski T, Herrmann A, Zimmermann B, Zachowski A, Muller P. An improved assay for measuring the transverse redistribution of fluorescent phospholipids in plasma membranes. Chem Phys Lipids. 1995;77:139–46.

Pons-Rejraji H, Artonne C, Sion B, Brugnon F, Canis M, Janny L, et al. Prostasomes: inhibitors of capacitation and modulators of cellular signalling in human sperm. Int J Androl. 2011;34:568–80.

Porter ME, Sale WS. The 9 + 2 axoneme anchors multiple inner arm dyneins and a network of kinases and phosphatases that control motility. J Cell Biol. 2000;151:F37–42.

Primakoff P, Hyatt H, Tredick-Kline J. Identification and purification of a sperm surface proteinwith a potential role in sperm–egg membrane fusion. J Cell Biol. 1987;104:141–9.

Quill TA, Sugden SA, Rossi KL, Doolittle LK, Hammer RE, Garbers DL. Hyperactivated sperm motility driven by CatSper2 is required for fertilization. Proc Natl Acad Sci U S A. 2003;100:14869–74.

Rejraji H, Sion B, Prensier G, Carreras M, Motta C, Frenoux JM, et al. Lipid remodeling of murine epididymosomes and spermatozoa during epididymal maturation. Biol Reprod. 2006;74:1104–13.

Robert M, Gagnon C. Semenogelin I: a coagulation forming multifunctional seminal vesicle protein. Cell Mol Life Sci. 1999;55:944–60.

Rodgers RJ, Irving-Rodgers HF, Russell DL. Extracellular matrix of the developing ovarian follicle 4. Reproduction. 2003;126:415–24.

Roldan ERS, Murase T, Shi QX. Exocytosis in spermatozoa in response to progesterone and zona pellucida. Science. 1994;266:1578–81.

Rubinstein E, Ziyyat A, Wolf JP, Le Naour F, Boucheix C. The molecular players of sperm egg fusion in mammals. Semin Cell Dev Biol. 2006;17:254–63.

Sakata Y, Saegusa H, Zong S, Osanai M, Murakoshi T, Shimizu Y, et al. Ca(v)2.3 (alpha 1E) Ca^{2+}channel participates in the control of sperm function. FEBS Lett. 2002;516:229–33.

Sathananthan AH, Tatham B, Dharmawardena V, Grills B, Lewis I, Trounson A. Inheritance of sperm centrioles and centrosomes in bovine embryos. Arch Androl. 1997;38:37–48.

Sato K, Blandau RJ. Time and process ofsperm penetration into cumulus-free mouse eggs fertilizedin vitro. Gamete Res. 1979;1:39–46.

Saunders CM, Larman MG, Parrington J, Cox LJ, Royse J, Blayney LM, et al. PLC zeta: a sperm –specific trigger of Ca(2+) oscillations in egg and embyo development. Development. 2002;129:3533–44.

Schatten G. The centrosome and its mode ofinheritance: the reduction of the centrosome during gametogenesisand its restoration during fertilization. Dev Biol. 1994;165:299–335.

Serres C, Feneux D, Jouannet P. Abnormal distribution of the periaxonemal structures in a human sperm flagellar dyskinesia. Cell Motil Cytoskel. 1984a;6:8–76.

Serres C, Feneux D, Jouannet P, David G. Influence of the flagellar wave development and propagationon the human sperm movement in seminal plasma. Gamete Res. 1984b;9:183–95.

Setchell BP, Maddocks S, Brooks DE. Anatomy, vasculature, innervation, and fluids of the male reproductive tract. In: Knobil E, Neill JD, editors. The physiology of reproduction, vol. 1. New York: Raven Press; 1994. p. 1063–175.

Shi QX, Roldan ER. Bicarbonate/CO2 is not required for zona pellucida- or progesterone-induced acrosomal exocytosis of mouse spermatozoa but is essential for capacitation. Biol Reprod. 1995;52:540–6.

Shur BD, Rodenhefer C, Ensslin MA, Lyng R, Raymond A. Identification of novel gamete receptors that mediate sperm adhesion to the egg coat. Mol Cell Endocrinol. 2006;250:137–48.

Si Y, Okuno M. Role of tyrosine phosphorylation of flagellar proteins in hamster sperm hyperactivation. Biol Reprod. 1999;61:240–6.

Simons K, Vaz WL. Model systems, lipid rafts, and cell membranes. Annu Rev Biophys Biomol Struct. 2004;33:269–95.

Smith E, Yang P. The radial spokes and central apparatus: mechano-chemical transducers that regulate flagellar motility. Cell Motil Cytoskeleton. 2004;57:8–17.

Stauss CR, Votta TJ, Suarez SS. Sperm motility hyperactivation facilitates penetration of the hamster zona pellucida. Biol Reprod. 1995;53:1280–5.

Sudhof TC, Rizo J. Synaptotagmins: C2-domainproteins that regulate membrane traffic. Neuron. 1996;17:379–88.

Sun F, Bahat A, Gakamsky A, Girsh E, Katz N, Giojalas LC, et al. Human sperm chemotaxis: both the oocyte and its surrounding cumulus cells secrete sperm chemoattractants. Hum Reprod. 2005;20:761–7.

Sutovsky P. Sperm-egg adhesion and fusion in mammals. Expert Rev Mol Med. 2009;11:e11.

Sutovsky P, Moreno RD, Ramalho-Santos J, Dominko T, Simerly C, Schatten G. Ubiquitinated spermmitochondria, selective proteolysis, and the regulation of mitochondrial inheritance in mammalian embryos. Biol Reprod. 2000;63:582–90.

Sutovsky P, Manandhar G, Wu A, Oko R. Interactions of the sperm perinuclear theca with the oocyte: implications for oocyte activation, anti-polyspermy defense and assisted reproduction. Microsc Res Tech. 2003;61:362–78.

Sutovsky P, van Leyen K, McCauley T, Day BN, Sutovsky M. Degradation of the paternal mitochondria after fertilization: implications for heteroplasmy, ART and mtDNA inheritance. Reprod Biomed Online. 2004;8:24–33.

Swann K. Acytosolic sperm factor stimulates repetitive calcium increases and mimics fertilization in hamster egg. Development. 1990;110:1295–302.

Takada S, Wilkerson CG, Wakabayashi K, Kamiya R, Witman GB. The outer dynein arm-docking complex: characterization of a subunit (Oda1) necessary for outer arm assembly. Mol Biol Cell. 2002;13:1015–29.

Talbot P, Shur BD, Myles DG. Cell adhesion and fertilization: steps in oocyte transport, sperm zona pellucida interactions and sperm egg fusion. Biol Reprod. 2003;68:1–9.

Tanaka H, Iguchi N, Egydio de Carvalhho C, Tadokoro Y, Yomogida K, Nishimune Y. Novel actin-like proteins T-ACTIN 1 and T-ACTIN 2 are differentially expressed in the cytoplasm and nucleus of mouse haploid germ cells. Biol Reprod. 2003;69:475–82.

Tanii I, Oh-oko T, Yshingag K, Toshimori K. A mouse acrosomal cortical matrix protein, MC41, has ZP2-binding activity and forms a complex with a 75-kDa serine protease. Dev Biol. 2001;238:332–41.

Tash JS. Protein phosphorylation: the second messenger signal transducer of flagellar motility. Cell Motil Cytoskeleton. 1989;14:332–9.

Tesarik J, Drahorad J, Testart J, Mendoza C. Acrosin activation follows its surface exposure and precedes membrane fusion in human sperm acrosome reaction. Development. 1990;110:391–400.

Thimon V, Frenette G, Saez F, Thabet M, Sullivan R. Protein composition of human epididymosomes collected during surgical vasectomy reversal: a proteomic and genomic approach. Hum Reprod. 2008;23:1698–707.

Tombes RM, Shapiro BM. Metabolite channeling: a phosphorylcreatine shuttle to mediate high energy phosphate transport between sperm mitochondrion and tail. Cell. 1985;41:325–34.

Toshimori K, Higashi R, Oura C. Distribution of intramembranous particles and filipin-sterol complexes in mouse sperm membranes: polyene antibiotic filipin treatment. Am J Anat. 1985;174:455–70.

Toshimori K, Tanii I, Araki S, Oura C. Characterization of the antigen recognized by a monoclonal antibody MN9: unique transport pathway to the equatorial segment of sperm head during spermiogenesis. Cell Tissue Res. 1992;270:459–68.

Turner RM. Tales from the tail: what do we really know about sperm motility? J Androl. 2003;24:790–803.

Turner TT, Giles RD. A sperm motility inhibiting factor in the rat epididymis. Am J Physiol. 1982;242:R199–203.

Ungar D, Hughson FM. SNARE protein structureand function. Annu Rev Cell Dev Biol. 2003;19:493–517.

Urner F, Sakkas D. Protein phosphorylation in mammalian sperm. Reproduction. 2003;125:17–26.

Visconti PE, Muschietti JP, Flawia MM, Tezon JG. Bicarbonate dependence of cAMP accumulation induced by phorbol esters in hamster spermatozoa. Biochim Biophys Acta. 1990;1054:231–6.

Visconti PE, Moore GD, Bailey JL, Leclerc P, Connors SA, Pan D, et al. Capacitation of mouse spermatozoa. II Protein tyrosine phosphorylation and capacitation are regulated by a cAMP-dependent pathway. Development. 1995;121:1139–50.

Visconti PE, Olds-Clarke P, Moss SB, Kalab P, Travis AJ, de las Heras M, et al. Properties and localization of a tyrosine phosphorylated form of hexokinase in mouse sperm. Mol Reprod Dev. 1996;43:82–93.

Visconti PE, Ning X, Fornes MW, Alvarez JG, Stein P, Connors SA, et al. Cholesterol efflux-mediated signal transduction in mammalian sperm: cholesterol release signals an increase in protein tyrosine phosphorylation during mouse sperm capacitation. Dev Biol. 1999;214:429–43.

Wassarman PM. Zona pellucida glycoproteins. Annu Rev Biochem. 1988;57:415–42.

Wassarman PM, Jovine L, Litscher ES. A profile of fertilization in mammals. Nat Cell Biol. 2001;3:E59–64.

Watanabe S, Tanaka A, Fujii S, Mizunuma H, Fukui A, Fukuhara R, et al. An investigation of the potential effect of vacuoles in human sperm on DNA damage using a chromosome assay and the TUNEL assay. Hum Reprod. 2011;26(5):978–86.

Wennemuth G, Carlson AE, Harper AJ, Babcock DF. Bicarbonate actions on flagellar and Ca2+−channel responses: initial events in sperm activation. Development. 2003;130:1317–26.

Whitaker M. Calcium at fertilization and in early development. Physiol Rev. 2006;86:25–88.

Wolf DE, Cardullo RA. Physiological properties of the mammalian sperm plasma membrane. In: Baccetti B, editor. Comparative spermatology 20 years after. New York: Raven Press; 1991. p. 599–604.

Wolkowicz JJ, Naaby-Hansen S, Gamble AR, Reddi PP, Flickinger CJ, Herr JC. Tektin B1 demonstrates flagellar localization in human sperm. Biol Reprod. 2002;66:241–50.

Woolley DM. Evidence for twisted plane undulation in golden hamster sperm tails. J Cell Biol. 1977;75:851–65.

Woolley DM, Osborn IW. Three-dimensional geometry of motile hamster spermatozoa. J Cell Sci. 1984;67:159–70.

World Health Organization. WHO laboratory manual for the examination and processing of human semen. 5th ed. Geneva: World Health Organization; 2010.

Yamashita Y, Shimada M, Okazaki T, Maeda T, Terada T. Production of progesterone from de novosynthesized cholesterol in cumulus cells and its physiological role during meiotic resumption of porcine oocytes. Biol Reprod. 2003;68:1193–8.

Yanagimachi R. The movement of golden hamster spermatozoa before and after capacitation. J Reprod Fertil. 1970;23:193–6.

Yanagimachi R. Zona-free hamster eggs: their use is assessing fertilization capacitity and examining chromosomes of human spermatozoa. Gamete Res. 1984;10:178–232.

Yang CH, Srivastava PN, Williams WL. Purification and properties of aryl sulfatases from rabbit sperm acrosome. Proc Soc Exp Biol Med. 1974;145:721–5.
Yeung CH, Cooper TG. Developmental changes in signalling transduction factors in maturing sperm duling epididymal transit. Cell Mol Biol. 2003;49:341–9.
Yeung CH, Woolley DM. Three-dimensional bend propagation in hamster sperm models and the direction of roll in free-swimming cells. Cell Motil. 1984;4:215–26.
Yoshinaga K, Toshimori K. Organization and modifications of sperm acrosomal molecules during spermatogenesis and epididymal maturation. Microsc Res Tech. 2003;61:39–45.
Yunes R, Michaut M, Tomes C, Mayorga LS. Rab3A triggers the acrosome reaction in permeabilized human spermatozoa. Biol Reprod. 2000;62:1084–9.
Zahler WL, Doak GA. Isolation of the outer acrosomal membrane from bull spermatozoa. Biochim Biophys Acta. 1975;406:479–88.
Zamboni L, Stefanini M. The fine structure of the neck of mammalian spermatozoa. Anat Rec. 1971;169:155–72.
Zaneveld LJD, Polakoski KL, Schumacher GFB. Properties of acrosomal hyaluronidase from bull spermatozoa. Evidence for its similarity to testicular hyaluronidase. J Biol Chem. 1973;248:564–70.
Zeng Y, Clark EN, Florman HM. Sperm membrane potential: hyperpolarization during capacitation regulates zona pellucida-dependent acrosomal secretion. Dev Biol. 1995;171:554–63.
Zeng Y, Oberdorf JA, Florman HM. pH regulation in mouse sperm: identification of Na(+)-, Cl(−)-, and HCO3(−)-dependent and arylaminobenzoate-dependent regulatory mechanisms and characterization of their roles in sperm capacitation. Dev Biol. 1996;173:510–20.

Sperm Function Tests

13

Manmohan Misro

Learning Objectives
- Routine semen analysis
- Rationale behind sperm function tests
- Laboratory tests of sperm function
- Modified protocols of specialized tests

13.1 Introduction

The penultimate goal of the male gamete is to release the paternal genetic material closest to the female counterpart leading to a successful completion of the physiological process called fertilization. Thus, from the stages of formation in the testis till its final union with the oocyte in the oviduct, the main function of the male gamete is much alike a vector protecting the genetic material by all means and delivering the same at the target site. This is being achieved by a variety of meaningful modifications in the sperm morphology or characteristics during different stages of its production in the testes and maturation in the epididymis, finally ending up with its fusion with the oocyte after completing its sojourn in the female reproductive tract. At each stage, the sperm has to undergo modifications, both topographically and biochemically. Any minor or major deviations in the process during such modifications render the sperm incompetent for fertilization or result in the production of a functionally defective sperm. In assisted reproduction procedure, such as

M. Misro, PhD
Reproductive Biomedicine, National Institute of Health and Family Welfare,
New Delhi, India
e-mail: mm_misro@yahoo.com

A. Kumar, M. Sharma (eds.), *Basics of Human Andrology*,
DOI 10.1007/978-981-10-3695-8_13

intracytoplasmic sperm injection (ICSI), many such necessary physiological steps are bypassed since the sperm is injected directly into the fully matured oocyte triggering nuclear fusion and fertilization. However, under normal physiological circumstances, all such stages with specific modifications are very much relevant to sperm function. Deficiencies in sperm characteristics are directly associated with sperm underperformance as sperm factor contributes a significant percentage (>70%) of all forms of male infertility.

Infertility in male is a complex problem affecting approximately 5% of eligible couples. Conventionally, male infertility is diagnosed following a simple evaluation of semen parameters like sperm count, motility and morphology. Additional tests on germ cell cytology and other semen biochemical tests mostly provide information on genital tract patency, accessory gland secretory efficiency or problems related to ejaculations or emissions. But, all such tests reveal very little about the functional competency of sperm, and their physiological significance changes with the passage of sperm from male to female reproductive tract. Starting from its precursor cell, the round spermatid, till its subsequent maturation into a full-fledged specialized sperm, there are subtle alterations not only in shape but also in chromatin chemistry which makes the later resistant to varied physiological fluctuations and exigencies. Such a process does not always ensure uniformity leading to the production of only "good" sperm in an ejaculate. As long as the proportion of defective creation remains small, it is mostly taken care of by sheer number of sperm as millions are being produced. However, subtle changes in sperm function in an ejaculate become apparent only when the defective population exceeds a particular threshold. The physiological significance of all such site-specific modifications and the appropriate test protocol for examining its associated function in the full-fledged mature sperm under in vitro conditions merit consideration since the same cannot be investigated and tested in vivo. However, sperm function testing has to be done in addition to routine analysis of sperm which cannot be dispensed with. Therefore, before we discuss the sperm function and associated testing protocols in detail, it is prudent to briefly analyse the laboratory methodology that is routinely followed under male infertility investigations.

13.2 Routine Analysis of Semen

It is an accepted fact that initial laboratory investigation concerning male fertility should start from the basic semen analysis, the method of which along with standard references has now been updated in the recent World Health Organization (WHO) manual. However, conventional semen analysis, as such, does not provide any details on sperm function. Instead, the results provide some basic but crucial information on the functional status of the germinal epithelium, epididymis and accessory sex glands which is directly linked to the physiological capacity and function of these organs. Besides the assessment of sperm concentration, the results of this analysis further yield full details on two important sperm characteristics, sperm motility and morphology, which have direct implications on fertilizing ability of sperm. Usually, two consecutive semen analyses 15 days apart from the same subject are carried out to categorize a semen sample as normal or subnormal (WHO 2010).

13.2.1 Assessment of Sperm Concentration

Concentration of sperm in the ejaculate is determined under a microscope using Neubauer's haemocytometer. Sperm concentration is determined on the basis of sperm counted per unit volume of semen utilized, multiplied by number of times it is diluted for examination. Total sperm count in the ejaculate is subsequently obtained after multiplying sperm concentration with semen volume (WHO 2010).

13.2.2 Assessment of Sperm Motility

There has been a little updating in the categorization of sperm motility in the new WHO manual. Earlier, forward progressive motility used to be classified under two heads, fast forward and slow forward. Individuals often found this method difficult to delineate these two categories without bias (Cooper and Yeung 2006). Now, sperm motility has been graded as progressive, nonprogressive or immotile as described below:

- Progressively motile (PR) – refers to those sperm which are moving fast, linearly or in large circles
- Nonprogressively motile (NP) – refers to those sperm which are static in spite of depicting active or docile flagellar movement
- Immotile (IM) – those sperm which are static and show no flagellar movement

Accordingly, PR and NP sperm are considered to calculate total sperm motility (WHO 2010).

13.2.3 Assessment of Sperm Morphology

Recent WHO guidelines utilize strict criteria given by Kruger for the assessment of sperm morphology. Kruger's criteria have been described in the chapter, "The Sperm". Using this criterion morphologically, normal sperm (3–5%) have been successfully used in in vitro fertilization, intrauterine insemination and spontaneous pregnancies (Coetzee et al. 1998; Van Waart et al. 2001; Van der Merwe et al. 2005). Accordingly, values lower than the above cut-off are classified as subnormal. The

Table 13.1 Reference values for semen parameters, as published in consecutive WHO manuals (Esteves et al. 2011)

Semen parameters	WHO, 1992	WHO, 1999	WHO, 2010[1]
Volume	≥2 ml	≥2 mL	1.5 mL
Sperm concentration/mL	≥20 × 10^6/mL	≥20 × 10^6/mL	15 × 10^6/mL
Total sperm count	≥40 × 10^6	≥40 × 10^6	39 × 10^6
Total motility (% motile sperm)	≥50%	≥50%	40%
Progressive motility[2]	≥25% (grade a)	≥25% (grade a)	32% (grade a + b)
Vitality (% alive)	≥75%	≥75%	58%
Morphology	≥30%[3]	14%[4]	4%[5]
Leukocyte count	<1.0 × 10^6/mL	<1.0 × 10^6/mL	<1.0 × 10^6/mL

WHO World Health Organization
[1]Lower reference limit obtained from the lower fifth centile value
[2]Grade a = rapid progressive motility (0.25 mm/s), grade b = slow/sluggish progressive motility (5–25 mm/s), normal = 50% motility (grades a + b) or 25% rapid progressive motility (grade a) within 60 min of ejaculation
[3]Arbitrary value
[4]No actual value given, but multicentric studies refer >14% (strict criteria) for in vitro fertilization (IVF)
[5]Normal shaped sperm according to Tygerberg (Kruger) strict criteria

cut-off is obtained from the data collected from volunteers of eight countries on the basis of their female partners achieving normal pregnancy in ≤12 months. The same data was then utilized to generate reference distributions of all other semen parameters. Table 13.1 provides the information of all the reference ranges in comparison to WHO guidelines of 1992, 1999 and 2010 (Esteves et al. 2011). Based on the lower reference limits given by WHO (2010), different nomenclatures has been used for the semen quality (see Table 13.2).

13.3 Rationale for Sperm Function Testing

Since basic infertility investigations in approximately 20% of couples yield normal semen parameters, etiologically they are classified as idiopathic. Other additional tests are, therefore, necessary to determine specific functional disorders associated with sperm abnormalities difficult to be detected in conventional semen analysis. Besides, sperm factor as a result of subnormal sperm function has been reported to be an indicator for couples with recurrent pregnancy loss (Gopalkrishnan et al. 2000; Saxena et al. 2008). The use of sperm from sperm banks further necessitates the use of such testing protocols for quality control before they are utilized in procedures associated with assisted reproduction for the treatment of infertility.

Sperm function test, ideally, should reveal the following crucial clinical information:

Table 13.2 Nomenclature used to indicate quality of semen parameters

Nomenclature	Total sperm count	Progressive motility (%)	Normal morphology (%)	Remarks
Normozoospermia	≥39 × 10^6 per ejaculate	≥32	≥4	
Oligozoospermia	<39 × 10^6 per ejaculate	–	–	
Asthenozoospermia	–	<32	–	
Teratozoospermia	–	–	<4	
Oligoasthenozoospermia	<39 × 10^6 per ejaculate	<32	–	
Oligoteratozoospermia	<39 × 10^6 per ejaculate	–	<4	
Asthenoteratozoospermia	–	<32	<4	
Oligoasthenoterato-zoospermia	<39 × 10^6 per ejaculate	<32	<4	
Azoospermia	No sperm	–	–	
Cryptozoospermia	Very low	–	–	Sperm few in number, identified only after centrifugation
Necrozoospermia	–	<32	–	High % of dead and immotile sperm
Aspermia	No semen			
Haemospermia (haematospermia)	Presence of erythrocytes in the ejaculate			
Leucospermia (leucocytospermia, pyospermia)	Presence of leukocytes in the ejaculate >1.0 × 10^6/ml			

(i) Able to detect a specific sperm dysfunction (Muller 2000)
(ii) Able to predict fertilization and pregnancy rates
(iii) Able to suggest appropriate therapeutic treatment for the same

Irrespective of requirement of specialized infrastructure and expertise, procedures of sperm function tests have now been recognized as essential and separately grouped under research procedures in the new WHO manual 2010 (Lamb 2010). However, more initiatives are needed to develop these tests in the form of routine protocols which can be performed along with regular semen analysis. Some of the related developments along with recently developed test protocols are described in the subsequent sections.

13.4 Tests of Sperm Function

13.4.1 Acrosome and Its Function

Under normal physiological circumstances, activation of acrosomal matrix to release its contents follows capacitation; both the events occur in the female reproductive tract and are considered prerequisites for successful fertilization. Only the acrosome-reacted sperm are able to release the acrosomal contents, aiding in penetrating the cumulus mass and zona pellucida (ZP). In fact, it is the sperm-zona binding that triggers the acrosome reaction resulting in release of the proteolytic enzymes designed to facilitate both ZP penetration and remodelling of the sperm surface in preparation for oocyte fusion (Aitken 2006). Though the clinical application of the acrosome reaction bioassay has been questioned by many, calcium ionophore-induced acrosome reaction testing has been found to predict the fertilizing potential of human sperm in vitro (Cummins et al. 1991).

Acrosome reaction has been reported to be induced in vitro through biochemical agents such as calcium ionophore A23187 and progesterone (Oehninger et al. 1994; Yovich et al. 1994; Perry et al. 1995). It opens up the possibility of monitoring acrosome reaction in vitro with the help of these agents. However, the test protocols based on this are complex that utilize several specific antibodies to different acrosomal components (Kallajoki et al. 1986; Sanchez et al. 1991).

There is another way of analysing acrosome function by assessing the acrosome status of the sperm. This is done by incubating sperm with fluorescent lectins that inherently bind to outer acrosomal membrane (*Arachis hypogaea* agglutinin) or to acrosomal contents (plasma sativum agglutinin). After binding with probe, the sperm depicts a fluorescent acrosomal cap or a band around its equatorial segment when acrosome reaction occurs (Aitken 2006).

Acrosome reaction in sperm can also be induced by binding through ZP (Menkveld et al. 1991; Liu and Baker 1992a, b). It is strongly supported by the fact that sperm binding to zona is directly correlated with normal morphology of acrosome in sperm (Liu and Baker 1988; Menkveld et al. 1996). The capacity of sperm acrosome to react in the presence of solubilized ZP has now been developed as a therapeutic tool using either the partner's own sperm or those available from other subjects (Franken et al. 1997). The procedure utilizes stripping of zona from oocytes by aspirating vigorously using a pipette with 90 μm inner diameter and incubating the same with sperm for inducing acrosome reaction. This is otherwise known as zona-induced acrosome reaction (ZIAR) test and has been recognized as an indicator of sperm dysfunction among men presenting normal semen parameters. The finding of such a study on in vitro fertilization outcome of 35 couples has been described in Table 13.3 (Franken et al. 2007).

Further simplification of test protocol on acrosome status of sperm comes from the assay of gelatin digestion following the release of acrosomal contents of sperm (Gopalkrishnan 1995). Though the assay has a wider acceptability, it suffers from the fact that its completion time is very long, about two and half hours, and sometimes fails to be reproducible even with experienced technical hands. The protocol

Table 13.3 Semen parameters versus acrosome response against the outcome of in vitro fertilization (IVF) of 35 couples as per results of zona-induced acrosome reaction (ZIAR) test

ZIAR	Mean sperm concentration (10^6 cells/ml)	Morphology (% normal)	Motility (%)	HZI	ZIAR	No. of oocytes	Fertilized group retrieved
<15%	55.8 ± 16	12.9 ± 4	54.5 ± 13	53 ± 13	3.7 ± 3[a]	8.9 ± 6	10.3 ± 21[c]
(n = 20)	(22–100)	(8–19)	(40–70)	(40–70)	(0–8)	(2–22)	(0–60)
>15%	66.4 ± 24	12.6 ± 3	64.7 ± 11	67 ± 12	22.2 ± 3[b]	8.2 ± 4	85.2 ± 11[d]
(n = 15)	(22–100)	(7–18)	(51–80)	(49–88)	(16–27)	(4–18)	(60–100)

Fisher's exact t-test: a vs b, $P = 0.001$; c vs d, $P = 0.001$

was further updated in the form of a kit which at present is being developed in the form of a commercial kit to be available in the Indian market soon. The protocol as described in the original protocol and the modified patented version are described below (Gopalkrishnan 1995; Misro and Chaki 2005a, b, c).

13.4.1.1 Test for Acrosome Status and Function (Gopalkrishnan 1995)

Time requirement: 2 h 40 min

The acrosome of sperm contains a number of proteases, which play a crucial role in the penetration of sperm through outer investments of oocyte.

The basis of this method is that acrosomal enzymes once released will dissolve protein (e.g. gelatin) through which light can pass, leading to the observation of halo structure around the sperm head when examined under a microscope. The protocol is described below.

Preparation of reagents:

Preparation of gelatin slides:

- Prepare 5% gelatin in distilled water (dist. H_2O) at 50°C.
- Take 40 µl of gelatin solution (warm) and spread on pre-cleaned slide and keep the gelatin-coated slides horizontally for 24 h at 4°C.
- Fix the gelatin slides in 0.05% glutaraldehyde in phosphate buffer saline (PBS) for 2 min.
- Wash the fixed gelatin slides twice in dist. H_2O.
- Gelatin slides are stored at 4°C vertically. The slides can be used up to 4 weeks.

Phosphate-buffered saline (PBS):
Sodium chloride (NaCl) = 0.2 gm
Disodium phosphate (Na_2HPO_4) = 0.14 gm
Potassium dihydrogen phosphate (KH_2PO_4) = 0.0023 gm
Dist. H_2O = 100 ml
PBS-glutaraldehyde solution:
25% glutaraldehyde = 0.2 ml
PBS = 100 ml

PBS-D-glucose (pH 7.8) solution:
PBS = 100 ml
D-Glucose =3 gms

Protocol:

- Dilute the semen sample 1:20 with PBS-D-glucose.
- Equilibrate for 30 min at 37°C.
- Bring the gelatin-coated slides to room temperature.
- Take 20 μl of diluted semen and gently smear on the gelatin slide.
- Keep for 5–10 min at room temperature to remove excess moisture.
- Incubate for 2 h at 37°C with wet filter paper on the sides in a petri dish.
- Observe the slide under microscope (400×).
- Count the sperm with halos against those without halos and calculate the percentage.

Normal: >50% sperm with halos.

Significance: This test evaluates the functional status of sperm acrosome and is a good indicator of sperm's ability to penetrate the oocyte investments.

Clinical relevance: Loss of acrosome, absence of acrosome or irregular acrosome may indicate the possibility of low or nil fertilization.

13.4.1.2 Lab Test Kit for Acrosome Status and Function (Misro and Chaki 2005a)

Time required: 40 min

Reagents provided in the kit: Gelatin-coated slides (black colour) and acrosome reaction solution (blue colour). Both are stable at room temperature.

Apparatus required: Any light microscope, an incubator maintained at 50°C, a micropipette and a moisture chamber (moisture chamber can be made using petri dish and blotting paper).

Protocol:

- Take 500 μl acrosome reaction solution in any small tube.
- Add 50 or100 μl of liquefied semen sample.
- Incubate at room temperature for 5 min.
- Make a smear smoothly on coated slide.
- Air-dry excess liquid on the slide. Do not do overdrying.
- Place the slide at 50°C in a moisture chamber and incubate for 30 min. Moisture chamber should be prewarmed at 50°C.
- Let it air-dry and observe under microscope.

Count the percentage of sperm with halos surroundings their heads. Normal acrosome status if >50% sperm with halos surrounding their head are found (see Fig. 13.1).

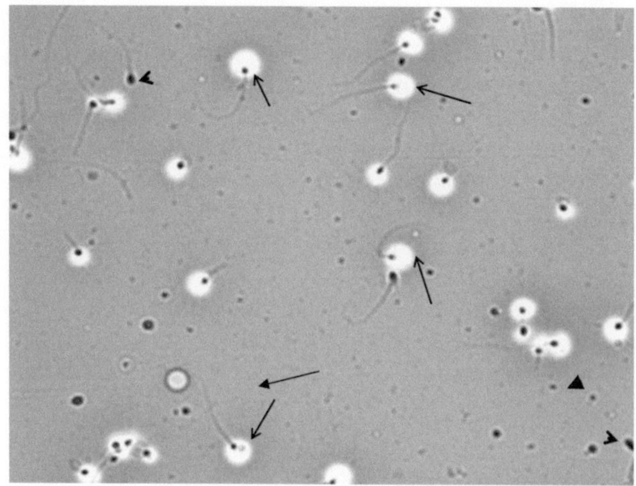

Fig. 13.1 Acrosome status: *Arrows* indicate sperm with halo (acrosome positive), and arrow heads indicate sperm without halo (acrosome negative)

13.4.2 Test for Sperm Plasma Membrane Integrity

Hypoosmotic swelling (HOS) of the sperm tail demonstrates the functional integrity of the sperm plasma membrane. The rationale of the test is based on the assumption that an undamaged sperm tail membrane permits the free flow of fluid into the cell under hypoosmotic conditions. The resultant swelling is manifested in the form of various types of sperm tail coiling that can easily be observed under light microscopy. The effect of hypoosmotic challenge is identical in all mammalian sperm (Drevius and Eriksson 1966). The hypoosmotic solution most commonly used for the purpose is composed of sodium citrate dihydrate ($Na_3C_6H_5O_7.2H_2O$; 0.735 g) and fructose (1.351 g) in 100 mL of dist. H_2O, with an osmolarity of 150 mOsm/litre. The same has been used in human sperm to test the plasma membrane integrity (Jeyendran et al. 1984, 1992; WHO 1992). Despite divergent degrees of correlation reported between the HOS test and different sperm evaluation parameters, the test has been recognized as the predictor of human sperm fertility (Okada et al. 1990). Though it is a very simple and sensitive test, it has never been routinely practised due to its lengthy duration of approximately 60 min, as per the WHO protocol. The protocol has been appropriately modified further and has been developed in the form of a laboratory kit (Misro and Chaki 2005a, b, c). The detailed procedure of the original as well the modified version is included here for better handling and reproducibility.

Reagents required: Hypoosmotic solution

Dissolve 0.735 g sodium citrate $Na_3C_6H_5O_7.2H_2O$ and 1.351 g fructose in 100 ml dist. H_2O. Store aliquots of this solution frozen at −20°C. Thaw and mix well before use.

Method (Jeyendran et al. 1984):

Time required: *60 min*

- Warm 1 ml HOS solution in a closed Eppendorf tube at 37°C for about 5 min.
- Add 0.1 ml liquefied semen to the solution and mix gently.
- Keep at 37°C for 100 min, but not longer than 120 min.
- Examine the sperm cells with phase-contrast microscope. Swelling of sperm is identified as changes in the shape of the tail as shown in Fig. 13.2. Repeat twice the score of swollen sperm in a total of 100 sperm, and calculate the mean percentage.

Modified method (Misro and Chaki 2005a, b, c):
Time required: *10 min*
Reagents provided: HOS solution (patented).
Storage condition: Room temperature.
Apparatus required: Any ordinary microscope and micropipette (10–100 μl).

Protocol:

- Take 500 μl of HOS solution in a small tube.
- Add 50 or 100 μl (depending on high/low sperm count) of liquefied semen sample to it.
- Mix gently and incubate at room temperature for 5 min.
- At the end of the incubation time, add 50 μl of colour stop solution and mix gently.
- Place a small drop of the mixture on a clean glass slide and cover it with cover slip.
- Observe it under a microscope and count the percentage of sperm with coiled tail (see Fig. 13.2). If the sample shows bent tail before the test, reduce the number after the test to get the actual result.

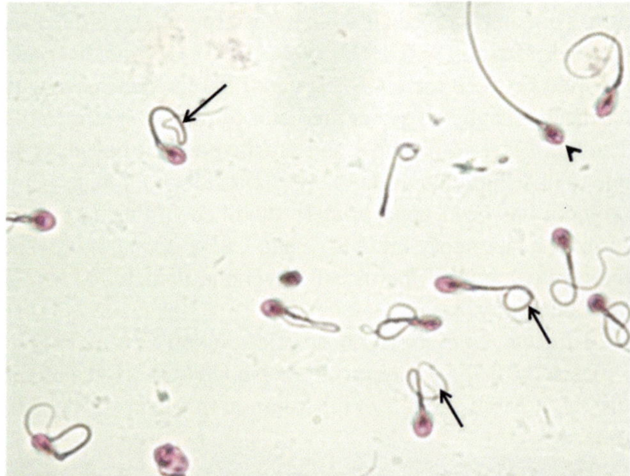

Fig. 13.2 The *arrows* indicate sperm positively responding to HOS test, whereas the arrow head indicates abnormal sperm (normal range, ≥60% with coiled tail)

13.4.3 Nuclear Chromatin Decondensation Test

Sperm chromatin is highly condensed, and the process of condensation starts right from the round spermatid stage at the level of the testis where they are formed. This is facilitated by the replacement of histone protein with protamine in the haploid chromatin network. However, the process of nuclear chromatin condensation continues during the transit of sperm from the testis and is completed during epididymal maturation. Since the condensation process involves millions of sperm and is indirectly dependent on the association of protamines in the sperm DNA, it is likely that all the sperm might not be condensed to the extent required. If a significant percentage of sperm is either too condensed or loosely condensed, this may reflect in their functional ability during fertilization. This is supported by the fact that during normal fertilization process following oocyte penetration, sperm chromatin decondenses, swells and forms the pronucleus prior to the fusion with oocyte nuclei. Thus, sperm chromatin decondensation is an essential prerequisite for successful fertilization, and the way this is tested in vitro using a semen sample is described below. The procedure as reported in the literature is again time consuming (Gopalkrishnan 1995). Therefore, a modified method to facilitate quick assessment of chromatin decondensation was developed, and the test can be completed within 10 min (Misro and Chaki 2005a, b, c). Both the procedures are described below for comparison and usability.

Method (Gopalkrishnan 1995):

Time required: 40 min

1. Mix sperm in 0.1 ml semen and 0.5 ml insemination medium, and then wash with 10 ml 0.9% sodium chloride and centrifuged at 2000 rpm for 10 min.
2. Suspended pellet in 0.1 ml 50 mmol/L sodium borate buffer (pH 9.0) containing 2 mmol/L DTT prepared daily.
3. Incubate suspension at room temperature (18–20°C) for 30 min.
4. Add 0.1 ml of 1% (w/v) SDS in sodium borate buffer and mix the suspension gently and incubate at room temperature for 2 min.
5. Stop the reaction and fix sperm by addition of 0.2 ml of 2.5% glutaraldehyde in sodium borate buffer.
6. Mix suspension (15 μl) with 5 μl of 0.8% rose bengal in 0.9% sodium chloride on a glass slide and cover with a square (22x22 mm) coverslip and examine under a phase-contrast microscope.

Method (Drevius and Eriksson 1966; Jeyendran et al. 1984; Misro and Chaki 2005c):

Reagents provided in the kit: One sachet containing chemical ingredients and a stop solution (patented).

Time required: 10 min

Storage condition: Once prepared, the solution is stable for 3–4 days in a refrigerator at 4°C.

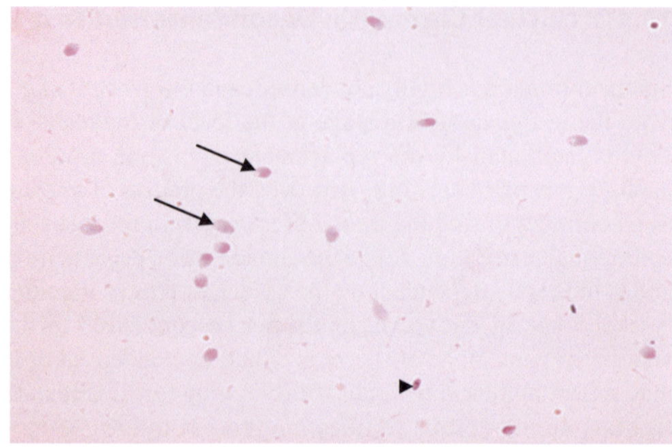

Fig. 13.3 Nuclear chromatin decondensation test. *Arrows* showing positive whereas the arrowhead showing negative NCD test (normal range, ≥ 70% with enlarged heads)

Apparatus required: A small centrifuge, microscope, micropipette and incubator maintained at 50°C.

Protocol:

Dissolve the whole content of a sachet in 10 ml dist. H_2O to make NCD (nuclear chromatin decondensation) solution.

- Take 500 μl of NCD solution (prewarmed at 50°C) in a small tube.
- Add 50–100 μl (depending on high/low sperm count) of liquefied semen sample to it. Mix gently and incubate at 50°C temperature for 5 min.
- Stop the reaction by adding 100 μl of stop solution.
- Place a small drop on a clean glass slide and cover with coverslip.
- Observe under the microscope and count % of decondensed (enlarged heads) sperm (see Fig. 13.3).

Besides the three easy-to-do methods of sperm function tests as described above, there are a few other tests that are very much specialized and require similar infrastructure and expertise to perform. In the following sections, therefore, their significance and usefulness are discussed without describing the method as such in detail.

13.4.4 Zona-Free Hamster Penetration Assay

The assay is based on the observation that following removal of the zona, the hamster egg becomes promiscuous and allows sperm from several species including humans to penetrate (Yanagimachi et al. 1976). This comprehensive bioassay examines the ability of human sperm to capacitate, to undergo the acrosome reaction and to undergo nuclear decondensation identical to the initial steps of fertilization after penetrating the egg. Though the test has gone several modifications since its development, its predictive value still remains to be controversial. A significantly higher correlation to in vitro fertilization outcome combined with low false positives and unacceptable levels of false negatives has been reported (Perreault and Rogers 1982; Aitken et al. 1991). The test is no longer commonly followed today because of the fact that it is labour intensive, technically demanding and extremely difficult to standardize.

13.4.5 Hemizona/Zona Pellucid (ZP) Binding Assay

ZP binding assay is based on the physiological observation that sperm-zona interactions are crucial leading to subsequent steps of fertilization. ZP is an acellular glycoprotein that surrounds the oocyte and serves as the site for sperm-egg recognition and induction of acrosome reaction. The glycoprotein ZP3 is reported to mediate such recognition and activations of sperm, while another glycoprotein ZP2 facilitates the binding of the acrosome-reacted sperm to the zona matrix. The assay developed for the purpose outlined the bioassay conditions and oocyte retrieval procedures followed in the hemizona assay and competitive intact ZP binding test (Burkman et al. 1988; Liu and Baker 1992a, b). Both bioassays provide functional homologues to physiological steps of sperm binding to ZP and provide a meaningful and distinct comparison between infertile versus fertile semen samples. Though the test has the advantage of providing information on sperm-zona interaction and associated defects, it is no longer widely offered. The reason behind this is that zona for the test is usually collected from excess, unutilized eggs or those eggs which did not fertilize. Moreover, eggs are to be bisected which require micromanipulation skills. The bisected ova are then incubated with sperm samples and examined following a period of incubation. Novel biomaterials may provide an alternate standard matrix to detect sperm-zona binding dysfunction in future and have the potential to develop as new laboratory protocols (Lamb 2010).

13.4.6 Test for Sperm DNA Damage

It is recognized that apparently sperm with normal morphology in an infertile semen sample may also have some DNA fragmentation which cannot be detected in routine examinations (Avendano et al. 2009). Sperm DNA fragmentation assessment in vitro has been increasingly recognized as a tool to assess the fertilizing potential combined with better diagnostic and prognostic prospects than the conventional assessment of parameters like sperm morphology, concentration and motility. When used along with assisted reproductive techniques, it has the risk of ending with early pregnancy loss leading to a low rate of carry home live births. Under normal circumstances, relatively small but not the large DNA damages are repaired during pre- and postreplication processes. Fertilization with such defective sperm carrying DNA damages may result in pregnancy loss or birth of offsprings with major or minor congenital malformations (Samsi et al. 2008). However, many andrologists still are not sure whether or not sperm DNA assessment should be routinely used in the male infertility evaluation though a variety of tests are available for the purpose. Since tests for whole sperm chromatin are already discussed earlier, tests which are typically developed to assess sperm DNA fragmentation defects are briefly described below, though these tests are seldom performed unless and until specifically recommended.

13.4.6.1 Terminal Deoxynucleotidyl Transferase dUTP Nick End Labelling

Terminal deoxynucleotidyl transferase dUTP nick end labelling (TUNEL) is a diagnostic technique to identify DNA strand breaks. It incorporates biotinylated deoxyuridine (dUTP) to 3'OH at DNA strand breaks, both single and double. Therefore, the intensity of signal increases with the increase in DNA strand breaks. During the development protocol, biotinylated dUTP is utilized to label the 3'ends of the fragmented DNA using recombinant deoxynucleotidyl transferase (TdT) enzyme. Sperm with normal DNA have capped telomeres at the 3'OH end. Thus, they do not fluoresce under a fluorescence microscope. But, those with DNA strand breaks having multiple 3'OH ends provide bright fluorescence (Lopes et al. 1998). Alternative use of TUNEL includes enzymatic detection of fragmented DNA. This is generally carried out using the enzyme horseradish peroxidase labelled with peroxidase streptavidin (streptavidin-HRP) which is in turn linked to biotinylated nucleotides. Hydrogen peroxide is used as the substrate and diaminobenzidine (DAB) as the chromogen, and the detection of the strand break is determined following the development of a deep brown stain.

The advantages of TUNEL assay are that it can detect both single- and double-strand breaks and provides useful information on the extent of DNA damage of sperm in a semen sample. It is, however, mostly a qualitative assay in the sense that the extent of DNA stand breaks in a single sperm cannot be quantified.

13.4.6.2 Comet Assay

The comet assay is also known as single gel electrophoresis assay, first described by Ostling and Johnson in 1984. It was later used with microgels which were utilized for electrophoresis under highly alkaline conditions (Singh et al. 1988). Alkaline conditions help in the uncoiling of DNA where the free ends migrate towards anode under the applied electric field. Following staining with acridine orange, the sperm head containing the DNA appears "comet"-like comprising a distinct head and tail. While the head represents intact DNA, the tail consists of strands of degraded DNA. In this way, both single- and double-strand breaks are easily analysed. This method was developed to measure low levels of strand breaks. Since then, a number of advancements have greatly increased the flexibility and utility of this technique for detecting various forms of DNA damage (e.g. single- and double-strand breaks, oxidative DNA base damage and DNA-DNA/DNA-protein/DNA-drug cross linking) and DNA repair in virtually any eukaryotic cell. The assay works on the principle that smaller fragmented DNAs migrate at a faster rate to anode as compared to the non-fragmented, intact DNAs. The protocol can be carried out both in neutral and acidic pH conditions.

To describe the method briefly, a suspension of sperm is mixed with low-melting agarose and layered onto a precoated agarose slide. The slide containing sperm is incubated inside a Coplin jar containing high salt concentration to release the DNA following removal of cellular proteins. The DNA then unwinds under alkaline conditions, and the alkali-labile sites are considered as single-strand breaks in the liberated DNA. Treatment with nucleases under controlled conditions is sometimes considered to experimentally induce strand breaks in the DNA for maintaining a parallel internal control. The slide containing sperm is then placed within an electrophoretic chamber and electrophoresed under alkaline (pH > 13) conditions that forces broken DNA strands to migrate close to anode. The released sperm DNA is stained with ethidium bromide and scored. The assay is very helpful to study various forms of DNA damage by altering steps of lysis and electrophoresis under controlled laboratory conditions. Though the method has acquired scientific validity, it requires sophisticated laboratory set-up. Therefore, it is not widely practised in andrology laboratories unless specifically recommended.

13.4.7 Test for Assessment of Sperm Oxidative Stress

Oxidative stress impairs all the functional capacities of sperm. Oxidative stress represents the imbalance between pro- and antioxidants present in any cellular system. Oxidative stress arises when pro-oxidants like reactive oxygen species (ROS), a class of free radicals, are generated in excess than what the cellular system can neutralize through antioxidants that normally limit the damaging effects. Free radicals are also generated in the normal course of oxidative metabolism. These free radicals are useful to some extent in the normal physiological processes like sperm capacitation, hyperactivation, acrosome reaction and sperm-oocyte fusion. However, generated in excess, they induce pathological effects leading to loss of function, apoptosis and cell death (see Fig. 13.4; Kothari et al. 2010).

It is imperative to find out the source of ROS generation. The external source of ROS is mostly represented by the leucocytes that are present in almost every ejaculate, and the internal source is represented by sperm themselves (Wolff 1995). It is reported that extrinsic ROS generated by leucocytes predominately affect the sperm plasma membrane leading to reduced motility (Henkel et al. 2005). In contrast, intrinsic ROS induces DNA damage (Franken and Oehninger 2012).

Estimation of sperm ROS can be easily measured with the help of a chemo-luminometer. Multiple factors affect chemiluminescent reactions. These include the concentration of reaction mixture, sample volume, temperature control and background luminescence. A brief procedure of measurement of sperm ROS using a chemo-luminometer is described below (Agarwal et al. 2008):

1. Centrifuge semen samples at 300 × g for 5 min, and remove the seminal plasma.
2. Suspend the sperm pellet in 3 mL of Dulbecco's PBS solution and wash again at 300 × g for 5 min.

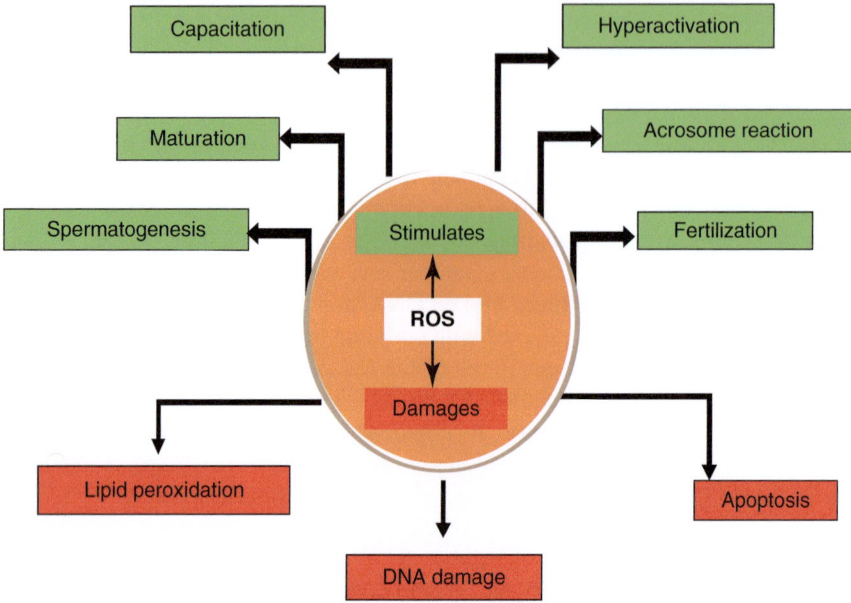

Fig. 13.4 Physiological roles of reactive oxygen species (ROS) include maturation, capacitation, hyperactivation, spermatogenesis, acrosome reaction and fertilization

3. The sperm concentration is adjusted to 20 × 106/mL before ROS measurement. Measure ROS by a chemiluminescence assay using 5 μL of luminol (5 mM, 5-amino-2, 3-dihydro-1, 4-phthalazinedione, Sigma Chemical Company, St. Louis, MO, USA).
4. Measure chemiluminescence in the integration mode using chemo-luminometer at 37°C for 15 min after adding luminol. ROS production is expressed as counted photons per minute (cpm)/20 × 10^6 sperm.

Several independent studies have also demonstrated that chemiluminescent assays of redox activity reflect the fertilizing potential of human sperm both in vivo and in vitro (Zorn et al. 2003; Said et al. 2004). However, in spite of their potential as diagnostic tools, these are only very limitedly utilized.

Key Questions

- What is the rationale of doing sperm function tests in the laboratory?
- Write the lower reference limit of various semen parameters updated by WHO in 2010.
- What is the principle of comet assay?
- Describe the test to be used for analysing sperm nuclear chromatin condensation.
- Write the principles of the tests used to examine acrosome function.

References

Agarwal A, Cocuzza M, Abdelrazik H, Sharma RK. Oxidative stress measurement in patients with male or female factor infertility. In: Popov I, Lewin G, editors. Handbook of chemiluminescent methods in oxidative stress assessment. Trivandrum: Transworld Res. Network; 2008. p. 198–218.

Aitken RJ. Sperm function tests and fertility. Int J Androl. 2006;29:69–75.

Aitken RJ, Irvine DS, Wu FC. Prospective analysis of sperm oocyte fusion and reactive oxygen species generation as criteria for diagnosis of infertility. Am J Obstet Gynecol. 1991;164:542–51.

Avendano C, Franchi A, Taylor S, Morshedi M, Bocca S, Oehninger S. Fragmentation of DNA in morphologically normal human spermatozoa. Fertil Steril. 2009;91:1077–84.

Burkman LJ, Coddington CC, Franken DR, Kruger T, Rosenwaks Z, Hogen GD. The hemizona assay (HZA): development of a diagnostic test for the binding of human spermatozoa to human hemizona pellucid to predict fertilization potential. Fertil Steril. 1988;49:688–95.

Coetzee K, Kruger TF, Lombard CJ. Predictive value of normal sperm morphology: a structured literature review. Hum Reprod Update. 1998;4:73–82.

Cooper TG, Yeung CH. Computer-aided evaluation of assessment of 'grade a' spermatozoa by experienced technicians. Fertil Steril. 2006;85:220–4.

Cummins JM, Pember SM, Jequier AM, Yovich JL, Hartmann PE. A test of the human sperm acrosome reaction following ionophore challenge (ARIC). Relationship to fertility and other seminal parameters. J Androl. 1991;12:98–103.

Drevius LO, Eriksson H. Osmotic swelling of mammalian spermatozoa. Exp Cell Res. 1966;42:136–56.

Esteves SC, Miyaoka R, Agarwal A. An update on the clinical assessment of the infertile male. Clinics. 2011;66:691–700.

Franken DR, Oehninger S. Semen analysis and sperm function testing. Asian J Androl. 2012;14:6–13.

Franken DR, Bastiaan HS, Kidson A, Wranz P, Habenicht UF. Zona pellucida mediated acrosome reaction and sperm morphology. Andrologia. 1997;29:311–7.

Franken DR, Esterhuizen A, Oehninger SC. Diagnostic impact of the zona pellucida induced acrosome reaction in an assisted reproductive programme. Embryo Talk. 2007;2:3–9.

Gopalkrishnan K. Standardized procedures in human sperm analysis. Curr Sci. 1995;68:353–62.

Gopalkrishnan K, Padwal V, Meharji PK, Gokral JS, Shah R, Juneja HS. Poor quality of sperm as it affects repeated early pregnancy loss. Arch Androl. 2000;45:111–7.

Henkel R, Kierspel E, Stalf T, Mehnert C, Menkveld R, Tinneberg HR, et al. Effect of reactive oxygen species produced by spermatozoa and leukocytes on sperm functions in nonleukocytospermic patients. Fertil Steril. 2005;83:635–42.

Jeyendran RS, Van der Ven HH, Perez-Pelaez M, Crabo BG, Zaneveld IJD. Development of an assay to assess the functional integrity of the human sperm membrane and its relationship to other semen characteristics. J Reprod Fertil. 1984;70:219–28.

Jeyendran RS, Van Der Van HH, Zaneveld LJD. The hypoosmotic swelling test: an update. Arch Androl. 1992;29:105–16.

Kallajoki M, Virtanen I, Suominen J. The fate of acrosomal staining during the acrosome reaction of human spermatozoa as revealed by a monoclonal antibody and PNA-lectin. Int J Androl. 1986;9:181–94.

Kothari S, Tompson A, Aggarwal A, du Plessis SS. Free radicals: their beneficial and detrimental effects on sperm function. Indian J Exp Biol. 2010;48(5):425–35.

Lamb DJ. Semen analysis in 21st century medicine: the need for sperm function testing. Asian J Androl. 2010;12:64–70.

Liu DY, Baker HW. The proportion of human sperm with poor morphology but normal intact acrosomes detected with *Pisum sativum* agglutinin correlates with fertilization in vitro. Fertil Steril. 1988;50:288–93.

Liu DY, Baker HW. Morphology of spermatozoa bound to the zona pellucida of human oocyte that failed to fertilize in vitro. J Reprod Fertil. 1992a;94:71–84.

Liu DY, Baker HW. Tests of human sperm function and fertilization in vitro. Fertil Steril. 1992b;58:465–83.

Lopes S, Sun JG, Jurisicova A, Meriano J, Casper RF. Sperm deoxyribonucleic acid fragmentation is increased in poor quality semen samples and correlates with failed fertilization in a cytoplasmic sperm injection. Fertil Steril. 1998;69:528–32.

Menkveld R, Franken DR, Kruger TF, Oehninger S, Hodgen GD. Sperm selection capacity of the human zona pellucida. Mol Reprod Dev. 1991;30:346–52.

Menkveld R, Rhemrev JP, Franken DR, Vermeiden JP, Kruger TF. Acrosomal morphology as a novel criterion for male fertility diagnosis: relation with acrosin activity, morphology (strict criteria), and fertilization in vitro. Fertil Steril. 1996;65:637–44.

Misro MM, Chaki SP. A method for evaluating sperm acrosome status. Indian patent application. No.3237/DEL/2005a

Misro MM, Chaki SP. Modified HOS test solution and a method there of. Patent filed on 9th November 2005b. No.2991/DEL/2005.

Misro MM, Chaki SP. Composition for assessment of human sperm function and method thereof. Patent filed on 28th December 2005c. No.3508/DEL/2005.

Muller CH. Rationale, interpretation, andrology lab corner validation, and uses of sperm function tests. J Androl. 2000;21:10–30.

Oehninger S, Blackmore P, Morshedi M, Sueldo C, Acosta AA, Alexander NJ. Defective calcium influx and acrosome reaction (spontaneous and progesterone-induced) in spermatozoa of infertile men with severe teratozoospermia. Fertil Steril. 1994;61:349–54.

Okada A, Inomata K, Matsuhashi M, Fujio K, Miura K, Shiina K, et al. Correlation of the hypo-osmotic swelling test, semen score and the zona-free hamster egg penetration assay in humans. Int J Androl. 1990;13:337–43.

Ostling O, Johanson KJ. Microelectrophoretic study of radiation-induced DNA damages in individual mammalian cells. Biochem Biophys Res Commun. 1984;123:291–8.

Perreault SD, Rogers BJ. Capacitation pattern of human spermatozoa. Sperm penetration assay. Fertil Steril. 1982;38:258–60.

Perry RL, Naeeni M, Barratt CL, Warren MA, Cooke ID. A time course study of capacitation and the acrosome reaction in human spermatozoa using a revised chlortetracycline pattern classification. Fertil Steril. 1995;64:150–9.

Said TM, Agarwal A, Sharma RK, Mascha E, Sikka SC, Thomas Jr AJ. Human sperm superoxide anion generation and correlation with semen quality in patients with male infertility. Fertil Steril. 2004;82:871–7.

Samsi MB, Kumar R, Dada R. Evaluation of nuclear DNA damage in human spermatozoa in men opting for assisted reproduction. Indian J Med Res. 2008;127:115–23.

Sanchez R, Toepfer-Petersen E, Aitken RJ, Schill WB. A new method for evaluation of the acrosome reaction in viable human spermatozoa. Andrologia. 1991;23:197–203.

Saxena P, Misro MM, Chaki SP, Chopra K, Roy S, Nandan D. Is abnormal sperm function an indicator among couples with recurrent pregnancy loss? Fertil Steril. 2008;90:1854–8.

Singh NP, McCoy MT, Tice RR, Schneider EL. A simple technique for quantitation of low levels of DNA damage in individual cells. Exp Cell Res. 1988;175:184–91.

Van der Merwe FH, Kruger TF, Oehninger SC, Lombard CJ. The use of semen parameters to identify the subfertile male in the general population. Gynecol Obstet Invest. 2005;59:86–91.

Van Waart J, Kruger TF, Lombard CJ, Ombelet W. Predictive value of normal sperm morphology in intrauterine insemination (IUI): a structured literature review. Hum Reprod Update. 2001;7:495–500.

Wolff H. The biologic significance of white blood cells in semen. Fertil Steril. 1995;63:1143–57.

World Health Organization. WHO manual for examination and processing of human semen. 5th ed. Geneva: World Health Organization; 2010.

Yanagimachi R, Yanagimachi H, Rogers BJ. The use of zona-free animal ova as a test-system for the assessment of the fertilizing capacity of the human spermatozoa. Biol Reprod. 1976;15:471–6.

Yovich JM, Edirisinghe WR, Yovich JL. Use of the acrosome reaction to ionophore challenge test in managing patients in an assisted reproduction program: a prospective, double-blind, randomized controlled study. Fertil Steril. 1994;61:902–10.

Zorn B, Vidmar G, Meden-Vrtovec H. Seminal reactive oxygen species as predictors of fertilization, embryo quality and pregnancy rates after conventional in vitro fertilization and intracytoplasmic sperm injection. Int J Androl. 2003;26:279–85.

Part IV

Reproductive Stages

Male Puberty

14

Amit Satapathy and Vandana Jain

Learning Objectives
- Age at onset of puberty
- Physiology
- Nutritional factors
- Physical and hormonal changes
- Delayed puberty
- Precocious puberty
- Endocrine disruptors and puberty

14.1 Introduction

The term "puberty" is derived from the Latin word *pubescere*, which means "to reach physical maturity, growth of body hair, and attain manhood." Puberty represents the stage of transition from the sexually immature child to the potentially fertile adolescent. Over the last few decades, advances in biochemistry, physiology, and hormone assay techniques have helped in elucidating the complex processes involved in the initiation and progression of pubertal development.

A. Satapathy, MD • V. Jain, MD (✉)
Department of Pediatrics, All India Institute of Medical Sciences, New Delhi, India
e-mail: amitkumar.satapathy@yahoo.co.in; drvandanajain@gmail.com

A. Kumar, M. Sharma (eds.), *Basics of Human Andrology*,
DOI 10.1007/978-981-10-3695-8_14

14.2 Age at Onset of Puberty in Males and Secular Trend

Timing of puberty is affected by genetic as well as environmental factors. Fetal nutrition, childhood dietary habits, obesity, and physical activity have a bearing on age of onset of puberty. Though age of attainment of menarche in girls has declined over the past decades, such secular trend has not been well documented for boys. In the USA, the median age at puberty in boys as per the National Health and Nutrition Examination Survey (NHANES III), from 1988 to 1994, is reported to be 10.1 years in Caucasian Americans and 9.3 years in African Americans, which is lower than that reported previously as 11.9 years (Reynolds and Wines 1951; Tanner and Davies 1985; Herman-Giddens et al. 2001; Karpati et al. 2002; Sun et al. 2002). Boys in more recent surveys are taller at younger ages and attain mature height earlier than in the past, indirectly suggesting advancement in the age of puberty (Herman-Giddens et al. 2001; Karpati et al. 2002).

14.3 The Hypothalamic–Pituitary–Gonadal Axis

The hypothalamus secretes gonadotropin-releasing hormone (GnRH) which stimulates the release of the gonadotropins – luteinizing hormone (LH) and follicle-stimulating hormone (FSH) – from the anterior pituitary, resulting in release of testosterone from testes (see Fig. 14.1).

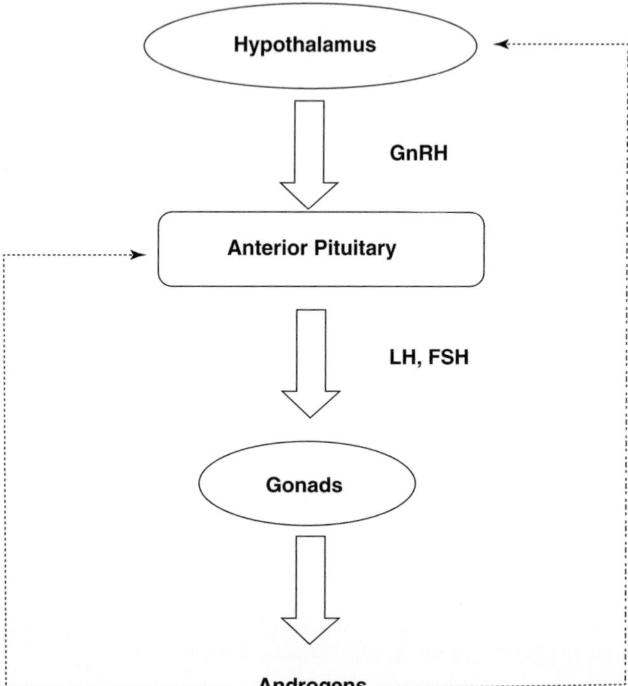

Fig. 14.1 The hypothalamic–pituitary–gonadal axis. *GnRH* gonadotropin-releasing hormone, *LH* luteinizing hormone, *FSH* follicle-stimulating hormone

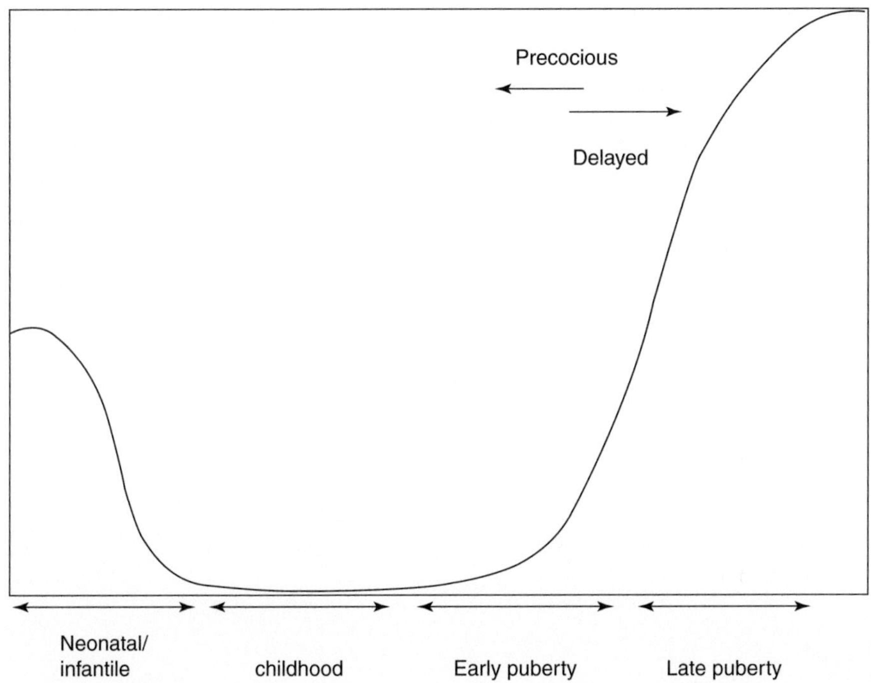

Fig. 14.2 The activity of gonadotropin-releasing hormone (GnRH) pulse generator during different phases of life

The pulsatile secretion of GnRH stimulates the production of LH and FSH, which stimulate the testes resulting in maturation of spermatogonia and increased secretion of testosterone. There is, in turn, a negative feedback loop as well, i.e., the gonadotropins inhibit the secretion of GnRH, and testosterone inhibits the release of gonadotropins as well as GnRH. The hypothalamic-pituitary-gonadal (HPG) axis is active in the fetal period, becomes quiescent by 7–9 months after birth, and remains so till the onset of puberty, when the GnRH pulse generator is reactivated – first during sleep and then during the day as well (see Fig. 14.2).

14.4 Physiology of Initiation of Puberty

The activation of the HPG axis at puberty is regulated by complex interactions among multiple neuronal pathways. The increased pulsatility of the GnRH secretion is central to initiation of puberty. GnRH release in the hypothalamus is controlled by excitatory/inhibitory amino acids, gamma aminobutyric acid (GABA)-ergic, and glutamatergic neurons, neurons responsive to leptin, kisspeptin, neuropeptide Y (NPY)-producing neurons, and prostaglandins (Mitsushima et al. 1994; Terasawa et al. 1999; Gottsch et al. 2004; Irwig et al. 2004). GABAergic neurons have inhibitory influence, restraining GnRH secretion during prepubertal period. However, the exact roles of glutamate and GABA in the initiation of puberty have not been elucidated. For example, no association of functional mutations or polymorphisms in GABA receptor gene with idiopathic precocious puberty (PP) is observed (Brito et al. 2006). Similarly, association between age at onset of puberty and sequence variation or polymorphisms in GnRH and GnRH receptor (GnRHR) genes has not been established too (Sedlmeyer et al. 2005). The role of kisspeptin, protein encoded by the KISS1 gene (1q32-q41), has been demonstrated as a potent activator of GnRH neurons in animals and humans (Shahab et al. 2005; Plant et al. 2006). After identification of mutation in KISS1R gene encoding GPR54 and in families with isolated hypogonadotropic hypogonadism (HH) and constitutional delay of growth and puberty (CDGP), the role of kisspeptin and its receptor GPR54 in

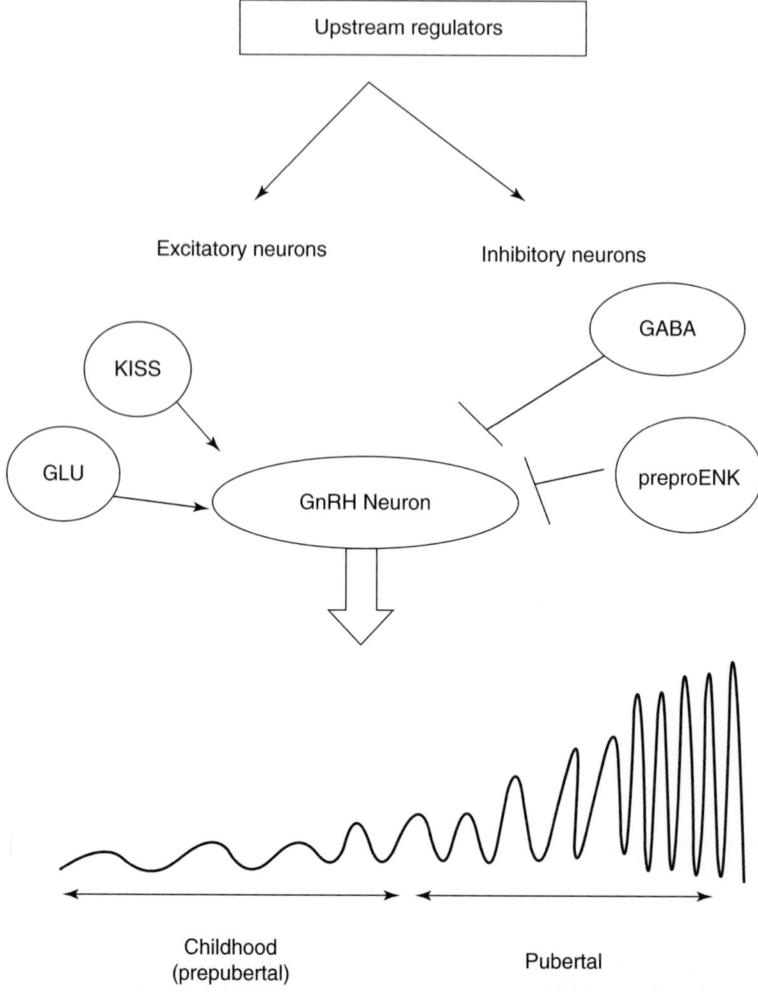

Fig. 14.3 Regulation of onset of puberty by the hypothalamic neurons. *GLU* glutamate, *GABA* gamma aminobutyric acid, *preproENK* preproenkephalin, *GnRH* gonadotropin-releasing hormone

regulation of onset of puberty has been studied in detail. Subsequently, identification of activating mutation in KISS1R gene in a patient with idiopathic precocious puberty (IPP) further confirms its role in puberty too. Though leptin does not affect GnRH secretion but is a permissive factor in onset and progression of mammalian puberty (Garcia et al. 2002; Quennell et al. 2009). A schematic representation of the regulation of onset of puberty by the hypothalamic neurons is shown in Fig. 14.3.

14.5 Role of Nutritional Factors in Timing of Puberty

Environmental factors play a major role in maturation of the HPG axis responsible for initiation of puberty. Pathological conditions, such as chronic illness, psychological distress, and intense physical training, delay pubertal maturation, whereas intracranial hamartoma, glioma, hydrocephalus, and low-dose cranial irradiation for malignancy are commonly associated with early or precocious puberty (Tahirovic 1998; Georgopoulos et al. 1999; Pozo and Argente 2002; Toogood 2004). Differences in timing of puberty in rural versus urban children and earlier maturation in improved health and nutrition condition favor a role for environmental factors (deMuinck et al. 2001).

In chronic childhood undernutrition, pubertal maturation and growth is delayed, in conjunction with lower levels of gonadotropins in blood (Satyanarayana and Naidu 1979; Kulim et al. 1984; Leensatra et al. 2005). Obesity has been shown to be associated with advanced pubertal development in many studies (Guo et al. 1997; Kindblom et al. 2006; Silventoinen et al. 2008). Gain in basal metabolic index (BMI) has been associated with early puberty. Although childhood growth rate is higher, but adult final height remains reduced because of poor height gain during growth spurt (He and Karlberg 2001).

Onset of puberty is often noted to occur at a younger age in children born small for gestational age due to intrauterine growth retardation (IUGR); however, those children with the absence of catch-up growth after birth mature late (Cooper et al. 1996). Rapid catch-up growth in infancy in IUGR children predisposes them to develop greater adiposity during later childhood (Lienhardt et al. 2002). The adiposity-related increase in leptin levels triggers earlier onset of puberty. Increased risk for advanced maturation has been documented in internationally adopted children with early malnutrition and subsequent catch-up growth after reaching new country with better nutrition (Proos et al. 1991; Ong et al. 2000).

In obesity, not only the levels of testosterone and estrogen are elevated, but the aromatization of androgens and bioavailability of sex steroids, secondary to low levels of sex hormone-binding globulin, increase as well. Adrenal gland function is also enhanced (Teilmann et al. 2006). The changes in hormonal milieu may contribute to earlier onset and/or accelerated tempo of maturation (deRidder et al. 1992). Early puberty in African American in comparison to Caucasian Americans may be explained partly by obesity as the mean body fat, as well as higher leptin levels in former population (Herman-Giddens et al. 1997; Biro et al. 2006). The secular trends of early puberty in children with obesity are more marked in girls as compared to boys probably because of lack of definitive pubertal event such as menarche in girls (Kaplowitz et al. 2001).

14.6 Physical Changes of Puberty

The physical changes of puberty can be broadly divided into development of secondary sexual characteristics and changes in anthropometric measures (bone age, height, weight, and body fat).

Secondary sexual characteristics are described using the five stages of development (Tanner et al. 1976). "Tanner stages" range from stage 1 (prepubertal) to stage 5 (postpubertal) and delineate the growth of pubic hair and genital development in boys. The Tanner staging or sexual maturity rating for boys is presented in Table 14.1 and illustrated in Fig. 14.4.

Increased testicular size is the first physical evidence of puberty. Pubertal testicular enlargement is considered to have begun when the longitudinal measurement of the testis is greater than 2.5 cm or volume $\geq$ 4 ml as assessed by orchidometer (see Fig. 14.5). This enlargement is predominantly due to increase in Sertoli cells and seminiferous tubules. The mean age of achieving a testicular volume of 4 ml has been reported to be 11.5–12 years in various studies (Tanner et al. 1976). Completion of genital development takes average time of 3 years (range 2–4.7 years) although testes reach adult volume over 5–7 years (Karlberg and Taranger 1976).

The phallus is more accurately measured in the stretched, flaccid state. The stretched penile length averages 6.2 cm in the prepubertal stage. During puberty it increases to 12.4 ± 2.7 cm (mean ± SE) in white adults, in black men the mean length is 14.6 cm, and in Asians it is 10.6 cm (Sutherland et al. 1996). There is gradual increase in male larynx and cricothyroid cartilage too resulting in change in quality of voice (Abbassi 1998).

Table 14.1 Tanner staging or sexual maturity rating in boys (Marshall and Tanner 1970)

Tanner staging: genitalia
G1: Penis and testis are of the same size during childhood; no scrotal thinning or reddening
G2: Enlargement of the testis to >4 ml; thinning and reddening of the scrotum
G3: Increase in penile length; further growth of testis and scrotum
G4: Further increase in penile length and width and darkening of the scrotal skin
G5: Adult size and configuration of genitalia
Tanner staging: pubic hair development
PH1: No pubic hair
PH2: Sparse long, slightly pigmented, downy hair, chiefly near the base of the penis
PH3: Darker, coarser, curly pubic hair joining sparsely in the midline over the symphysis pubis
PH4: Adult-type hair, but in a smaller area than in adults; no spread to the medial thighs
PH5: Adult-type hair in the classic inverse triangle distribution; might spread to the medial thighs

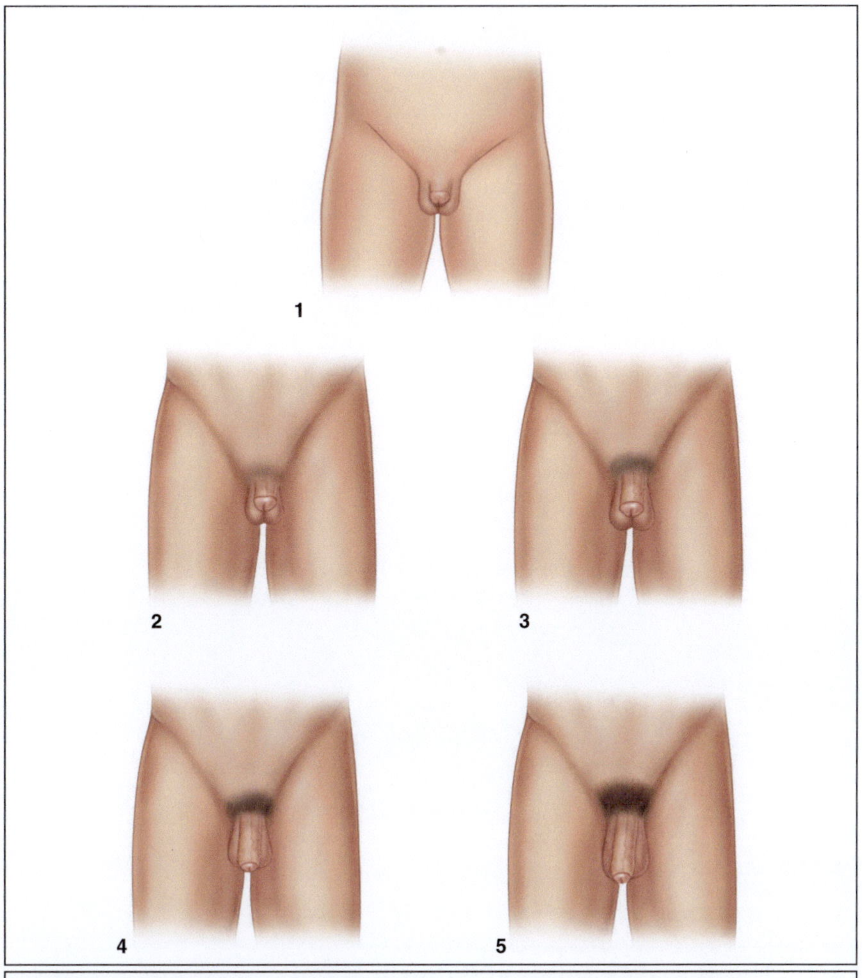

Stage 1: prepubertal with no pubic hair
Stage 2: Sparse light coloured hair at the base of penis, scrotal skin textured
Stage 3: pubic hair spread across public symphysis with increase in size of testis and scrotum, increase in penile length
Stagr 4: abundant pubic hair with curling, further, enlargemant of scrotum and testis, darkening of skin, increase in penile length and diameter as that of adult type
Stage 5: pubic hair extending up to medial aspect of thigh and umbilicus, scrotum and testis become adult size

Fig. 14.4 Tanner staging (Adapted and modified from Marshall and Tanner 1970)

Boys reach peak height velocity (PHV) approximately 2 years later than girls, at a mean age of 13.5 years (Delemarre-Van De Waal et al. 1991). PHV occurs at Tanner pubic hair stages 3 or 4 of puberty in most boys. The total height gain in boys during puberty averages 31 cm with mean difference between adult height in men and women being 12.5 cm (Boyar et al. 1974).

Fig. 14.5 Orchidometer

14.7 Hormonal Changes at Puberty

14.7.1 Gonadotropins

LH and FSH levels are low at birth but there occurs a surge in gonadotropins in early neonatal period termed as "mini puberty" before returning to the low level at which they stay till puberty. The onset of pubertal hormonal changes is marked by increase in pulsatile secretion of GnRH resulting in increased release of LH and increased frequency of LH pulses during sleep. As puberty progresses, GnRH pulse becomes more persistent throughout the day (August et al. 1972). LH causes increased sex hormone production by the testes, whereas spermatogonia maturation is stimulated by FSH.

14.7.2 Testosterone

Prepubertal boys have low plasma testosterone level (<0.3 nmol/l or 0.1 ng/ml), except during the first 3–5 months of life. Serum testosterone level is elevated only during night time at onset of puberty. Sustained elevation of testosterone is detected in daytime when testicular volume becomes 4 ml and continues to increase thereafter during puberty (Knorr et al. 1974) with peak at stages 2 and 3 of puberty.

14.7.3 Estrogens

Estradiol levels are low in prepubertal stage and rise during puberty but decrease thereafter. Aromatization of testosterone and androstenedione is the predominant source of estradiol in males. This aromatized estradiol not only helps in skeletal maturation but also helps in acquisition of bone mass (Weinstein et al. 1974).

14.7.4 Adrenal Androgens

The plasma levels of dehydroepiandrosterone (DHEA) and its sulfated form (DHEAS) begin to increase by the age of 7–8 years and remain elevated till early adulthood. Though there is no sex difference in secretion of adrenal androgen in early stage puberty, but later on it is higher in males. In some children, pubic hair and axillary hair may develop without appearance of the other sexual maturation signs and symptoms. They may be associated with advanced bone age too. Dissociation of adrenarche and gonadarche is noticed in premature adrenarche and central/true precocious puberty (CPP; Sklar et al. 1980).

14.7.5 Growth Hormone

Serum growth hormone (GH) and insulin-like growth factor-1 (IGF-1) levels rise during puberty in both boys and girls (Veldhuis et al. 2000).

14.8 Delayed Puberty

Delayed puberty is defined as the absence of secondary sexual characters by the age of 14 years in boys. It is more common in boys as compared to girls with overall prevalence of 2.5% in adolescent children. The causes are classified into temporary delay, which is common, and permanent sexual infantilism, which can be hypogonadotropic–hypothalamic–pituitary in origin or hypergonadotropic testicular failure.

14.8.1 Temporary Delay

14.8.1.1 Constitutional Delay of Growth and Puberty

Constitutional delay of growth and puberty (CDGP) is the commonest cause of delayed puberty in boys. The patients have a delayed tempo of growth and development. Endocrine function is appropriate for the stage of physiological development, but not for chronological age. The typical boy with constitutional delay of puberty presents at 14–15 years of age with complaints of being shorter compared to peers and lacking secondary sexual characters. Boys with CDGP present more commonly than girls probably because of higher psychosocial concern. They may have history of delayed puberty in siblings and other family members. Bone age, though delayed, usually correlates well with the stage of sexual maturation. Prognosis is good and final adult height remains unaffected.

14.8.1.2 Chronic Illness

Chronic illness may delay puberty by disturbing normal physiology or due to prolonged undernutrition. Long-term use of glucocorticoids, anticancer chemotherapy, or radiation therapy delays growth or sexual maturation. Adolescents with chronic conditions should be monitored for pubertal development. In milder conditions, reassurance is sufficient. However, pathologic causes of delayed puberty must be detected early and should be treated timely to attain catch-up growth. Hormone replacement with exogenous sex steroids may be considered. Nutritional rehabilitation is essential in children with malabsorption (e.g., inflammatory bowel disease, cystic fibrosis, celiac disease) causing delayed puberty. Children with chronic illness should be evaluated for other causes also if the primary disease fails to fully explain the delayed puberty.

14.8.1.3 Hypothyroidism

Thyroid hormone is essential for normal puberty. Thyroid hormone deficiency delays the onset and retards the progress of pubertal maturation by interfering with gonadotropin secretion.

14.8.1.4 Hyperprolactinemia

Hyperprolactinemia with or without a pituitary microadenoma or galactorrhea can delay the onset or progression of puberty.

14.8.2 Permanent Sexual Infantilism

14.8.2.1 Hypogonadotropic Hypogonadism

Delayed onset of puberty with low serum gonadotropin levels is defined as hypogonadotropic hypogonadism. It may be secondary to hypothalamus or pituitary dysfunction. The defect in gonadotropin secretion may be absolute or partial. It may be isolated deficiency or present in combination with deficiency of other pituitary hormones. Undescended testes and gynecomastia are commonly seen in these children. A list of some of the causes is given in the Table 14.2.

14.8.2.2 Isolated Gonadotropin Deficiency

Stature is normal till adolescence in patients with isolated gonadotropin deficiency. After reaching pubertal age, the deficiency of sex steroids leads to slowing of growth. However, growth continues beyond the usual age because of delayed epiphyseal fusion, so that the final height is normal or above normal. Body proportion is eunuchoid, i.e., limbs are longer, and the upper to lower body segment decreases (below 0.9).

Table 14.2 Causes of hypogonadotropic hypogonadism in males

Genetic: Kallmann syndrome, multiple pituitary hormone deficiency, isolated luteinizing hormone deficiency, X-linked adrenal hypoplasia congenital, Prader–Willi syndrome, Laurence–Moon–Bardet–Biedl syndrome
CNS tumors: germinoma, hypothalamic glioma, craniopharyngioma, etc.
Infiltrative diseases: Langerhans cell histiocytosis, sarcoidosis, hemochromatosis
Trauma: pituitary stalk transaction, skull fracture
Infection: tuberculosis, human immunodeficiency virus (HIV), fungus
Miscellaneous: chronic systemic illness, hypothyroidism, Cushing's disease, hyperprolactinemia, eating disorders, psychogenic disease

The commonest form of isolated gonadotropin deficiency is Kallmann syndrome. The clinical features consist of hyposmia or anosmia (from aplasia or hypoplasia of the olfactory lobes) associated with gonadotropin deficiency. Microphallus is present at birth in about 50% of the affected boys. Though most cases are sporadic but familial cases are also reported. It may have X-linked, autosomal dominant, or recessive inheritance with variable penetrance. There is heterogeneity of presentation. It is caused by mutations in the KAL gene, which is normally responsible for the typical pattern of migration of GnRH and olfactory neurons (Schwanzel-Fukuda and Pfaff 1989). Unilateral or bilateral absence of olfactory bulbs and sulci is noted on magnetic resonance imaging (MRI) in a majority of patients. Recently, mutations in other genes have also been implicated in Kallmann syndrome. These are FGFR1, prokineticin 2 (PROK2), and its receptor (PROKR2) genes (Dode et al. 2003, 2006).

Mutations in the GNRH gene (8p11.2-p21) have not been identified in humans, but a number of autosomal recessive mutations in the GNRHR are known to cause isolated hypogonadotropic hypogonadism (Layman et al. 1998).

14.8.2.3 Congenital or Acquired Hypopituitarism

Hypopituitarism is associated with deficiency of gonadotropin along with other pituitary hormones and can be congenital or secondary to other systemic illness. Isolated untreated GH deficiency also presents with delayed puberty, and till the achievement of pubertal bone age, it is difficult to determine whether gonadotropin deficiency is also present. Mutations in HESX-1 and PROP-1 genes, which lead to abnormal development of the pituitary, are associated with congenital hypopituitarism. Acquired causes of hypopituitarism include CNS tumors, commonest being craniopharyngioma, tuberculosis, trauma, and histiocytosis X. Hydrocephalus and cranial irradiation are conditions that may lead to either precocious or delayed puberty.

Syndromes such as Prader–Willi syndrome and Bardet–Biedl syndrome are also associated with hypogonadotropic hypogonadism.

14.8.2.4 Hypergonadotropic Hypogonadism

Bilateral testicular failure results in markedly elevated concentrations of serum gonadotropins, due to loss of feedback inhibition. The commonest cause of primary testicular failure is Klinefelter syndrome.

14.8.2.5 Klinefelter Syndrome

Males with Klinefelter syndrome usually have fibrotic testes and poorly developed external genitalia with gynecomastia. Because of the tall stature, children usually present late despite of delay in puberty onset. Borderline intellectual disability and behavioral difficulties are often present. Although genotype is typically 46 XXY, genetic variability and mosaicism also occur. Testes remain small with adult size of less than 6 ml. Though FSH, LH, and testosterone remain to be normal at beginning of puberty but as it progresses, testosterone tends to decline and FSH and LH levels increase.

14.8.2.6 Other Causes of Congenital Gonadal Failure

Noonan syndrome is caused by delayed sexual maturation, cryptorchidism, testicular atrophy, or anorchia. "Vanishing testes syndrome" is characterized by 46 XY karyotype and presents as delayed puberty with the absence of the testis. They can present as male with unilateral nonpalpable testis or as isolated micropenis depending on the time of assault during sexual maturation in utero. Other rare causes of congenital gonadal failure include defects in steroid hormone synthesis such as cholesterol desmolase complex and 3-beta-hydroxysteroid dehydrogenase that usually present with ambiguous genitalia.

14.8.2.7 Acquired Causes of Gonadal Failure

Acquired forms of primary hypogonadism may have postinfectious, autoimmune, traumatic, or metabolic etiology. Mumps orchitis occurring during adolescence or adulthood can cause gonadal failure. It can also occur after long-term use of chemotherapy like cyclophosphamide, radiation therapy or surgery.

14.8.3 Principles of Diagnostic Evaluation

Evaluation of delayed puberty is undertaken if puberty has not begun, i.e., testicular size is <4 ml or 2.5 cm in longitudinal axis, after 14 years of age or there is pubertal arrest after initiation. Detailed history, including history of any chronic illness, medication, and pubertal timing of siblings should be asked. Detailed anthropometric evaluation should be done including height, weight, arm span and upper to lower segment ratio, in addition to BMI. The presence of any dysmorphic features should be carefully noticed for syndromic association. Sexual maturation rating as given by Tanner should be done after detailed examination. Phallic length should be measured accurately. Other secondary sexual characteristics including facial and body hair, acne, and voice quality should be noted. The presence of gynecomastia should be noted as it may be indicative of prolactinoma or Klinefelter syndrome.

Diagnostic investigations are guided by the clues from history and examination. X-ray for bone age is useful as it correlates best with pubertal staging. Measurement of serum gonadotropins – LH, FSH, and testosterone – should be done, to differentiate between hyper- and hypogonadotropic hypogonadism. Complete blood count, renal function, liver function, and thyroid function test should be done to rule out associated systemic illness on the basis of clinical clues from history. If there is short stature or poor growth velocity, screening IGF-1, followed by dynamic testing for GH levels, may also be suggested to rule out GH deficiency. Karyotype is suggested if clinical features are suggestive of Klinefelter syndrome. If CNS causes are suspected, MRI of the brain is indicated.

Since it is difficult to distinguish CDGP from permanent hypogonadotropic hypogonadism (HH) as both present with low levels of LH, FSH, and testosterone, longitudinal observation till 17–18 years of age is needed to resolve the issue. An early morning testosterone levels of >0.7 nmol/L (20 ng/dL) in a sensitive assay have been stated to indicate secondary sexual development over the next 12–15 months. The GnRH stimulation test using GnRH or GnRH agonist has very low diagnostic value in distinguishing CDGP from hypogonadotropic hypogonadism due to significant variability in response. A more robust response to GnRH is observed in CDGP, whereas very low, flat responses are seen in HH. GnRH stimulation test using buserelin has shown a positive predictive value of 89% in distinguishing HH from CDGP (Wilson et al. 2006). The increment in testosterone after human chorionic gonadotropin (hCG) stimulation test has been reported to be >9 nmol/L in CDGP, compared to <3 nmol/L in permanent hypogonadotropic hypogonadism (Degros et al. 2003). An algorithmic approach to evaluation of delayed puberty is shown in Fig. 14.6.

14.8.4 Management

The treatment depends on the diagnosis and whether the pubertal delay is temporary or permanent. In patients with CDGP, reassurance of attainment of puberty at a later age than usual is generally sufficient. However, if there is considerable psychological distress and parental anxiety for delayed puberty, low-dose testosterone may be administered for a short period. Usually, 100 mg of testosterone enanthate is given intramuscularly every 4 weeks for 3 months. This not only helps in growth spurt and development of secondary sexual characters but also primes for spontaneous puberty. Repeat course of monthly injection of testosterone can be tried if there is no spontaneous progression to puberty after initial treatment (Soliman and De Sanctis 2012).

Permanent hypogonadotropic hypogonadism is managed with intramuscular injections of testosterone every 4 weeks. The starting dose is 100 mg, initiated at the normal age of pubertal onset. The dose is increased to 200–300 mg, which is continued indefinitely. Dermal patches or androgen gel preparations can also be used in place of injectable testosterone. In boys with concomitant GH deficiency, testosterone is started at a later age in lower doses to maximize height gain before epiphyseal fusion.

14.9 Precocious Puberty

Puberty in boys is considered as precocious if secondary sexual characters appear before 9 years of age. Precocious puberty (PP) is much less common in boys than girls. PP may be gonadotropin dependent which can be central/true precocious puberty (CPP) or gonadotropin independent that can be pseudo- or peripheral

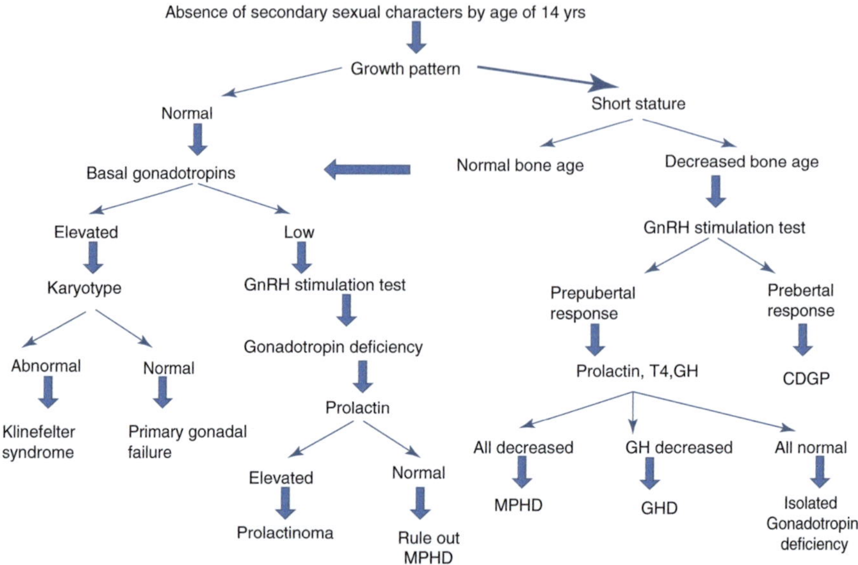

Fig. 14.6 Flowchart showing an approach to evaluate a boy with delayed puberty. *GnRH* gonadotropin-releasing hormone, *GHD* growth hormone deficiency, *CDGP* constitutional delay of growth and puberty, *MPHD* multiple pituitary hormone deficiency, *GH* growth hormone, *T4* thyroxine

precocious puberty (PPP). The HPG axis is active in central PP whereas there is no such activation of HPG axis in pseudo- or peripheral precocious puberty (Lee and Kerrigan 2004). The etiological classification of PP is given in Table 14.3.

14.9.1 Central/True Precocious Puberty (CPP)

Precocious activation of the HPG axis is defined as CPP. This condition is more common in girls as compared to boys. Boys with CPP have basal and GnRH-stimulated LH and testosterone levels as that in puberty but at an early age. Patients have features of puberty, such as development of sexual hair and other secondary sex characteristics, bilaterally symmetrical enlarged testes along with accelerated linear growth, and advanced bone age with early closure of epiphysis. Though the affected children are tall in childhood, but have a short final height as the epiphyseal fusion occurs early. Organic causes of CPP are more common in boys in comparison to girls. Hypothalamic hamartoma is the most common pathology seen in boys with CPP.

Hypothalamic hamartomas are nonneoplastic, tumor like lesions, composed of ectopic hypothalamic tissue, attached to the hypothalamus. These usually contain GnRH neurons, but in some tumors, transforming growth factor-alpha (TGF-α) may be found (Feuillan et al. 1999). Hamartomas are typically diagnosed by MRI. Some

Table 14.3 Causes of precocious puberty

1. Central/true precocious puberty (gonadotropin dependent)
(a) Idiopathic
(b) CNS abnormalities:
(i) Tumors: hypothalamic hamartoma, astrocytoma, craniopharyngioma
(ii) Infection: tuberculosis, abscess, encephalitis, trauma
(iii) Congenital malformation: hydrocephalus, arachnoid cyst
(iv) Others: cranial irradiation, chemotherapy
2. Pseudo- or peripheral precocious puberty (gonadotropin independent)
(a) Genetic: McCune–Albright syndrome
Familial testotoxicosis (activating mutation of LH receptor)
Congenital virilizing adrenal hyperplasia
(b) Tumors: Leydig cell tumor
Adrenal functional adenoma/carcinoma
CNS: chorioepithelioma
Hepatoma
(c) Others: primary hypothyroidism, iatrogenic
3. Combined (secondary) precocious puberty
(a) McCune–Albright syndrome
(b) Congenital adrenal hyperplasia

may present with gelastic – laughing/giggling – seizures that are resistant to anti-convulsant treatment. CPP due to hamartoma is usually diagnosed at an earlier age as compared to idiopathic and other organic etiologies. Medical management with GnRH agonists is the mainstay, although gamma knife surgery may be required in children with refractory seizures (Barajas et al. 2005). Figure 14.7 shows the MRI finding of hypothalamic hamartoma.

Other causes of CPP in boys include other CNS tumors such as glioma, pineal tumors, neurofibroma, astrocytoma, and rarely craniopharyngioma. History of cranial irradiation in children with malignancy, history of CNS infections such as tuberculosis, and other CNS conditions such as hydrocephalus are also associated with CPP. There may be associated GH deficiency in neurogenic CPP resulting in delayed rather than advanced bone age, as expected in precocious puberty.

Clinical photograph of a child with CPP due to neurofibromatosis type 1 is shown in Fig. 14.8.

Idiopathic CPP is a diagnosis of exclusion. With advances in the understanding of genetic determinants of pubertal onset, especially GPR54 receptor gene, its ligand kisspeptin, and GnRH receptor gene, the etiopathogenesis of more and more "idiopathic" cases is likely to be unraveled in the future.

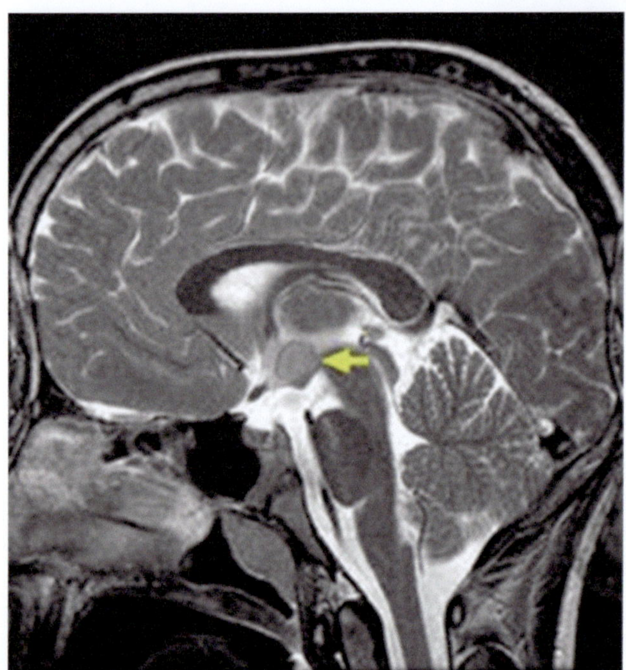

Fig. 14.7 Magnetic resonance imaging of brain, sagittal view showing hypothalamic hamartoma (*arrow*)

14.9.2 Pseudo- or Peripheral Precocious Puberty (PPP) or Gonadotropin-Independent Precocious Puberty

PPP accounts for up to 25% of all cases of precocious puberty in boys, whereas in girls PPP is very uncommon. The testosterone levels are usually in pubertal range, while gonadotropin levels are low and do not show a rise after stimulation with GnRH agonists.

Adrenal causes: Boys with simple virilizing type of congenital adrenal hyperplasia (CAH) due to 21-hydroxylase deficiency often present with precocious puberty. Usually, CAH is diagnosed late due to the absence of salt-wasting features in early infancy. Patients typically present with enlargement of the penis, and pubic hair development, without any enlargement of testes. Patients have tall stature for age with advanced bone age. Boys with inadequately treated CAH also present similarly. Some of these boys may develop secondary CPP as well with activation of HPG axis once the bone age is advanced to greater than 11–12 years.

CAH due to 11-β-hydroxylase deficiency may present with precocious puberty with hypertension. Children with adrenal tumors also present with pubic hair development and enlargement of the penis, in association with other features of adrenal excess, such as hirsutism, hypertension, and cushingoid facies.

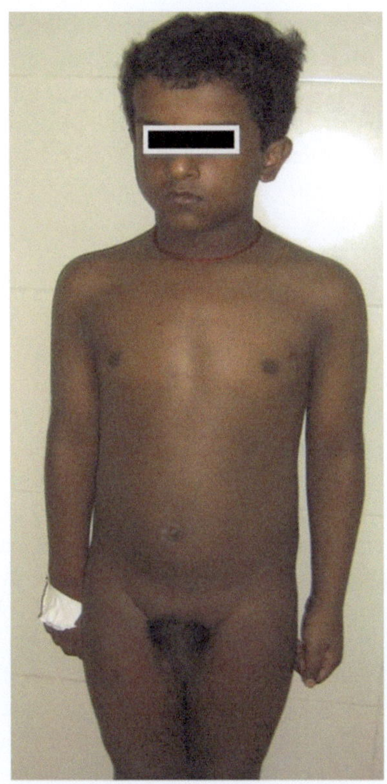

Fig. 14.8 Central/true precocious puberty in a 7-year-old boy with neurofibromatosis 1 (Note the muscular appearance and enlarged testis)

Familial testotoxicosis or male-limited gonadotropin-independent precocious puberty is an autosomal dominant inherited rare condition caused by activating mutations in the LH receptor. LH levels are low but testosterone levels are high, and there is symmetrical testicular enlargement with increased spermatogenesis, even though the HPG axis is not activated (Gondos et al. 1985).

Another cause of PPP in boys is hCG-secreting tumor and may be localized to organs other than gonads such as the liver, brain, or mediastinum. The α-subunit of hCG is identical to that of LH. Because of this similarity, hCG stimulates production of testosterone by Leydig cells resulting in testicular enlargement. Though the basal LH levels may be high due to cross-reactivity with hCG, but the response to GnRH is prepubertal. Serum alpha-fetoprotein (AFP) levels are usually elevated in patients with these tumors.

Conditions such as Leydig cell tumors and McCune–Albright syndrome are other rare causes of sexual precocity in boys.

14.9.3 Premature Adrenarche/Pubarche

Onset of sexual hair before the age of 9 years in boys, in the absence of development of other secondary sexual characters, is labeled as premature adrenarche/pubarche.

Premature adrenarche/pubarche is due to the increase of adrenal androgens as a result of increase in the activity of 17, 20 lyase, and 17 alpha hydroxylase enzymes. It may manifest only as a small increase in growth velocity in some children or as pubarche, i.e., development of genital and axillary hair. It requires close follow-up at regular intervals of 4–6 months as in some cases children with early adrenarche have early gonadarche as well.

14.9.4 Principles of Diagnostic Evaluation

Management of precocious puberty depends on identifying the underlying etiology. Detailed history and clinical examination is essential to establish an etiology. History includes information about the onset and progression of puberty and the presence of other secondary sexual characteristics. Family history of PP provides a clue in the diagnosis of familial testotoxicosis.

The presence of pubic hair and increase in penile size, in the absence of testicular enlargement, point toward adrenal etiology. Asymmetrical enlargement of the testis is more commonly seen in McCune–Albright syndrome and Leydig cell tumor cases, whereas symmetrical enlargement is observed in CPP, hCG-secreting tumors, and testotoxicosis. Children having only pubarche with normal bone age are considered normal variants and need to be followed-up. They have increased risk of having PP later on.

Baseline estimation of LH, FSH, testosterone, and DHEAS is essential in evaluation of precocious puberty, followed by GnRH or GnRH agonist stimulation test. On basal measurements, LH and FSH are suppressed in PPP cases, while pubertal level of LH is observed in CPP patients. Basal plasma testosterone is high in CPP as well as PPP, but the level is much higher in PPP. In GnRH stimulation test, single bolus of GnRH of 2.5 mcg/kg (maximum of 100 mcg) is administered followed by measurement of LH and FSH at 0, 30, and 60 min. For GnRH agonist stimulation test, leuprolide acetate 20 mcg/kg, to a maximum of 500 mcg, is administered intravenously, and LH and FSH are measured at 0, 60, 120, and 180 min. Only two measurements at 0 and 180 min may also be taken subsequently. The cutoffs for diagnosis of CPP vary according to the test protocol used. For the GnRH agonist test, a peak LH of >7 IU/L is generally considered as diagnostic of CPP (Partsch et al. 2002). The other endocrinological evaluations may include thyroid profile, basal and ACTH stimulated serum 17-hydroxyprogesterone level, serum hCG and α-fetoprotein measurement depending on the clinical signs.

Bone age analysis should be done in all cases of PP for advanced bone age. MRI for evaluating hypothalamus and pituitary should be performed in all boys with CPP. Other imaging investigations can be performed as guided by clinical and hormonal assays.

Approach to evaluation of precocious puberty in boys is given in Fig. 14.9.

14.9.5 Management

Boys with isolated adrenarche do not require any treatment, but regular follow-up is warranted. Hypothyroidism is treated with adequate dose of thyroxine. PP due to CAH

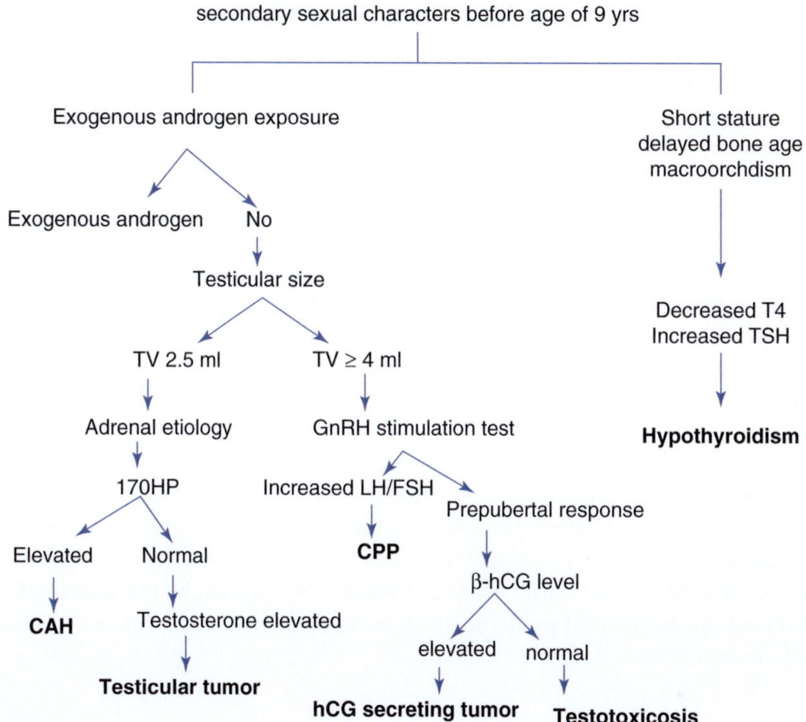

Fig. 14.9 Flowchart showing an approach to evaluate a boy with precocious puberty. *TV* testicular volume, *17OHP* 17-hydroxyprogesterone, *CAH* congenital adrenal hyperplasia. *β-hCG* beta human chorionic gonadotropin, *CPP* central/true precocious puberty, *GnRH* gonadotropin-releasing hormone, *T4* thyroxine, *LH* luteinizing hormone, *FSH* follicle-stimulating hormone, *TSH* thyroid-stimulating hormone

is managed with hydrocortisone and fludrocortisone. Though, there is no definite consensus regarding treatment for boys with CPP, but the decision is mostly based on the presence of psychological/behavioral concerns and prognosis regarding final height.

14.9.5.1 Central/True Precocious Puberty (CPP)

The treatment of choice for CPP is GnRH agonists. These are basically D-amino acid synthetic analogs with enhanced receptor occupancy. This results in receptor downregulation and suppression of the HPG axis in contrast to the physiological pulsatile exposure (Kaplan and Grumbach 1989). Depot preparations of GnRH agonists are given intramascularly on a once monthly or quarterly basis. The most commonly used preparations in our country are depot leuprolide acetate and depot triptorelin (Neely et al. 1992; Carel et al. 1995). Clinical response is assessed by monitoring pubertal stage, height velocity, and skeletal maturation. Plasma LH level < 2 IU/L, when measured 1 h after the dose of GnRH agonist depot, is also indication of adequate suppression. GnRH agonists are safe and well-tolerated drugs, but may have adverse effect of bone mineral accrual.

14.9.5.2 Pseudo- or Peripheral Precocious Puberty (PPP)

Treatment of PPP is difficult. Testolactone, spironolactone, ketoconazole, flutamide, cyproterone, and medroxyprogesterone are the available options used in various circumstances.

Treatment of PP with GnRH agonist should be discontinued when child has reached the usual age of puberty. The gonadotropin secretion is expected to resume within 4 months of stopping medication (Manasco et al. 1988).

14.10 Endocrine Disruptors and Puberty

Exogenous environmental chemicals that mimic or block the actions of endogenous hormones are known as endocrine disruptors (EnD; Rasier et al. 2008). EnD can be estrogenic, antiestrogenic, androgenic, antiandrogenic, or direct stimulant of GnRH production. They can alter hormonal signaling by targeting the hypothalamic–pituitary axis. EnDs with androgenic potential cause virilization phenotypes. EnDs with estrogenic activity can accelerate pubertal onset through kisspeptins, whereas those with antiandrogenic effects via suppression of testicular steroidogenesis and androgen receptor blockade cause delayed puberty. Dichlorodiphenyltrichloroethane (DDT), a commonly used pesticide, is an example of EnDs causing early puberty due to its estrogenic effect.

Key Questions

- Describe the environmental and nutritional factors influencing onset of puberty.
- At what age will you consider evaluation of a boy for delayed puberty?
- What are the causes of hyper- and hypogonadotropic hypogonadism?
- Write approach to a boy with precocious puberty.
- What are the treatment options for central/true precocious puberty in boys?

References

Abbassi V. Growth and normal puberty. Pediatrics. 1998;102:507–11.

August GP, Grumbach MM, Kaplan SL. Hormonal changes in puberty. 3. Correlation of plasma testosterone, LH, FSH, testicular size, and bone age with male pubertal development. J Clin Endocrinol Metab. 1972;34:319–26.

Barajas MA, Ramírez-Guzman MG, Rodríguez-Vázquez C, Toledo-Buenrostro V, Cuevas-Solórzano A, Rodríguez-Hernández G. Gamma knife surgery for hypothalamic hamartomas accompanied by medically intractable epilepsy and precocious puberty: experience in Mexico. J Neurosurg. 2005;102(suppl):53–5.

Biro FM, Huang B, Crawford PB, Lucky AW, Striegel-Moore R, Barton BA, et al. Pubertal correlates in black and white girls. J Pediatr. 2006;148:234–40.

Boyar RM, Rosenfeld RS, Kapen S, Finkelstein JW, Roffwarg HP, Weitzman ED, et al. Human puberty. Simultaneous augmented secretion of luteinizing hormone and testosterone during sleep. J Clin Invest. 1974;54:609–18.

Brito VN, Mendonca BB, Guilhoto LMFF, Freitas KCM, Arnhold IJP, Latronico AC. Allelic variants of the gamma-aminobutyric acid-A receptor alpha1-subunit gene (GABRA1) are not associated with idiopathic gonadotropin-dependent precocious puberty in girls with and without electroencephalographic abnormalities. J Clin Endocrinol Metab. 2006;91:2432–6.

Carel JC, Lahlou N, Guazzarotti L, Joubert-Collin M, Roger M, Colle M, et al. Treatment of central precocious puberty with leuprorelin. French Leuprorelin Trial Group. Eur J Endocrinol. 1995;132:699–704.

Cooper C, Kuh D, Egger P, Wadsworth M, Barker D. Childhood growth and age at menarche. BJOG. 1996;103:814–7.

Degros V, Cortet-Rudelli C, Soudan B, Dewailly D. The human chorionic gonadotropin test is more powerful than the gonadotropin-releasing hormone agonist test to discriminate male isolated hypogonadotropic hypogonadism from constitutional delayed puberty. Eur J Endocrinol. 2003;149:23–9.

Delemarre-Van De Waal HA, Wennink JM, Odink RJ. Gonadotrophin and growth hormone secretion throughout puberty. Acta Paediatr Scand Suppl. 1991;372:26–31.

de Muinck Keizer SM, Mul D. Trends in pubertal development in Europe. Hum Reprod Update. 2001;7:287–91.

de Ridder CM, JHH T, Bruning PF, van den Brande JL, Zonderland ML, WBM E. Body fat mass, body fat distribution, and pubertal development: a longitudinal study of physical and hormonal sexual maturation of girls. J Clin Endocrinol Metab. 1992;7:442–6.

Dodé C, Levilliers J, Dupont J-M, De Paepe A, Le Dû N, Soussi-Yanicostas N, et al. Loss-of-function mutations in FGFR1 cause autosomal dominant Kallmann syndrome. Nat Genet. 2003;33:463–5.

Dodé C, Teixeira L, Levilliers J, Fouveaut C, Bouchard P, Kottler ML, et al. Kallmann syndrome: mutations in the genes encoding prokineticin-2 and prokineticin receptor-2. PLoS Genet. 2006;2:1648–52.

Feuillan PP, Jones JV, Barnes K, Oerter-Klein K, Cutler Jr GB. Reproductive axis after discontinuation of gonadotropin releasing hormone analog treatment of girls with precocious puberty: long term follow up comparing girls with hypothalamic hamartoma to those with idiopathic precocious puberty. J Clin Endocrinol Metab. 1999;84:44–9.

Garcia MR, Amstalden M, Williams SW, Stanko RL, Morrison CD, Keisler DH, et al. Serum leptin and its adipose gene expression during pubertal development, the estrous cycle, and different seasons in cattle. J Anim Sci. 2002;80:2158–67.

Georgopoulos N, Markou K, Theodoropoulou A, Paraskevopoulou P, Varaki L, Kazantzi Z, et al. Growth and pubertal development in elite female rhythmic gymnasts. J Clin Endocrinol Metab. 1999;84:4525–30.

Gondos B, Egli CA, Rosenthal SM, Grumbach MM. Testicular changes in gonadotropin independent familial sexual precocity. Familial testotoxicosis. Arch Pathol Lab Med. 1985;109:990–5.

Gottsch ML, Cunningham MJ, Smith JT, Popa SM, Acohido BV, Crowley WF, et al. A role for kisspeptins in the regulation of gonadotropin secretion in the mouse. Endocrinology. 2004;145:4073–7.

Guo SS, Chumlea WC, Roche AF, Siervogel RM. Age- and maturity-related changes in body composition during adolescence into adulthood: the Fels Longitudinal Study. Int J Obes (Lond). 1997;21:1167–75.

He Q, Karlberg J. BMI in childhood and its association with height gain, timing of puberty, and final height. Pediatr Res. 2001;49:244–51.

Herman-Giddens ME, Slora EJ, Wasserman RC, Bourdony CJ, Bhapkar MV, Koch GG, et al. Secondary sexual characteristics and menses in young girls seen in office practice: a study from the pediatric research in office settings network. Pediatrics. 1997;99:505–12.

Herman-Giddens ME, Wang L, Koch G. Secondary sexual characteristics in boys: estimates from The National Health and Nutrition Examination Survey III, 1988-1994. Arch Pediatr Adolesc Med. 2001;155:1022–8.

Irwig MS, Fraley GS, Smith JT, Acohido BV, Popa SM, Cunningham MJ, et al. Kisspeptin activation of gonadotropin releasing hormone neurons and regulation of Kiss-1 mRNA in the male rat. Neuroendocrinology. 2004;80:264–72.

Kaplan SL, Grumbach MM. True precocious puberty: treatment with GnRH agonists. In: Delemarre vander-waal H, Plant TM, Vanrees JP, editors. Control of onset of puberty. Amsterdam: ELSEVIER; 1989. p. 357–73.

Kaplowitz PB, Slora EJ, Wasserman RC, Pedlow SE, Herman-Giddens ME. Earlier onset of puberty in girls: relation to increased body mass index and race. Pediatrics. 2001;108:347–53.

Karlberg P, Taranger J. The somatic development of children in a Swedish urban community. Acta Paediatr Scand Suppl. 1976;258:1–48.

Karpati AM, Rubin CH, Kieszak SM, Marcus M, Troiano RP. Stature and pubertal stage assessment in American boys: the 1988-1994 Third National Health and Nutrition Examination Survey. J Adolesc Health. 2002;30:205–12.

Kindblom JM, Lorentzon M, Norjavaara E, Lönn L, Brandberg J, Angelhed JE, et al. Pubertal timing is an independent predictor of central adiposity in young adult males: the Gothenburg osteoporosis and obesity determinants study. Diabetes. 2006;55:3047–52.

Knorr D, Bidlingmaier F, Butenandt O, Fendel H. Plasma testosterone in male puberty. I. Physiology of plasma testosterone. Acta Endocrinol. 1974;75:181–94.

Kulim HE, Bwido N, Mutie D, Santner SJ. Gonadotropin excretion during puberty in malnourished children. J Pediatr. 1984;105:325–8.

Layman LC, Cohen DP, Jin M, Xie J, Li Z, Reindollar RH, et al. Mutations in gonadotropin-releasing hormone receptor gene cause hypogonadotropic hypogonadism. Nat Genet. 1998;18:14–5.

Lee PA, Kerrigan JR. Precocious puberty. In: Pescovitz OH, Eugster EA, editors. Textbook of pediatric endocrinology. Philadelphia: Lip-pincott Williams & Wilkins; 2004. p. 316–30.

Leenstra T, Petersen LT, Kariuki SK, Oloo AJ, Kager PA, FO tK. Prevalence and severity of malnutrition and age at menarche; cross-sectional studies in adolescent schoolgirls in western Kenya. Eur J Clin Nutr. 2005;59:41–8.

Lienhardt A, Carel J-C, Preux P-M, Coutant R, Chaussain J-L. Amplitude of pubertal growth in short stature children with intrauterine growth retardation. Horm Res. 2002;57:88–94.

Manasco PK, Pescovitz OH, Feuillan PP, Hench KD, Barnes KM, Jones J, et al. Resumption of puberty after long term leutinizing hormone releasing hormone agonist treatment of central precocious puberty. J Clin Endocrinol Metab. 1988;67:368–72.

Marshall WA, Tanner JM. Variations in the pattern of pubertal changes in boys. Arch Dis Child. 1970;45(239):13–23.

Mitsushima D, Hei DL, Terasawa E. Gamma-aminobutyric acid is an inhibitory neurotransmitter restricting the release of luteinizing hormone-releasing hormone before the onset of puberty. Proc Natl Acad Sci U S A. 1994;91:395–9.

Neely EK, Hintz RL, Parker B, Bachrach LK, Cohen P, Olney R, et al. Two-year result of treatment with depot leuporide acetate for central precocious puberty. J Pediatr. 1992;121:634–40.

Ong KKL, Ahmed ML, Emmett PM, Dunger DB. Association between postnatal catch-up growth and obesity in childhood: prospective cohort study. BMJ. 2000;320:967–71.

Partsch CJ, Heger S, Sippell WG. Management and outcome of central precocious puberty. Clin Endocrinol (Oxf). 2002;56:129–48.

Plant TM, Ramaswamy S, DiPietro MJ. Repetitive activation of hypothalamic G protein-coupled receptor 54 with intravenous pulses of kisspeptin in the juvenile monkey (Macaca mulatta) elicits a sustained train of gonadotropin-releasing hormone discharges. Endocrinology. 2006;147:1007–13.

Pozo J, Argente J. Delayed puberty in chronic illness. Best Pract Res Clin Endocrinol Metab. 2002;16:73–90.

Proos LA, Hofvander Y, Tuvemo T. Menarcheal age and growth pattern of Indian girls adopted in Sweden. II. Catch-up growth and final height. Indian J Pediatr. 1991;58:105–14.

Quennell JH, Mulligan AC, Tups A, Liu X, Phipps SJ, Kemp CJ, et al. Leptin indirectly regulates gonadotropin-releasing hormone neuronal function. Endocrinology. 2009;150:2805–12.

Rasier G, Parent AS, Gerard A, Denooz R, Lebrethon MC, Charlier C, et al. Mechanisms of interaction of endocrine-disrupting chemicals with glutamate-evoked secretion of gonadotropin-releasing hormone. Toxicol Sci. 2008;102:33–41.

Reynolds EL, Wines JV. Physical changes associated with adolescence in boys. AMA Am J Dis Child. 1951;82:529–47.

Satyanarayana K, Naidu N. Nutrition and menarche in rural Hyderabad. Ann Hum Biol. 1979;6:163–5.

Schwanzel-Fukuda M, Pfaff DW. Origin of luteinizing hormone-releasing hormone neurons. Nature. 1989;338:161–4.

Sedlmeyer IL, Pearce CL, Trueman JA, Butler JL, Bersaglieri T, Read AP, et al. Determination of sequence variation and haplotype structure for the gonadotropin-releasing hormone (GnRH) and GnRH receptor genes: investigation of role in pubertal timing. J Clin Endocrinol Metab. 2005;90:1091–9.

Shahab M, Mastronardi C, Seminara SB, Crowley WF, Ojeda SR, Plant TM. Increased hypothalamic GPR54 signaling: a potential mechanism for initiation of puberty in primates. Proc Natl Acad Sci U S A. 2005;102:2129–34.

Silventoinen K, Haukka J, Dunkel L, Tynelius P, Rasmussen F. Genetics of pubertal timing and its associations with relative weight in childhood and adult height: the Swedish young male twins study. Pediatrics. 2008;121:e885–91.

Sklar CA, Kaplan SL, Grumbach MM. Evidence for dissociation between adrenarche and gonadarche: studies in patients with idiopathic precocious puberty, gonadal dysgenesis, isolated gonadotropin deficiency, and constitutionally delayed growth and adolescence. J Clin Endocrinol Metab. 1980;51:548–56.

Soliman AT, De Sanctis V. An approach to constitutional delay of growth and puberty. Indian J Endocrinol Metab. 2012;16(5):698–705.

Sun SS, Schubert CM, Chumlea WC, Roche AF, Kulin HE, Lee PA, et al. National estimates of the timing of sexual maturation and racial differences among US children. Pediatrics. 2002;110:911–9.

Sutherland RS, Kogan BA, Baskin LS, Mevorach RA, Conte F, Kaplan SL, et al. The effect of prepubertal androgen exposure on adult penile length. J Urol. 1996;156:783–7.

Tahirovic HF. Menarchal age and the stress of war: an example from Bosnia. Eur J Pediatr. 1998;157:978–80.

Tanner JM, Davies PS. Clinical longitudinal standards for height and height velocity for North American children. J Pediatr. 1985;107:317–29.

Tanner JM, Whitehouse RH, Marubini E, Resele LF. The adolescent growth spurt of boys and girls of the Harpenden growth study. Ann Hum Biol. 1976;3:109–26.

Teilmann G, Pedersen CB, Skakkebaek NE, Jensen TK. Increased risk of precocious puberty in internationally adopted children in Denmark. Pediatrics. 2006;118:e391–9.

Terasawa E, Luchansky LL, Kasuya E, Nyberg CL. An increase in glutamate release follows a decrease in gamma aminobutyric acid and the pubertal increase in luteinizing hormone releasing hormone release in the female rhesus monkeys. J Neuroendocrinol. 1999;11:275–82.

Toogood AA. Endocrine consequences of brain irradiation. Growth Horm IGF Res. 2004;14:S118–24.

Veldhuis JD, Roemmich JN, Rogol AD. Gender and sexual maturation-dependent contrasts in the neuroregulation of growth hormone secretion in prepubertal and late adolescent males and females—a general clinical research center-based study. J Clin Endocrinol Metab. 2000;85:2385–94.

Weinstein RL, Kelch RP, Jenner MR, Kaplan SL, Grumbach MM. Secretion of unconjugated androgens and estrogens by the normal and abnormal human testis before and after human chorionic gonadotropin. J Clin Invest. 1974;53:1–6.

Wilson DA, Hofman PL, Miles HL, Unwin KE, CE MG, Cutfield WS. Evaluation of the buserelin stimulation test in diagnosing gonadotropin deficiency in males with delayed puberty. J Pediatr. 2006;148:89–94.

Male Sexual Function

15

Mona Sharma and Anand Kumar

Learning Objectives

- Normal penile dimensions
- Structure of the penis
- Phases of male sexual response cycle
- Physiology of penile erection
- Neurotransmitters, hormones, and molecular pathways mediating penile erection
- Disorders of male sexual function and their management

15.1 Introduction

One of the major concerns in reproductive medicine is the problem of erectile dysfunction (ED) which develops progressively with age and even affects younger people. The current understanding of male sexual function comprises of neuroendocrine and vascular events causing penile erection. Alterations in these events lead to ED. This chapter begins with understanding of the structure of male copulatory organ, the penis, followed by details of sexual response cycle, sexual dysfunction, and recent treatment modalities.

M. Sharma, MD, DNB, MAMS (✉) • A. Kumar, MD, FAMS
Department of Reproductive Biology, All India Institute of Medical Sciences, New Delhi, India
e-mail: dr.mona18sharma@gmail.com

A. Kumar, M. Sharma (eds.), *Basics of Human Andrology*,
DOI 10.1007/978-981-10-3695-8_15

15.2 Normal Penile Dimensions in Adult Males

A study has been conducted on 80 men in the age group of 21–82 years on penile size of the adult human male (Wessels et al. 1996). Penile length is measured from pubopenile skin junction to the urethral opening in flaccid penis, in flaccid but stretched penis, and in erect penis. The penile circumference is taken at mid-shaft. The fat pad depth is measured by pushing the prepubic fat with tape onto the pubic bone and measuring the distance from the bone surface to the surface of unpushed skin. Functional penile length is calculated by adding erect length and fat pad depth (see Fig. 15.1). The results of this study show that the stretched flaccid length and erect length are the same (see Table 15.1). The authors recommend penile augmentation in patients having erect length less than 7.5 cm or 2.95 inches. The study shows that pubic fat pad depth is increased in obese men. The actual penile length is affected by the fat pad depth. In such cases, weight loss and suprapubic liposuction can be the better alternatives for penile lengthening (Wessells et al. 1996). Penile lengthening procedures are described further in this chapter.

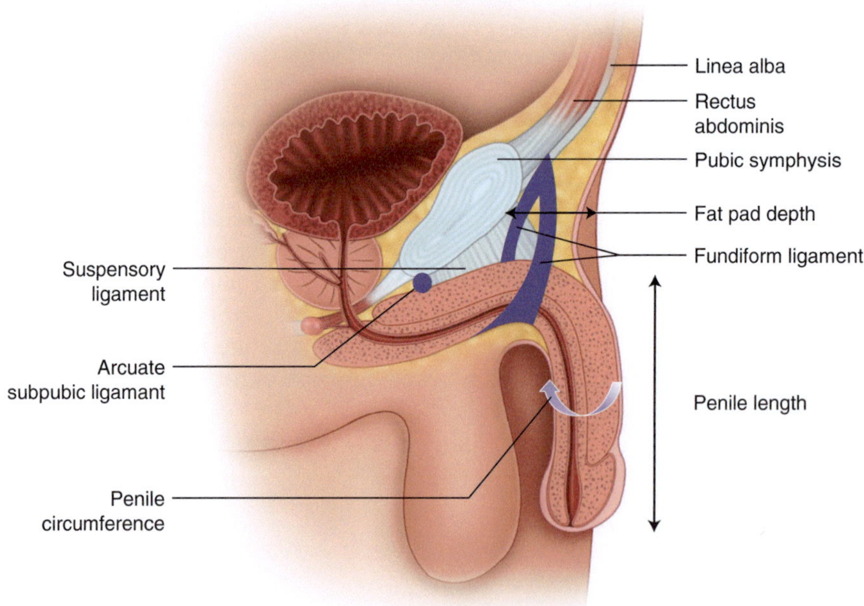

Fig. 15.1 Schematic diagram showing measurements of the penis and suspensory apparatus

Table 15.1 Normal penile dimensions (Adapted and modified from Wessells et al. 1996)

Dimensions	Range
Flaccid length	5.0–15.5 cm or 1.96–6.10 inches
Stretched length	7.5–19 cm or 2.95–7.48 inches
Erect length	7.5–19 cm or 2.95–7.48 inches
Functional length	21.25–87 cm or 8.36–34.25 inches
Flaccid circumference	6.5–13 cm or 2.55–5.11 inches
Erect circumference	9.0–16.0 cm or 3.54–6.29 inches
Fat pad depth	0.5–7.5 cm or 0.19–2.95 inches
Flaccid volume	21.25–87.0 ml
Erect volume	33.75–152.0 ml

15.3 Structure of the Penis

The penis consists of body and a root. The body is the dangling part covered by skin, whereas the root is the attached part lying in the perineum. The body of penis consists of three erectile masses of tissues termed as corporal bodies. Each contains irregular endothelium-lined vascular spaces which are called cavernous spaces or sinusoids. Paired corpora cavernosa (CC) lie dorsally and the median corpus spongiosum (CS) lies ventrally. The paired crura and a bulb form the root of the penis. Each crus is a blunt and elongated structure attached to the ischiopubic ramus. Both crura converge and together form CC of the body of the penis. Between the two crura lies the bulb of the penis which is firmly attached to the inferior surface of the perineal membrane. The bulb continues anteriorly into the CS (see Fig. 15.2a, b).

The CC and CS are surrounded by a dense fascial sheath known as the tunica albuginea (Standring 2005). The CS lies below the CC and it is the structure which is traversed by the urethra. It gradually expands toward the end of the penis and forms glans penis (see Fig. 15.2). The penis is covered by a superficial loose penile fascia. The deepest fibers condense to form Buck's fascia which anchors the penis

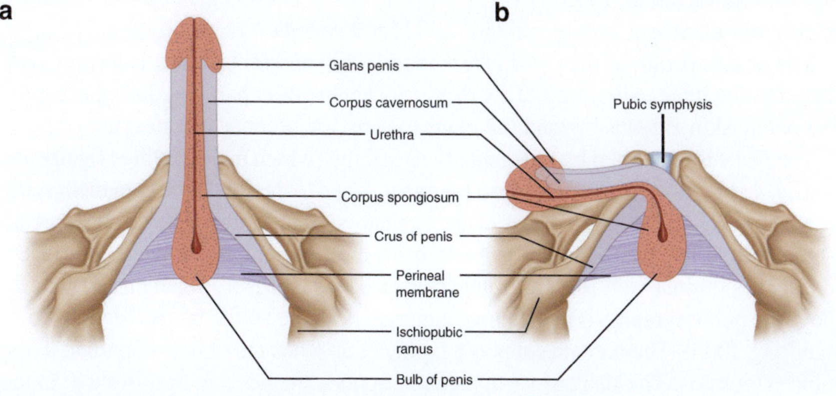

Fig. 15.2 Schematic diagram showing different parts of the penis in (**a**) erect and (**b**) flaccid state

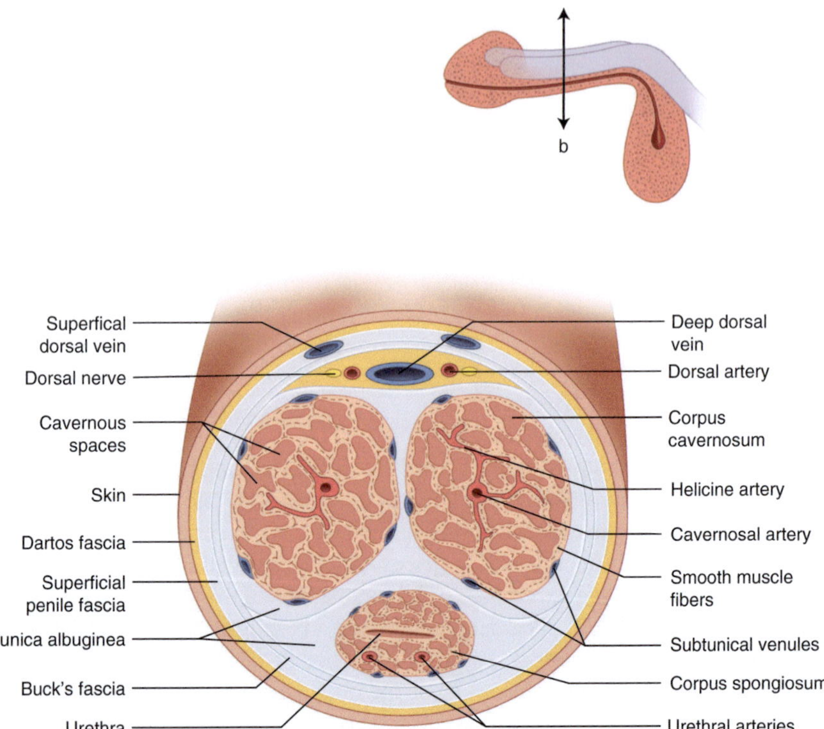

Fig. 15.3 Cross section across plane ab showing internal structures of the penis

to the symphysis pubis. This fascia compresses the subtunical plexus and circumflex veins during erection and limits venous drainage (Kirby 1999). Microscopically, smooth muscle fibers, extracellular matrix, network of endothelial cell-lined sinuses or cavernous spaces, and helicine arteries form CC as shown diagrammatically in Fig. 15.3 (Lerner et al. 1993). The CS also consists of endothelial cell-lined sinuses, smooth muscle fibers, collagen fibers, and blood vessels.

The skin covering of the penis is remarkably thin and mobile as it is loosely connected to the tunica albuginea. This makes it capable of expanding during erection. The penile skin extends beyond the glans to form the prepuce or foreskin.

The penis is supported by "suspensory apparatus" which includes three ligaments. The most superficial is the fundiform ligament. Deep to the fundiform ligament is the suspensory ligament and deepest is the arcuate subpubic ligament (Hoznek et al. 1998). The fundiform ligament extends from the lower fibers of linea alba in the anterior abdominal wall to the scrotal septum. The suspensory ligament is attached above the pubic symphysis and, below, it merges with the penile fascia (see Fig. 15.1; Standring 2005). The arcuate subpubic ligament attaches the tunica albuginea to the pubic symphysis. The suspensory apparatus supports the penis in flaccid state. In the erect penis, the suspensory ligament keeps the base of the penis in front of the pubis thereby supporting the free part of the penis during erection (Hoznek et al. 1998).

15.4 Male Sexual Response Cycle

Male sexual response has been described under different classifications. One of the classifications is the *psychosexual cycle* which was first defined by the team of William Masters and Virginia Johnson. Their work was published in the book *Human Sexual Response* (Masters and Johnson 1966). They divided this cycle into four phases: excitement, plateau, orgasm, and resolution. Each of these phases consists of two penile events as represented in Fig. 15.4. Another classification is a *functional classification* which is considered clinically most relevant. It consists of five phases: sexual desire, erection, ejaculation, orgasm, and detumescence (Walsh and Wilson 1987). Figure 15.4 is the schematic representation of psychosexual phases which does not exactly correspond to the phases of functional classification, but some overlapping can be observed. In functional classification of sexual cycle, phase of *sexual desire* corresponds to excitement phase of psychosexual cycle; phase of erection forms part of plateau phase of psychosexual cycle; phase of *ejaculation* forms part of orgasm phase of psychosexual cycle; *orgasm* phase is a separate phase in functional sexual cycle; and phase of *detumescence* forms part of resolution phase of psychosexual cycle. As the functional classification is clinically important, so is described next in detail. The phase of ejaculation usually occurs with orgasm, so both the phases will be described as one phase.

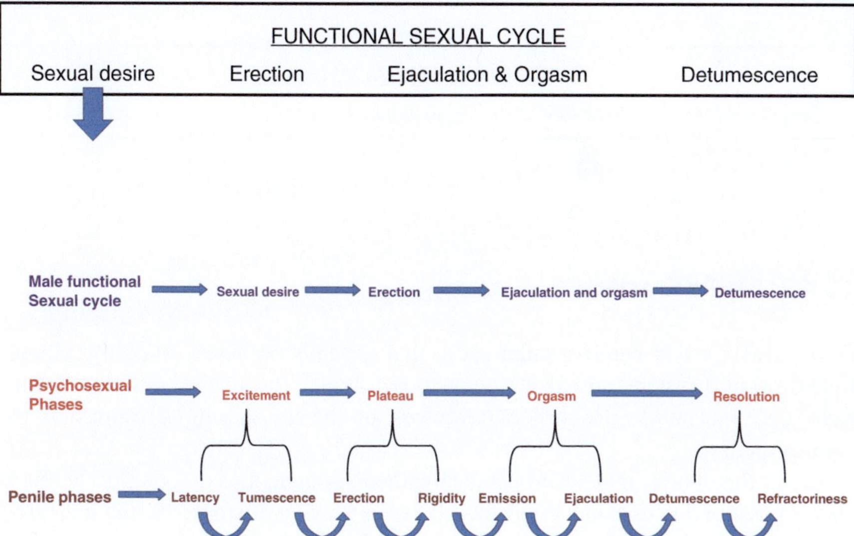

Fig. 15.4 Schematic diagram showing phases of male sexual cycle and corresponding phases of psychosexual cycle and penile activity phases

15.4.1 Sexual Desire

Upon sexual stimulation, integration of various sensory stimuli such as visual, tactile, olfactory, auditory, and cerebral results in activation of sacral parasympathetic pathway. This leads to smooth muscle relaxation of corporal bodies and helicine arterioles. This causes increased blood flow into corporal bodies and enlargement of the penis. This event is known as penile tumescence.

Positron emission tomography (PET) and functional MRI (fMRI) studies have described various brain regions which get activated during sexual arousal. Based on these studies, a neurobehavioral model of brain processes involved during visually evoked sexual arousal has been established (Stoleru et al. 2003; Redoute et al. 2000). The model consists of four components: (1) *Cognitive* component assesses visual stimuli as sexual and involves activation of the bilateral inferior temporal cortex, orbitofrontal cortex, superior and inferior parietal lobules, and left anterior cingulate cortex. (2) *Motivational* component processes the visual stimuli and directs behavior to a sexual goal. This component involves activation of the left cingulate cortex. (3) *Neuroendocrine and autonomic* component coordinates the pathways regulating cardiovascular, respiratory, and genital responses. This component involves activation in the left anterior cingulate cortex and hypothalamus. (4) *Emotional* component includes pleasure associated with sexual arousal and involves activation of the right insula and sensory cortex (see Fig. 15.5). The details of the brain areas are given in chapter "Male Behaviors I: Brain Areas Regulating Male Behaviors."

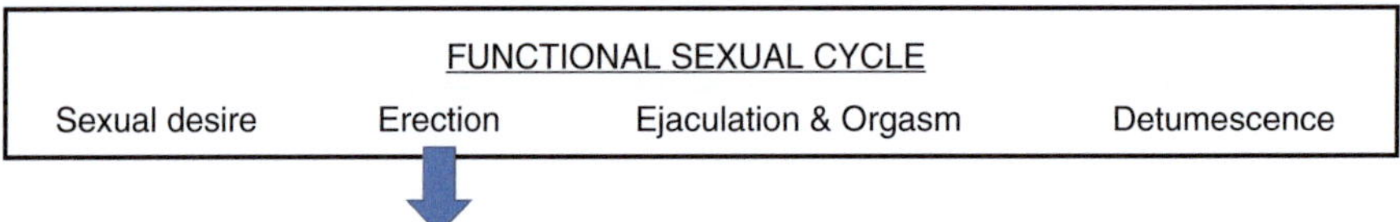

15.4.2 Penile Erection

Penile erection is a neurovascular event that depends on neural integrity, a functional vascular system, and healthy cavernosal tissue. The smooth muscles of arterioles and sinusoids in the corpora cavernosa are the key structures responsible for penile erection.

Under the influence of sacral parasympathetic system, the smooth muscle relaxation of corporal bodies and arterioles allows expansion of sinusoids and increased blood flow into the sinusoids, respectively (see Fig. 15.6). Penile sinusoids engorge, but further engorgement is curtailed by the fibrous tunica albuginea which does not stretch, resulting in increased intracorporeal pressure of around 100 mm Hg. The veins between the tunica albuginea and corporal bodies are compressed and blocked because of increased intracorporeal pressure. This does not allow venous blood to return, thereby maintaining rigidity of organ (Dean and Lue 2005). Erection thus

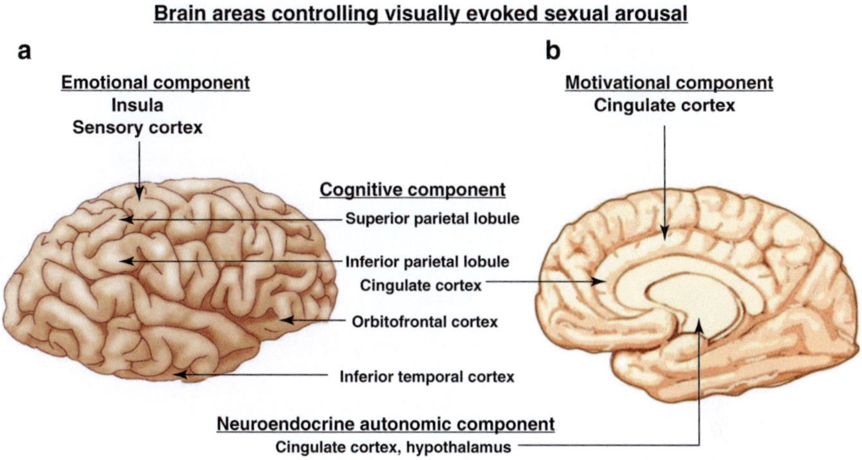

Fig. 15.5 Brain areas involved during visually evoked sexual arousal (**a**, lateral view; **b**, medial view)

comprises of three important events: sinusoidal relaxation, arteriolar dilatation, and venous occlusion.

During erection, blood flow increases in the corpus spongiosum in the same manner as in the corpora cavernosa, but the rise in pressure is not high. This is because the tunical covering over the spongiosum is very thin and, therefore, venous occlusion is minimal. During full erection, the corpus spongiosum functions as arteriovenous shunts. Ischiocavernosus and bulbospongiosus muscles compress the spongiosal veins that further raises pressure and leads to rigidity.

15.4.2.1 Types of Erection

At least three kinds of erections can be distinguished in man: psychogenic, reflexogenic, and nocturnal types (Dean and Lue 2005). *Psychogenic* erection is a result of audio or visual stimuli producing sexual arousal and stimulating spinal centers controlling erection. The tactile stimuli to genitals produce sensations reaching the spinal erection centers, thereby producing *reflexogenic* erections. *Nocturnal penile* erections or tumescence is a physiologic phenomenon in normal healthy males and occurs several times during rapid eye movement (REM) phase of sleep (Schmidt and Schmidt 2004). Although not very reliable, many clinical centers still use nocturnal penile tumescence test to monitor penile erection, mainly to differentiate organic vs psychogenic ED. Nocturnal penile tumescence remains present in psychogenic ED but is absent in organic ED.

15.4.2.2 Neuroendocrine Regulation of Erection

Penile erection is regulated by coordination between the central autonomic and peripheral somatic pathways. Both the pathways are controlled by various neurotransmitters and hormones. The central control of penile erection is described in chapter “Male Behaviors II: Neural Pathways Regulating Behaviors.”

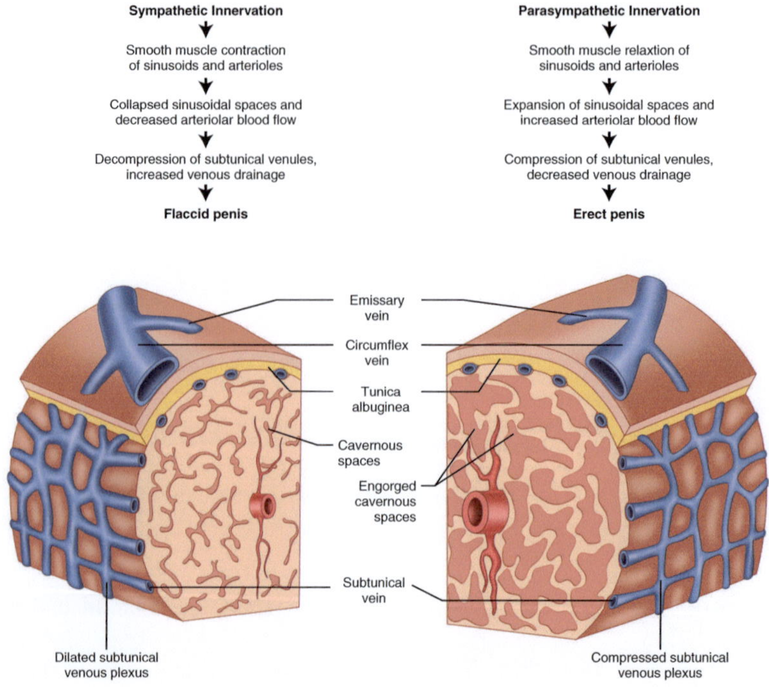

Fig. 15.6 Schematic diagram showing the hemodynamic changes during penile erection

Peripheral Control of Penile Erection

The penis is innervated by both sympathetic and parasympathetic nerves. Depending on the different neurotransmitters or modulators released, nerves can be adrenergic, cholinergic, and non-adrenergic noncholinergic (NANC). In the next section, various agents mediating relaxation and contraction of penile vessels and smooth muscles are described.

Relaxation Mediating Agents

Acetylcholine

Penile cavernosal smooth muscles and endothelium receive rich cholinergic innervations (Dail 1993; Hedlund et al. 1999). The released acetylcholine (Ach) from parasympathetic cholinergic nerves acts on the muscarinic receptors – M_1–M_4 – present in the penile cavernosum (Traish et al. 1995). M_2 receptors are present on the smooth muscles, whereas M_3 receptors are present on the endothelium (Toselli et al. 1994; Traish et al. 1995). The mechanism of Ach-induced cavernosal smooth muscle and arteriolar relaxation is indirect. This is either by inhibiting the release of contractant- noradrenaline (NA), by stimulating muscarinic receptors on adrenergic nerve terminals or by release of relaxation-producing factor, nitric oxide (NO), from the endothelium (Klinge and Sjostrand 1977; Andersson and Wagner 1995).

Nitric Oxide (NO)

The role of NO in penile erection is widely accepted (Burnett 1997). The endothelium and the NANC nerves are the source of NO in the cavernosal tissue. NO is a small gaseous, lipophilic signaling molecule that is produced by nitric oxide synthase (NOS) in NANC neurons and endothelium. NO induces relaxation of CC smooth muscles, resulting in sinusoidal filling and penile erection (Ignarro et al. 1990). Three distinct isoforms of NOS have been identified. The neuronal isoform (nNOS) is present in the nerve endings, and the endothelial (eNOS) is mainly membrane-bound in endothelial cells of penile blood vessels and trabecular tissue (Dail et al. 1995; Stanarius et al. 1999). Both isoforms require calcium (Ca^{2+}) and calmodulin for activity. The third inducible isoform (iNOS) is Ca^{2+}-independent and is also expressed in human cavernosal smooth muscle cells during pathological conditions such as penile fibrosis (Rajasekaran et al. 1998; Gonzalez-Cadavid et al. 1999). But, it does not have a direct physiological role in penile erection. NO derived from nNOS in the nitrergic nerves appears to be responsible for the initiation of the smooth muscle relaxation, whereas NO from eNOS contributes to the maintenance of erection. The mechanism of NO-mediated smooth muscle relaxation and erection is described further in this chapter.

Prostanoids

The major prostanoids synthesized in the human cavernosum are PGD_2, PGE_2, $PGF_{2\alpha}$, PGI_2, and thromboxane A_2 (Miller and Morgan 1994). These prostanoids mediate their effects by binding to the G protein-coupled receptors, namely, DP, EP, FP, IP, and TP receptors, respectively (Pierce et al. 1995; Narumiya et al. 1999). Prostanoids $PGF_{2\alpha}$ and thromboxane A_2 are involved in contraction of erectile tissue, whereas PGE_2 is responsible for smooth muscle relaxation (Lee et al. 1999).

Molecular Mechanism of Penile Vascular and Cavernosal Smooth Muscle Relaxation

The two important mechanisms causing penile smooth muscle relaxation are the guanyl cyclase (GC)/cyclic guanosine monophosphate (cGMP) and adenyl cyclase (AC)/cyclic adenosine monophosphate (cAMP) pathways. Parasympathetic stimulation releases Ach and inhibits NA release from nerve endings in the penis. Ach binds to muscarinic receptors on endothelial cells which increase inositol triphosphate (IP_3). This leads to increase in intracellular Ca^{2+} that stimulates eNOS and production of NO. The released NO from endothelial cells and from NANC nerve endings diffuses into adjacent smooth muscle cells of corpora cavernosum. There, it binds to GC enzyme which converts guanyl triphosphate (GTP) into cGMP (Dean and Lue 2005).

cGMP further activates protein kinase G that phosphorylates certain proteins which lead to opening of potassium channels and hyperpolarization, inhibition of Ca^{2+} channels thereby blocking Ca^{2+}entry into cell, and inhibition of release of Ca^{2+} by the endoplasmic reticulum (ER; Saenz de Tejada et al. 2004). All these events lead to decreased intracellular Ca^{2+} due to which myosin detaches from actin filament. The actin-myosin dissociation causes smooth muscle relaxation (Walsh 1991). cGMP also blocks activation of Rho-kinase pathway. This pathway causes smooth muscle contraction, and therefore, blocking of this pathway is an important step for muscle relaxation and penile erection. Rho-kinase pathway is described further in this chapter.

Smooth muscle relaxation is also mediated by prostaglandins (Alexandre et al. 2007). PGE_1 binds to the receptors present on the smooth muscle cells and activates AC that converts adenosine triphosphate (ATP) into cAMP.

cAMP activates protein kinase A which further reduces intracellular Ca^{2+} release similar to cGMP (see Fig. 15.7). Phosphodiesterase (PDEs) converts these cyclic nucleotides into inactive metabolites 5'GMP and 5'AMP, respectively. PDE-5 is expressed predominantly in CC and is a key enzyme in the NO/cGMP pathway causing smooth muscle relaxation and penile erection (Leite et al. 2007).

Contraction Mediating Agents

Noradrenaline (NA)

The sympathetic innervations to the penis via adrenergic supply keep it in flaccid state. The released NA from the sympathetic nerve endings binds to α-adrenergic receptors (α-AdR) in the penile vessels and cavernosal smooth muscles. The binding predominantly to α_1AdR and α_2AdR causes vessel as well as muscle contractions (Andersson and Wagner 1995). The stimulation of α_2AdR in horse penile

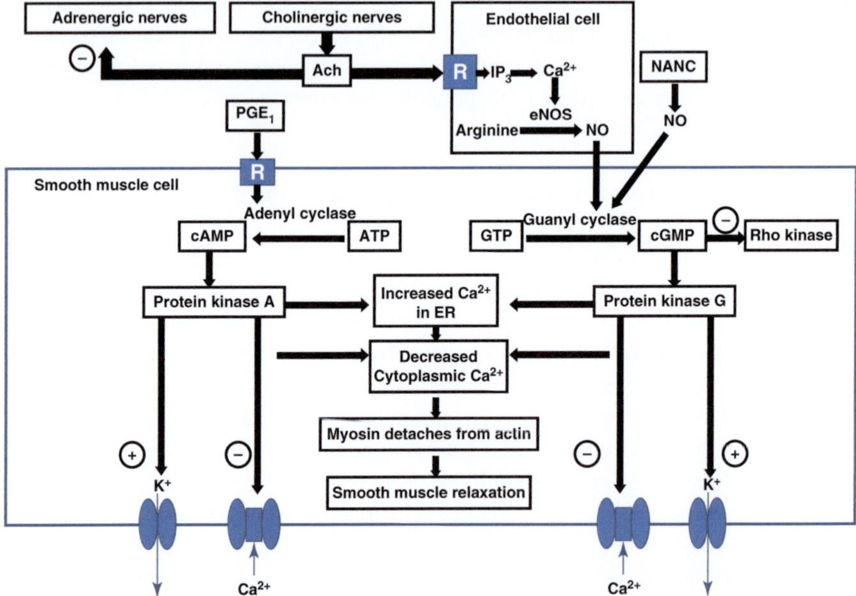

Fig. 15.7 Mechanism of penile vascular and cavernosal smooth muscle relaxation. *Ach* acetylcholine, *PGE1* prostaglandin E_1, *IP3* inositol triphosphate, *Ca2+* calcium, *K+* potassium, *eNOS* endothelial nitric oxide synthase, *NANC* noradrenergic noncholinergic, *NO* nitric oxide, *R* receptor, *ATP* adenosine triphosphate, *cAMP* cyclic adenosine monophosphate, *GTP* guanosine triphosphate, *cGMP* cyclic guanosine monophosphate, *ER* endoplasmic reticulum

vessels is associated with inhibition of NANC nerves and their transmitter release (Simonsen et al. 1997). This has been suggested that NA-mediated inhibition of release of NO maintains detumescence as blockade of AdR increases release of NO (Saenz de Tejada et al. 2000).

Endothelins

Endothelins are the vasoconstricting peptides primarily synthesized in the endothelium of blood vessels. The expression of ET 1-mRNA has been observed in the human cavernosum (Saenz de Tejada et al. 1991). The receptors for endothelins are ET_A and ET_B. These have been found in corporal smooth muscle membranes (Christ et al. 1995). It has also been suggested that ET-1 produces long-lasting contractions in smooth muscles of penile vessels and cavernous spaces (Becker et al. 2000). The other contraction mediating endothelins are ET-2 and ET-3 which are less potent than ET-1 (Saenz de Tejada et al. 1991). ET-1 causes contractions by opening of Ca^{2+} channels and Ca^{2+} influx (Holmquist et al. 1990).

Angiotensins

The increased levels of angiotensin II have been observed in cavernous blood during detumescent stage. The human cavernosal tissue produces angiotensin II which mediates contractions in the cavernosal smooth muscles in humans (Becker et al. 2001). This effect is increased on inhibiting NOS (Comiter et al. 1997). The intracavernosal administration of angiotensin II receptor blocker, losartan, causes smooth muscle relaxation leading to erection (Kifor et al. 1997).

Molecular Mechanism of Penile Vascular and Cavernosal Smooth Muscle Contraction

Penile sympathetic stimulation releases NA from adrenergic nerve endings along with the release of endothelins and PGF2α from endothelial cells (Granchi et al. 2002). All these bind to the receptors present on the smooth muscle cells. This binding stimulates phospholipase C (PLC) activity (Web 2003). PLC upregulates the activity of phosphatidylinositol 4,5 bisphosphate (PIP_2). PIP_2 further catalyzes the formation of two second messengers, inositol triphosphate (IP_3), and diacylglycerol (DG). IP3 binds to the receptors on ER resulting in the release of Ca^{2+} into cytosol. DG with released Ca^{2+} activates protein kinase C (PKC). PKC promotes phosphorylation of Ca^{2+} channels that leads to increase Ca^{2+} influx through these channels. This increases intracellular Ca^{2+} concentration (Walsh 1991). Ca^{2+} binds to protein calmodulin and this complex activates myosin light chain kinase (MLCK) enzyme. MLCK phosphorylates myosin light chain (MLC). The MLC phosphorylation allows cycling of myosin cross bridges with actin that leads to muscle contraction. Therefore, MLC phosphorylation is a prerequisite for muscle contraction. MLC phosphorylation is also regulated by MLC phosphatase (MLCP). MLCP has MLC-binding subunit and it dephosphorylates MLC thereby causing muscle relaxation. The phosphorylation of myosin-binding subunit of MLCP inactivates its enzymatic activity which enables MLC to remain phosphorylated (see Fig. 15.8). The phosphorylated state of MLC promotes muscle contraction (Somlyo and Somlyo 2000).

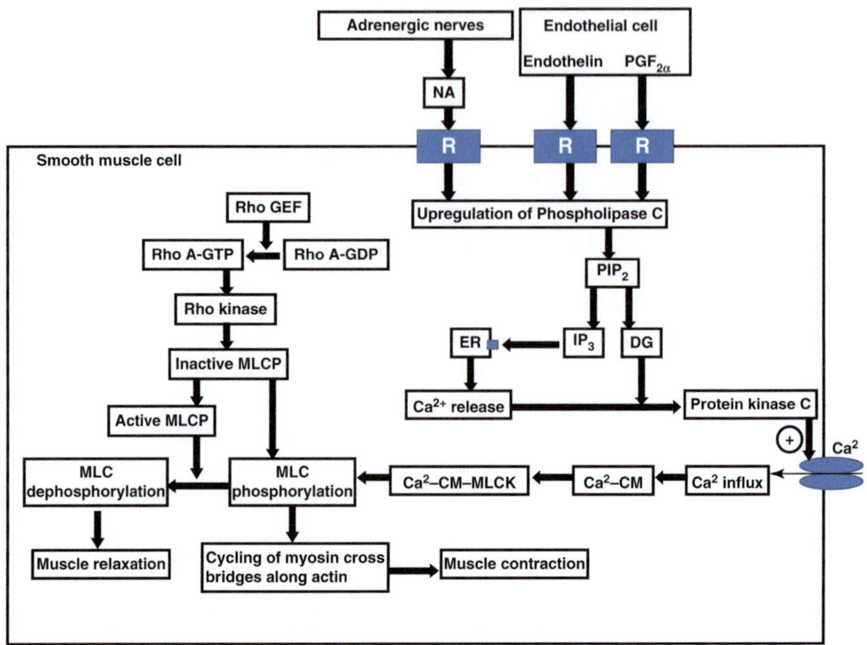

Fig. 15.8 Mechanism of penile vascular and cavernosal smooth muscle contraction. *NA* noradrenaline, *PGF2α* prostaglandin $F_{2α,}$ *R* receptor, *Rho A* Ras homolog gene family member A, *RhoGEF* Rho guanine nucleotide exchange factor, *GTP* guanosine triphosphate, *GDP* guanosine diphosphate, *MLC* myosin light chain, *MLCP* myosin light chain phosphatase, *MLCK* myosin light chain kinase, *PIP2* phosphatidylinositol 4,5 bisphosphate, *IP3* inositol triphosphate, *DG* diacylglycerol, *CM* calmodulin, *Ca2+* calcium, *K+* potassium

Increase in intracellular Ca^{2+} concentration is transient and further contraction is maintained by Ca^{2+}-sensitizing mechanism which gets initiated simultaneously with PLC activation (Hirano 2007). The mechanism includes a protein, Ras homolog gene family member A protein (Rho A), which is a small GTPase protein that regulates cytoskeletal dynamics, cell cycle, and cell polarity (Boureux et al. 2007; Bustelo et al. 2007). Rho A exists in two forms: inactive GDP-bound and an active GTP-bound form. The Rho guanine nucleotide exchange factor (RhoGEF) upregulates the exchange of GDP for GTP, thereby generating the active form of Rho A (Rho A-GTP). Rho A-GTP upregulates activity of Rho-kinase. Rho-kinase phosphorylates myosin-binding subunit of MLC phosphatase and inhibits its activity. This enables MLC to remain phosphorylated thereby promoting contraction (Hirano 2007).

15.4.2.3 Recent Insights into the Molecular Pathways of Erection Physiology

The emerging molecular pathways of erection physiology are important as these may be the underlying mechanisms causing ED. Among these pathways are the renin-angiotensin system, mitogen-activated protein kinases, tumor necrosis factor-α (TNF-α), and arginase II. These pathways can be the target of future drug therapies for ED.

Renin-Angiotensin System

It has been proved that renin-angiotensin system (RAS) exists in CC and is involved in erectile function (Becker et al. 2001). Angiotensin II (Ang II) is the main active peptide produced locally from RAS (Bader and Ganten 2008). It acts on angiotensin 1 (AT_1) receptors present on the smooth muscle and endothelial cells and activates Rho A/Rho-kinase pathway (Ryan et al. 2004). The Rho-kinase activation leads to inactivation of MLC phosphatase. The mechanism downstream of activation of Rho A/Rho-kinase pathway causing smooth muscle contraction has been discussed in the previous section of this chapter. The activation of Rho A/Rho-kinase pathway by Ang II eventually leads to ED (Ying et al. 2006). Another role of AT_1 in RAS is the activation of nicotinamide adenine dinucleotide phosphate (NADPH) oxidase which increases reactive oxygen species (ROS) production (Jin et al. 2004). ROS rapidly reacts with NO and reduces its bioavailability. This leads to diminished or absent smooth muscle relaxation that leads to ED.

Mitogen-Activated Protein Kinases

Mitogen-activated protein kinases (MAPK) belong to group of serine/threonine protein kinases, usually involved in stress response, programmed cell death, and immune mechanisms (Liu et al. 2007). The extracellular signal-regulated kinase 1/2 (ERK 1/2) is one of the most important MAPK pathways involved in ED. ERK regulate eNOS expression in endothelial cells. It stimulates phosphorylation of eNOS and inactivates eNOS for further production of NO (Bernier et al. 2000). The ERK-mediated inhibition of eNOS has been observed in human CC. Moreover, patients with ED show increased ERK expression in CC (Sommer et al. 2002).

Tumor Necrosis Factor-Alpha (TNF-α)

TNF-α is a proinflammatory cytokine involved in various cardiovascular disorders (Meldrum et al. 1998; Berk et al. 2001). The major target for TNF-α action is the endothelium. It impairs endothelium-mediated vasodilatation by decreasing release of NO (Chia et al. 2003). Endothelium dysfunction has been the key pathological event in ED. Oxidative stress and inflammatory conditions predispose to endothelial dysfunction. TNF-α activates Rho A/Rho-kinase pathway and increases Ca^{2+} sensitivity in CC (Mong et al. 2008; Hunter et al. 2003). TNF-α downregulates the expression of nNOS in CC (Carneiro et al. 2009). The upregulation of Rho A/Rho-kinase pathway and decreased nNOS lead to decreased smooth muscle relaxation and ED.

Arginase

Another recent addition in the pathological molecules of ED is the arginase. This enzyme exists in two isoforms- I and II, and both isoforms exist in CC (Cox et al.

1999; Mori and Gotoh 2000). Arginase II is majorly involved in diabetes and age-related ED (Bivalacqua et al. 2001; Numao et al. 2007). Enhanced arginase activity is reported in oxidative stress and inflammatory conditions (Bivalacqua et al. 2007). Arginase catalyzes the conversion of L-arginine into ornithine and urea. L-Arginine is also the substrate for NOS (Boucher et al. 1999). Therefore, increased arginase activity decreases overall production of NO that eventually leads to decreased smooth muscle relaxation and ED.

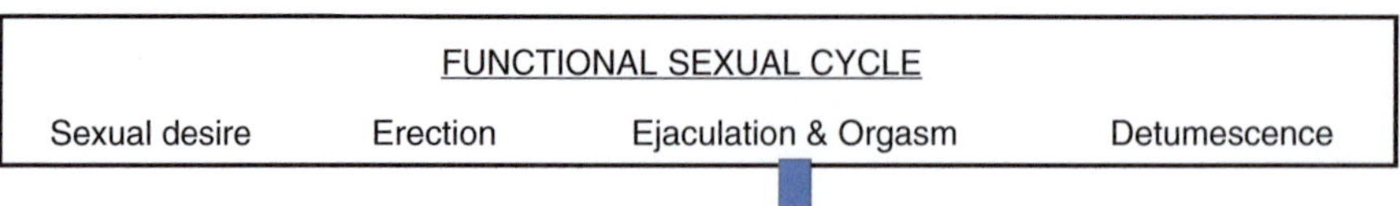

15.4.3 Ejaculation and Orgasm

Ejaculation constitutes the next phase of the sexual response cycle and is usually accompanied by orgasm. Ejaculation consists of two sequential events: emission and expulsion. Both the events are under the control of sympathetic nervous system innervating reproductive organs taking part in ejaculation. Two groups of anatomical structures are specifically involved and distinguished in each event. The organs involved during emission are the epididymis, vas deferens, seminal vesicles, prostate gland, prostatic urethra, and bladder neck. The urethra and bladder neck along with pelvic striated muscles participate during expulsion of ejaculate (Giuliano and Clement 2005).

Emission is ejection of sperm and secretions of accessory sex glands into the prostatic urethra. During the emission phase, sperm are transported by peristaltic contractions of smooth muscles of the vas deferens. The vas deferens joins the duct of the seminal vesicles at its ampullar ending and forms the ejaculatory duct. The ejaculatory duct travels through the prostate gland and enters the urethra at the level of verumontanum.

Expulsion is ejection of sperm along with glandular secretions from the urethra to the meatal opening. Adequate expulsion of ejaculate requires a synchronized relaxation of the external urinary sphincter with simultaneous bladder neck closure during rhythmic contractions of the striated muscles of the pelvic floor, including bulbocavernosus, ischiocavernosus, and levator ani muscles (Giuliano and Clement 2005). The rhythmic contraction of these muscles facilitates movement of seminal fluid through the urethra and, finally, its expulsion from the urethral meatus (Gerstenburg et al. 1990). The neurophysiology of ejaculation has been discussed in the chapter, Neurovascular Supply and Lymphatic Drainage of Male Reproductive Organs.

The brain areas activated during ejaculation process are the medial preoptic area (MPOA), paragigantocellular nucleus of pons (nPGi), parvocellular subparafascicular nucleus of the thalamus (SPF_P), bed nucleus of stria terminalis (BNST), paraventricular nucleus (PVN), and medial amygdala (MeA). The neurotransmitters regulating ejaculation are serotonin (5-HT), dopamine (DA), oxytocin (OXT), NO, NA, and Ach. Among these neurotransmitters, 5-HT plays a primary role in physiology of ejaculation. Most of the 5-HT neurons are located in the brain stem raphe nuclei. 5-HT neuronal terminals have been found in the MPOA, BNST, and MeA in rats (Descarries et al. 1975; Azamitia and Gannon 1983). The thoracolumbar sympathetic and spinal motor neurons are involved in ejaculation. The autonomic components innervate pelvic organs such as the prostate gland, seminal vesicles, and bulbospongiosus muscle that take part in ejaculation.

The spinal interneurons in lumbar (L) 3–4 segments in rats have been identified which are involved in ejaculation (Truitt and Coolen 2002). These interneurons are named as lumbar spinothalamic cells (LSt) or spinal generator for ejaculation. The sensory inputs from genital areas reach the LSt cells through pudendal nerve fibers. The sensory fibers of the pudendal nerve terminate close to LSt cells although the evidence of direct communication between LSt cells and pudendal nerve terminals is still lacking (McKenna and Nadelhaft 1986). The LSt cells project to the autonomic neurons innervating pelvic organs such as the prostate gland and seminal vesicles and to the motor neurons innervating the bulbospongiosus muscle (Xu et al. 2005). LSt cells also send projections to SPF_P (Coolen et al. 2003). SPF_P sends output fibers to MPOA, BNST, and MeA (Yasui et al. 1991; Canteras et al. 1995). Neurons from MPOA project to PVN from where the output fibers project directly to the autonomic neurons and pudendal motor neurons in rats (Luiten et al. 1985; McKenna and Nadelhaft 1986). MPOA has reciprocal connections with other brain areas involved in ejaculation such as PVN, nPGi, BNST, and MeA (Simerly and Swanson 1988; Rizvi et al. 1992; Murphy et al. 1999). The fibers from nPGi project to the L3–L5 spinal segments. The nerve terminals reaching L3–L5 spinal segments release 5-HT which inhibits ejaculation (Marson and McKenna 1992, 1996). The pathway involved in ejaculation has been schematically represented in Fig. 15.9.

The ejection of semen is usually associated with orgasm. Orgasm is the objective of sexual stimulation and is perceived by the sensory cortex as pleasurable. After the start of orgasm, pulses of semen begin to flow from the urethra, reach a peak discharge, and then diminish in flow. The typical orgasm consists of 10–15 contractions. Once the first contraction has taken place, ejaculation will continue to completion as an involuntary process. At this stage, ejaculation cannot be stopped. Initial contractions occur at an average interval of 0.6 s with an increment of 0.1 s per contraction (Newman et al. 1991). The rate of contractions gradually slows during orgasm (Gerstenburg et al. 1990). The defects in male sexual cycle may lead to orgasm without ejaculation and even ejaculation in the absence of orgasm.

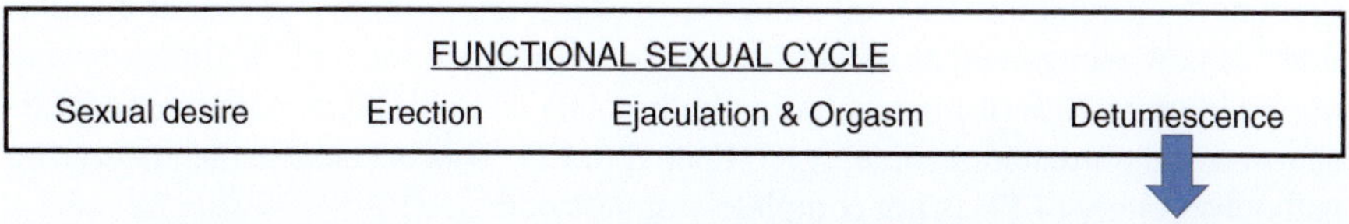

FUNCTIONAL SEXUAL CYCLE

Sexual desire Erection Ejaculation & Orgasm Detumescence

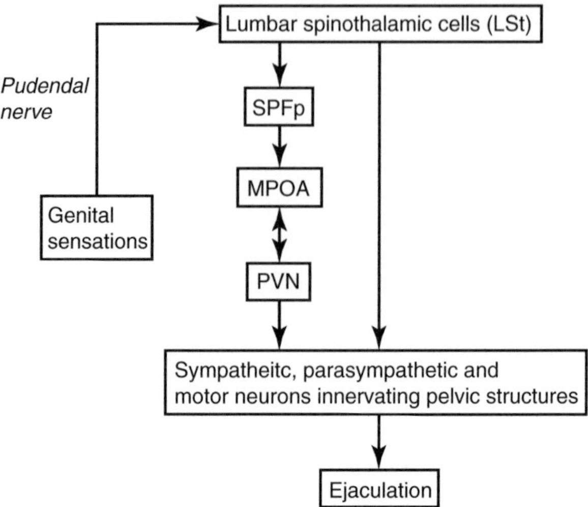

Fig. 15.9 Schematic diagram showing pathway involved in ejaculation. *SPFp* parvocellular part of the subparafascicular thalamus, *MPOA* medial preoptic area, *PVN* paraventricular nucleus

15.4.4 Detumescence

The balance between contraction and relaxation factors in the penis determines the state of penile rigidity or flaccidity. Detumescence is the result of cessation of release of neurotransmitters mediating penile relaxation, the breakdown of second messengers such as cGMP by PDEs, or sympathetic discharge following ejaculation. Contraction of the trabecular smooth muscle reopens the venous channels, the trapped blood is expelled, and flaccidity returns. The maintenance of the intracorporeal smooth muscle in a contracted state results from three factors: intrinsic myogenic activity, adrenergic neurotransmission, and endothelium-derived contraction mediating factors such as $PGF_{2\alpha}$ and endothelin (Andersson and Wagner 1995). The mechanism of smooth muscle relaxation has been discussed in the previous section of this chapter.

15.5 Disorders of Male Sexual Function

15.5.1 Premature Ejaculation

Premature ejaculation (PE), also known as "early ejaculation" or "rapid ejaculation," is now recognized as one of the most common male sexual dysfunctions and occurs in up to 30% of men including both young and old. PE remains an underdiagnosed and untreated disorder as compared to ED. This is because the underlying pathophysiology of PE is not completely understood.

Masters and Johnson, the research pioneers in the field of human sexual behavior, explained PE as a condition where a man is unable to delay his ejaculation until his partner got sexually satisfied. Recently, the American Urological Association Guideline on the Pharmacologic Management of Premature Ejaculation defined PE as “ejaculation that occurs sooner than desired, either before or shortly after penetration, causing distress to either one or both partners” (Montague et al. 2004). For clinical assessment and research purpose, the “intravaginal ejaculation latency time” (IELT) is considered the most objective parameter (Waldinger et al. 1994). IELT is the time taken from the vaginal penetration to the start of intravaginal ejaculation. Based on this parameter, PE can be classified as subjective or objective type. *Subjective* PE is defined as loss of voluntary control over ejaculation and is associated with male or both partner distresses. *Objective* PE can be severe when ejaculation occurs before penetration or with IELT ≤ 15 s, moderate with IELT ≤ 1 min, and mild with IELT < 2 min (Jannini et al. 2005).

There are many pathological conditions associated with PE such as psychorelational when there are partner relation disturbances and performance anxiety; neurobiological in cases of multiple sclerosis, spina bifida, and tumors of the spinal cord; urological like chronic prostatitis; and hormonal such as hyperthyroidism (Jannini et al. 2006). The underlying mechanism of PE in these conditions is largely unknown because of inadequate experimental support.

Current efforts focus on developing combined psychopharmacological strategies to reduce the attached stigma and improve education on this very common sexual dysfunction (Rowland et al. 2004). At this time, no drugs have been approved by regulatory agencies for treatment of PE (Montague et al. 2004). Based on their side effect of delayed ejaculation, selective serotonin reuptake inhibitors (SSRIs) such as paroxetine and fluoxetine are being used as off-label drugs to treat PE (Waldinger et al. 1998; McMahon and Touma 1999).

15.5.2 Erectile Dysfunction (ED)

ED is a widespread problem of men affecting all reproductive age groups. According to the International Society of Impotence Research, ED is defined as the inability to achieve or maintain penile erection sufficient enough to acquire a satisfactory sexual performance (Lizza and Rozen 1999). The same society classified ED into organic and psychogenic types with further subtypes of each (see Fig. 15.10).

15.5.2.1 Organic Causes of ED

Vasculogenic

Arteriogenic

Generalized atherosclerotic processes are associated with diabetes, hyperlipidemia, and coronary artery diseases, all of which produce arterial insufficiency and, therefore, impair penile perfusion that leads to ED. The underlying mechanisms causing arteriogenic ED are the systemic vasoconstriction and failure of endothelium-dependent vasodilatation (Michal and Ruzbarsky 1980; Rosen et al. 1990; Junemann et al.1991).

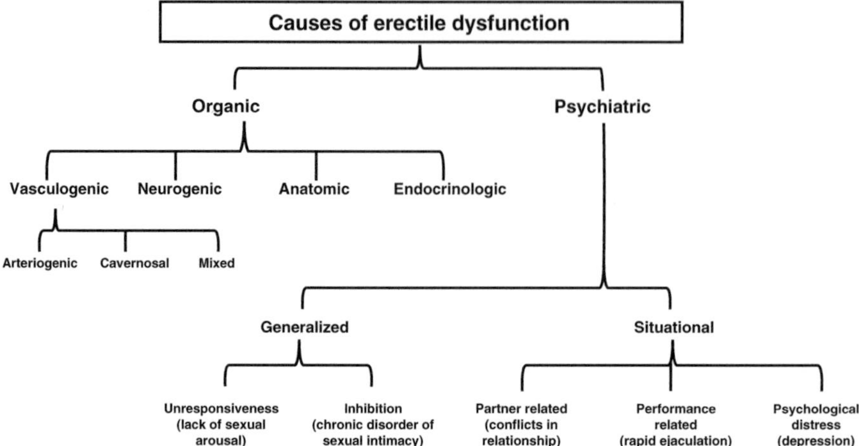

Fig. 15.10 Causes of erectile dysfunction

Cavernosal

ED can occur due to inadequate venous occlusion which is a pathological event associated with various degenerative diseases such as Peyronie's disease, diabetes, and old age (Metz et al. 1983). The mechanism underlying these degenerative diseases is altered fibroelastic components of sinusoids, smooth muscles, and the endothelium. Increase in collagen fibers and decreases in elastic fibers in the tunica albuginea impair veno-occlusive mechanisms (Rajfer et al. 1988). Excessive adrenergic tone in anxious individuals may lead to insufficient smooth muscle relaxation causing inadequate sinusoidal expansion and venular compression (Christ et al. 1990). All these events lead to ED.

Neurogenic

Erection is a neurovascular event. Any disease affecting the MPOA, PVN, and hippocampus can cause neurogenic ED as these are the main integration centers for male sexual function. The disorders affecting these areas are Parkinson's disease, encephalitis, and temporal lobe epilepsy and are usually associated with ED (Wermuth and Stenager 1992). Other neural disorders associated with ED are Alzheimer's disease, spina bifida, tumors, and disc herniation (Eardley and Kirby 1991). Iatrogenic damage to cavernous nerves during radical prostatectomy and other perineal surgeries may also lead to ED (Yeager and Van Heerden 1980; Finkle and Taylor 1981).

Anatomic

The structural malformations of the penis may cause ED. One of them is Peyronie's disease (PD). It is a connective tissue disorder that is characterized by localized fibrotic plaques in the tunica albuginea, most commonly on the dorsal surface of the penis, that result in penile bending and often pain (Taylor and Levine 2007). PD occurs due to chronic inflammation of the tunica albuginea (CITA). This inflammatory process itself is self-limited and the pain typically resolves with time (Gelbard et al. 1990). In cases of progressive disease, medical treatment is given to reduce fibrosis. Antioxidants such as vitamin E, receptor blockers of transforming growth factors such as tamoxifen citrate, and Ca^{2+} channel blockers such as verapamil are also used topically which reduce fibroblast activity and enhance local collagenase activity (Teloken et al. 1999; Martin et al. 2002; Safarinejad et al. 2007). But, the medical treatment does not give good results. Unfortunately, by progression, penile deformity remains in 90–95% of patients. Surgery is the sole conclusive treatment for this sequel. In addition, PD is frequently associated with ED; in a recent retrospective study of 1,001 patients with PD, 58.1% of patients reported having ED (Mulhall et al. 2006; Kadioglu et al. 2011).

Chordee is a congenital malformation of the penis where the tip of the penis curved downward due to the very short fibrous urethra that pulls the penis downward. This condition usually accompanies hypospadias where the urethra opens underneath the penis. It causes recurrent infection and inflammation in the urethra causing fibrosis and curved penis which leads to ED (Mieusset and Soulié 2005).

Endocrinologic

Effects of androgens on male sexual behavior are well known. Testosterone enhances libido and frequency of nocturnal erections (Mulligan and Schmitt 1993). Men on androgen ablation therapy for prostate cancer present with decreased libido and ED. Increased prolactin levels also result in sexual dysfunction by inhibiting the secretion of gonadotropin-releasing hormone that leads to decreased testosterone levels. Both the hyperthyroid and hypothyroid conditions lead to ED.

15.5.2.2 Psychogenic Causes of ED

Psychogenic ED is mediated by few mechanisms – by direct inhibition of the spinal erection centers by the brain, by increased sympathetic outflow, and by increased peripheral catecholamine secretions which prevent smooth muscle relaxation (Steers 1990; Diederichs et al. 1991; Bancroft 2000). The common causes are sexual unresponsiveness or inhibition leading to lack of arousability which can also be age related, partner related such as poor relations and sexual preferences, performance related such as anxiety and rapid ejaculation, and psychological distress such as depression.

15.5.2.3 Principles of Management of ED

Historically, ED management was largely empirical and comprised of aphrodisiacs, herbal supplements to achieve erectile enhancement despite meager academic support for their use. Early scientific progress was achieved in psychoanalysis, sex therapy, and endocrine treatments. Initial mechanical and device concepts included penile splints, vacuum constrictive devices, and now improved surgically inserted penile prosthetics. See Fig. 15.11 for management options of ED.

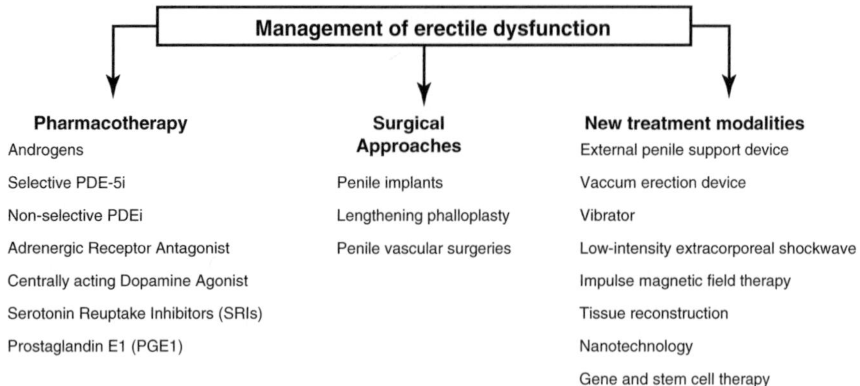

Fig. 15.11 Management of erectile dysfunction

Pharmacotherapy

Androgens

Testosterone therapy is encouraged in patients with ED associated with hypogonadism. In eugonadal patients with ED, testosterone treatment activates sexual behavior without enhancing erectile capacity (Schiavi et al. 1997).

Selective Phosphodiesterase-5 Inhibitors

Selective phosphodiesterase-5 inhibitors (PDE-5i) appears to be the most important enzyme in the control of normal penile erectile activity. Sildenafil, commercially available as Viagra, tadalafil, and vardenafil are selective PDE-5i (Boswell-Smith et al. 2006). It increases concentration of cGMP in penile tissue causing increased smooth muscle relaxation and better erection. Avanafil is also a selective PDE-5i having an advantage of rapid onset of action (Kedia et al. 2013).

Nonselective Phosphodiesterase Inhibitors (PDEi)

Papaverine is a nonselective PDEi which increases concentrations of both cAMP and cGMP in erectile tissue (Jeremy et al. 1997). It has been found effective in neurogenic and psychogenic ED but is not a promising therapy in vasculogenic ED.

Adrenergic Receptor Antagonists

Adrenergic stimulation leads to smooth muscle contraction of cavernosal arteries causing decreased blood flow and smooth muscle contraction of sinusoids leading to their collapse. These events lead to detumescence. Phentolamine is an α_1- and α_2-adrenergic receptor blocker and causes smooth muscle relaxation and penile erection (Gwinup 1988). This drug has been found effective in mild to moderate ED.

Centrally Acting Dopamine Agonist

Apomorphine is a nonselective DA agonist and acts on D1 and D2 receptors in PVN which is a brain area regulating male sexual function. Apomorphine enhances erectile efficiency through dopaminergic and oxytocinergic pathways. Dopaminergic receptors are located in the oxytocinergic and dopaminergic neurons in PVN. Apomorphine binds to these receptors and activates these neurons which project to the spinal centers mediating erection (Heaton et al. 1995).

Selective Serotonin Reuptake Inhibitors (SSRIs)

Serotonin is a negative regulator of erection. Trazodone is a serotonin antagonist and reuptake inhibitor which is used as antidepressant. Its use in ED has shown some positive results but still, more clinical trials are needed to prove its proerectile benefits (Lance et al. 1995).

Prostaglandin E1

The role of prostaglandin E1 (PGE1) in penile erection is very well established. Alprostadil is a synthetic PGE1 and has been used for ED. The drug is used as transurethral therapy and is found beneficial (Williams et al. 1998). Intracavernous therapy has also been used in ED where alprostadil is given in combination with papaverine and phentolamine (Stief and Wetterauer 1988; Linet and Neff 1994; Fallon 1995).

Surgical Approaches

Penile Implants

Latest penile implants are developed for delivering vasodilator drugs to CC. This device has two parts: a canula connected to a scrotal reservoir. The scrotal pump on squeezing delivers drug into CC (Lim 2003). Further trials and research are needed to establish benefits of this device. Another implant which is yet to be available commercially, is made up of nickel-titanium alloy (Le et al. 2013). This device changes its configuration between flaccid and erect type with application of heat.

Lengthening Phalloplasty

These procedures are good options for patients with ED due to short penis in cases of Peyronie's disease or trauma. The procedure includes division of the suspensory ligament that leads to increase in apparent length of the penis. The procedure is followed by penile augmentation by inserting silicon prosthesis in the space created after ligament excision to prevent reattachment. In obese men, suprapubic fat pad is also excised in order to increase the apparent length of the penis (Li et al. 2006).

Penile Vascular Surgeries

There are two types of penile vascular surgeries done in ED. First is the venous ligation where penile veins such as dorsal penile, circumflex, emissary, and cavernous veins are ligated or selectively embolized (Lue 1989). Another vascular surgery is the arterial revascularization where the inferior epigastric artery is anastomosed with the dorsal penile vein (Virag et al. 1981). Apart from anastomosing, balloon dilatation of stenosed arteries is also done (Babaev and Jhaveri 2012). These surgeries lead to increased arterial blood flow and venous engorgement that enhances erectile function.

Mechanical Devices and Other Treatment Modalities

External Penile Support Device

This is a recently introduced mechanical device called "Erektor." It is worn during sexual intercourse. This device provides length and rigidity to the penis. It consists of two rings: the base ring and an expandable ring, attached at the ends of flexible tubes (see Fig. 15.12). The penile shaft is placed in the two rings keeping the rod on the ventral side of the shaft. The "penile cast" is another device having a shape similar to the penile shaft. This device is also worn during intercourse and is available in different sizes. There are no clinical trials available to prove efficacy of these devices. But these can provide good noninvasive therapeutic options to men with ED (Stein et al. 2014).

Vacuum Erection Devices

These devices are commercially available since the 1980s. These have three components: vacuum pump, a cylinder, and constriction rings. The rings are placed at the open end of cylinder. The cylinder is placed over the penis and pump is operated to create a vacuum (see Fig. 15.13). Once penile engorgement occurs, rings are transferred at the base of the penis to maintain erection during intercourse (Nadig et al. 1986). Human trials have been done which show good therapeutic potential of these devices in ED after radical prostatectomy (Tal et al. 2011).

Vibrator

The first approved penile stimulatory vibrator is called "Viberect." The device has circular pads which are placed in contact with the penile shaft and a touch pad on the dorsal side to customize the vibration intensity (see Fig. 15.14). The vibrator stimulates branches of the pudendal nerve that further activates parasympathetic pathway and NANC fibers in the penis. NANC fibers release NO as neurotransmitter that eventually leads to penile smooth muscle relaxation and erection (Tajkarimi and Burnett 2011). The use of this device demands clinical trials to show its benefits in ED.

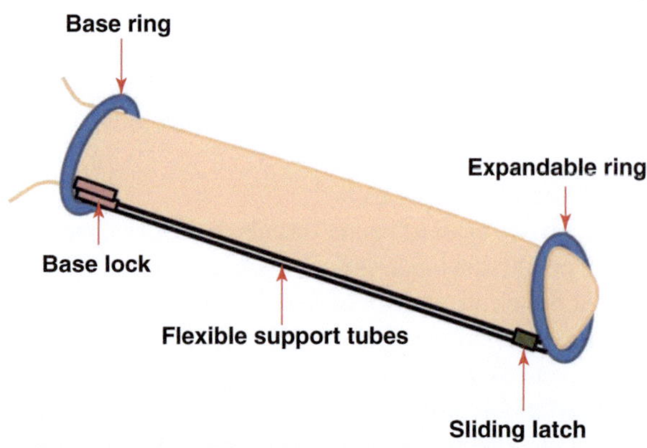

Fig. 15.12 Schematic diagram showing different parts of Erektor

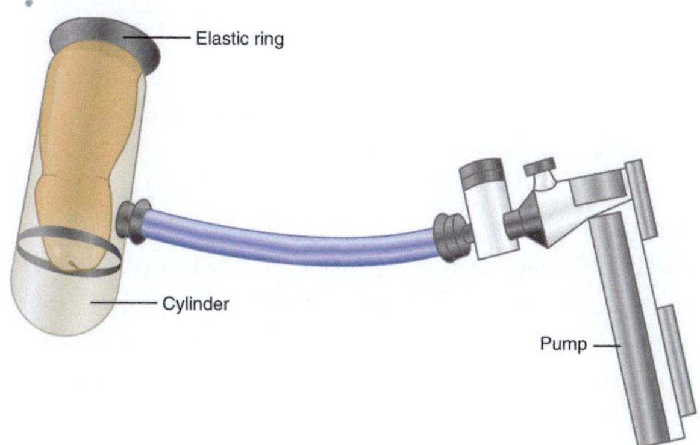

Fig. 15.13 Schematic diagram showing different parts of vacuum erection device

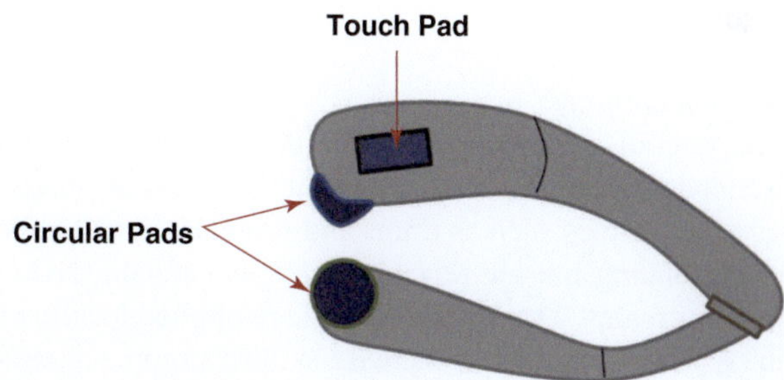

Fig. 15.14 Schematic diagram showing different parts of Viberect

Low-Intensity Extracorporeal Shockwave

Penile low-intensity extracorporeal shockwave (LI-ESW) is a recently developed novel modality for the treatment of ED. LI-ESW are the acoustic waves which carry energy while propagating through a medium and generate pressure impulses. When these waves are applied to any organ, waves interact with deeper tissues and generate mechanical stress. This causes release of angiogenic and endothelial proliferating factors such as vascular endothelial growth factor (VEGF) and proliferating cell nuclear antigen, respectively (Wang et al. 2003). This leads to neovascularization and increased endothelial function that increases blood flow to the target organ. One of the underlying causes of organic ED is poor cavernosal arterial flow. Therefore, LI-ESW-induced neovascularization improves arterial blood flow which in turn enhances penile erection.

The use of LI-ESW has been done in clinical trials. The results have shown improved penile hemodynamics and erectile functions (Vardi et al. 2010, 2012).

Impulse Magnetic Field Therapy

Magnetic stimulation produces alternating currents in cellular electrolyte balance. This affects various cellular functions such as membrane permeability, mitochondrial function, and energy production (Pelka et al. 2002). It also increases oxygen uptake by cells and blood circulation. The use of impulse magnetic field therapy has been done in humans by various groups. During this therapy, a magnetic coil is applied on the penis followed by measuring the intracorporeal pressure. The results show increase in intracorporeal pressure and improved penile erection in patients with ED (Shafik et al. 2000).

Tissue Reconstruction

The first known penile reconstruction was done in 1936. In this, the bone cartilage was used to reconstruct artificial penis (Bretan 1989). Other options used for tissue reconstruction are cartilage rods and collagen matrices (Yoo et al. 1998, 1999; Chen et al. 2010). These kinds of tissue reconstruction procedures can be used in patients with ED due to penile trauma or cancers. Still, more human studies are needed to assess its therapeutic efficacy. Tissue reconstruction procedures for Peyronie's disease consist of plaque excision and pericardial patch or saphenous vein patch grafting (Carson 1997).

Nanotechnology

The use of nanotechnology in drug delivery system is spreading very fast. Nanoparticles have size ≤100 nm and have larger surface to bind, adsorb, and carry other molecules such as drugs and proteins. For ED, nanoparticles are synthesized encapsulating muscle relaxants such as tadalafil, NO, and PDE-5i in the form of gels or creams. These preparations are applied locally on the penile shaft and glans in rats. The results show increase in intracorporeal pressure in rats (Han et al. 2010). There are no clinical trials done in humans to assess the efficacy and potential toxicity associated with the use of topical nanoparticle creams for ED.

Gene and Stem Cell Therapy

The role of NO as a potent vasodilator and having proerectile functions is very well established. It is produced by NOS enzyme. The activity of this enzyme decreases with age and in diseases such as diabetes. The use of gene therapy has been examined by injecting recombinant adenovirus containing eNOS gene into corpora cavernosa of aged rats. This study shows increase in cGMP levels and subsequent increase in intracorporeal pressure in rats (Champion et al. 1999). Similarly, in diabetic rat model, the use of RhoA inhibitor gene therapy restores erectile function (Bivalacqua et al. 2004a). Therefore, gene therapy may prove a new therapeutic intervention in patients with ED. The efficacy of eNOS gene therapy is also tested by first transferring the eNOS genes into rat mesenchymal cells (MSCs) and then these transduced MSCs are injected into the penis of aged rats. This study shows increased eNOS protein expression, NOS activity, cGMP levels, and enhanced erectile response. These studies suggest stem cell and gene therapy as a potential therapeutic modality in ED (Bivalacqua et al. 2004b). NO levels are also regulated by arginases apart from NOS enzyme. Inhibition of arginase enzyme can upregulate NO-mediated smooth muscle relaxation in the penile tissue. The transfer of antisense arginase gene in aged rats shows decreased arginase protein levels and increased cGMP levels along with enhanced erectile function (Bivalacqua et al. 2007).

Key Questions

- How organic erectile dysfunction can be differentiated from psychogenic erectile dysfunction?
- What are the agents that cause relaxation and contraction of penile vessels and smooth muscles?
- Describe the organic causes of erectile dysfunction.
- Write the principles of medical management of erectile dysfunction.
- What are the functional principles of aids to overcome erectile disability?

References

Alexandre B, Lemaire A, Desvaux P, Amar E. Intracavernous injections of PGE1 for erectile dysfunction. Patient satisfaction and quality of sex life on long term treatment. J Sex Med. 2007;4:426–31.

Andersson KE, Wagner G. Physiology of penile erection. Physiol Rev. 1995;75(1):191–236.

Azamitia E, Gannon P. The ultrastructural localization of serotonin immunoreactivity in myelinated and unmyelinated axons within the medial forebrain bundle of rat and monkey. J Neurosci. 1983;3:2083–90.

Babaev A, Jhaveri R. Angiography and endovascular revascularization of pudendal artery atherosclerotic disease in patients with medically refractile erectile dysfunction. J Invasive Cardiol. 2012;24:236–40.

Bader M, Ganten D. Update on tissue renin-angiotensin systems. J Mol Med (Berl). 2008;86:615–21.

Bancroft J. Lecture 4: psychogenic erectile dysfunction-a theoretical approach. Int J Impot Res. 2000;12(Suppl 3):S46–8.

Becker AJ, Uckert S, Stief CG, Truss MC, Hartman U, Sohn M, et al. Systemic and cavernous plasma levels of endothelin 1 in healthy males during different functional conditions of the penis. World J Urol. 2000;18:227–31.

Becker AJ, Uckert S, Stief CG, Truss MC, Machtens S, Scheller F, et al. Possible role of bradykinin and angiotensin II in the regulation of penile erection and detumescence. Urology. 2001;57(1):193–8.

Berk BC, Abe JI, Min W, Surapisitchat J, Yan C. Endothelial atheroprotective and anti-inflammatory mechanisms. Ann N Y Acad Sci. 2001;947:93–109.

Bernier SG, Haldar S, Michel T. Bradykinin-regulated interactions of the mitogenactivated protein kinase pathway with the endothelial nitric-oxide synthase. J Biol Chem. 2000;275:30707–15.

Bivalacqua TJ, Burnett AL, Hellstrom WJ, Champion HC. Overexpression of arginase in the aged mouse penis impairs erectile function and decreases eNOS activity: influence of in vivo gene therapy of anti-arginase. Am J Physiol Heart Circ Physiol. 2007;292:H1340–51.

Bivalacqua TJ, Champion HC, Usta MF, Cellek S, Chitaley K, Webb RC, et al. Rho A/Rho kinase suppresses endothelial nitric oxide synthase in the penis: a mechanism for diabetes associated erectile dysfunction. Proc Natl Acad Sci U S A. 2004a;101:9121–6.

Bivalacqua TJ, Deng W, Champion HC, Hellstrom WJ, Kadowitz PJ. Gene therapy techniques for the delivery of endothelial nitric oxide synthase to the corpora cavernosa for erectile dysfunction. Methods Mol Biol. 2004b;279:173–85.

Bivalacqua TJ, Hellstrom WJ, Kadowitz PJ, Champion HC. Increased expression of arginase II in human diabetic corpus cavernosum: in diabetic-associated erectile dysfunction. Biochem Biophys Res Commun. 2001;283:923–7.

Boswell-Smith V, Spina D, Page CP. Phosphodiesterase inhibitors. Br J Pharmacol. 2006;147:S252–7.

Boucher JL, Moali C, Tenu JP. Nitric oxide biosynthesis, nitric oxide synthase inhibitors and arginase competition for L-arginine utilization. Cell Mol Life Sci. 1999;55:1015–28.

Boureux A, Vignal E, Faure S, Fort P. Evolution of Rho family of RAS like GTPases in eukaryotes. Mol Biol Evol. 2007;24(1):203–16.

Bretan P. History of prosthetic treatment of impotence. Urol Clin North Am. 1989;16:1–5.

Burnett AL. Nitric oxide in the penis: physiology and pathology. J Urol. 1997;157:320–4.

Bustelo XR, Sauzeau V, Berenjeno IM. GTP binding proteins of Rho/Rac family: regulation, effectors and functions in vivo. Bioessays. 2007;29(4):356–70.

Canteras NS, Simerly RB, Swanson LW. Organization of projections from medial nucleus of the amygdala: a PHAL study in the rat. J Comp Neurol. 1995;360:213–45.

Carneiro FS, Sturgis LC, Giachini FR, Carneiro ZN, Lima VV, Wynne BM, et al. TNF-alpha knockout mice have increased corpora cavernosa relaxation. J Sex Med. 2009;6:115–25.

Carson CC. Peyronie's disease: etiology diagnosis and treatment. In: Mulcahy JJ, editor. Diagnosis and management of male sexual dysfunction. New York: Igagu-Shoin; 1997. p. 164–81.

Champion HC, Bivalacqua TJ, Hyman AL, Ignarro LJ, Hellstrom WJ, Kadowitz PJ. Gene transfer of endothelial nitric oxide synthase to the penis augments erectile response in the aged rats. Proc Natl Acad Sci U S A. 1999;96:11648–52.

Chen K, Ebreli D, Yoo J, Atala A. Bioengineered corporal tissue for structural and functional restoration of the penis. Proc Natl Acad Sci U S A. 2010;107:3346–50.

Chia S, Qadan M, Newton R, Ludlam CA, Fox KA, Newby DE. Intra-arterial tumor necrosis factor-alpha impairs endothelium-dependent vasodilatation and stimulates local tissue plasminogen activator release in humans. Arterioscler Thromb Vasc Biol. 2003;23:695–701.

Christ GJ, Lerner SE, Kim DC, Melman A. Endothelin-1 as a putative modulator of erectile dysfunction: I. Characteristics of contraction of isolated corporal tissue strips. J Urol. 1995;153:1998–2003.

Christ GJ, Maayani S, Valcic M, Melman A. Pharmacological studies of human erectile tissue: characteristics of spontaneous contractions and alterations in alpha-adrenoceptor responsiveness with age and disease in isolated tissues. Br J Pharmacol. 1990;101:375–81.

Comiter CV, Sullivan MP, Yalla SV, Kifor I. Effect of angiotensin II on corpus cavernosum smooth muscle in relation to notric oxide environment: in vitro studies in canines. Int J Impot Res. 1997;9:135–40.

Coolen LM, Veening JG, Wells AB, Shipley MT. Afferent connections of parvocellular subparafascicular thalamic nucleus in the rat: evidence for functional subdivisions. J Comp Neurol. 2003;463:132–56.

Cox JD, Kim NN, Traish AM, Christianson DW. Arginase-boronic acid complex highlights a physiological role in erectile function. Nat Struct Biol. 1999;6:1043–7.

Dail WG, Barba V, Leyba L, Galindo R. Neural and endothelial nitric oxide synthase activity in rat penile erectile tissue. Cell Tissue Res. 1995;282:109–16.

Dail WG. Autonomic innervation of male reproductive genitalia. In: Maggi CA, editor. The autonomic nervous system, nervous control of the urogenital system, vol. 6. London: Harwood Academic Publishers; 1993. p. 69–101.

Dean RC, Lue TF. Physiology of penile erection and pathophysiology of erectile dysfunction. Urol Clin North Am. 2005;32:379–95.

Descarries L, Beaudet A, Watkins KC. Serotonin nerve terminals in adult rat neocortex. Brain Res. 1975;100:563–88.

Diederichs W, Stief CG, Benard F, Bosch R, Lue TF, Tanagho EA. The sympathetic role as an antagonist of erection. Urol Res. 1991;19:123–6.

Eardley I, Kirby R. Neurogenic Impotence. In: Kirby R, Carson C, Webster G, editors. Impotence: diagnosis and management of male erectile dysfunction. Oxford: Butterworth-Heinemann; 1991. p. 227–31.

Fallon B. Intracavernous injection therapy for male erectile dysfunction. Urol Clin North Am. 1995;22:833–45.

Finkle AL, Taylor SP. Sexual potency after radical prostatectomy. J Urol. 1981;125:350–2.

Gelbard MK, Dorey F, James K. The natural history of Peyronie's disease. J Urol. 1990;144:1376–9.
Gerstenburg TC, Levin RJ, Wagner G. Erection and ejaculation in man. Assessment of the electromyographic activity of the bulbocavernosus and ischiocavernosus muscles. Br J Urol. 1990;65:395–402.
Giuliano F, Clement P. Physiology of ejaculation: emphasis on serotonergic control. Eur Urol. 2005;48:408–17.
Gonzalez-Cadavid NF, Ignarro LJ, Rajfer J. Nitric oxide and the cyclic GMP system in the penis. Mol Urol. 1999;3:51–9.
Granchi S, Vanelli GB, Vignozzi L, Crescioli C, Ferruzzi P, Mancina R, et al. Expression and regulation of endothelin-1 and its receptors in human penile smooth muscle cells. Mol Hum Reprod. 2002;8:1053–64.
Gwinup G. Oral phentolamine in nonspecific erectile insufficiency. Ann Intern Med. 1988;109:162–3.
Han G, Tar M, Kuppam D, Freidman A, Melman A, Freidman A, et al. Nanoparticles as a novel delivery vehicle for therapeutics targeting erectile dysfunction. J Sex Med. 2010;7:224–33.
Heaton JP, Morales A, Adams MA, Johnston B, el-Rashidy R. Recovery of erectile function by the oral administration of apomorphine. Urology. 1995;45:200–6.
Hedlund P, Alm P, Andersson KE. NO synthase in cholinergic nerves and NO-induced relaxation in the rat isolated corpus cavernosum. Br J Pharmacol. 1999;127:349–60.
Hirano K. Current topic in the regulatory mechanism underlying the Ca2+ sensitization of the contractile apparatus in vascular smooth muscle. J Pharmacol Sci. 2007;104:109–15.
Holmquist F, Andersson K-E, Hedlund H. Actions of endothelin on isolated corpus cavernosum from rabbit and man. Acta Physiol Scand. 1990;139:113–22.
Hoznek A, Rahmouni A, Abbou C, Delmas V, Colombel M. The suspensory ligament of the penis: anatomic and radiological description. Surg Radiol Anat. 1998;20:413–7.
Hunter I, Cobban HJ, Vandenabeele P, MacEwan DJ, Nixon GF. Tumor necrosis factoralpha-induced activation of RhoA in airway smooth muscle cells: role in the Ca2 + sensitization of myosin light chain20 phosphorylation. Mol Pharmacol. 2003;63:714–21.
Ignarro LF, Bush PA, Buga GM, Wood KS, Fukuto JM, Rajfer J. Nitric oxide and cyclic GMP formation upon electrical field stimulation cause relaxation of corpus cavernosum smooth muscle. Biochem Biophys Res Commun. 1990;170:843–50.
Jannini EA, Carosa E, Pepe M, Lombardo F, Lenzi A. Update on pathophysiology of premature ejaculation: the bases for new pharmacological treatments. EAU-EBU Update series 4. 2006. p. 141–9.
Jannini EA, Lombardo F, Lenzi A. Correlation between ejaculatory and erectile dysfunction. Int J Androl. 2005;28:40–5.
Jeremy JY, Ballard SA, Naylor AM, Miller MA, Angelini GD. Effects of sildenafil, a type-5 cGMP phosphodiesterase inhibitor, and papaverine on cyclic GMP and cyclic AMP levels in the rabbit corpus cavernosum in vitro. Br J Urol. 1997;79:958–63.
Jin L, Ying Z, Webb RC. Activation of Rho/Rho kinase signaling pathway by reactive oxygen species in rat aorta. Am J Physiol Heart Circ Physiol. 2004;287:H1495–500.
Junemann KP, Aufenanger J, Konrad T, Pill J, Berle B, Persson-Junemann C, et al. The effect of impaired lipid metabolism on the smooth muscle cells of rabbits. Urol Res. 1991;19: 271–5.
Kadioglu A, Sanli O, Akman T, Canguven O, Aydin M, Akbulut F, et al. Factors affecting the degree of penile deformity in Peyronie disease: an analysis of 1001 patients. J Androl. 2011;32:502–8.
Kedia GT, Uckert S, Assadi-Pour F, Kuczyk MA, Albrecht K. Avanafil for the treatment of erectile dysfunction: initial data and clinical key properties. Ther Adv Urol. 2013;5(1):35–41.
Kifor I, Williams GH, Vickers MA, Sullivan MP, Jodbert P, Dluhy RG. Tissue angiotensin II as a modulator of erectile function. I. Angiotensin peptide content, secretion and effects in the corpus cavernosum. J Urol. 1997;157:1920–5.

Kirby RS. An Atlas of erectile dysfunction: an illustrated textbook and reference for clinicians. New York: The Parthenon Publishing Group; 1999.

Klinge E, Sjöstrand NO. Suppression of the excitatory adrenergic neurotransmission; a possible role of cholinergic nerves in the retractor penis muscle. Acta Physiol Scand. 1977;100:368–76.

Lance R, Albo M, Costabile RA, Steers WD. Oral trazodone as empirical therapy for erectile dysfunction: a retrospective review. Urology. 1995;46:117–20.

Le B, Colombo A, Mustoe T, McVary K. Evaluation of a Ni-Ti shape memory alloy for use in a novel penile prosthesis. J Urol. 2013;189(Suppl):e502.

Lee SW, Wang HZ, Zhao W, Ney P, Brink PR, Christ GJ. Prostaglandin E1 activates the large-conductance KCa channel in human corporal smooth muscle cells. Int J Impot Res. 1999;11:189–99.

Leite R, Giachini FR, Carneiro FS, Nunes KP, Tostes RC, Webb RC. Targets for treatment for erectile dysfunction: is NO/cGMP still the answer? Recent Pat Cardiovasc Drug Discov. 2007;2:119–32.

Lerner SE, Melman A, Christ GJ. A review of erectile dysfunction: new insights and more questions. J Urol. 1993;149:1246–55.

Li CY, Kayes O, Kell PD, Christopher N, Minhas S, Ralph DJ. Penile suspensory ligament division for penil augmentation: indications and results. Eur Urol. 2006;49:729–33.

Lim P. Recent advances and research updates. 2003. ISSN-0972-4689.

Linet OI, Neff LL. Intracavernous prostaglandin E1 in erectile dysfunction. Clin Investig. 1994;72:139–49.

Liu Y, Shepherd EG, Nelin LD. MAPK phosphatases – regulating the immune response. Nat Rev Immunol. 2007;7:202–12.

Lizza EF, Rozen RC. Definition and classification of erectile dysfunction: report of nomenclature committee of the impotence society of impotence research. Int J Impot Res. 1999;11(3):141–3.

Lue TF. Penile venous surgery. Urol Clin North Am. 1989;16:607–11.

Luiten PG, ter Horst GJ, Karst H, Steffens AB. The course of paraventricular hypothalamic efferents to autonomic structures in medulla and spinal cord. Brain Res. 1985;329:374–8.

Marson L, McKenna KE. A role for 5-hydroxytryptamine in descending inhibition of spinal sexual reflexes. Exp Brain Res. 1992;88:313–20.

Marson L, McKenna KE. CNS cell groups involved in the control of the ischiocavernosus and bulbospongiosus muscles: a transneuronal tracing study using pseudorabies virus. J Comp Neurol. 1996;374:161–79.

Martin DJ, Badwan K, Parker M, Mulhall JP. Transdermal application of verapamil gel to the penile shaft fails to infiltrate the tunica albuginea. J Urol. 2002;168:2483–5.

Masters WH, Johnson VE. Human sexual response. 1st ed. Boston: Little, Brown and Company Boston; 1966.

McKenna KE, Nadelhaft I. The organization of pudendal nerve in the male and female rats. J Comp Neurol. 1986;248:532–49.

McMahon CG, Touma K. Treatment of premature ejaculation with paroxetine hydrochloride as needed: 2 single-blind placebo controlled crossover studies. J Urol. 1999;161:1826–30.

Meldrum DR, Meng X, Dinarello CA, Ayala A, Cain BS, Shames BD, et al. Human myocardial tissue TNF alpha expression following acute global ischemia in vivo. J Mol Cell Cardiol. 1998;30:1683–9.

Metz P, Ebbehoj J, Uhrenholdt A, Wagner G. Peyronie's disease and erectile failure. J Urol. 1983;130:1103–4.

Michal V, Ruzbarsky V. Histological changes in the penile arterial bed with aging and diabetes. In: Zorgniotti A, Rossi G, editors. Vasculogenic impotence: proceedings of the first international conference on corpus cavernosum revascularization. Springfield: Charles C Thomas; 1980. p. 113–9.

Mieusset R, Soulié M. Hypospadias: psychosocial, sexual and reproductive consequences in adult life. J Androl. 2005;26:163–8.

Miller MA, Morgan RJ. Eicosanoids, erections and erectile dysfunction. Prostaglandins Leukot Essent Fatty Acids. 1994;51:1–9.

Mong PY, Petrulio C, Kaufman HL, Wang Q. Activation of Rho kinase by TNF-alpha is required for JNK activation in human pulmonary microvascular endothelial cells. J Immunol. 2008;180:550–8.

Montague DK, Jarow J, Broderick GA, Dmochowski RR, Heaton JP, Lue TF, et al. AUA guideline on the pharmacologic management of premature ejaculation. J Urol. 2004;172:290–4.

Mori M, Gotoh T. Regulation of nitric oxide production by arginine metabolic enzymes. Biochem Biophys Res Commun. 2000;275:715–9.

Mulhall JP, Schiff J, Guhring P. An analysis of the natural history of Peyronie's disease. J Urol. 2006;175:2115–8.

Mulligan T, Schmitt B. Testosterone for erectile failure. J Gen Intern Med. 1993;8:517–21.

Murphy AZ, Rizvi TA, Ennis M, Shipley MT. The organization of preoptic- medullary circuits in the male rats: evidence for interconnectivity of neural structures involved in reproductive behavior, antinociception and cardiovascular regulation. Neuroscience. 1999;91:1103–16.

Nadig PW, Ware JC, Blumoff R. Non invasive device to produce and maintain an erection like state. Urology. 1986;27:126–31.

Narumiya S, Sugimoto Y, Ushikubi F. Prostanoid receptors: structures, properties, and functions. Physiol Rev. 1999;79:1193–226.

Newman HF, Reiss H, Northrup JD. Physical basis of emission, ejaculation, and orgasm in the male. Urology. 1991;19:341–50.

Numao N, Masuda H, Sakai Y, Okada Y, Kihara K, Azuma H. Roles of attenuated neuronal nitric-oxide synthase protein expression and accelerated arginase activity in impairing neurogenic relaxation of corpus cavernosum in aged rabbits. BJU Int. 2007;99:1495–9.

Pelka R, Jaenicke C, Gruenwald J. Impulse magnetic field therapy for erectile dysfunction: a double blind, placebo controlled study. Adv Ther. 2002;19:53–60.

Pierce KL, Gil DW, Woodward DF, Regan JW. Cloning of human prostanoid receptors. Trends Pharmacol Sci. 1995;16:253–6.

Rajasekaran M, Mondal D, Agrawal K, Chen IL, Hellstrom W, Sikka S. Ex vivo expression of nitric oxide synthase isoforms (eNOS/iNOS) and calmodulin in human penile cavernosal cells. J Urol. 1998;160:2210–5.

Rajfer J, Rosciszewski A, Mehringer M. Prevalence of corporeal venous leakage in impotent men. J Urol. 1988;140:69–71.

Redoute J, Stoleru S, Gregoire MC, Costes N, Cinotti L, Lavenne F, et al. Brain processing of visual sexual stimuli in human males. Hum Brain Mapp. 2000;11(3):162–77.

Rizvi TA, Ennis M, Shipley MT. Reciprocal connections between the medial preoptic area and the midbrain periaqueductal gray in rat: a WGA-HRP and PHA-L study. J Comp Neurol. 1992;315:1–15.

Rosen MP, Greenfield AJ, Walker TG, Grant P, Guben JK, Dubrow J, et al. Arteriogenic impotence: findings in 195 impotent men examined with selective internal pudendal angiography. Young Investigator's Award. Radiology. 1990;174:1043–8.

Rowland DL, Perelman MA, Althof SE, Barada J, Mc Cullough A, Bull S, et al. Self-reported premature ejaculation and aspects of sexual functioning and satisfaction. J Sex Med. 2004;1:225–32.

Ryan MJ, Didion SP, Mathur S, Faraci FM, Sigmund CD. Angiotensin II-induced vascular dysfunction is mediated by the AT1A receptor in mice. Hypertension. 2004;43:1074–9.

Saenz de Tejada I, Carson MP, de las Morenas A, Goldstein I, Traish AM. Endothelin: localization, synthesis, activity, and receptor types in human penile corpus cavernosum. Am J Physiol. 1991;261:H1078–85.

Saenz de Tejada I, Kim NN, Goldstein I, Traish AM. Regulation of pre-synaptic alpha adrenergic activity in the corpus cavernosum. Int J Impot Res. 2000;12(Suppl 1):S20–5.

de Tejada S, Angulo J, Cellek S, Gonzalez-Cadavid N, Heaton J, Pickard R, et al. Physiology of erectlie function. J Sex Med. 2004;1:254–65.

Safarinejad MR, Hosseini SY, Kolahi AA. Comparison of vitamin E and propionyl-L-carnitine, separately or in combination, in patients with early chronic Peyronie's disease: a double-blind, placebo controlled, randomized study. J Urol. 2007;178(4 Pt 1):1398–403.

Schiavi RC, White D, Mandeli J, Levine AC. Effect of testosterone administration on sexual behavior and mood in men with erectile dysfunction. Arch Sex Behav. 1997;26:231–41.

Schmidt MH, Schmidt HS. Sleep related erections: neural mechanisms and clinical significance. Curr Neurol Neurosci Rep. 2004;4:170–8.

Shafik A, El-Sibai O, Shafik A. Magnetic stimulation of the cavernous nerve for the treatment of erectile dysfunction in humans. Int J Impot Res. 2000;12:137–41.

Simerly RB, Swanson LW. Projections of medial preoptic nucleus: a Phaseolus vulgaris leucoagglutinin anterograde tract tracing study in the rat. J Comp Neurol. 1988;270:209–42.

Simonsen U, Prieto D, Hernandez M, Saenz de Tejada I, Garcia-Sacristan A. Prejunctional alpha 2-adrenoceptors inhibit nitrergic neurotransmission in horse penile resistance arteries. J Urol. 1997;157:2356–60.

Somlyo AP, Somlyo AV. Signal transduction by G-proteins, rho kinase and protein phosphatase to smooth muscle and non- muscle myosin II. J Physiol. 2000;522(pt 2):177–85.

Sommer F, Klotz T, Steinritz D, Schmidt A, Addicks K, Engelmann U, et al. MAP kinase 1/2 (Erk 1/2) and serine/threonine specific protein kinase Akt/PKB expression and activity in the human corpus cavernosum. Int J Impot Res. 2002;14:217–25.

Stanarius A, U¨ckert S, Machtens SA, Stief CG, Wolf G, Jonas U. Immunocytochemical distribution of endothelial nitric oxide synthase (eNOS) in human corpus cavernosum (Abstract 849). J Urol. 1999;161(Suppl):221.

Standring S. Penis. Gray's anatomy. 39th ed: Elsevier/Churchill Livingstone: Edinburgh: New York; 2005. p. 1315–7.

Steers WD. Neural control of penile erection. Semin Urol. 1990;8:66–79.

Stein MJ, Lin H, Wang R. New advances in erectile technology. Ther Adv Urol. 2014;6(1):15–24.

Stief CG, Wetterauer U. Erectile responses to intracavernous papaverine and phentolamine: comparison of single and combined delivery. J Urol. 1988;140:1415–6.

Stoleru S, Redoute J, Costes N, Lavenne F, Bars DL, Dechoud H, et al. Brain processing of visual sexual stimuli in men with hypoactive sexual desire disorder. Psychiatry Res. 2003;124(2):67–86.

Tajkarimi K, Burnett A. Viberect device use by men with erectile dysfunction: safety, ease of use, tolerability and satisfaction survey. J Sex Med. 2011;8:441.

Tal R, Teloken P, Mulhall J. Erectile function rehabilitation after radical prostatectomy: practice patterns among AUA members. J Sex Med. 2011;8:2370–6.

Taylor FL, Levine LA. Peyronie's disease. Urol Clin North Am. 2007;34:517–34.

Teloken C, Rhoden EL, Grazziotin TM, Ross CT, Sogari PR, Souto CA. Tamoxifen versus placebo in the treatment of Peyronie's disease. J Urol. 1999;162:2003–5.

Toselli P, Moreland R, Traish AM. Detection of m2 muscarinic acetylcholine receptor mRNA in human corpus cavernosum by in-situ hybridization. Life Sci. 1994;55:621–7.

Traish AM, Palmer MS, Goldstein I, Moreland RB. Expression of functional muscarinic acetylcholine receptor subtypes in human corpus cavernosum and in cultured smooth muscle cells. Receptor. 1995;5:159–76.

Truitt WA, Coolen LM. Identification of a potential ejaculation generator in the spinal cord. Science. 2002;297:1566–9.

Vardi Y, Appel B, Jacob G, Massawi O, Gruenwald I. Can low intensity extracorporeal shock wave therapy improve erectile function? A 6- month follow up pilot study in patients with organi erectile dysfunction. Eur Urol. 2010;58:243–8.

Vardi Y, Appel B, Kilchevsky A, Gruenwald I. Does low intensity extracorporeal shock wave therapy have a physiological effect on erectile function? Short term results of a randomized, double blind, sham controlled study. J Urol. 2012;187:1769–75.

Virag R, Zwang G, Dermange H, Legman M. Vasculogenic impotence: a review of 92 cases with 54 surgical operations. Vasc Surg. 1981;15:9–16.

Waldinger MD, Hengeveld MW, Zwinderman AH, Oliveier B. Effect of SSRI antidepressants on ejaculation: a double-blind, randomized, placebo-controlled study with fluoxetine, fluvoxamine, paroxetine, and sertraline. J Clin Psychopharmacol. 1998;18:275–81.

Waldinger MD, Hengeveld MW, Zwinderman AH. Paroxetine treatment of premature ejaculation: a double blind, randomized, placebo controlled study. Am J Psychiatry. 1994;151:1377–9.

Walsh MP. Calcium dependent mechanism of smooth muscle contraction. Biochem Cell Biol. 1991;69:771–800.

Walsh PC, Wilson JD. Impotence and infertility in men. In: Braunwald E, Isselbacher KJ, Petersdrop RG, editors. Harrison's principles of internal medicine. 11th ed. New York: Mc Grew Hill Book company; 1987. p. 217–20.

Wang CJ, Wang FS, Yang KD, Weng LH, Hsu CC, Huang CS, et al. Shock wave therapy induces neovascularization at the tendon bone junction: a study in rabbits. J Orthop Res. 2003;21:984–9.

Web RC. Smooth muscle contraction and relaxation. Adv Physiol Educ. 2003;27:201–6.

Wermuth L, Stenager E. Sexual aspects of Parkinson's disease. Semin Neurol. 1992;12:125–7.

Wessells H, Lue TF, McAninch JW. Penile length in the flaccid and erect states: guidelines for penile augmentation. J Urol. 1996;156:995–7.

Williams G, Abbou CC, Amar ET, Desvaux P, Flam TA, Lycklama a Nijeholt GA, et al. Efficacy and safety of transurethral alprostadil therapy in men with erectile dysfunction. Br J Urol. 1998;81:889–94.

Xu C, Yaici ED, Conrath M, Blanchard P, Leclerc P, Benoit G, et al. Galanin and neurokinin 1 receptor immunoreactivity spinal neurons controlling the prostate and bulbospongiosus muscle identified by transsynaptic labeling in the rat. Neuroscience. 2005;134:1325–41.

Yasui Y, Saper CB, Cechetto DF. Calcitonin gene related peptide (CGRP) immunoreactive projections from the thalamus to the striatum and amygdala in the rat. J Comp Neurol. 1991;308:293–310.

Yeager ES, Van Heerden JA. Sexual dysfunction following proctocolectomy and abdominoperineal resection. Ann Surg. 1980;191:169–70.

Ying Z, Jin L, Palmer T, Webb RC. Angiotensin II up-regulates the leukemia-associated Rho guanine nucleotide exchange factor (RhoGEF), a regulator of G protein signaling domain-containing RhoGEF, in vascular smooth muscle cells. Mol Pharmacol. 2006;69:932–40.

Yoo J, Lee I, Atala A. Cartilage rods as a potential material for penile reconstruction. J Urol. 1998;160:1164–8.

Yoo J, Park H, Lee I, Atala A. Autologous engineered cartilage rods for penile reconstruction. J Urol. 1999;162:1119–21.

Andropause

16

A.C. Ammini[†], Shivaprakash Somasundaram, and Ravinder Goswami

Learning Objectives

- Biological basis and criteria of defining male andropause
- Symptoms of andropause
- Confounders in male andropause
- Principles of diagnostic evaluation and management

16.1 Introduction

Andropause or male menopause is the term used to describe a set of symptoms that are seen in some of the ageing men. It is a state characterized by significant decline in the production of testosterone, dehydroepiandrosterone and other hormones like growth hormone (GH). The first description of this state was given by Heller and Myers in 1944. They called it male climacteric and observed that subnormal testosterone and symptoms like loss of libido, nervousness, irritability, poor concentration, depression, impaired memory, fatigue and insomnia improved following

[†] Deceased

A.C. Ammini, MD, DM • S. Somasundaram, MD, DM
R. Goswami, MD, DM, FASc, FNASc, FAMS (✉)
Department of Endocrinology and Metabolism, All India Institute of Medical Sciences, New Delhi 110029, India
e-mail: gosravinder@hotmail.com

A. Kumar, M. Sharma (eds.), *Basics of Human Andrology*,
DOI 10.1007/978-981-10-3695-8_16

testosterone replacement. In contrast to menopause in which there is abrupt cessation of fertility and marked decline in estrogen production, andropause is a slow transition with preserved fertility (Yeap 2009).

16.2 Terminologies Used for Andropause

In males, the age-related reproductive changes have been described by various terms such as viropause, partial androgen deficiency of the ageing male (PADAM), androgen deficiency of the ageing male (ADAM), testosterone deficiency syndrome and late-onset hypogonadism (Wang et al. 2008; Bhasin et al. 2010). The association of declining testosterone levels coupled with hypogonadal symptoms has triggered treatment with testosterone replacement at an alarming rate and large-scale marketing of testosterone formulations (Harrison 2011).

16.3 Biological Basis and Criteria of Defining Andropause

The biologically active, 'bioavailable testosterone' is free and albumin-bound testosterone. With ageing, the 'bioavailable testosterone' declines, testicular volume decreases, sperm become less motile and the number of Sertoli and Leydig cells decreases to around half of the numbers seen in young men (Neaves et al. 1984; Anderson 2000; Well et al. 2007). Ageing-related reduction in androgen availability is characterized by aberrant gonadotropin-releasing hormone (GnRH) secretion, normal luteinizing hormone (LH) response to stimulation with GnRH, impaired Leydig cell testosterone production in response to LH (gonadal ageing), loss of circadian rhythmicity in testosterone levels and enhanced androgen-negative feedback on LH secretion (Harman and Tsitouras 1980; Harman et al. 1982). There is a possibility of decreased number and sensitivity of brain androgen receptors with ageing. This might also account for occurrence of symptoms suggestive of male andropause with relatively normal circulating testosterone concentration (Wylie and Froggatt 2010). The effects of decreased testosterone levels on brain ageing and neuroanatomical changes of ageing brain are described in the chapter Male Behaviors IV: Ageing Brain.

Normally, 58% of plasma testosterone is bound to the sex hormone-binding globulin (SHBG) and 40% to albumin. Serum testosterone values tend to decrease after 40 years of age with average fall of approximately 1–2% per year during the fourth to seventh decade (Davidson et al. 1983; Simon et al. 1992; Kaufman and Vermeulen 2005; Sartorius et al. 2012; Kim et al. 2012). Interestingly, serum SHBG also shows a rise with ageing. Therefore, there is inverse relation between serum testosterone and SHBG levels with ageing. This would result in a proportionately faster decline in bioavailable testosterone by 2–3% per year (Feldman et al. 2002).

16.4 Symptoms of Andropause and Confounders

Multiple factors associated with ageing can have confounding effects on testosterone levels. Obesity, frailty, diabetes mellitus, inflammatory arthritis, renal failure, cardiovascular disease, chronic obstructive pulmonary disease and chronic use of steroid and opiates are more common causes of low testosterone than ageing per se. Also, the SHBG levles are affected by several of these comorbid conditions resulting in lower values. Hepatic cirrhosis, hepatitis, hyperthyroidism, human immunodeficiency virus (HIV) infection, anticonvulsants and estrogen therapy are other causes of elevated SHBG levels. Androgen receptor-CAG repeat polymorphisms can also affect the circulating androgens and andropausal symptoms (Harman et al. 2001; Mohr et al. 2005; Zitzmann 2009; Chia Liu et al. 2012). Availability of reliable testosterone assays, lack of normative ranges for bioavailable testosterone in apparently healthy aged males and the absence of consensus on the type of testosterone to be measured are important issues in the diagnosis of andropause (de Ronde et al. 2006; Rosner et al. 2007). Due to these reasons, it is difficult to comment on the accurate prevalence of late-onset hypogonadism (Araujo et al. 2007).

The association between hypogonadal symptoms and serum testosterone is weak (Bhasin et al. 2001). This is probably due to confounding effects of a number of disease states in ageing males which can give rise to andropausal symptoms. In the European Male Ageing Study (EMAS), criteria are set for late-onset hypogonadism in the general population, and thresholds are determined for symptoms such as infrequent morning erection, erectile dysfunction and decreasing sexual thoughts with lower serum testosterone values (Wu et al. 2010).

Late-onset hypogonadism is defined as a combination of at least three sexual symptoms associated with a total testosterone level of <11 nmol/L (3.2 ng /ml) and a free testosterone of <220 pmol /L. However, the association between the sexual symptoms and a testosterone level becomes less significant when adjustment is made for the chronological age, body mass index and the coexisting medical illness. These facts highlight the possibility of nonspecific symptoms of ageing manifesting as late-onset hypogonadism. The accrual of fat with ageing is considered to be one of the manifestations of estrogen deficiency (Finkelstein et al. 2013). Secondary hypogonadism is associated with obesity, independent of age (Tajar et al. 2010).

Ageing is associated with decline in GH and insulin-like growth factor (IGF) secretion, commonly termed as somatopause. There is also a deficiency of dehydroepiandrosterone (DHEA) and its sulphated ester DHEAS during ageing which is known as adrenopause. There is data available on the effect of androgen replacement on levels of GH in ageing males (Gentili et al. 2002). On the contrary, in adults with GH deficiency, GH replacement causes fall in serum testosterone mainly because of decline in SHBG (Giavoli et al. 2006). Oral DHEA supplementation in dose of 50 mg/day for 6 months in ageing males between 40 and 70 years results in only marginal increase in serum androstenedione.

According to the Endocrine Society Clinical Practice Guidelines, hypogonadism in men is defined as a syndrome resulting from inability of the testis to produce physiologically sufficient testosterone concentration and sperm due to disruption of hypothalamic-pituitary-testicular axis. The diagnosis of androgen deficiency is recommended in men with symptoms and unequivocally subnormal serum testosterone values. Some of the specific symptoms and signs of androgen deficiency are delayed sexual development, poor libido, decreased frequency of erections, eunuchoid proportion, etc. Gynaecomastia, loss of axillary and pubic hair, reduced shaving frequency, small testes, infertility, oligo or azoospermia, fracture on trivial trauma and osteoporosis are other manifestations of androgen deficiency. In the presence of these symptoms and signs, it is recommended to measure total serum testosterone values in the morning as an initial diagnostic test followed by confirmation by repeat measurement. In cases where serum total testosterone is in the lower range, SHBG measurement may be made to have estimates of bioavailable testosterone.

Subnormal serum testosterone levels are linked with visceral adiposity, risk of diabetes mellitus, coronary artery disease and fractures. Testosterone in Older Men with Mobility Limitations (TOM) trial assessed the effects of testosterone administration on lower extremity strength and physical function in older men with limitations in mobility (Basaria et al. 2010). Adverse cardiovascular events are observed with testosterone gel application. The impact of testosterone therapy on the risk of prostate cancer is an area of concern (Spitzer et al. 2013).

The concept of late-onset hypogonadism is drawn mainly from comparison with classical hypogonadism in young men. Moreover, clinical syndrome of androgen deficiency in young men cannot be extrapolated to the older population. The commonly used symptom-based questionnaires which help to diagnose andropause are Ageing Males' Symptoms Rating Scale and Massachusetts Male Ageing Study Scale. Several issues remain unresolved such as testosterone cut off to define hypogonadism, whether free or total testosterone is to be measured, selection of testosterone assay and exact concentration of testosterone to diagnose androgen deficiency.

16.5 Principles of Diagnostic Evaluation and Treatment

With the available evidence, the decision to diagnose and treat late-onset hypogonadism has to be done only with unequivocal clinical and biochemical evidence of hypogonadism. The current diagnosis of andropause is based on the strict definition of serum total testosterone less than 8 nmol/L (2.3 ng/ml) and the presence of decreased morning erections. The diagnostic criteria may not be accurate especially if the androgen receptor sensitivity is affected with ageing and may demand further biomarkers of androgen activity in ageing males. Nevertheless, whenever testosterone therapy is contemplated in ageing males, it would require extra vigil. Absolute contraindications to testosterone therapy include prostate and male breast cancer. Role of testosterone is implicated in prostate cancer as it activates various genes controlling cell division. Testosterone aromatizes to estrogen and therefore, testosterone therapy is contraindicated in patients with breast cancer. Any suspicion towards side effects including prostate and breast carcinoma would require discontinuation of the therapy.

Relative contraindications are serum prostate-specific antigen (PSA) > 4 ng/ml, significant lower urinary tract symptoms, haematocrit >50%, untreated or poorly controlled congestive heart failure and sleep apnoea syndrome. All elderly subjects put on testosterone need to be monitored with haematocrit and haemoglobin as testosterone increases erythropoietin production. PSA and per rectal digital examination should also be part of monitoring. The basis of testosterone association with sleep apnoea is not clearly known. Short-acting testosterone is preferred whenever androgen therapy is planned in ageing males. 17-Methyltestosterone is best avoided because of liver toxicity. There is no clear evidence for the use of other preparations like dihydrotestosterone, DHEAS or androstenedione. Well-designed placebo controlled trials using objective criteria for improvement are required before testosterone replacement therapy can be considered for andropause.

Key Questions

- What is the biological basis of andropause?
- Describe the current criteria of diagnosing andropause.
- What are the contraindications of androgen therapy in hypogonadal males?

References

Anderson RA. Sertoli cell function in the ageing male. Clin Endocrinol (Oxf). 2000;53:139–40.

Araujo AB, Esche GR, Kupelian V, O'Donnell AB, Travison TG, Williams R, et al. Prevalence of symptomatic androgen deficiency in men. J Clin Endocrinol Metab. 2007;92:4241–7.

Basaria S, Coviello AD, Travison TG, Storer TW, Farwell WR, Jette AM, et al. Adverse events associated with testosterone administration. N Engl J Med. 2010;363:109–22.

Bhasin S, Woodhouse L, Casaburi R, Singh AB, Bhasin D, Berman N, et al. Testosterone dose-response relationships in healthy young men. Am J Physiol Endocrinol Metab. 2001;281:E1172–E81.

Bhasin S, Cunningham GR, Hayes FJ, Matsumoto AM, Snyder PJ, Swerdloff RS, et al. Testosterone therapy in men with androgen deficiency syndromes: an Endocrine Society Clinical Practice Guideline. J Clin Endocrinol Metab. 2010;95:2536–59.

Chia Liu CC, Lee YC, Wang CJ, Yeh HC, Li WM, Wu WJ, et al. The impact of androgen receptor cag repeat polymorphism on andropausal symptoms in different serum testosterone levels. J Sex Med. 2012;9:2429–37.

Davidson JM, Chen JJ, Crapo L, Gray GD, Greenleaf WJ, Catania JA. Hormonal changes and sexual function in ageing men. J Clin Endocrinol Metab. 1983;57:71–7.

de Ronde W, van der Schouw YT, Pols HA, Gooren LJ, Muller M, Grobbee DE, et al. Calculation of bioavailable and free testosterone in men: a comparison of 5 published algorithms. Clin Chem. 2006;52:1777–84.

Feldman HA, Longcope C, Derby CA, Johannes CB, Araujo AB, Coviello AD, et al. Age trends in the level of serum testosterone and other hormones in middleaged men: longitudinal results from the Massachusetts male ageing study. J Clin Endocrinol Metab. 2002;87:589–98.

Finkelstein JS, Lee H, Burnett-Bowie SA, Pallais JC, Yu EW, Borges LF, et al. Gonadal steroids and body composition, strength, and sexual function in men. N Engl J Med. 2013;369:1011–22.

Gentili A, Mulligan T, Godschalk M, Clore J, Patrie J, Iranmanesh A, et al. Unequal impact of short-term testosterone repletion on the somatotropic axis of young and older men. J Clin Endocrinol Metab. 2002;87:825–34.

Giavoli C, Ferrante E, Ermetici F, Bergamaschi S, Ronchi CL, Lania AG, et al. Effect of recombinant hGH (rhGH) replacement on gonadal function in male patients with organic adult-onset GH deficiency. Clin Endocrinol (Oxf). 2006;65:717–21.

Harman SM, Tsitouras PD. Reproductive hormones in ageing men: I. Measurement of sex steroids, basal luteinizing hormone and Leydig cell response to human chorionic gonadotropin. J Clin Endocrinol Metab. 1980;51:35–40.

Harman SM, Tsitouras PD, Costa PT, Blackman MR. Reproductive hormones in ageing men. II. Basal pituitary gonadotropins and gonadotropin responses to luteinizing hormone- releasing hormone. J Clin Endocrinol Metab. 1982;54:547–51.

Harman SM, Metter EJ, Tobin JD, Pearson J, Blackman MR. Longitudinal effects of ageing on serum total and free testosterone levels in healthy men. Baltimore Longitudinal Study of Ageing. J Clin Endocrinol Metab. 2001;86:724–31.

Harrison J. Talking about my generation: a state-of-the-art review of health information for men in the andropause. Health Info Libr J. 2011;28:161–70.

Heller CG, Myers GB. The Male climacteric: its symptomatology, diagnosis and treatment. JAMA. 1944;126:472–7.

Kaufman JM, Vermeulen A. The decline of androgen levels in elderly men and its clinical and therapeutic implications. Endocr Rev. 2005;26:833–76.

Kim YS, Hong D, Lee DJ, Joo NS, Kim KS. Total testosterone may not decline with ageing in Korean men aged 40 years or older. Clin Endocrinol (Oxf). 2012;77:296–301.

Mohr BA, Guay AT, O'Donnell AB, McKinlay JB. Normal, bound and nonbound testosterone levels in normally ageing men: results from the Massachusetts Male Ageing Study. Clin Endocrinol (Oxf). 2005;62:64–73.

Neaves WB, Johnson L, Porter JC, Parker CR, Petty CS. Leydig cell numbers, daily sperm production, and serum gonadotropin levels in ageing men. J Clin Endocrinol Metab. 1984;59:756–63.

Rosner W, Auchus RJ, Azziz R, Sluss PM, Raff H. Utility, limitations, and pitfalls in measuring testosterone: an Endocrine Society position statement. J Clin Endocrinol Metab. 2007;92:405–13.

Sartorius G, Spasevska S, Idan A, Turner L, Forbes E, Zamojska A, et al. Serum testosterone, dihydrotestosterone and estradiol concentrations in older men self-reporting very good health: the healthy man study. Clin Endocrinol (Oxf). 2012;77:755–63.

Simon D, Preziosi P, Barrett-Connor E, Roger M, Saint-Paul M, Nahoul K. The influence of ageing on plasma sex hormones in men: the Telecom Study. Am J Epidemiol. 1992;135:783–91.

Spitzer M, Huang G, Basaria S, Travison TG, Bhasin S. Risks and benefits of testosterone therapy in older men. Nat Rev Endocrinol. 2013;9:414–24.

Tajar A, Forti G, O'Neill TW, Lee DM, Silman AJ, Finn JD, et al. Characteristics of secondary, primary, and compensated hypogonadism in ageing men: evidence from the European Male Ageing Study. J Clin Endocrinol Metab. 2010;95:1810–8.

Wang C, Nieschlag E, Swerdloff R, Behre HM, Hellstrom WJ, Gooren LJ, et al. Investigation, treatment and monitoring of late-onset hypogonadism in males: ISA, ISSAM, EAU, EAA and ASA recommendations. Eur J Endocrinol. 2008;159:507–14.

Well D, Yang H, Houseni M, Iruvuri S, Alzeair S, Sansovini M, et al. Age-related structural and metabolic changes in the pelvic reproductive end organs. Semin Nucl Med. 2007;37:173–84.

Wu FC, Tajar A, Beynon JM, Pye SR, Silman AJ, Finn JD, et al. Identification of late-onset hypogonadism in middle-aged and elderly men. N Engl J Med. 2010;363:123–35.

Wylie K, Froggatt N. Late onset hypogonadism, sexuality and fertility. Hum Fertil (Camb). 2010;13:126–33.

Yeap BB. Testosterone and ill-health in ageing men. Nat Clin Pract Endocrinol Metab. 2009;5:113–21.

Zitzmann M. The role of the CAG repeat androgen receptor polymorphism in andrology. Front Horm Res. 2009;37:52–61.

Part V

Neurophysiology of Male Behaviors

Male Behaviors I: Brain Areas Regulating Male Behaviors

17

Mona Sharma, Rakesh Kumar Chadda, P.K. Anish, and Anand Kumar

Learning Objectives

- Cortical and subcortical brain areas regulating behaviors
- Neural connections and functions

17.1 Introduction

The basic structural organization of brain areas and their neural connections is important to understand the biological mechanisms of behaviors. In this chapter, we shall first discuss the most visible areas on the surface of the brain, the cerebral cortex. The cortical areas receive sensory inputs, control motor outputs and also various behaviors. The cortical areas are the frontal, parietal and temporal lobes. Another cortical area which is deeply seated is the insula. The cortical area which is visualized on the medial surface of brain is the limbic lobe. The brain areas forming the limbic lobe are the hippocampus, parahippocampus and cingulate cortex. The deeper subcortical brain areas regulating behavior are the amygdala (AMG), hypothalamus, ventral pallidum (VP) and septal region. The subcortical brain areas along with the limbic lobe constitute the limbic system. In the last part of this chapter, few structures in the brainstem such as the ventral tegmental area

M. Sharma, MD, DNB, MAMS (✉) • A. Kumar, MD, FAMS
Department of Reproductive Biology, All India Institute of Medical Sciences, New Delhi, India
e-mail: dr.mona18sharma@gmail.com; anandkumarrepbiol@hotmail.com

R.K. Chadda, MD, FAMS, FRC • P.K. Anish, MD
Department of Psychiatry, All India Institute of Medical Sciences, New Delhi, India
e-mail: drrakeshchadda@gmail.com; anishpk9@gmail.com

A. Kumar, M. Sharma (eds.), *Basics of Human Andrology*,
DOI 10.1007/978-981-10-3695-8_17

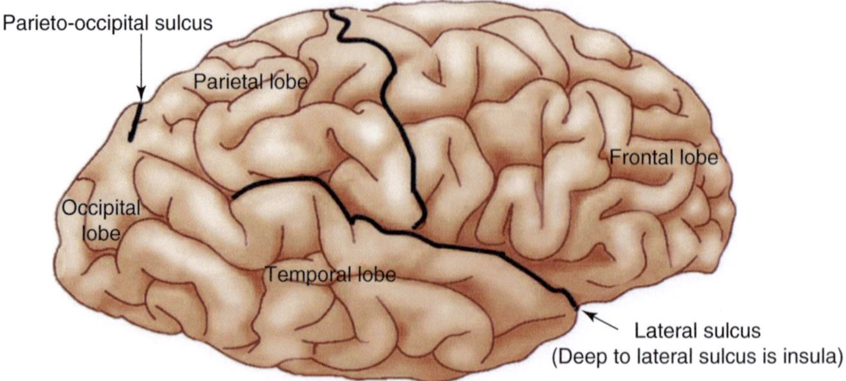

Fig. 17.1 Lateral surface of the brain showing cortical areas

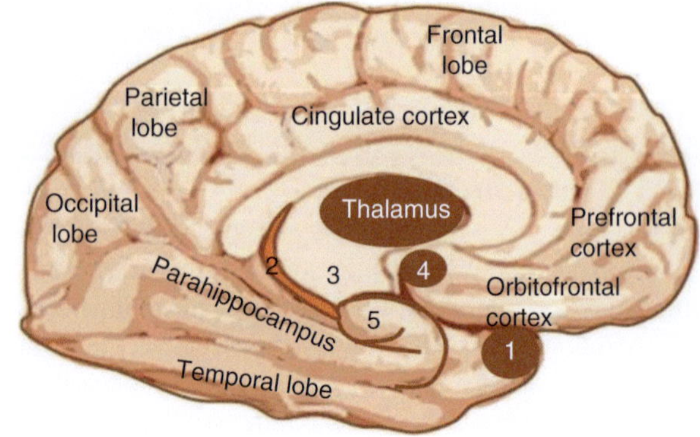

Fig. 17.2 Medial surface of the brain showing cortical and subcortical areas. *1*, amygdala; *2*, fornix; *3*, hypothalamus; *4*, septal region; *5*, hippocampus

(VTA) and substantia nigra (SN) are also discussed (see Figs. 17.1, 17.2 and 17.3). The discussions will be based on various studies done in animals and few studies in humans. The neural connections are described in terms of incoming projections or afferents and outgoing projections or efferents. The functions and neurotransmitters of all the neural connections of brain areas are not yet clear. But the defined functions of the neural connections that regulate behaviors have been mentioned in this chapter.

17.2 Frontal Lobes

The frontal lobe is the most anterior part of the cerebral cortex and lies in front of central sulcus. The important subdivisions of the frontal lobe are the primary motor cortex, premotor cortex, frontal eye field, prefrontal cortex (PFC) and anterior part of paracentral lobule on the medial surface (see Figs. 17.1, 17.3 and 17.4).

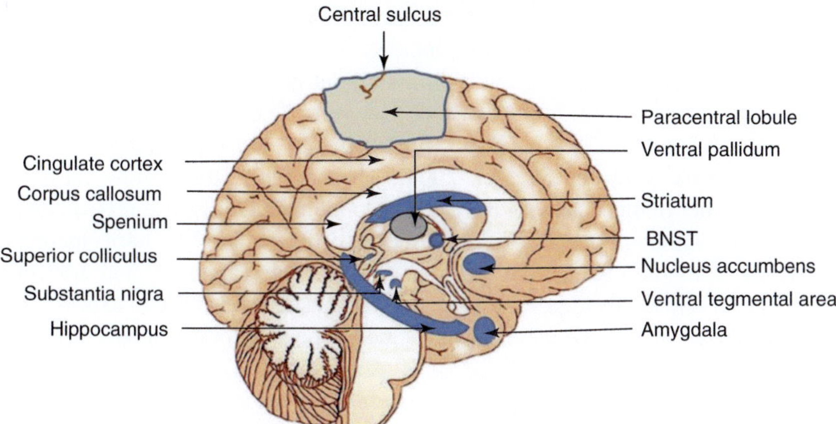

Fig. 17.3 Medial view of the brain showing cortical and subcortical areas. *BNST* bed nucleus of the stria terminalis

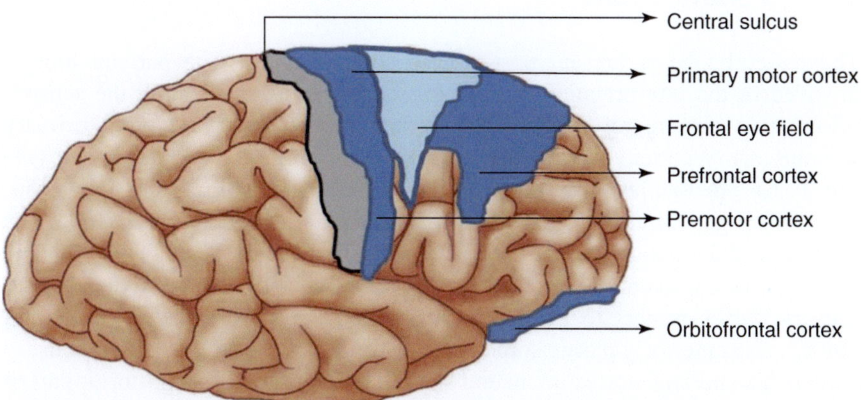

Fig. 17.4 Lateral surface of the brain showing various functional areas of the frontal lobe

The primary motor cortex receives afferents from the somatosensory cortex, premotor cortex and cerebellum. The efferents project to the spinal cord motor neurons, brainstem nuclei and basal ganglia (Parent 1996; Standring 2005a). Motor cortices function in control of movements. Motor behavior is an integral part of reproduction starting from viewing of sexual stimuli to the final sexual act. The premotor cortex selects the movements to be performed, whereas the primary motor cortex is responsible for final execution of movements. Paracentral lobule provides motor nerve supply to genitals (Standring 2005a).

Frontal eye field gets afferents from the midbrain superior colliculus, SN and visual sensory areas. The efferents project to the brainstem nuclei and premotor cortex. Frontal eye field is involved in control of voluntary eye movements (Parent 1996; Standring 2005a).

PFC is further subdivided into lateral, medial and orbital frontal cortices. The medial part of PFC forms the anterior cingulate cortex. PFC has bidirectional connections with premotor, frontal eye field, temporal cortex and VTA. It also gets serotonergic afferents from brainstem raphe neurons. The efferents also project to the nucleus accumbens (NAc), AMG, hypothalamus and hippocampus (Berger et al. 1991; Jankowski and Seasack 2004). The details of these brain areas are described further in this chapter. PFC is the execution area for goal-directed behaviors. The various components of goal-directed behavior are cognition, memory, motivation, emotions, judgement and decision-making (Dembrow and Johnston 2014). PFC is also involved in sexual behavior. Sexual arousal is found to be associated with activation of prefrontal regions including the anterior cingulate cortex and orbitofrontal cortex (Redoute et al. 2000). Hypersexual behavior along with expression of increased sexual drive in public has been observed after damage to the orbitofrontal region (Miller et al. 1986). See Fig. 17.5 for the connections and functions of the frontal lobe.

17.3 Parietal Lobes

The parietal lobe is present posterior to central sulcus. The parietal lobe is involved in the interpretation of sensory stimuli and transforming the sensory information into motor output. The parietal cortex is divided into primary somatosensory cortex, superior parietal lobule and inferior parietal lobule (see Fig. 17.6). The lobules form the posterior parietal cortex (PPC). The somatosensory cortex gets afferents from the thalamus, midbrain raphe nuclei and locus coeruleus in the pons. The locus coeruleus is a nucleus in the pons which is the major site of noradrenaline synthesis in the brain. The efferents from the somatosensory cortex project to the brainstem, spinal cord and striatum (Standring 2005a). The anteriorly placed somatosensory cortex is the area receiving sensory stimuli. The parietal cortex on medial surface of the brain forms posterior part of paracentral lobule which is considered as the genital sensory cortex (see Fig. 17.3). The anterior part of paracentral lobule is formed by the frontal cortex controlling motor innervations to the genitals. Paracentral lobule gets activated during self-stimulation of genitals (Baird 2007). Penile erection induced by visually evoked stimuli involved activation of paracentral lobule (Moulier et al. 2006). The seizures arising from parietal cortex may be associated with increased genital sensations (Toone 1991).

PPC gets afferents from the thalamus and primary sensory cortex and sends efferent projections to the premotor cortex, cingulate cortex and insula (Standring 2005a). PPC is involved in cognitive functions such as selective attention, sequential planning of events and memory retrieval (Behrmann et al. 2004; Wagner et al. 2005; Cui 2014). Selective attention is a process by which a few inputs are selected and processed further to bring about a preferred motor output and behavior. See Fig. 17.7 for connections and functions of the parietal lobe.

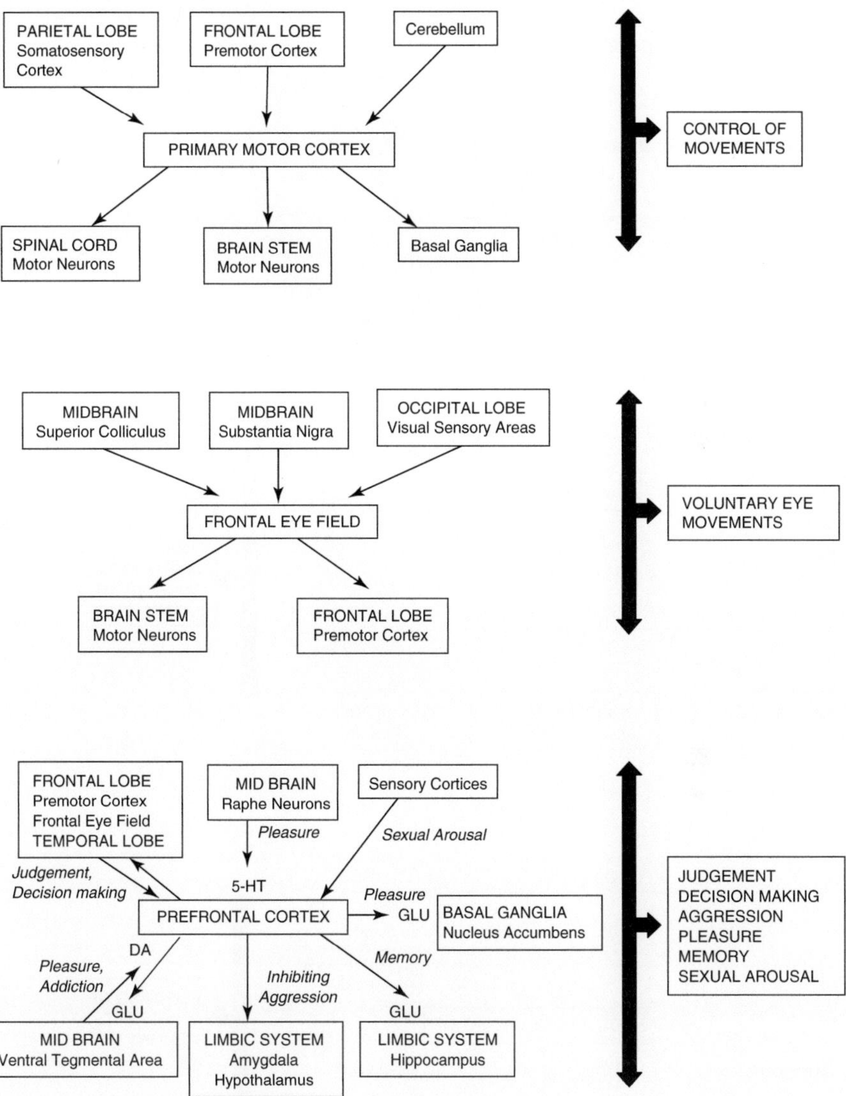

Fig. 17.5 Connections and functions of the frontal lobe. *5-HT* serotonin, *DA* dopamine, *GLU* glutamate

17.4 Temporal Lobes

The temporal lobe is present below the lateral sulcus. The temporal cortex is divided into three cortical areas—superior, middle and inferior temporal cortices (see Fig. 17.8). Superior temporal cortex forms part of auditory cortex and inferior temporal cortex forms visual association cortex. Superior temporal cortex has

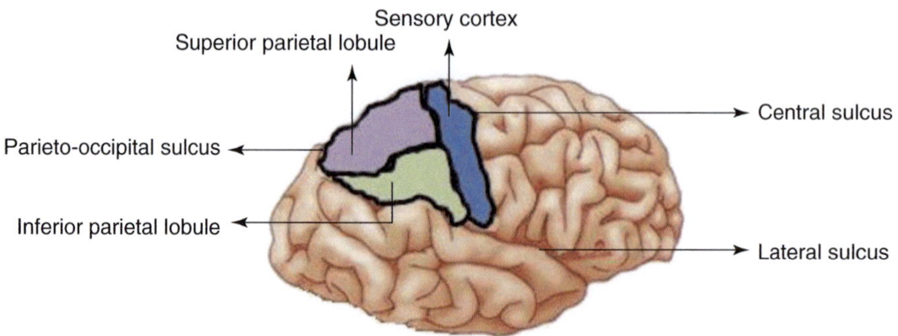

Fig. 17.6 Lateral surface of the brain showing various functional areas of the parietal lobe

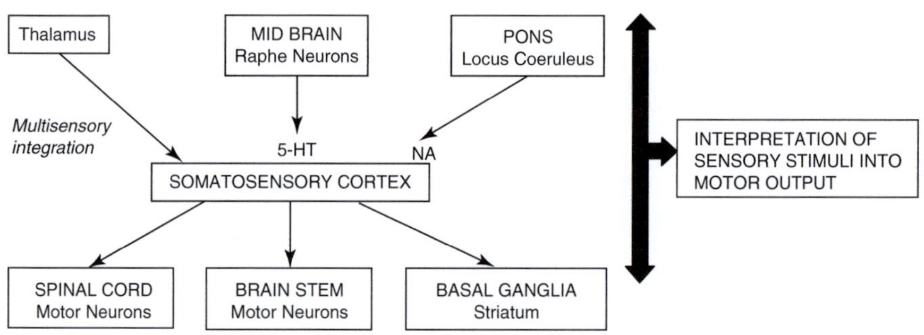

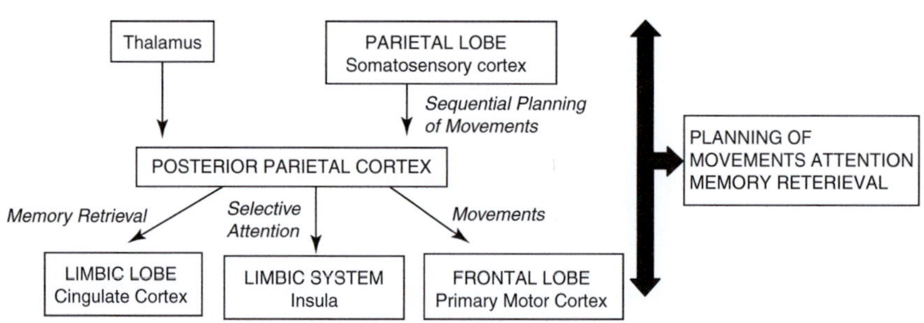

Fig. 17.7 Connections and functions of the parietal lobe. *5-HT* serotonin, *NA* noradrenaline

bidirectional connections with the thalamus and also sends efferents to the PFC and insula. The middle temporal cortex gets afferents from the auditory cortex, somatosensory cortex and visual association cortices. The efferents from the middle temporal cortex project to the PFC. Inferior temporal cortex gets afferents from primary visual cortex in the occipital lobe and sends efferents to the PFC and limbic lobe (Standring 2005a). The temporal lobe functions in spatial detection of

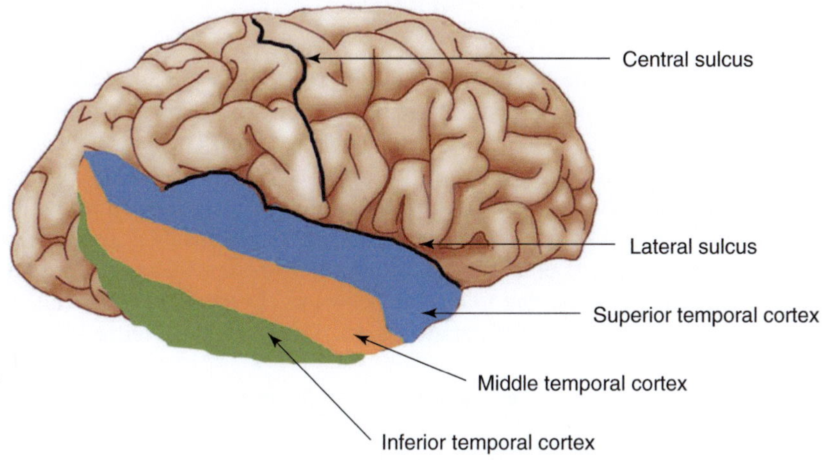

Fig. 17.8 Lateral surface of the brain showing various functional areas of the temporal lobe

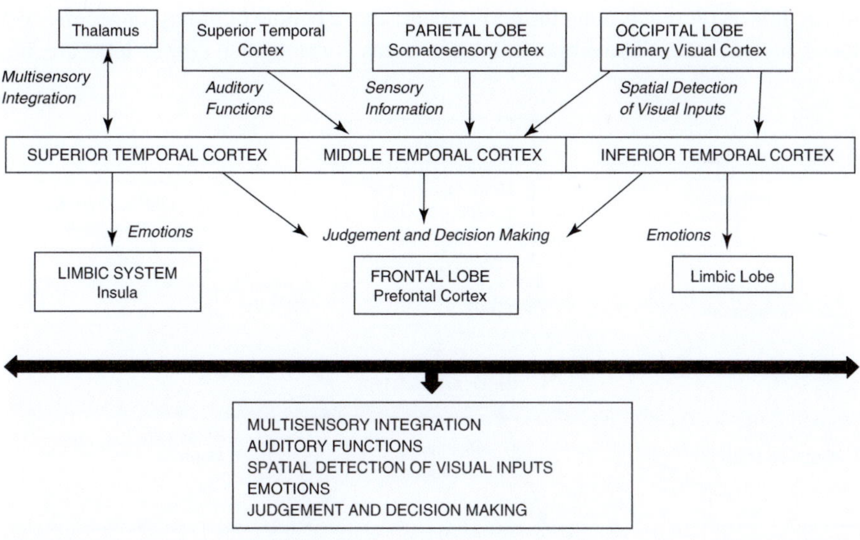

Fig. 17.9 Connections and functions of the temporal lobe

visual inputs and emotional behavior and has auditory processing. The role of the temporal lobe is also implicated in sexual behavior. In temporal lobe epilepsy, hyposexuality has been identified as one of the most common sexual dysfunction. Hypersexuality or increased sexual urge has been reported in some patients after unilateral temporal lobectomy (Herzog et al. 1986). The anterior commissure which is a bundle of neuronal fibres connecting two temporal lobes influences sexual orientation (Baird 2007). See Fig. 17.9 for connections and functions of the temporal lobe.

17.5 Insula

The insula is the part of the cerebral cortex buried deep to the lateral sulcus and it is overlapped by parietal and temporal lobes. It can be visualized after retracting lateral sulci apart (see Fig. 17.1). The insula has bidirectional connections with the auditory cortex, PFC, somatosensory cortex, cingulate cortex, AMG, basal ganglia, brainstem and thalamus. The insula is the site for affective integration of sensorimotor, autonomic and limbic functions (Flynn et al. 1999). The insula is also involved in conscious awareness of emotions, detection of errors and memory. Conscious awareness of errors is important to change behaviors. Deficits in insular function may lead to dysfunction in emotional awareness. The condition is termed as alexithymia (Taylor 2000).

See Fig. 17.10 for connections and functions of the insula.

17.6 Limbic Lobe

The term limbic lobe is derived from a Latin word limbus which means rim. The limbic lobe is present around the corpus callosum, a bundle of fibres connecting two lobes of the brain. The important brain areas forming the limbic lobe are the

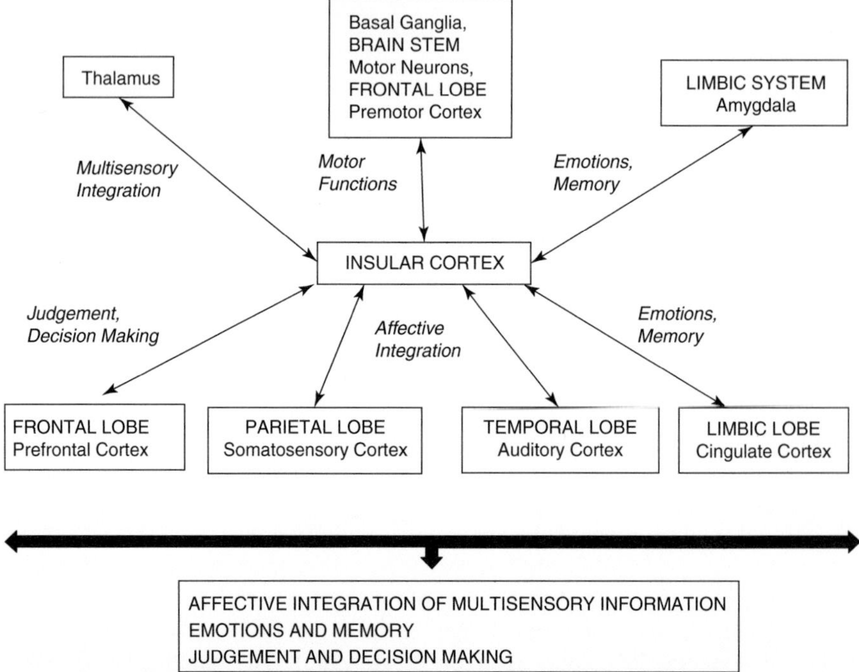

Fig. 17.10 Connections and functions of the insula

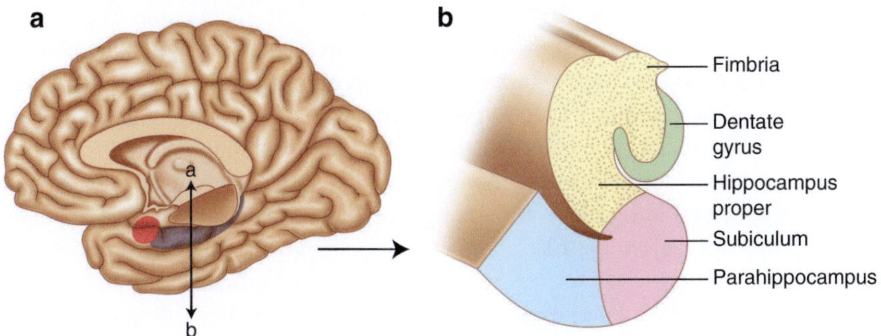

Fig. 17.11 (**a**) Plane ab for coronal section of the hippocampus; (**b**) schematic diagram showing different parts of the hippocampus in coronal section

hippocampus, parahippocampus and cingulate cortex. The limbic lobe structures are easily seen on the medial surface of each lobe of the brain (see Fig. 17.2). The forebrain structures like the orbitofrontal cortex, pallidum, septal region, AMG, thalamus, hypothalamus and olfactory cortex along with the limbic lobe structure constitute 'the limbic system' (Purves et al. 2004).

The hippocampus extends from AMG anteriorly to the splenium posteriorly. The splenium is the last and most posterior part of the corpus callosum (see Fig. 17.3). The hippocampus is located in the floor of the temporal extension of the lateral ventricle. The lateral ventricle is a cavity found within each lobe of the brain. The hippocampus consists of three different parts—the dentate gyrus, the hippocampus proper and the subiculum (Parent 1996). The shape of the hippocampus resembles a sea horse and so was given the name as in Greek, hippocampus means seahorse (see Fig. 17.11a, b). The dentate gyrus is buried deep under the most medial part of the temporal lobe. The dentate gyrus continues into the hippocampus proper which is a complex mass situated deep in the temporal lobe. The next part is the subiculum which receives information from the hippocampal neurons and transmits it forward through the fimbria. The fimbria later continues as the fornix (Parent 1996).

The major afferents to the hippocampus project from the entorhinal cortex (ERC) (Amaral et al. 1987). The hippocampus also gets dopaminergic afferents from VTA and serotonergic afferents from the midbrain raphe nuclei (Amaral and Insausti 1990). The reciprocal connections with ERC are implicated in memory process (Lavenax and Amaral 2000).

The hippocampus sends the efferent projections through the fornix. In humans, the fornix consists of about 1.2 million fibres (Powell et al. 1957). The fornix consists of neuronal axons of the subiculum and hippocampus proper. The axons form the fimbria which later continues as the fornix. The fornix projects to the septal nuclei and mammillary body of the hypothalamus (Nauta 1956; Swanson and Cowan 1977). The hippocampus also sends efferents to the cingulate cortex and parahippocampus and glutamatergic efferents to NAc (Groenewegen et al. 1987; Amaral and Insausti 1990; Brog et al. 1993). The role of the hippocampus is implicated in recent memory, pleasure, love and other emotions such as fear and anxiety (Green and Shimamoto 1953; Papez 1937). It has been proved that the limbic lobe and hypothalamus are interconnected. Emotions are organized in the hippocampus and then, are expressed through mammillary bodies in the hypothalamus under its autonomous influence. See Fig. 17.12 for connections and functions of the hippocampus.

The parahippocampus is situated along the medial surface of temporal lobes, and its role is implicated in memory processing (see Fig. 17.2). The parahippocampus contains a distinct region, the ERC. The parahippocampus is connected to the cingulate cortex, hippocampal formation, various sensory cortical areas and frontal lobe (Hendelman 2006). Parahippocampal atrophy has been associated with memory impairments (Kohler et al. 1998).

The cingulate cortex is present above the corpus callosum on the medial surface of each cerebral lobe (see Figs. 17.2 and 17.3). The cingulate cortex has bidirectional connections with the parahippocampus (Hendelman 2006). The cingulate cortex is involved in the forebrain networks that process acute pain and emotion and in the initiation of motor activities (Vogt 2005). The cingulate cortex also regulates various affective responses like fear and aggression. Excessive activity in the cingulate gyrus may present in the form of behavioral abnormalities like obsessive-compulsive disorders. Decreased cingulate activity is associated with depression and reduced responses to pain (Devinsky et al. 1995).

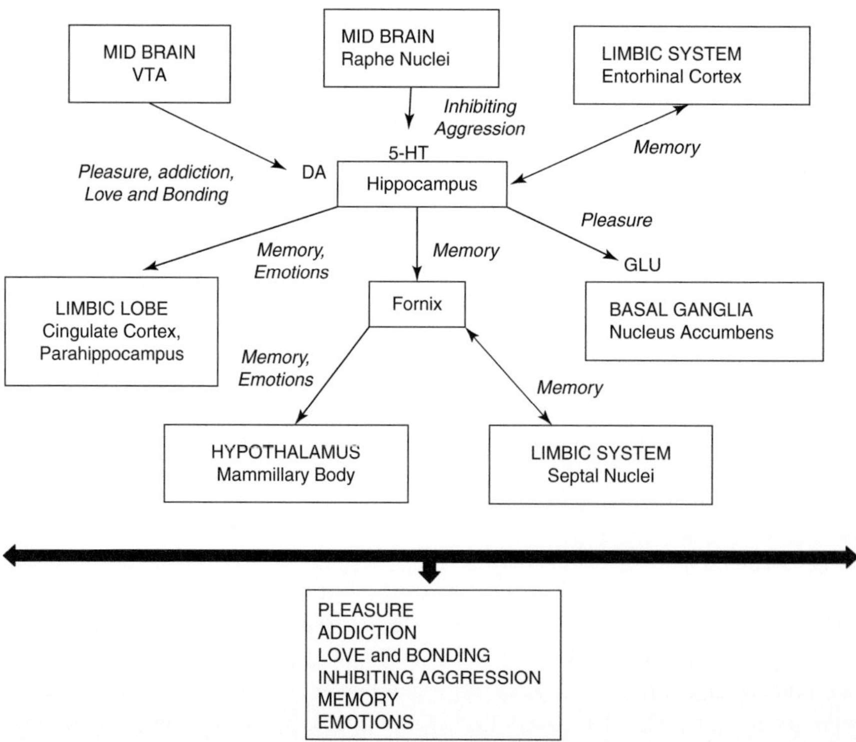

Fig. 17.12 Connections and functions of the hippocampus. *DA* dopamine, *5-HT* serotonin, *GLU* glutamate, *VTA* ventral tegmental area

17.7 Amygdala (AMG)

AMG is a mass of neuronal bodies situated in the anteromedial part of each temporal lobe (see Figs. 17.2 and 17.3). There are different nuclear groups in the AMG which are connected to different brain areas. The nuclear groups are the corticomedial group and basolateral group, each having subnuclei (Crosby and Humphrey 1941). The corticomedial group of nuclei is majorly connected to the olfactory cortex, whereas the basolateral group of nuclei is connected to the PFC. Another group of nuclei implicated in behaviors is the bed nucleus of the stria terminalis (BNST) that forms part of extended AMG group of nuclei (Alheid et al. 1995).

AMG has bidirectional connections with the medial preoptic area (MPOA) of the hypothalamus, sensory cortices, brainstem raphe nuclei, BNST, VTA and SN of the midbrain. AMG sends efferent projections through the stria terminalis. Some of the fibres of the stria terminalis relay in BNST, and some its fibres are projections from BNST to AMG. A few fibres of the stria terminalis pass through BNST and then relay into the hypothalamus and septal nuclei (Standring 2005b). The efferents from AMG also project directly to the hippocampus, striatum and brainstem autonomic neurons (Heimer and Nauta 1969; McBride and Sutin 1977; Groenewegen et al. 1980; Mehler 1980; Amaral et al. 1992; Aggleton 2001; Swanson and Cowan 2004).

Major function of AMG is to control autonomic influences on emotional behavior. AMG gives major outputs to the autonomic nuclei of the brainstem and hypothalamus by which it influences autonomic responses related to behavioral expression of emotions (Purves et al. 2004). AMG has role in expression of fear, modulation of memory and mediation of social recognition (Fleming et al. 1994). Emotional stimuli such as fearful faces activate AMG. AMG is probably a critical component of a system that evaluates the environment for potential dangers (Amaral 2003). It may also be associated with decision-making and thus, guide or even drive human behaviors like appraisal of the emotional state of others and value judgements in complex social situations (Trimble and George 2010). Functional neuroimaging studies have demonstrated activation of AMG during appetitive sexual behavior such as viewing of sexual stimuli leading to sexual arousal. Consummatory sexual behaviors, erection and orgasm correspond with decreased AMG activity (Beauregard et al. 2001; Hamann et al. 2004). See Fig. 17.13 for connections and functions of AMG. BNST is the group of nuclei forming extended AMG, and it is situated in the basal and medial part of the forebrain (see Fig. 17.3). BNST controls emotions such as fear and anxiety through its connections with the hippocampus, AMG and hypothalamus in non-human primates as well as in humans (Dong et al. 2001; Dong and Swanson 2004a, b; Fox et al. 2008; Alvarez et al. 2011). The role of BNST is also implicated in pleasure, motivation, sexual behavior and addiction through its connections with NAc, VTA and MPOA (Weller and Smith 1982; Dong et al. 2001; O'Daly et al. 2012; Silberman and Winder 2013).

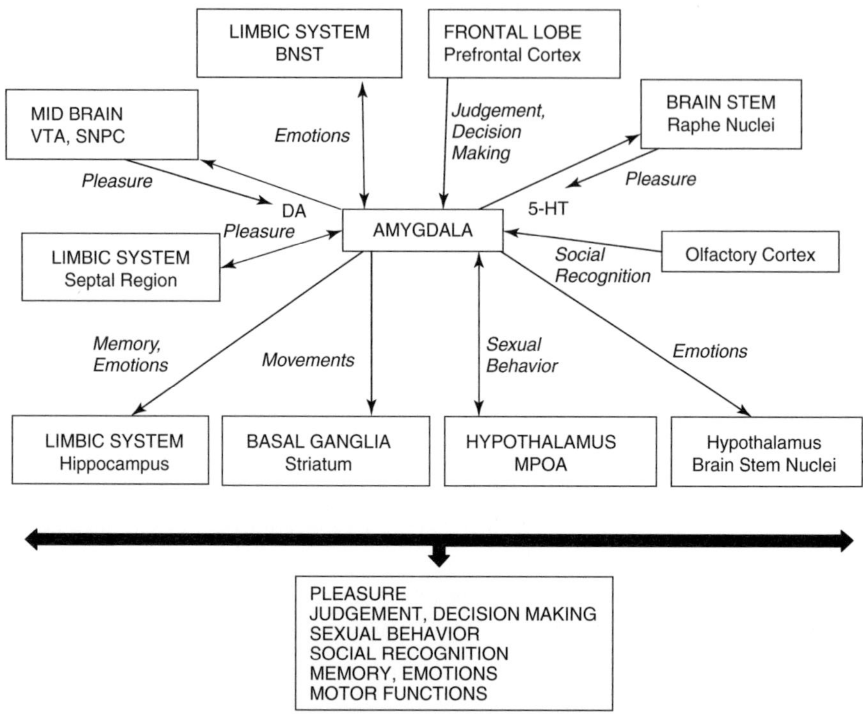

Fig. 17.13 Connections and functions of the amygdala. *DA* dopamine, *5-HT* serotonin, *VTA* ventral tegmental area, *SNPC* substantia nigra pars compacta, *BNST* bed nucleus of the stria terminalis, *MPOA* Medial preoptic area

17.8 Hypothalamus

The hypothalamus is a part of diencephalon which contains various nuclei that regulate emotions, autonomic functions, endocrine functions and some somatic functions (see Fig. 17.14). The hypothalamus is interconnected with various visceral and somatic nuclei of the brainstem and spinal cord. The hypothalamus regulates hunger, thirst and emotional and sexual behavior through its connections with AMG and hippocampus.

Amongst the hypothalamic nuclei, MPOA is the key neural structure regulating male sexual behavior. The initial phase of erection has been reported with stimulation of MPOA, indicating that the hypothalamus may serve to trigger an overt sexual response (Ferreti et al. 2005). MPOA has been implicated in both affective and consummatory aspects of male sexual behavior. A lesion in MPOA impairs copulation in male rats and in a large number of other mammalian species (Hull et al. 2006). MPOA has bidirectional connections with AMG, BNST, midbrain raphe and cingulate cortex (Muller-Preuss and Jurgens 1975; Parent et al. 1981; Weller and Smith 1982; Canteras et al. 1995). MPOA also sends efferents to spinal cord motor neurons (Hull and Dominguez 2006). See Fig. 17.15 for connections and functions of MPOA. Supraoptic and paraventricular nuclei (PVN) of the hypothalamus

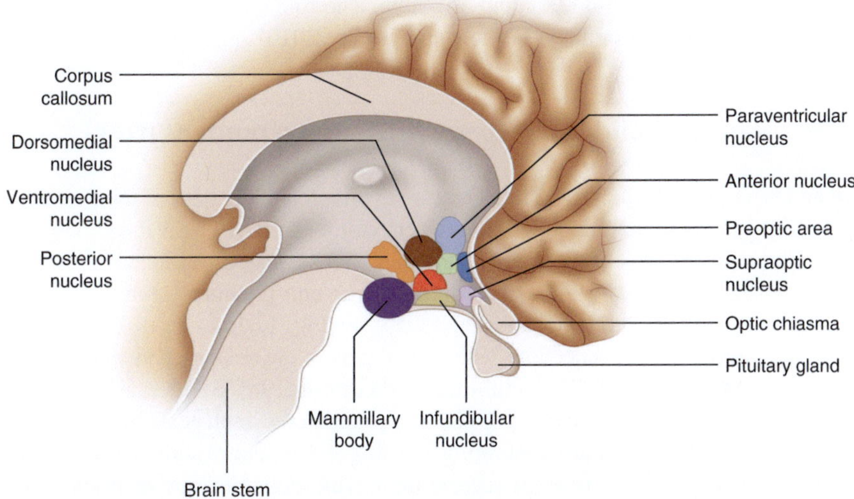

Fig. 17.14 Medial surface of the brain showing hypothalamic nuclei

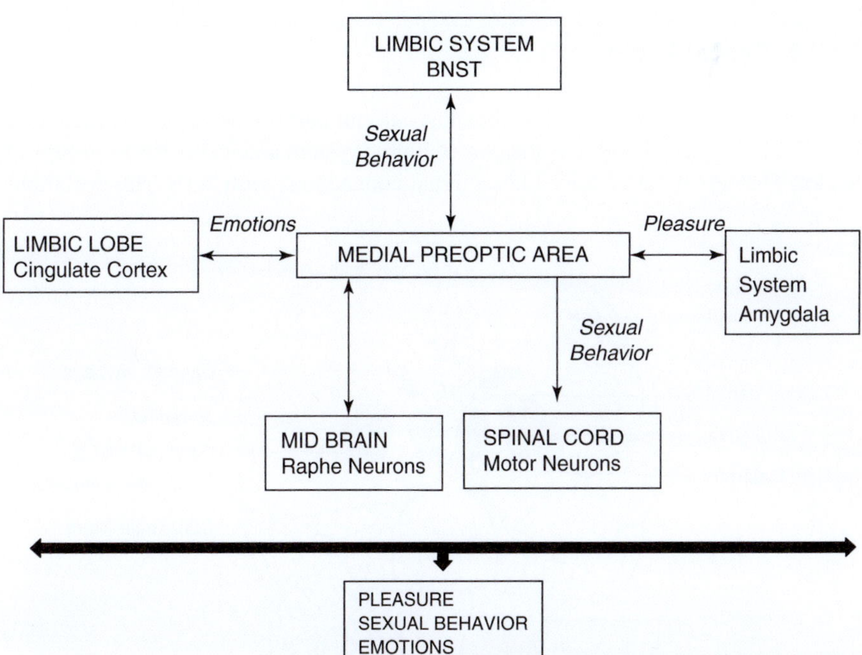

Fig. 17.15 Connections and functions of the medial preoptic area. *BNST* bed nucleus of the stria terminalis

synthesize oxytocin (OXT) and vasopressin (VP), and their axons pass into the posterior pituitary and release OXT and VP there. The role of OXT and VP is implicated in love and pair bonding (Trimble and George 2010).

17.9 Ventral Pallidum (VP) and Nucleus Accumbens (NAc)

VP and NAc are the subcortical nuclear masses situated at the base of the forebrain and control motivational aspect of human behavior (see Fig. 17.3). VP and NAc are the components of the basal ganglia. The other nuclear masses of the basal ganglia are the striatum consisting of the caudate nucleus and putamen, globus pallidus, subthalamic nucleus and SN (Standring 2005b; see Fig. 17.16).

VP has bidirectional connections with reward-generating areas such as PFC, AMG, VTA and NAc (Haber et al. 1985; Phillipson and Griffiths 1985; Smith et al. 2009). VP acts as the final pathway for reward or pleasure generation (Smith et al. 2009).

The NAc is another reward-generating area that also mediates addictive actions of drugs of abuse. The NAc neurons receive input from dopaminergic neurons of the VTA and SN, from the glutamatergic neurons of the hippocampus, AMG and medial PFC. Efferents from NAc project to VTA, SN, VP and PFC (Fallon and Moore 1978; Mogenson et al. 1980; Phillison and Griffith 1985; Heimer et al. 1991, 1995; Salgado and Kaplitt 2015). See Fig. 17.17 for connections and functions of VP and NAc.

17.10 Septal Region

The septal region is present below the most anterior part of the corpus callosum (see Fig. 17.2). The septal region is the part of limbic system and it consists of groups of nuclei. The septal nuclei have bidirectional connections with AMG, hippocampus

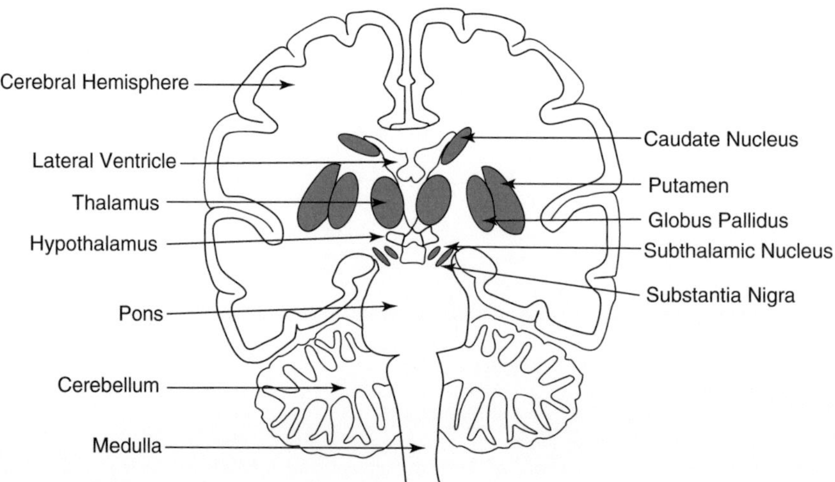

Fig. 17.16 Vertical section of the brain showing subcortical areas

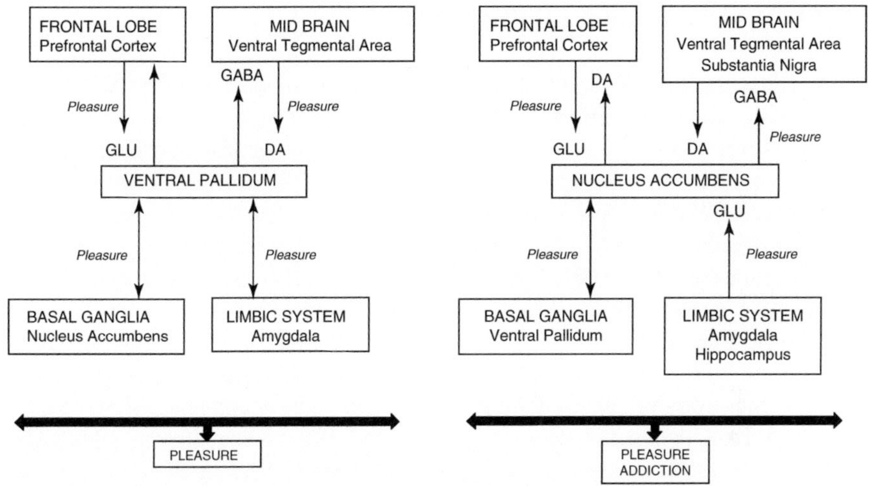

Fig. 17.17 Connections and functions of the ventral pallidum and nucleus accumbens. *GABA* gamma aminobutyric acid, *GLU* glutamate, *DA* dopamine

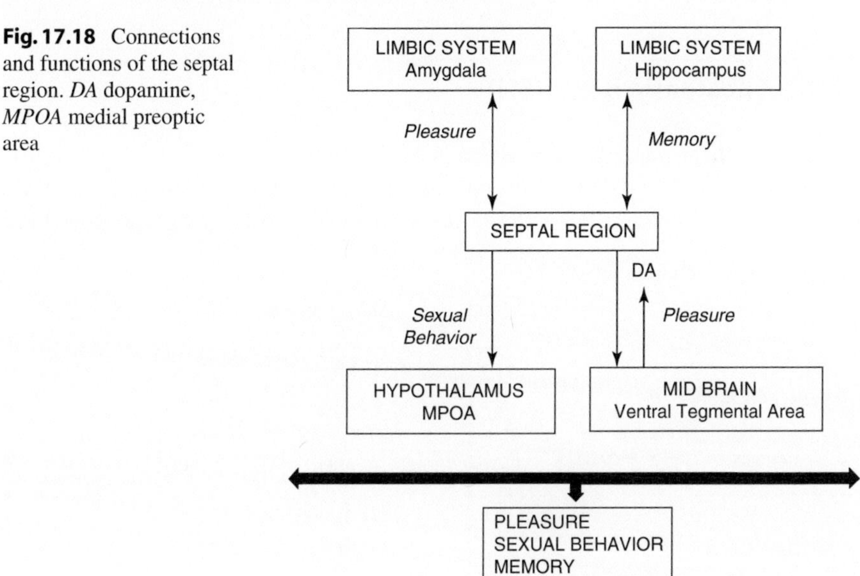

Fig. 17.18 Connections and functions of the septal region. *DA* dopamine, *MPOA* medial preoptic area

and VTA. Efferents from the septal nuclei also project to MPOA (Swanson and Cowan 2004; Luttgen et al. 2004; Hendelman 2006; Rutz et al. 2007). Septal region appears to be involved in memory, pleasurable response and orgasm. Septal stimulation elicits orgasm and a compulsion to masturbate. Septal injury leads to hypersexual behavior (Gorman and Cummings 1992). See Fig. 17.18 for connections and functions of the septal region.

17.11 Ventral Tegmental Area (VTA)

VTA is the collection of neurons located in the midbrain (see Fig. 17.3). VTA is the largest collection of dopaminergic neurons which give origin to mesocorticolimbic pathways mediating reward and motivation. The mesocorticolimbic pathways extend from dopaminergic neurons in VTA projecting to widespread areas such as PFC, NAc, AMG and hippocampus. In humans, 75% of dopaminergic neurons exist in the substantia nigra pars compacta (SNPC), 15% in VTA and 10% in the midbrain reticular formation (Hirsch et al. 1992; Francois et al. 1999; Duzel et al. 2009).

The VTA dopaminergic neurons receive GABAergic afferents from NAc and GABAergic interneurons of VTA (Spanagel and Weiss 1999). The glutamatergic afferents to VTA come from BNST and PFC (Georges and Aston-Jones 2002). The afferents from VP are GABAergic and, hence, inhibit dopamine (DA) release from the VTA dopaminergic neurons (Kalivas et al. 1993; Klitenick et al. 1992). The GABAergic efferent neurons in VTA project to PFC and NAc. GABAergic neurons regulate the activity of dopaminergic neurons in VTA by axonal collaterals (Steffensen et al. 1998). VTA sends dopaminergic efferents to PFC, AMG and hippocampus. Glutamatergic connection of VTA with BNST is involved in addiction (Georges and Aston-Jones 2002). See Fig. 17.19 for connections and functions of VTA.

17.12 Substantia Nigra (SN)

SN in Latin means dark substance and is a pigmented nuclear mass present in the midbrain (see Fig. 17.3). SN forms part of the basal ganglia. The main parts of SN are the posterior zone, pars compacta (PC); an anterior zone, pars reticulata (PR);

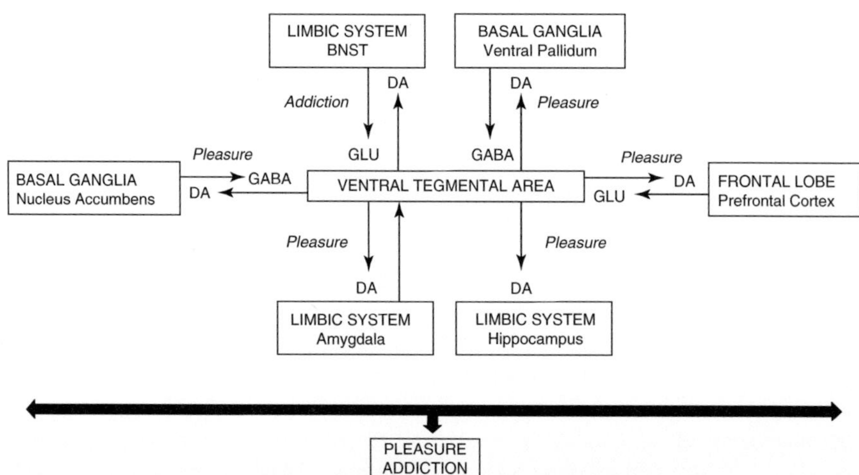

Fig. 17.19 Connections and functions of the ventral tegmental area. *BNST* bed nucleus of the stria terminalis, *DA* dopamine, *GLU* glutamate, *GABA* gamma aminobutyric acid

and the pars lateralis (PL). PC is a dense cellular zone made up of dopaminergic neurons. The neuromelanin pigment present in the dopaminergic neurons of PC gives it dark pigmented appearance (Rabey and Hefti 1990). The PR contains majorly the GABAergic neurons. PL is considered as the lateral extension of PC containing dopaminergic neurons (Poirier et al. 1983).

Various animal studies have identified the afferent and efferent projection fibres of SN. PC receives GABAergic afferents from the striatum and glutamatergic afferents from PFC (Somogyi et al. 1981; Naitao and Kita 1994). The efferent projections of PC are majorly dopaminergic to the striatum. PC has a role in reward- or pleasure-mediated behaviors.

PR receives GABAergic afferents from the striatum and globus pallidus (Somogyi et al. 1981; Smith et al. 1998; Parent et al. 1991). The glutamatergic afferents to PR come from subthalamic nucleus and PFC (Kita and Kitai 1987; Naitao and Kita 1994). PR along with the globus pallidus forms the major output of the basal ganglia. PR sends GABAergic efferent projections to the thalamus and superior colliculus (Rinvik 1975; Rinvik et al. 1976; Beckstead et al. 1979). The thalamic fibres then, project to the primary and supplementary motor cortical areas for control of movements. See Fig. 17.20 for connections and functions of SN.

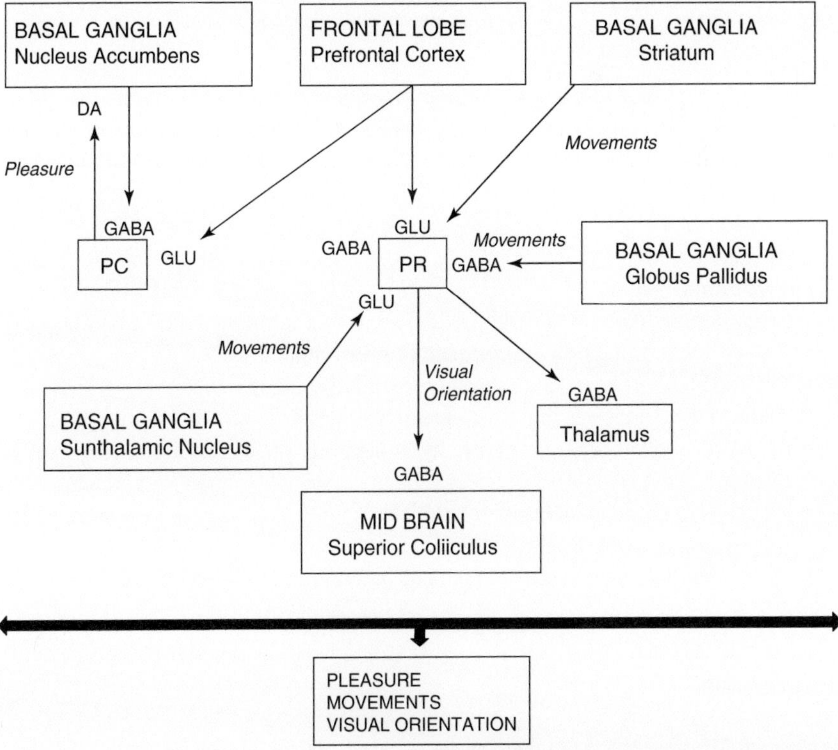

Fig. 17.20 Connections and functions of the substantia nigra. *PC* pars compacta, *PR* pars reticulata, *GLU* glutamate, *DA* dopamine, *GABA* gamma aminobutyric acid

Table 17.1 Brain areas and functions

Brain areas	Functions
Frontal lobe	Control of movements, judgement, decision-making, aggression, pleasure, memory, sexual arousal
Parietal lobe	Interpretation of sensory stimuli into motor output, sequential planning of movements, attention, memory retrieval
Temporal lobe	Multisensory integration, auditory functions, spatial detection of visual inputs, emotions, judgement, decision-making
Insula	Affective integration of multisensory information, emotions, memory, judgement decision-making
Hippocampus	Aggression, pleasure, addiction, love and bonding, memory, emotions
Amygdala	Pleasure, judgement, decision-making, movements, memory, emotions, sexual behavior
Bed nucleus of the stria terminalis	Pleasure, addiction, emotions, sexual behavior
Medial preoptic area	Pleasure, sexual behavior, emotions
Ventral pallidum	Pleasure
Nucleus accumbens	Pleasure, addiction
Septal region	Pleasure, sexual behavior, memory
Ventral tegmental area	Pleasure, addiction
Substantia nigra	Pleasure, movements, visual orientation

See Table 17.1 for brain areas and their functions. Brain controls behaviors. Male behaviors such as aggression, pleasure, love and sex are related to reproduction. Various neural pathways which regulate male behaviors are discussed in the subsequent chapter.

Key Questions

- What are the cortical and subcortical areas that regulate male behaviors?
- Describe the functional importance of prefrontal cortex.
- Name the brain areas constituting the limbic system.
- Draw a diagram showing the connections and functions of the amygdala.
- Draw a labelled diagram of the hypothalamus showing different nuclear groups.
- What are the connections by which the nucleus accumbens regulate pleasure?
- Describe the different parts of the substantia nigra with their functions.

References

Aggleton JP. The amygdala: a functional analysis. 2nd ed. London: Oxford UP; 2001.

Alheid GF, De Olmos J, Beltramino CA. Amygdala and extended amygdala. In: Paxinos G, Orlando FL, editors. The rat nervous system. San Diego: Academic; 1995. p. 495–578.

Alvarez RP, Chen G, Bodurka J, Kaplan R, Grillon C. Phasic and sustained fear in humans elicits distinct patterns of brain activity. Neuroimage. 2011;55:389–400.

Amaral DG, Insausti R, Cowan WM. The entorhinal cortex of the monkey: I. Cytoarchitectonic organization. J Comp Neurol. 1987;264:326–55.

Amaral DG, Insausti R. Hippocampal formation. In: Paxinos G, editor. The human nervous system, Ch 21. New York: Academic Press; 1990. p. 711–55.

Amaral DG, Price JL, Pitkanen A, Carmichael ST. Anatomical organization of primate amygdaloid complex. In: Aggleton JP, editor. The amygdala: neurobiological aspects of emotion, memory and mental dysfunction. New York: Wiley-Liss; 1992. p. 1–66.

Amaral DG. The amygdala, social behavior, and danger detection. Ann N Y Acad Sci. 2003;1000:337–47.

Baird AD, Wilson SJ, Bladin PF, Saling MM, Reutens DC. Neurological control of human sexual behavior: insights from lesion studies. J Neurol Neurosurg Psychiatry. 2007;78:1042–9.

Beauregard M, Leve'sque J, Bourgouin P. Neural correlates of conscious self regulation of emotion. J Neurosci. 2001;21:1–6.

Beckstead RM, Domesick VB, Nauta WJH. Efferent connections of the substantia nigra and ventral tegemntal area in the rat. Brain Res. 1979;175:191–217.

Behrmann M, Geng JJ, Shomstein S. Parietal cortex and attention. Curr Opin Neurobiol. 2004;14:212–7.

Berger B, Gasper P, Verney C. Dopaminergic innervations of the cerebral cortex: unexpected differences between rodents and primates. Trends Neurosci. 1991;14:21–7.

Brog JS, Salyapongse A, Deutch AY, Zahm DS. The patterns of afferent innervation of the core and shell in the "accumbens" part of the rat ventral striatum: immunohistochemical detection of retrogradely transported fluoro-gold. J Comp Neurol. 1993;338:255–78.

Canteras NS, Simerly RB, Swanson LW. Organization of projections from medial nucleus of amygdala: a PHAL study in the rat. J Comp Neurol. 1995;360:213–45.

Crosby EC, Humphrey T. Studies of the vertebrate telencephalon. II. The nuclear pattern of the anterior olfactory nucleus, tuberculum olfactorum and the amygdaloid complex in adult man. J Comp Neurol. 1941;74:309–52.

Cui H. From intention to action: hierarchical sensorimotor transformation in the posterior parietal cortex. eNeuro. 2014;1(1):1–6.

Dembrow N, Johnston D. Subcircuit-specific neuromodulation in the prefrontal cortex. Front Neural Circuits. 2014;8(54):1–9.

Devinsky O, Morrell M, Vogt B. Contributions of anterior cingulate cortex to behavior. Brain. 1995;118:279–306.

Dong HW, Petrovich GD, Watts AG, Swanson LW. Basic organization of projections from the oval and fusiform nuclei of the bed nuclei of the stria terminalis in adult rat brain. J Comp Neurol. 2001;436:430–55.

Dong H-W, Swanson LW. Organization of axonal projections from the anterolateral area of the bed nuclei of the stria terminalis. J Comp Neurol. 2004a;468:277–98.

Dong H-W, Swanson LW. Projections from bed nuclei of the stria terminalis, posterior division: implications for cerebral hemisphere regulation of defensive and reproductive behaviors. J Comp Neurol. 2004b;471:396–433.

Düzel E, Bunzeck N, Guitart-Masip M, Wittmann B, Schott BH, Tobler PN. Functional imaging of the human dopaminergic midbrain. Trends Neurosci. 2009;32:321–8.

Fallon JH, Moore RY. Catecholamine innervations of the basal forebrain. IV. Topography of the dopamine projection to the basal forebrain and neostriatum. J Comp Neurol. 1978;180:545–80.

Ferretti A, Massimo C, Del Gratta C, Matteo RD, Merla A, Montorsi F, et al. Dynamics of male sexual arousal: distinct components of brain activation revealed by fMRI. Neuroimage. 2005;26:1086–96.

Fleming AS, Suh EJ, Korsmit M, Rusak B. Activation of Fos-like immunoreactivity in the medial preoptic area and limbic structures by maternal and social interactions in rats. Behav Neurosci. 1994;108:724–34.

Flynn FG, Benson DF, Ardila A. Anatomy of the insulafunctional and clinical correlates. Aphasiology. 1999;13:55–78.

Fox AS, Shelton SE, Oakes TR, Davidson RJ, Kalin NH. Trait-like brain activity during adolescence predicts anxious temperament in primates. PLoS One. 2008;3:e2570.

Francois C, Yelnik J, Tande D, Agid Y, Hirsch EC. Dopaminergic cell group A8 in the monkey:anatomical organization and projections to the striatum. J Comp Neurol. 1999;414:334–47.

Georges F, Aston-Jones G. Activation of ventral tegmental area cells by the bed nucleus of stria terminalis: a novel excitatory amino acid input to midbrain dopamine neurons. J Neurosci. 2002;22(12):5173–87.

Gorman DG, Cummings JL. Hypersexuality following septal injury. Arch Neurol. 1992;49:308–10.

Green JD, Shimamoto T. Hippocampal seizures and their propagation. Arch Neurol Psychiatry. 1953;70:687–702.

Groenewegen HJ, Becker NEHM, Lohman AHM. Subcotical afferents of nucleus accumbens septi in the cat, studied with retrograde axonal transport of horse raddish peroxidase and bisbenzimid. Neuroscience. 1980;5:1902–16.

Groenewegen HJ, Vermeulen-Van der Zee E, Te Kortschot A, Witter MP. Organization of the projections from the subiculum to the ventral striatum in the rat. A study using anterograde transport of Phaseolus vulgarus leucoagglutinin. Neuroscience. 1987;23:103–20.

Haber SN, Groenewegen HJ, Grove EA, Nauta WJ. Efferent connections of the ventral pallidum: evidence of a dual striato pallidofugal pathway. J Comp Neurol. 1985;235:322–35.

Hamann S, Herman RA, Nolan CL, Wallen K. Men and women differ in amygdala response to visual sexual stimuli. Nat Neurosci. 2004;7:411–6.

Heimer L, Nauta WJH. The hypothalamic distribution of the stria terminalis in the rat. Brain Res. 1969;13:284–97.

Heimer L, Zahm DS, Alheid GF. Basal ganglia. In: Paxinos G, editor. The rat nervous system. Academic Press: San Diego; 1995. p. 579–628.

Heimer L, Zahm DS, Churchill L, Kalivas PW, Wohltmann C. Specificity in the projection patterns of accumbal core and shell in the rat. Neuroscience. 1991;41:89–125.

Hendelman WJ. Limbic system. In: Atlas of functional neuroanatomy. 2nd ed. Boca Raton: CRC Press, Taylor and Francis Group; 2006. p. 202–38.

Herzog AG, Seibel MM, Schomer DL, Vaitukaitis JL, Geschwind N. Reproductive endocrine disorders in men with partial seizures of temporal lobe origin. Arch Neurol. 1986;43:347–50.

Hirsch EC, Mouatt A, Faucheux B, Bonnet AM, Javoy-Agid F, Graybiel A, et al. Dopamine tremor and Parkinson's disease. Lancet. 1992;340:125–6.

Hull EM, Dominguez JM. Getting his act together: roles of glutamate, nitric oxide, and dopamine in the medial preoptic area. Brain Res. 2006;1126:66–75.

Hull EM, Wood RI, McKenna KE. Neurobiology of male sexual behavior. In: Neill JD, editor. Knobil and neill's physiology of reproduction. 3rd ed. Elsevier Press: Amsterdam; 2006. p. 1729–824.

Jankowski MP, Sesack SR. Prefrontal cortical projections to the rat dorsal raphe nucleus: ultrastructural features and associations with serotonin and gamma-aminobutyric acid neurons. J Comp Neurol. 2004;468:518–29.

Kalivas PW, Churchill L, Klitenick MA. GABA and enkephalin projection from the nucleus accumbens and ventral pallidum to VTA. Neuroscience. 1993;57:1047–60.

Kita H, Kitai ST. Efferent projections of subthalamic nucleus in the rat: light and electron microscopic analysis with the PHA-L method. J Comp Neurol. 1987;260:435–52.

Klitenick MA, DeWitte P, Kalivas PW. Regulation of somatodendritic dopamine release in the ventral tegmental area by opioids and GABA. J Neurosci. 1992;12:2623–32.

Kohler S, Black SE, Sinden M, Szekely C, Kidron D, Parker JL, et al. Memory impairments associated with hippocampal versus parahippocampal gyrus atrophy: an MR volumetry study in Alzheimer's disease. Neuropsychologia. 1998;36(9):901–14.

Lavenex P, Amaral DG. Hippocampal neocortical interactions: a hierarchy of associativity. Hippocampus. 2000;10:420–30.

Luttgen M, Ove Ogren S, Meister B. Chemical identity of 5-HT2A receptor immunoreactive neurons of the rat septal complex and dorsal hippocampus. Brain Res. 2004;1010(1–2):156–65.

McBride RL, Sutin J. Amygdaloid and pontine projections to the ventromedial nucleus of the hypothalamus. J Comp Neurol. 1977;174:377–96.

Mehler WR. Subcortical afferent connections of amygdale in the monkey. J Comp Neurol. 1980;190:733–62.

Miller BL, Cummings JL, McIntyre H, Ebers G, Grode M. Hypersexuality or altered sexual preference following brain injury. J Neurol Neurosurg Psychiatry. 1986;49:867–73.

Mogenson GJ, Jones DL, Yim CY. From motivation to action: functional interface between the limbic system and the motor system. Prog Neurobiol. 1980;14:69–97.

Moulier V, Mouras H, Pelegrini-Issac M, Glutron D, Rouxel R, Grandjean B, et al. Neuroanatomical correlates of penile erection evoked by photographic stimuli in human males. Neuroimage. 2006;33(2):689–99.

Muller-Preuss P, Jurgens U. Projections from the cingular vocalization area in the squirrel monkey. Brain Res. 1975;103:29–43.

Naito A, Kita H. The coriconigral projections in the rat: an anterograde tracing study with biotinylated dextran amine. Brain Res. 1994;637:317–22.

Nauta WHJ. An experimental study of fornix in the rat. J Comp Neurol. 1956;104:247–72.

O'Daly OG, Trick L, Scaife J, Marshall J, Ball D, Phillips ML, et al. Withdrawal-associated increases and decreases in functional neural connectivity associated with altered emotional regulation in alcoholism. Neuropsychopharmacology. 2012;37:2267–76.

Papez JW. A proposed mechanism of emotion. Arch Neurol Psychiatry. 1937;38:725–43.

Parent A, Descarries L, Beaudet A. Organization of ascending serotonin systems in the adult rat brain. A radioautographic study after intraventricular administration of 5-hydroxytryptamine. Neuroscience. 1981;6:115–38.

Parent A, Hazrati LN, Lavoie B. The pallidum as a dual structure in primates. In: Bernardi G, Carpenter MB, Di Chiara G, Morelli M, Stanzione P, editors. The basal ganglia III. New York: Plenum; 1991. p. 81–8.

Parent A. Limbic system: Ch 18. In: Carpenter's human neuroanatomy. 9th ed. Baltimore: Williams and Wilkins; 1996. p. 744–94.

Phillipson OT, Griffiths AC. The topographic order of inputs to nucleus accumbens in the rat. Neuroscience. 1985;16:275–96.

Poirier LJ, Giguere M, Marchand R. Comparative morphology of the substantia nigra and ventral tegmental area in the monkey, cat and rat. Brain Res Bull. 1983;11:371–97.

Powell TPS, Guillery RW, Cowan WM. A quantitative study of the fornix-mammillo-thalamic system. J Anat. 1957;91:419–32.

Purves D, George JA, David F, William CH, Anthony SL, James OM, et al. Emotions. In: Neuroscience. 3rd ed. Sunderland: Sinauer Associates; 2004. p. 687–710.

Rabey JM, Hefti F. Neuromelanin synthesis in rat and human substantia nigra. J Neural Transm Park Dis Dement Sect. 1990;2(1):1–14.

Redoute J, Stolero S, Gregoire MC, Costes N, Cinotti L, Lavenne F, et al. Brain processing of visual sexual stimuli in human males. Hum Brain Mapp. 2000;11:162–77.

Rinvik E, Grofova I, Ottersen OP. Demonstration of nigrotectal and nigroreticular projections in the cat by axonal transport of protein. Brain Res. 1976;112:388–94.

Rinvik E. Demonstration of nigrothalamic connections in the cat by retrograde axonal transport of horse raddish peroxidase. Brain Res. 1975;90:313–8.

Rutz S, Riegert C, Rothmaier AK, Jackisch R. Presynaptic modulation of 5-HT release in the rat septal region. Neuroscience. 2007;146(2):643–58.

Salgado S, Kaplitt MG. Nucleus accumbens: a comprehensive review. Stereotact Funct Neurosurg. 2015;93:75–93.

Silberman Y, Winder DG. Emerging role for corticotropin releasing factor signaling in the bed nucleus of the stria terminalis at the intersection of stress and reward. Front Psych. 2013;4:42.

Smith KS, Tindell AJ, Aldridge JW, Berridge KC. Ventral pallidum roles in reward and motivation. Behav Brain Res. 2009;196(2):155–67.

Smith Y, Bevan MD, Shink E, Bolam JP. Microcircuitry of direct and indirect pathways of the basal ganglia. Neuroscience. 1998;86:353–87.

Somogyi P, Bolam JP, Totterdell S, Smith AD. Monosynaptic input from the nucleus accumbens-ventral striatum region to retrgradely labeled nigrostriatal neurons. Brain Res. 1981;217:245–63.
Spanagel R, Weiss F. The dopamine hypothesis of reward: past and current status. Trends Neurosci. 1999;22:521–7.
Standring S. Basal Ganglia. Ch 23. In: Gray's anatomy. 39th ed. Edinburgh: Elsevier Churchill Livingstone; 2005b. p. 419–30.
Standring S. Cerebral hemisphere. Ch 22. In: Gray's anatomy. 39th ed. Edinburgh: Elsevier Churchill Livingstone; 2005a. p. 387–417.
Steffensen SC, Svingos AL, Pickel VM, Henriksen SJ. Electrophysiological characterization of GABAergic neurons in the ventral tegmental area. J Neurosci. 1998;18:8003–15.
Swanson LW, Cowan WM. An autoradiographic study of the organization of the efferent connections of the hippocampal formation in the rat. J Comp Neurol. 1977;172:49–84.
Swanson LW, Cowan WM. Connections of the septal region in the rat. J Comp Neurol. 2004;186(4):621–55.
Taylor GJ. Recent developments in alexithymia theory and research. Can J Psychiatry. 2000;45:134–42.
Toone B. Sex, sexual seizures and the female with epilepsy. In: Trimble MR, editor. Women and epilepsy. West Sussex: John Wiley & Sons; 1991. p. 201–6.
Trimble MR, George MS. Brain behavior relationships. In: Trimble MR, George MS, editors. Biological psychiatry. 3rd ed. Chichester/Hoboken: John Wiley and Sons; 2010. p. 76–116.
Vogt BA. Pain and emotion interactions in subregions of the cingulate gyrus. Nat Rev Neurosci. 2005;6:533–44.
Wagner AD, Shannon BJ, Kahn I, Buckner RL. Parietal lobe contributions to episodic memory retrieval. Trends Cogn Sci. 2005;9:445–53.
Weller KL, Smith DA. Afferent connections to bed nucleus of stria terminalis. Brain Res. 1982;232:255–70.

Male Behaviors II: Neural Pathways Regulating Male Behaviors

18

Mona Sharma and Anand Kumar

Learning Objectives

- Neural pathway mediating aggression
- Motivation and pathways of pleasure
- Molecular mechanism of addiction
- Biochemical molecules and pathways regulating love and pair-bonding
- Neural and hormonal regulation of sexual behavior

18.1 Introduction

The brain regulates behaviors in coordination with various external influences. The brain areas regulating behaviors and their connections have been described in the previous chapter. The outgoing and incoming projections and their neurotransmitters interact and construct neural circuits. We shall now describe various neural pathways regulating male behaviors such as aggression, pleasure, love, pair-bonding, and sexual behavior, all of which influence reproduction.

M. Sharma, MD, DNB, MAMS (✉) • A. Kumar, MD, FAMS
Department of Reproductive Biology, All India Institute of Medical Sciences, New Delhi, India
e-mail: dr.mona18sharma@gmail.com; anandkumarrepbiol@hotmail.com

A. Kumar, M. Sharma (eds.), *Basics of Human Andrology*,
DOI 10.1007/978-981-10-3695-8_18

18.2 Aggression

Aggression is a sexually dimorphic behavior in most species. Aggression has important implications for the evolutionary survival and struggle for resources of the individual animal. Experimental mouse models have been used to discover underlying regulatory molecules of aggression. In almost all vertebrates including humans, males are considered more aggressive than females. The gender differences in the aggressive behavior are based on hormonal and social factors. Sex hormones are integral for neural circuit development and sexually dimorphic behaviors. Androgens are the main neuroendocrine mediators of aggressive behavior in rats and mouse (Simon et al. 1998). Testosterone with androgen receptor and aromatase are the key regulators of transition from genotype to phenotype (Cornill et al. 2006; Chang et al. 1995; Forlano et al. 2006; MacLusky and Naftolin 1981). The circulating testosterone binds to androgen receptors and gets converted into estrogen by aromatase enzyme. Aromatase-expressing neurons show extensive sexual dimorphism in number and projection patterns. Increased aromatase-expressing neurons with richer synapses are found in the bed nucleus of the stria terminalis (BNST) and medial group of nuclei in the amygdala (MeA) in male as compared to female mice (Wu et al. 2009).

Increase in testosterone levels in men is associated with more aggression (Zitzmann and Nieschlag 2001). Castration reduces androgen levels and, hence, aggression in mice. Testosterone treatment induces aggression in mice (Matochik et al. 1994). In males, estradiol and dihydrotestosterone induce aggressive behavior (Finney and Erpino 1976). Estrogen receptor alpha (ERα) knockout mice show decreased aggressive behavior (Ogawa et al. 1997). Steroid receptors regulate aggression while acting downstream on neurotransmitter pathways involving arginine vasopressin (AVP). AVP-mediated aggression is regulated by circulating levels of testosterone (De Vries et al. 1985). Castration decreased AVP immunoreactivity in rats along with decreased AVP-mRNA levels in brain areas controlling aggression such as BNST and MeA (Van Leeuwen et al. 1985). Testosterone replacement in these animals restores AVP levels. Vasopressinergic neurons stimulate glutamatergic neurons of BNST (see Fig. 18.1). The BNST stimulation leads to increased aggression in androgen-treated animals (Carrillo et al. 2011).

The other neurotransmitter systems implicated in the control of aggressive behavior are the serotonin and 5-hydroxytryptamine (5-HT). 5-HT is reported to have the opposite effect, and it diminishes aggressive behavior (Olivier and Mos 1990). Low 5-HT levels are associated with aggressiveness (Manuck et al. 2006). The use of 5-HT precursors, 5-HT reuptake inhibitors, and 5-HT_{1A} and 5-HT_{1B} agonists reduces aggressive behavior in rodents (Olivier et al. 1995; Miczek et al. 1998; Fish et al. 1999). 5-HT_{1B} receptors are found in brain areas controlling aggressive behavior such as the hippocampus and brain stem raphe nuclei (Olivier 2005). But, the effects on 5-HT_{1C}, 5-HT_2, and 5-HT_3 do not influence aggression (Simon et al. 1998). The aggressive stimuli reach the brain stem raphe nuclei through the ascending sensory tracts of the spinal cord. The raphe neurons express 5-HT receptors which get activated by the stimuli. The receptor activation leads to stimulation of

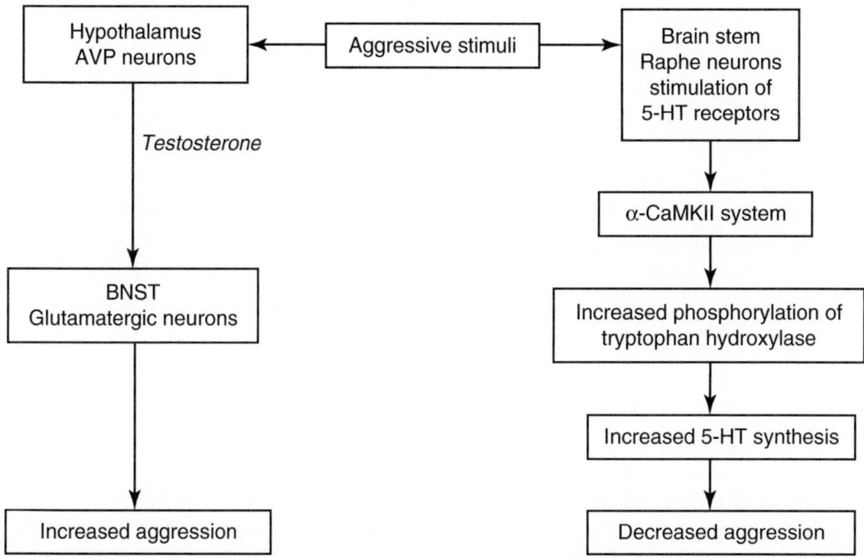

Fig. 18.1 Pathways of aggression. *AVP* arginine vasopressin, *BNST* bed nucleus of stria terminalis, *5-HT* serotonin, *α-CaMKII* alpha calcium–calmodulin-dependent kinase II

secondary signal transduction systems in the neurons. The most commonly implicated is the alpha calcium–calmodulin-dependent kinase II (α-CaMKII) system. The knockout rat models of CaMKII show increased aggressive behavior (Chen et al. 1994). The α-CaMKII signaling stimulates the phosphorylation of tryptophan hydroxylase which is the enzyme needed in 5-HT synthesis (see Fig. 18.1).

The other brain areas such as the amygdala (AMG), hippocampus, hypothalamus, prefrontal cortex (PFC), and medial preoptic area (MPOA) also regulate aggressive behavior. In humans, the impulsive aggressive behavior is associated with reduced activity of PFC. PFC inhibits aggression by sending inhibitory inputs to the hypothalamus and AMG (Davidson et al. 2000). Bilateral AMG lesions are associated with the development of Kluver–Bucy syndrome which is characterized by altered behavior with loss of aggressiveness, hyperorality, or tendency to examine the objects by the mouth and hypersexuality (Horel 1975).

18.3 Motivation

Motivation is a theoretical construct which explains initiation, intensity, and quality of a behavior (Maehr and Meyer 1997). Motivation can be appetitive or aversive. Appetitive motivation is directed toward goals that are associated with pleasurable events like food, sex, drugs, etc. Aversive motivation guards the unpleasant conditions (Bozarth 1994). The brain areas involved in motivation lie within the limbic system (Esch et al. 2004). The AMG assesses the experience, whether pleasurable or avoidable, and it also compares the present experience with the past (Nestler et al. 2001). Memories of pleasurable events are remembered through the hippocampus. PFC processes all the information and directs the final behavior. Activities in relation to sexual behavior can be typically divided into appetitive and consummatory. Motivation and associated behaviors in order to engage in sexual activity can be called appetitive, while the actual performance of a sexual act can be termed consummatory activity (Ball and Balthazart 2008). The experimental studies on male Japanese quail for neural regulation of appetitive and consummatory behaviors have revealed the role of different subregions of MPOA (Balthazart et al. 1998). Dopamine (DA) release in MPOA of male rats contributes to the sexual motivation and copulatory function (Hull et al. 1995).

18.4 Pleasure

The pleasure circuitry or pleasure system of the brain is highly interconnected and complex and drives nearly all behaviors including sex, emotions, and drug dependence (Schultz 2010). The basic drive of pleasure is motivation. Pleasures are the objects or goals that we seek. The circuitry evaluates pleasure-generating stimuli and guides decision-making and goal-directed behavior (Wise 2002).

The pleasure-generating brain areas are the AMG, PFC, hippocampus, nucleus accumbens (NAc), ventral tegmental area (VTA), substantia nigra (SN), and ventral pallidum (VP). The major neurochemical pathways forming the pleasure circuitry include DA-mediated mesolimbic and mesocortical pathways. DA acts on different receptors in pleasure circuitry to either stimulate or inhibit the production of cyclic adenosine monophosphate (cAMP). The DA receptors are the G-protein-coupled receptors usually consisting of two subfamilies. The D1-like receptors such as D1 and D5 activate adenylyl cyclase and increase production of cAMP. The increased cAMP levels activate protein kinase A (PKA) which leads to phosphorylation of various cellular substrates. The activated cellular substrates further increase neuronal excitability. D2-like receptors such as D2, D3, and D4 inhibit adenylyl cyclase activity and decrease production of cAMP resulting in reduced PKA activity and decreased neuronal excitability (Missale et al. 1998; Beaulieu and Gainetdinov 2011; Kebabian and Calne 1979).

The mesolimbic pathway extends from dopaminergic neurons present in VTA and SN to NAc through the medial forebrain bundle (MFB; Bozarth 1994; Routtenberg and Malsbury 1969). The MFB consists of dopaminergic, serotonergic, GABAergic, and glutamatergic neurons which are closely implicated in pleasure. The neurotransmitters interact with DA receptors at various points along the pleasure pathway and regulate dopaminergic release (Denise and Edward 2001). The next-stage neurons in NAc are GABAergic which project to VP (see Fig. 18.2). VP also sends GABAergic projections to NAc and VTA (Churchill and Kalivas 1994). The pathway from NAc to VP is implicated in motor response to pleasure-mediated goals by converting limbic motivation stimuli into motor behavior (Wise and Bozarth 1984). The GABAergic projections from NAc inhibit the GABAergic neurons of VP. Inhibition of GABAergic neurons in VP leads to decreased inhibition of dopaminergic neurons of VTA and, hence, increased pleasure response (Johnson and North 1992). VTA and NAc are also connected to the paraventricular nucleus (PVN) and MPOA which send oxytocinergic and dopaminergic projections, respectively, to spinal cord motor neurons responsible for eliciting motor response (Melis et al. 2007; Fallon and Loughlin 1995; Sofroniew 1983; Bjorklund et al. 1975; Skagerberg and Lindvall 1985). See Figs. 18.2 and 18.3 for pleasure circuitry and its neurotransmitters.

As the mesolimbic pathway is associated with expressions of pleasure and desire, the role of this pathway has been implicated in drug addiction, depression, and schizophrenia. The pathway of addiction is discussed next in this chapter.

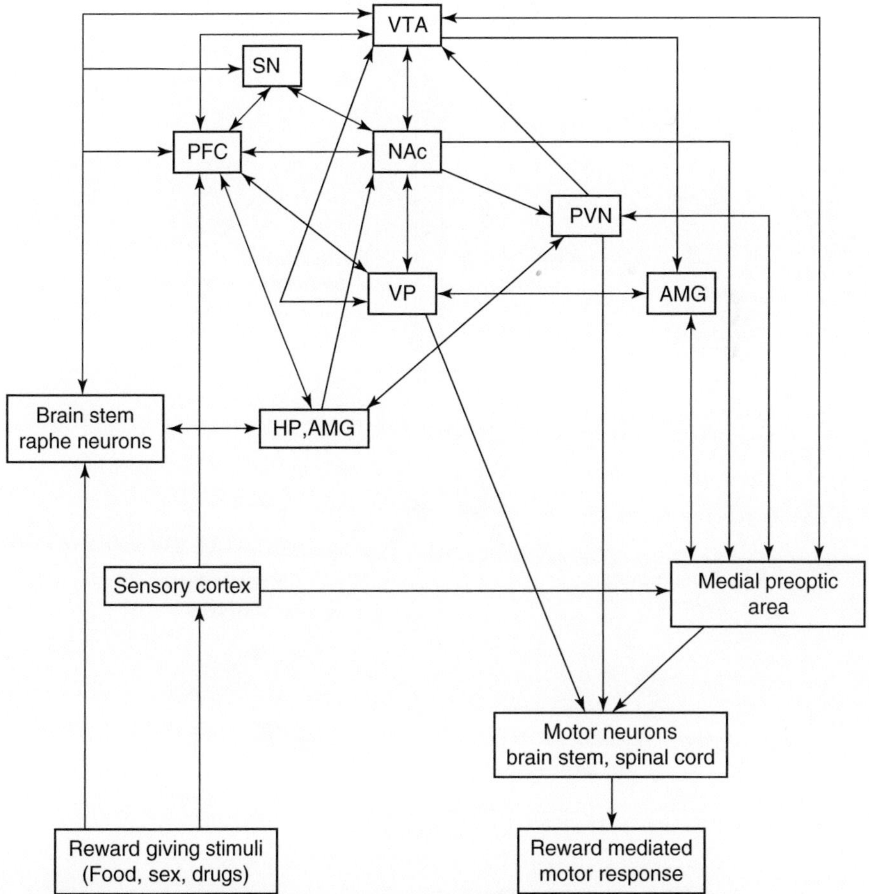

Fig. 18.2 Pleasure circuitry. *PFC* prefrontal cortex, *AMG* amygdala, *HP* hippocampus, *VTA* ventral tegmental area, *NAc* nucleus accumbens, *SN* substantia nigra, *PVN* paraventricular nucleus, *VP* ventral pallidum [A→B: outgoing projection from A to B; ↔: bilateral projections]

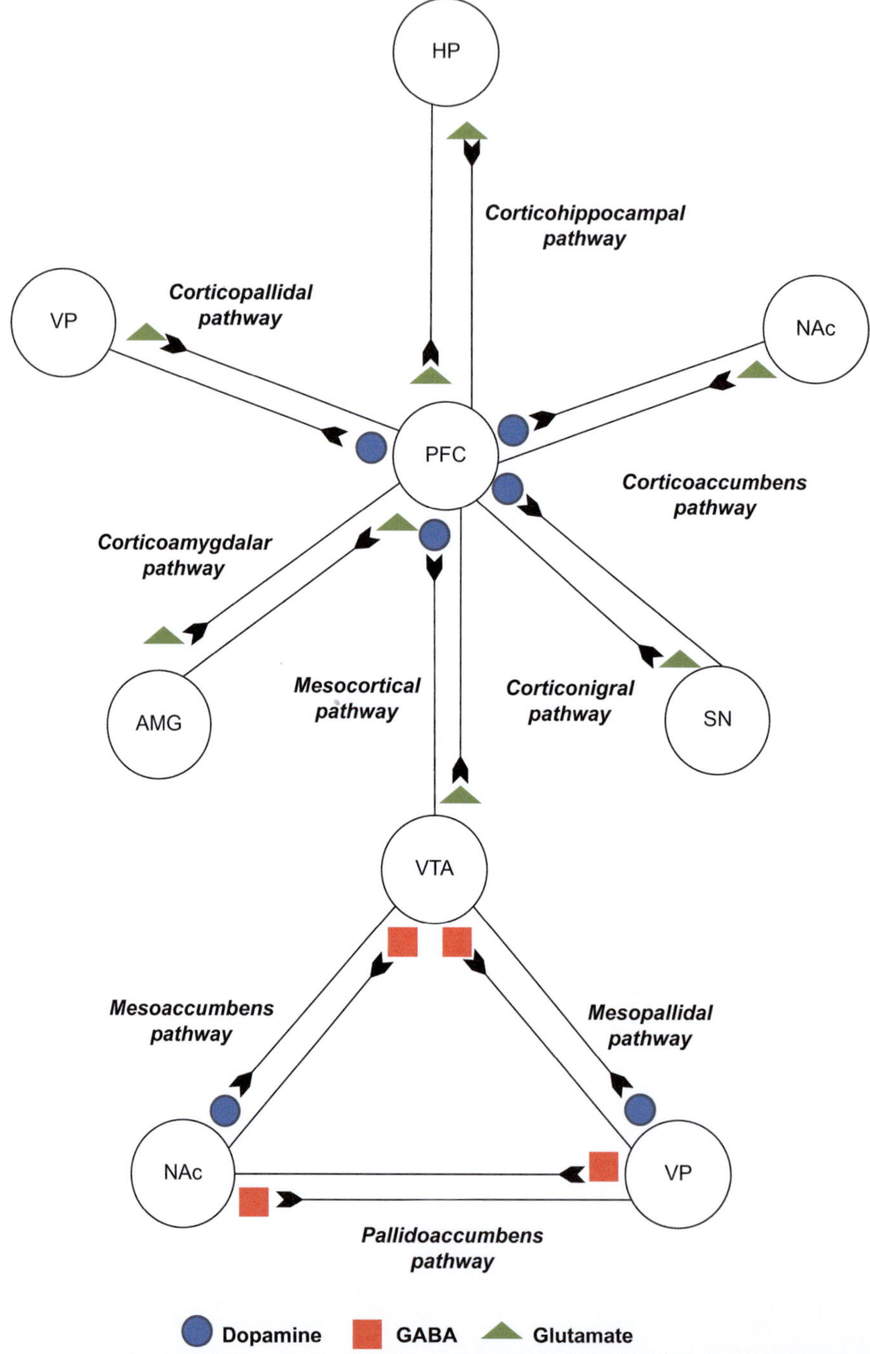

Fig. 18.3 Neurotransmitters and neural pathways of pleasure. *VTA* ventral tegmental area, *NAc* nucleus accumbens, *VP* ventral pallidum, *PFC* prefrontal cortex, *HP* hippocampus, *AMG* amygdala, *SN* substantia nigra, *GABA* gamma-aminobutyric acid A———B Outgoing projections from A to B

Another dopaminergic pathway in pleasure circuitry is the mesocortical pathway which connects VTA to PFC. The mesocortical pathway is involved in integrating emotions, cognition, and motivation which is essential for expression of pleasure-mediated behaviors. PFC controls emotional integrity by sending glutamatergic projections to the NAc, AMG, hippocampus, and VTA (see Figs. 18.2 and 18.3). All these areas send reciprocal projections to PFC (Tzschentke and Scmidt 2000; Groenewegen and Uylings 2000; Mc Donald 1996). The limbic areas such as the NAc, AMG, and hippocampus are innervated by dopaminergic projections of VTA. The hippocampus and AMG also send glutamatergic projections to NAc (Grace et al. 2007; Mora et al. 2008). NAc integrates contextual, emotional, and cognitive information given by the hippocampus, AMG, and PFC, respectively. NAc then mediates goal-directed motor response by sending projections to VP that further sends projections to brain stem motor neurons (Mogenson et al. 1980; Mogenson and Yang 1991).

The role of 5-HT in pleasure circuitry is based on the fact that pleasure-generating dopaminergic areas are interconnected with serotonergic neurons. The serotonergic neurons are located in the midline raphe nuclei of the brain stem (Michelsen et al. 2007). The most important of all midline raphe neurons are the dorsal raphe neurons (DRN) projecting to the dopaminergic areas related to pleasure processing (Parent et al. 1981). The serotonergic neurons regulate the DA transmission by acting on different 5-HT receptors present in DA-innervated areas. The 5-HT receptors consist of subfamilies from 5-HT_{1-4} with further receptor subtypes. The receptors like 5-HT_{1A}, 5-HT_{1B}, 5-HT_{2A}, 5-HT_3, and 5-HT_4 facilitate DA release, whereas 5-HT_{2B} and 5-HT_{2C} inhibit DA release (Alex and Pehek 2007; Di Giovanni et al. 1999). The DRN projects widely to various brain areas such as the cortex, AMG, VTA, SN, and hippocampus.

Brain areas projecting to DRN are the PFC, cingulate cortex, AMG, VTA, SN, and hypothalamus. The VTA and SN dopaminergic neurons project to DRN and activate them by D2-like receptor activation. The activity of DRN is regulated by 5-HT via 5-HT_{1A} receptors located on the neurons (Carey et al. 2004). The greatest number of neurons projecting to VTA also resides in DRN (Watabe-Uschida et al. 2012). DRN also includes nonserotonergic neurons implicated in pleasure circuitry, and these neurons majorly project to the VTA. The nonserotonergic neurons consist of glutamatergic, dopaminergic, and GABAergic neurons. The major projecting neurons of DRN provide glutamatergic excitatory inputs to VTA dopaminergic neurons (Mc Devitt et al. 2014). The excitatory inputs to VTA stimulate the pleasure circuitry (see Fig. 18.2).

18.5 Addiction

Pleasure is a feeling of enjoyment and satisfaction that one feels while doing a set of activities. Motivation- and pleasure-mediated activities are beneficial for survival, so are important components of normal behavior. The natural pleasures control behavior by enhancing decision-making. Loss of ability for making decisions leads to loss of control of behavior and addiction. This is called motivational toxicity (Bozarth 1994).

Hedonic effects of food, drugs, and sex can motivate their excessive use and experience resulting in addiction. The chronic use of abusive drugs, food, and excessive sexual behavior induces changes in the pleasure circuitry (Kelley 2004). The changes decrease the pleasurable effects by altering DA-mediated pleasure pathway. The decreased pleasure associated with pleasure-mediated activities leads to dependence and tolerance (Berridge and Robinson 1998).

The major pleasure centers are the VTA, AMG, hippocampus, and PFC. As discussed in the previous section, pleasure circuit extends from dopaminergic neurons of VTA projecting to the NAc neurons. The final behavior is assessed on pleasure scale by VTA–NAc pathway. The VTA–NAc pathway, through its connections to widespread areas, informs regarding the level of pleasure associated with an activity (Nestler and Malenka 2004). The more the level of pleasure associated with an activity, more likely the activity is repeated by an organism. Drugs acting as psychomotor stimulants activate pleasure system by affecting VTA–NAc through DA signaling. Similarly, natural pleasures such as food and sexual activity also activate brain's pleasure circuitry (Bozarth 1987).

The AMG, hippocampus, and PFC send glutaminergic projections to VTA and NAc (Bardo 1998; Nestler 2001). The chronic drug abuse or excessive use of natural pleasures leads to increase glutamate sensitivity in VTA and NAc neurons by increasing glutamate receptor synthesis or transition of receptors to the neuronal membrane. The glutamate sensitivity causes increase DA release and cAMP signaling in NAc (Tzschentke 2001). Increased cAMP signaling in NAc induces various transcription factors such as cyclic AMP response element-binding protein (CREB; see Fig. 18.4). The chronic use of drugs or any pleasure-associated activity leads to sustained activity of CREB.

CREB induces expression of gene for protein dynorphin in NAc. Dynorphin is produced by neurons in NAc that project to VTA and inhibit release of DA in VTA. The decreased levels of DA in VTA dampen the pleasure circuitry. Now, the same amount of drug or activity becomes less pleasurable, and an addict needs more amount of drug or activity to produce same level of pleasure associated before the chronic use of the same (Carlezon et al. 1998; Nestler 2001; Lu et al. 2003; Shaw-Lutchman et al. 2003).

18.6 Love and Bonding

Love is one of the core psychological emotions that cherishes and nurtures relationships. One of the basic driving forces for reproduction and social attachments is love and pair-bonding. Human monogamy, love, and affection have become important areas of neuroscience research. The latest interest is in the role of biochemical molecules of love. The various biochemicals regulating love and attachment are DA, oxytocin (OXT), AVP, and 5-HT (Young and Wang 2004; Zeki 2007). DA is a pleasure chemical responsible for euphoria, and it energizes relationships. OXT and AVP

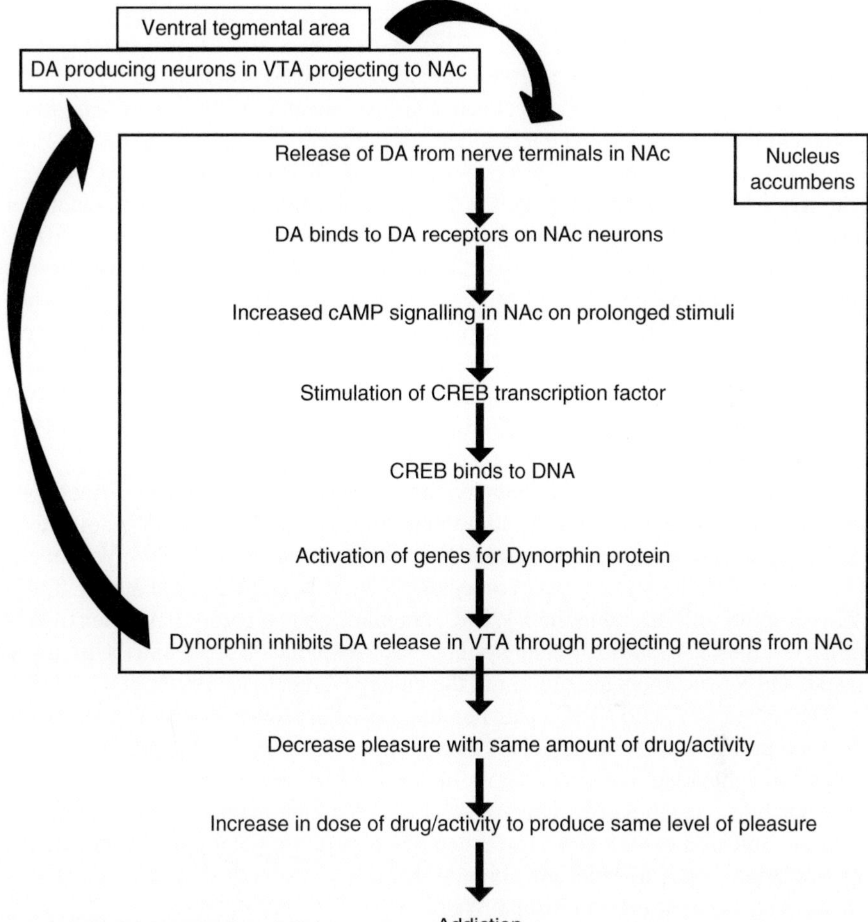

Fig. 18.4 Pathway of addiction. *VTA* ventral tegmental area, *NAc* nucleus accumbens, *CREB* cyclic AMP response element-binding protein, *DA* dopamine, *cAMP* cyclic adenosine monophosphate, *DNA* deoxyribonucleic acid

interact with their receptors in the dopaminergic areas such as NAc, VTA, and PFC and induce DA release, thereby transforming love into a pleasure experience (Young and Wang 2004). OXT and VP show opposite effects on amygdaloid output because both these neurotransmitters activate different neurons of the AMG (Debiec 2007). The role of the AMG is implicated in fear responses. OXT receptors are located in the lateral part of the central nucleus of the AMG. Here, they activate GABAergic neurons which project to medial part of the central nucleus of the AMG. The GABAergic neurons inhibit the amygdaloid output and hence, reduce fear and arise trust. On the other side, AVP directly activates neurons in the medial parts of the central nucleus of the AMG, thereby increasing amygdaloid output and arising fear (Huber et al. 2005). In both males and females during romantic love, the effect of OXT is stronger than AVP on decreasing the amygdaloid activity or fear reduction.

Low levels of 5-HT are seen in the early stages of love, depression, and anxiety disorders (Zeki 2007; Deboer et al. 2012). Early stages of love also show certain behavioral changes such as depression, anxiety, and obsessive thinking.

Various regions of the brain activated in love are the insular cortex, cingulate cortex, hippocampus, NAc and hypothalamus, whereas regions like AMG and PFC are deactivated (Deboer et al. 2012; Diamond and Dickenson 2012; Tarlaci 2012). The role of AMG in emotions such as fear, PFC in evaluation of other's emotions and intentions, is proven by deactivation of these areas in love (Deboer et al. 2012).

The brain areas regulating pair-bonding such as PFC, NAc, and VP are also the pleasure-generating areas, suggesting pair-bonding a pleasure-driven phenomenon. The role of OXT and AVP in pair-bonding and attachment has been well described. The two chemicals are released during sexual activity producing feeling of satisfaction and attachment. The various areas of increased OXT and AVP receptor density are the parts of pleasure system such as VTA, NAc, VP, and PFC (Lim and Young 2004).

OXT receptors are G-protein-coupled receptors located in the AMG; NAc; hypothalamic nuclei such as MPOA, supraoptic nucleus (SON), and PVN; olfactory bulbs; reproductive organs; thymus; and heart (Freundmercier et al. 1994; Yoshimura et al. 1993). AVP has three types of receptors: V1a, V1b, and V2 (Birnbaumer 2000). V1a receptors are located in the VP, PFC, cingulate cortex, entorhinal cortex, mammillary bodies, and cardiovascular system. V1b receptors are present in the pituitary gland, and V2 receptors are present in the kidney (Zingg et al. 1996).

The role of OXT and AVP in pair-bonding has been proved based on experiments done on prairie and montane voles (Young et al. 1999). Following mating, prairie voles form monogamous pairs, show biparental care, and stay together lifelong. Pair-bonding is established when voles spend more time in contact with their partners as compared to strangers. The montane voles present an opposite behavior in pair-bonding. Montane voles are nonmonogamous and do not show biparental care (Insel et al. 1995; Wang and Insel 1996).

During mating in prairie voles, OXT and AVP are released in pleasure-generating areas which facilitate pair-bonding (Carter et al. 1995). On inhibiting the release of OXT and AVP, the pair-bonding is no longer seen. The difference in the mating behavior of prairie and montane voles is due to the differences in the location of OXT receptors (OXTR) and AVP receptors (AVPR). The prairie voles possess higher densities of OXTR in DA-rich areas such as NAc and V1a receptors in AMG and VP as compared to the montane voles (Insel and Shapiro 1992; Insel et al. 1994). Montane voles have higher OXTR and V1a receptor density in lateral septal region (Insel et al. 1994). The gender differences in the role of OXT and AVP have been seen in pair-bonding (Lim and Young 2006). AVP induces partner preference in male prairie voles, whereas OXT does this in females (Young et al. 2001).

The formation of monogamous pair-bonds in female prairie voles is due to the OXT–DA interactions in NAc. Both OXT and DA exert similar effects on certain behaviors such as sexual and pair-bonding (Wang and Aragona 2004; Baskerville and Douglas 2008). Social and sexual interactions stimulate oxytocinergic neurons. Oxytocinergic neurons in the hypothalamus are present in MPOA, SON, and PVN (Freundmercier et al. 1994). The oxytocinergic neuron-rich areas are innervated by dopaminergic neurons projecting from zona incerta and have shown the presence of D2-like DA receptors (Decavel et al. 1987; Baskerville et al. 2009). Therefore, it can be suggested that DA regulates release of hypothalamic OXT. Oxytocinergic neurons in PVN release OXT in various brain areas such as the AMG, hippocampus, VTA, and spinal cord (Sofroniew 1983; Melis et al. 2007; Melis et al. 2009). Oxytocinergic neurons project from PVN and terminate into VTA close to dopaminergic neurons that project to the NAc, hippocampus, and AMG, suggesting neuroanatomical basis of OXT–DA interactions in these areas (Melis et al. 2007; Fallon and Loughlin 1995). NAc and PFC receive dopaminergic projections from VTA and oxytocinergic projections from PVN. Both PFC and NAc have receptors for DA and OXT (Smeltzer et al. 2006; Young 1999). PFC sends glutamatergic projections to VTA and NAc (Carr et al. 1999; Carr and Sesack 2000). DA in PFC regulates glutamatergic projections to NAc and so regulates DA release in NAc. Coactivation of DA and OXT receptors in NAc facilitates pair-bonding in female prairie voles (Liu and Wang 2003).

After mating in prairie voles, there is release of DA in NAc by its effect on D2 receptors. By D1 receptor stimulation, the pair-bonding is inhibited. After formation of pair-bonds in prairie voles, D1 receptor upregulation occurs which maintains the existing pair-bond and prevents formation of new pair-bonds (Aragona et al. 2003). Systemic or local administration of DA agonist, apomorphine, into NAc facilitates pair-bonding, whereas DA receptor antagonist haloperidol blocks partner preference and pair-bonding in prairie voles (Aragona et al. 2003; Gingrich et al. 2000). Administration of OXTR antagonist into NAc inhibits D2 receptor-induced pair-bonding. D2 receptor blockade in NAc inhibits pair-bonding induced by OXT (Liu and Wang 2003). The experiments suggest that OXT and DA interact in NAc and are important for pair-bonding in female prairie voles.

In male prairie voles, the role of AVP–DA interactions is implicated in pair-bonding. The roles of D2 receptors in NAc and V1a receptors in VP are important for pair-bonding (Lim and Young 2004; Aragona et al. 2003; Pitkow et al. 2001). In experiments, upregulating V1a receptor activity in VP facilitates pair-bonding in male prairie voles (Lindvall et al. 1974; Lim and Young 2004). Based on animal studies, neural pathway for pair-bonding can be proposed (see Fig. 18.5).

Sexual and olfactory stimuli reach MeA. MeA sends projections to BNST and VP (Meisel and Sach 1994). AVP neurons are present in BNST and MeA (DeVries and Buijs 1983). Vasopressinergic projections in VP stimulate dopaminergic neurons projecting to NAc (Mogenson et al. 1987). As VP and NAc are interconnected in pleasure pathway, DA and AVP also interact within pleasure pathway to regulate pair-bonding in males (see Fig. 18.5).

Apart from pair-bonding, the role of OXT is also implicated in mother–infant bonding (Kendrick et al. 1997). AVP also regulates male-related behaviors such as aggression (Goodson and Bass 2001; Ferris et al. 1984).

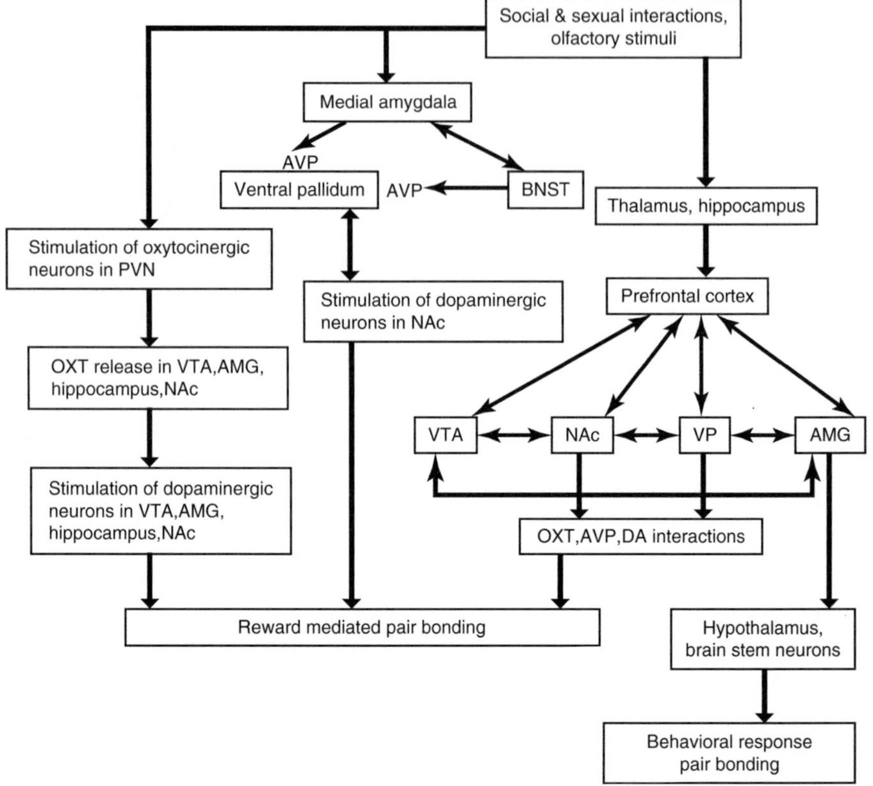

Fig. 18.5 Pathway of pair-bonding. *PVN* paraventricular nucleus, *VTA* ventral tegmental area, *NAc* nucleus accumbens, *BNST* bed nucleus of stria terminalis, *AMG* amygdala, *VP* ventral pallidum, *AVP* arginine vasopressin, *OXT* oxytocin, *DA* dopamine

18.7 Sexual Behavior

Male sexual behavior is primarily influenced by testosterone and its metabolites. Increase in testosterone at puberty is essential for increased sexual activity of maturing males. Testosterone is also responsible for decreasing fat mass, increasing lean mass and bone density, and improving sexual behavior in men with hypogonadism (Wang et al. 2000). The effect of gonadal steroids on sexual behavior is mediated through androgen and estrogen receptors. The receptors are found in selective areas in the brain such as MPOA, an area that is important for male sexual behavior (Hull and Rodriguez 2009).

18.7.1 Regulation of Sexual Behavior

A number of neurotransmitters like DA, nitric oxide (NO), gamma-aminobutyric acid (GABA), and many others are involved in regulation of sexual behavior.

18.7.1.1 Dopamine (DA)

The major dopaminergic pathways involved in male sexual behavior are the nigrostriatal, mesolimbic, and incertohypothalamic (see Table 18.1). The nigrostriatal pathway originates in the SN– pars compacta and projects to the striatum. The nigrostriatal pathway is implicated in the control of motor response to sexual stimuli. The mesolimbic pathway begins in the VTA and projects to the NAc. The mesolimbic pathway is important for motivation and pleasure associated with sexual activity. Incertohypothalamic pathway is an intrinsic dopaminergic system of the hypothalamus. The pathway originates in the zona incerta in the subthalamus and periventricular area and projects to the MPOA and PVN of the hypothalamus. The incertohypothalamic pathway is involved in coordination of genital reflexes necessary for erection (Ferriti et al. 2005).

The role of DA in enhancing male sexual behavior was first recognized when the use of L-dopa, precursor of DA, in patients with Parkinson's disease resulted in increased libido and sexual potency (Bowers et al. 1971). DA agonist apomorphine has been used to enhance erectile function (Heaton 1995). Dopaminergic neurons are present majorly in SN, VTA, and periventricular area of the hypothalamus. The dopaminergic neurons in the periventricular area send projections to the MPOA (Bjorklund et al. 1975). The neurons in MPOA project to the autonomic and somatic nuclei in the lumbosacral part of the spinal cord (Bjorklund et al. 1975; Skagerberg and Lindvall 1985). Stimulation of MPOA elicits erections, whereas ablation of MPOA impairs sexual behavior in rats (Giuliano et al. 1997). Both the areas, MPOA and PVN, have DA receptors (D1-, D2-like receptors). It has also been shown that stimulation of D1-like receptors in MPOA with DA agonist apomorphine and D2-like receptors in PVN facilitates erections (Hull et al. 1992; Andersson and Wagner 1995). Centrally acting DA antagonist haloperidol blocks the apomorphine-induced penile erections. The peripherally acting DA antagonist domperidone fails to inhibit erection in mice (Rampin et al. 2003). The experiments show that centrally acting DA facilitates erections. Also, injection of DA antagonist flupenthixol into MPOA decreases penile reflexes in male rats (Warner et al. 1991).

Release of DA in MPOA has been observed during copulation in male rats. The release of DA or increase in extracellular DA is regulated by release of glutamate in MPOA (Whitton 1997). The glutamatergic projections in MPOA come from locally present glutamatergic neurons, MeA, PVN, suprachiasmatic nucleus (SCN), and

Table 18.1 Dopaminergic pathways involved in male sexual behavior

Pathway	Origin	Reaches up to	Function
Nigrostriatal	Substantia nigra pars compacta	Striatum	Control of motor response to sexual stimuli
Mesolimbic	Ventral tegmental area of the midbrain	Nucleus accumbens in ventral striatum	Motivation and pleasure
Incertohypothalamic	Zona incerta and periventricular area	MPOA, PVN	Coordination of genital reflexes

mammillary nuclei of the hypothalamus (Kocsis et al. 2003). The areas get activated by sexually relevant stimuli. The released glutamate binds to N-methyl-D-aspartate (NMDA) receptors present in nitric oxide synthase (NOS) containing neurons in MPOA (Forrest et al. 1994). NOS is an enzyme that catalyzes the production of NO from L-arginine. The glutamate–NMDA binding opens calcium (Ca^{2+}) channels which lead to Ca^{2+} influx. The Ca^{2+} influx activates calmodulin (CaM), a Ca^{2+} sensor protein molecule that activates NOS (see Fig. 18.6). NOS-containing neurons are present near the dopaminergic neurons in MPOA (Du and Hull 1999). The released NO inhibits the DA transporter activity in MPOA that leads to increased extraneuronal DA levels (Pogun et al. 1994). The DA transporter transports DA into axonal terminals where it gets metabolized. The released NO from NOS-containing

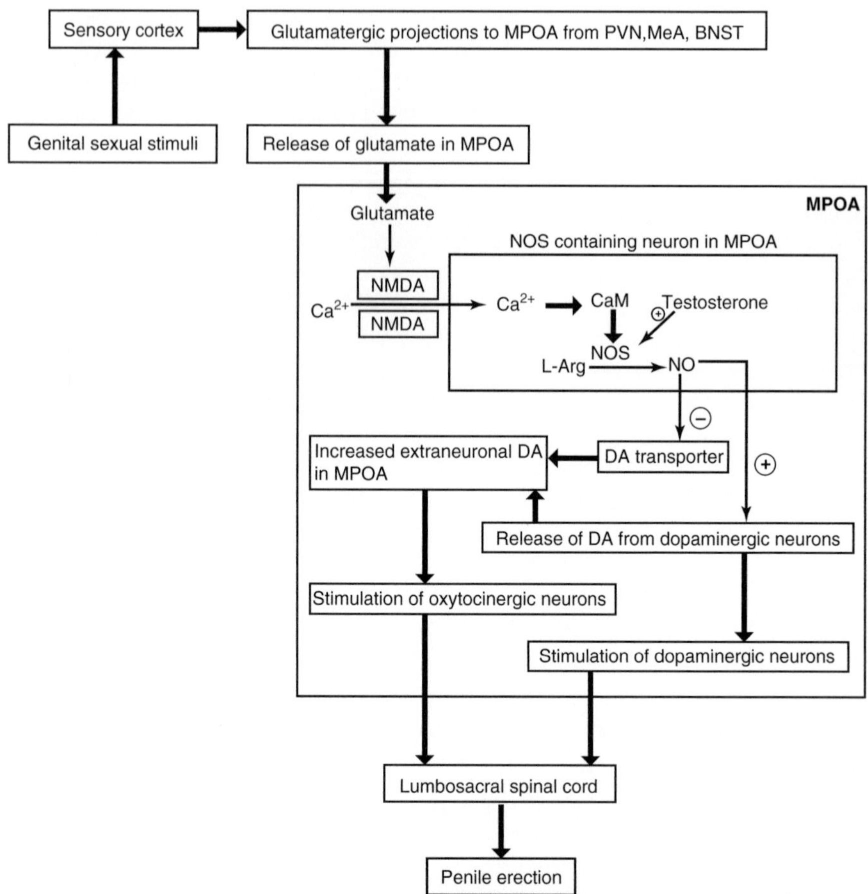

Fig. 18.6 Flowchart showing neural pathway of sexual behavior. *MPOA* medial preoptic area, *PVN* paraventricular nucleus, *BNST* bed nucleus of stria terminalis, *MeA* medial amygdala, *DA* dopamine, *Ca^{2+}* calcium, *NMDA* N-methyl-D-aspartate receptors, *NOS* nitric oxide synthase, *CaM* calmodulin, *L-Arg* arginine, *NO* nitric oxide

neurons diffuses into adjacent dopaminergic neurons and stimulates the release of DA from axonal terminals (Dominguez et al. 2004). The released DA binds to DA receptors present on the dopaminergic neurons in MPOA (see Fig. 18.6). Further, the dopaminergic neurons project to the autonomic spinal cord nuclei controlling penile erection (Bjorklund et al. 1975; Skagerberg and Lindvall 1985).

18.7.1.2 Androgens

Androgens are responsible for development of sexually dimorphic brain regions (MacLusky and Naftolin 1981). MPOA of the hypothalamus is one of the sexually dimorphic areas where androgen receptors are present. Testosterone binds to androgen receptors and activates copulatory behavior in most vertebrate species (Meisel and Sachs 1994; Jarzab et al. 1990). Testosterone is also essential for sexual desire in men (Mills et al. 1996). Testosterone administered to hypogonadal men increases libido and erectile function (Davidson et al. 1979). Spontaneous nocturnal penile erections are also impaired in androgen deficiency (Carani et al. 1992).

Action of DA is also testosterone dependent. Castration in rodents abolishes apomorphine-induced erections. Testosterone replacement in these animals restores erection (Heaton and Varin 1994; Melis et al. 1994). Castration reduces NOS-positive neurons in male hamster MPOA (Hadeishi and Wood 1996). Testosterone upregulates expression of NOS in MPOA in male rats (Du and Hull 1999). The upregulation of NOS in MPOA produces more NO which stimulates release of DA in MPOA (see Fig. 18.6).

18.7.1.3 Oxytocin (OXT)

The central OXT system is important in mediating DA-induced penile erections (Baskervill et al. 2009). Various studies have shown that oxytocinergic neurons are present in PVN, SON, and MPOA (Argiolas and Melis 2004). All these areas receive dopaminergic innervations from incertohypothalamic system (Bjorklund et al. 1975). The dopaminergic fibers impinge on the oxytocinergic neurons in PVN and SON (Decavel et al. 1987).

The oxytocinergic neurons in PVN, SON, and MPOA express D2-like receptors (Baskerville et al. 2009). DA binds to these receptors and activates intracellular signaling molecules such as signal-regulated kinases that lead to opening of Ca^{2+} channels and intracellular Ca^{2+} release (Moreland et al. 2004; Bitner et al. 2006). The intracellular Ca^{2+} activates NOS which produces NO (Schuman and Madison 1994). The released NO by autoactivation mechanisms activates oxytocinergic neurons (see Fig. 18.6). The activated oxytocinergic neurons project from PVN into the sacral 2–4 spinal segments and activate parasympathetic nuclei which stimulate erectile response (Argiolas et al. 1986; Argiolas 1992; Tang et al. 1998; Melis et al. 1998; Veronneau-Longueville et al. 1999; Melis and Argiolas 2003).

18.7.1.4 Prolactin

Increased prolactin levels decrease sexual behavior in men and depress genital reflexes in rats (de Groat and Booth 1993; Rehman et al. 2000). The mechanism of hyperprolactinemia affecting erectile function in men is currently not understood. It has been suggested that increased prolactin may act centrally by inhibiting dopaminergic neurons in MPOA which are involved in erectile function (Lookingland and Moore 1984). Prolactin also increases corporal smooth muscle contractility. The direct effect of hyperprolactinemia on reducing penile erection has been shown in dogs (Ra et al. 1996). It has also been shown that induced hyperprolactinemia in rats suppresses apomorphine-induced penile erections (Sato et al. 1997).

18.7.1.5 Serotonin (5-HT)

The role of 5-HT in human sexual behavior is well established. The serotonergic neurons and their receptors are located in brain stem raphe nuclei (Andersson and Wagner 1995). Serotonergic neurons then project to the MPOA. The neurons from MPOA project to autonomic spinal nuclei controlling erection (Tang et al. 1998). The use of 5-HT agonist p-chloroamphetamine in rats elicits penile erection. Pretreatment of rats with 5-HT synthesis inhibitor p-chlorophenylalanine depletes brain 5-HT and reduces p-chloroamphetamine-induced penile erections (Yonezawa et al. 2000). In another study, microinjections of 5-HT into the MPOA decrease erectile function (Verma et al. 1989). The effect of 5-HT on sexual behavior can be facilitatory or inhibitory depending upon stimulation of different types of 5-HT receptors (de Groat et al. 1993; Rehman et al. 1998). Penile erection is facilitated mainly through 5-HT_{2C} receptor activation, whereas 5-HT_{1A} receptor activation inhibits erectile responses in rats (Millan et al. 1997; Fernandez-Guasti et al. 1992). Selective reuptake inhibitors of 5-HT, which are used as antidepressants, reduce motivation and impair erectile functions (Clayton et al. 2006).

18.7.1.6 Opioid Peptides

Opioid peptides depress copulatory behavior in male rats. Morphine injection into PVN inhibits penile erection induced by either OXT or apomorphine administration. The action is mediated by stimulation of μ receptors present in PVN (Andersson and Wagner 1995). The μ receptor stimulation inhibits the NOS activity in PVN and decreases NO production. The decrease in NO production reduces OXT and DA release that leads to inhibition of penile erection (Melis et al. 1997b).

18.7.1.7 Adrenocorticotropic Hormone, α-Melanocyte-Stimulating Hormone

Melanocortins (MC) are the group of natural peptides which are derived from proopiomelanocortin (POMC) hormone. POMC undergoes cleavage and produces two MC- adrenocorticotropic hormone (ACTH) in anterior pituitary and α-melanocyte stimulating hormone (α-MSH) in the intermediate pituitary gland. Both the peptides exert their effects by binding to the MC receptors. The MC receptors are part of G-protein-coupled receptors family and have five subtypes named MC1–MC5 (Adan and Gispen 2000). Out of these receptors, MC3 and MC4 show abundant expression in central nervous system. MC3 receptors are found in the hypothalamus, brain stem, and cortex (Lindblom et al. 1998). MC4 receptors are present in all the regions in the brain and are the only MC receptors which are found in the spinal cord of rat (Kishi et al. 2003; Mountjoy and Wild 1998). Administration of ACTH and α-MSH into the hypothalamic periventricular region produces penile erection in rats (Argiolas et al. 2000). MC-induced erections are androgen dependent as these get abolished in castrated animals. Moreover, testosterone replacement in these animals restores erections (Bertolini et al. 2001). Intrathecal administration of melanotan, an α-MSH analogue, induces erection in rats (Wessels et al. 2003). The mechanism of erection induced by MC is through the activation of NO in the central nervous system. ACTH-induced penile erections are abolished by pretreatment with NOS inhibitor (Poggioli et al. 1995).

18.7.1.8 Nitric Oxide (NO)

The role of NO in central control of penile erection has been defined in many studies (Melis and Argiolas 1997; Lorrain et al. 1996). The major brain areas of NO action are the MPOA and PVN (Sato et al. 1998; Sato et al. 1999). Injection of nitroglycerine, which is an NO donor, in PVN produces erection in rats (Melis and Argiolas 1995). It has also been observed that production of NO also increases in PVN during erection and copulation (Melis et al. 1998). Various agents such as DA, OXT, ACTH, and 5-HT induce penile erection. Injection of NOS inhibitors into PVN prevents penile erection induced by DA, OXT, ACTH, and 5-HT (Melis and Argiolas 1993; Poggioli et al. 1995; Melis et al. 1997a). One of the physiological actions of NO is to activate guanylyl cyclase (GC) enzyme which converts 5-guanosine triphosphate (GTP) to 3,5-cyclic guanosine monophosphate (cGMP). The injection of GC inhibitors into PVN does not prevent erection induced by various agents. Moreover, erections are not induced on injecting cGMP into PVN which suggest that NO action is independent of activation of cGMP in the brain (Melis and Argiolas 1997).

18.7.2 Neural Pathway of Sexual Behavior

The genital sensory inputs reach the sensory cortical areas from where the glutamatergic fibers project through the thalamus and reach MeA, MPOA, PVN, and BNST. MeA and BNST are also connected with hypothalamic MPOA and PVN reciprocally. On sexual stimulation, there is release of glutamate in MPOA and PVN. Glutamate further stimulates the release of DA as described in the previous section. The D2-like receptors are involved in mediating erection via OXT (Baskerville and Douglas 2008). DA released in MPOA and PVN activates oxytocinergic neurons. The oxytocinergic fibers from MPOA and PVN project to the erectile centers in the lumbosacral spinal cord (see Fig. 18.6). The glutamatergic fibers from the AMG also project to VTA dopaminergic neurons and stimulate pleasure-mediated sexual behavior.

See Fig. 18.7 for various brain areas and neurotransmitters regulating male behaviors.

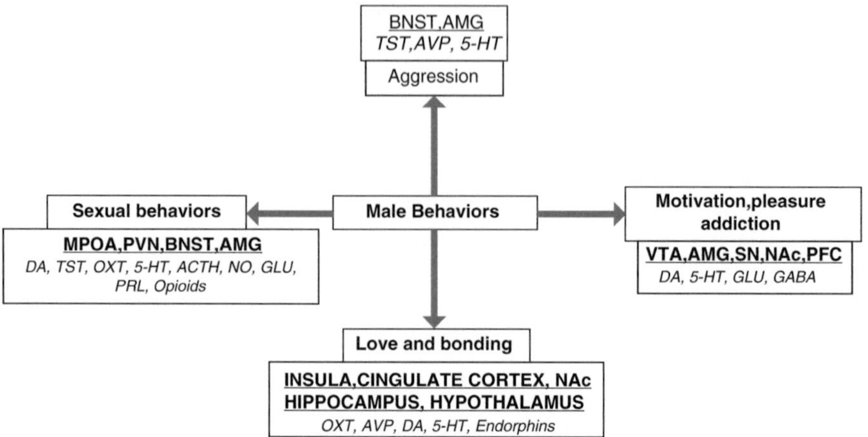

Fig. 18.7 Brain areas and neurotransmitters regulating male behaviors. *BNST* bed nucleus of stria terminalis, *AMG* amygdala, *VTA* ventral tegmental area, *SN* substantia nigra, *NAc* nucleus accumbens, *PFC* prefrontal cortex, *MPOA* medial preoptic area, *PVN* paraventricular nucleus, *TST* testosterone, *AVP* arginine vasopressin, *DA* dopamine, *5-HT* serotonin, *GLU* glutamate, *GABA* gamma-aminobutyric acid, *OXT* oxytocin, *ACTH* adrenocorticotropic hormone, *NO* nitric oxide, *PRL* prolactin

Key Questions

- Describe the signaling pathways and neurotransmitters mediating aggression.
- Draw and discuss the pleasure pathway.
- How pleasure can lead to motivational toxicity and addiction?
- Describe the role of oxytocin and vasopressin in pair-bonding.
- Discuss the interactions between dopamine, androgen, and oxytocin in eliciting sexual response.

References

Adan RA, Gispen WH. Melanocortins and the brain: from effects via receptors to drug targets. Eur J Pharmacol. 2000;405:13–24.

Alex KD, Pehek EA. Pharmacologic mechanisms of serotonergic regulation of dopamine neurotransmission. Pharmacol Ther. 2007;113:296–320.

Andersson KE, Wagner G. Physiology of penile erection. Physiol Rev. 1995;75(1):191–236.

Aragona BJ, Liu Y, Curtis JT, Stephen FK, Wang ZX. A critical role for nucleus accumbens dopamine in partner preference formation in male prairie voles. J Neurosci. 2003;23(8):3483–90.

Argiolas A. Oxytocin stimulation of penile erection. Pharmacology, site, and mechanism of action. Ann N Y Acad Sci. 1992;652:194–203.

Argiolas A, Melis MR. The role of oxytocin and paraventricular nucleus in the sexual behavior of male mammals. Physiol Behav. 2004;83(2):309–17.

Argiolas A, Melis MR, Gessa GL. Oxytocin: an extremely potent inducer of penile erection and yawning in male rats. Eur J Pharmacol. 1986;130:265–72.

Argiolas A, Melis MR, Murgia S, Schioth HB. ACTH and alpha MSH induced grooming, stretching, yawning and penile erection in male rats: site of action in the brain and role of melanocortin receptors. Brain Res Bull. 2000;51:425–31.

Ball GF, Balthazart J. How useful is the appetitive and consummatory distinction for our understanding of the neuroendocrine control of sexual behavior? Horm Behav. 2008;53:307–11.

Balthazart J, Absil P, Ge´ rard M, Appeltants D, Ball GF. Appetitive and consummatory male sexual behavior in Japanese quail are differentially regulated by subregions of the preoptic medial nucleus. J Neurosci. 1998;18(16):6512–27.

Bardo MT. Neuropharmacological mechanism of drug reward: beyond dopamine in nucleus accumbens. Crit Rev Neurobiol. 1998;12:37–67.

Baskerville TA, Douglas AJ. Interactions between dopamine and oxytocin in the control of sexual behavior. Prog Brain Res. 2008;170:277–90.

Baskerville TA, Allard J, Wayman C, Douglas AJ. Dopamine-oxytocin interactions in penile erection. Eur J Neurosci. 2009;30:2151–64.

Beaulieu JM, Gainetdinov RR. The physiology signaling and pharmacology of dopamine receptors. Pharmacol Rev. 2011;63(1):182–217.

Berridge K, Robinson T. What is the role of dopamine in reward: hedonic impact, reward learning or incentive silence? Brain Res Rev. 1998;28:309–69.

Bertolini A, Gessa GL, Ferrari C. Penile erections and ejaculation: a central effect of ACTH- like peptides in mammals. In: Sandler M, Gessa GL, editors. Sexual behavior: pharmacology and biochemistry. New York: Raven Press; 2001. p. 247–57.

Birnbaumer M. Vasopressin receptors. Trends Endocrinol Metab. 2000;11(10):406–10.

Bitner RS, Nikkel AL, Otte S, Martino B, Barlow EH, Bhatia P, et al. Dopamine D4 receptor signaling in the rat paraventricular hypothalamic nucleus: evidence of natural coupling involving immediate early gene induction and mitogen activated protein kinase phosphorylation. Neuropharmacology. 2006;50:521–31.

Bjorklund A, Lindvall O, Nobin A. Evidence of an incerto-hypothalamic dopamine neuron system in the rat. Brain Res. 1975;89(1):29–42.

Bowers MB, Woert MV, Davis L. Sexual treatment during L-dopa treatment for Parkinsonism. Am J Psychiatry. 1971;127(12):1691–3.

Bozarth MA. Ventral tegmental reward system. In: Oreland L, Engel J, editors. Brain reward system and abuse. New York: Raven Press; 1987. p. 1–17.

Bozarth MA. Pleasure systems in the brain. In: Warburton M, editor. Pleasure: the politics and the reality. New York: Wiley; 1994. p. 5–14.

Carani C, Bancroft J, Granata A, Del Rio G, Marrama P. Testosterone and erectile function, nocturnal penile tumescence and rigidity, and erectile response to visual erotic stimuli in hypogonadal and eugonadal men. Psychoneuroendocrinology. 1992;17(6):647–54.

Carey RJ, Depalma G, Damianopoulos E, Muller CP, Huston JP. The 5-HT1A receptor and behavioral stimulation in the rat: effects of 8-OHDPAT on spontaneous and cocaine-induced behavior. Psychopharmacology (Berl). 2004;177:46–54.

Carillo M, Ricci LA, Melloni RH. Glutamate vasopressin interactions and the neurobiology of anabolic steroid induced offensive aggression. Neuroscience. 2011;185:85–96.

Carlezon Jr WA, Thome J, Olson VG, Lane-Ladd SB, Brodkin ES, Hiroi N, et al. Regulation of cocaine reward by CREB. Science. 1998;18:2272–5.

Carr DB, Sesack SR. Projections from the rat prefrontal cortex to the ventral tegmental area: target specificity in the synaptic associations with mesoaccumbens and mesocortical neurons. J Neurosci. 2000;20:3864–73.

Carr DB, O'Donnell P, Card JP, Sesack SR. Dopamine terminals in the rat prefrontal cortex synapse on pyramidal cells that project to the nucleus accumbens. J Neurosci. 1999;19(24):11049–60.

Carter CS, DeVries AC, Getzz LL. Physiological substrates of mammalian monogamy: the prairie vole model. Neurosci Biobehav Rev. 1995;19(2):303–14.
Chang C, Saltzman A, Yeh S, Young W, Keller E, Lee HJ, et al. Androgen receptor: an overview. Crit Rev Eukaryot Gene Expr. 1995;5:97–125.
Chen C, Rainnie DG, Greene RW, Tonegawa S. Abnormal fear response and aggressive behavior in mutant mice deficient for calcium- calmodulin kinaseII. Science. 1994;288:291–4.
Churchill L, Kalivas PW. A topographically organized gamma-amino butyric acid projection from ventral pallidum to nucleus accumbens in the rat. J Comp Neurol. 1994;345(4):579–95.
Clayton A, Keller A, McGarvey EL. Burden of phase specific sexual dysfunction with SSRIs. J Affect Disord. 2006;91:27–32.
Cornill CA, Ball GF, Balthazart J. Functional significance of the rapid regulation of brain estrogen action: where do the estrogen come from? Brain Res. 2006;1126:2–26.
Davidson JM, Camargo CA, Smith ER. Effects of androgen on sexual behavior in hypogonadal men. J Clin Endocrinol Metab. 1979;48(6):955–8.
Davidson RJ, Putnam KM, Larson CL. Dysfunction in the neural circuitry of emotion regulation – a possible prelude to violence. Science. 2000;289:591–4.
De Groat WC, Booth AM. Neural control of penile erection. In: Maggi CA, editor. The autonomic nervous system. Nervous control of urogenital system, vol. 6. London: Harwood Academic Publishers; 1993. p. 465–524.
De Vries GJ, Buijs RM. The origin of the vasopressinergic and oxytocinergic innervation of the rat brain with special reference to the lateral septum. Brain Res. 1983;273:307–17.
De Vries GJ, Buijs RM, Van Leeuwen FW, Caffe AR, Swaab DF. The vasopressinergic innervations of the brain in normal and castrated rats. J Comp Neurol. 1985;233:236–54.
Debiec J. From affiliative behaviors to romantic feelings: a role of nanopeptides. FEBS Lett. 2007;581(14):2580–6.
Deboer A, Van Buel EM, Ter Horst GJ. Love is more than just a kiss: neurobiological perspective on love and affection. Neuroscience. 2012;201:114–24.
Decavel C, Geffard M, Calas A. Comparative study of dopamine- and noradrenaline-immunoreactive terminals in the paraventricular and supraoptic nuclei of the rat. Neurosci Lett. 1987;77(2):149–54.
Denise MT, Edward MS. Addiction and the brain: the role of neurotransmitters in the cause and treatment of drug dependence. CMAJ. 2001;164(6):817–21.
Di Giovanni G, De Deurwaerdere P, Di Mascio M, Di Matteo V, Esposito E, Spampinato U. Selective blockade of serotonin 2C/2B receptors enhances mesolimbic and mesostriatal dopaminergic function: a combined in vivo electrophysiological and microdialysis study. Neuroscience. 1999;91:587–97.
Diamond LM, Dickenson JA. The neuroimaging of love and desire: review and further directions. Clin Neuropsychiatry. 2012;9:39–46.
Dominguez JM, Muschamp JW, Schmich JM, Hull EM. Nitric oxide mediates glutamate-evoked dopamine release in the medial preoptic area. Neuroscience. 2004;125:203–10.
Du J, Hull EM. Effect of testosterone on neuronal nitric oxide synthase and tyrosine hydroxylase. Brain Res. 1999;836(1–2):90–8.
Esch T, Guarna M, Bianchi E, Zhu W, Stefano GB. Commonalities in the central nervous system's involvement with complementary medical therapies: Limbic morphinergic processes. Med Sci Monit. 2004;10:MS6–17.
Fallon JH, Loughlin SE. Substantia nigra. In: Paxinos G, editor. The rat nervous system. New York: Academic Press; 1995.
Fernandez-Guasti A, Escalante AL, Ahlenius S, Hillegaart V, Larsson K. Stimulation of 5-HT1A and 5-HT1B receptors in brain regions and its effects on male rat sexual behavior. Eur J Pharmacol. 1992;210:121–9.
Ferretti A, Caulo M, Del Gratta C, Di Matteo R, Merla A, Montorsi F. Dynamics of male sexual arousal: distinct components of brain activation revealed by fMRI. Neuroimage. 2005;26:1086–96.

Ferris CF, Albers HE, Wesolowski SM, Goldman BD, Luman SE. Vasopressin injected into the hypothalamus triggers a stereotypic behavior in golden hamsters. Science. 1984;224:521–3.

Finney HC, Erpino MJ. Synergistic effect of estradiol benzoate and dihydrotestosterone on aggression in mice. Horm Behav. 1976;7:391–400.

Fish EW, Faccidomo S, Miczek KA. Aggression heightened by alcohol or social instigation in mice: reduction by the 5-HT(1B) receptor agonist CP-94,253. Psychopharmacology (Berl). 1999;146:391–9.

Forlano PM, Schlinger BA, Bass AH. Brain aromatase: new lessons from no-mammalian model system. Front Neuroendocrinol. 2006;27:247–74.

Forrest D, Yuzaki M, Soares HD, Ng L, Luk DC, Sheng M, et al. Targeted disruption of NMDA receptor 1 gene abolishes NMDA response and results in neonatal death. Neuron. 1994;13:325–38.

Freundmercier MJ, Stoeckel ME, Klein MJ. Oxytocin receptors on oxytocin neurons—Histoautoradiographic detection in the lactating rat. J Physiol. 1994;480:155–61.

Gingrich B, Liu Y, Cascio C, Wang Z, Insel TR. Dopamine D2 receptors in the nucleus accumbens are important for social attachment in female prairie voles (Microtus ochrogaster). Behav Neurosci. 2000;114:173–83.

Giuliano F, Bernabe J, Brown K, Droupy S, Benoit G, Rampin O. Erectile response to hypothalamic stimulation in rats: role of peripheral nerves. Am J Physiol. 1997;273:1990–7.

Goodson JL, Bass AH. Social behavior functions and related anatomical characteristics of vasotocin/vasopressin systems in vertebrates. Brain Res Rev. 2001;35:246–65.

Grace AA, Floresco SB, Goto Y, Lodge D. Regulation of firing of dopaminergic neurons and control of goal-directed behaviors. Trends Neurosci. 2007;30:220–7.

Groenewegen HJ, Uylings HBM. The prefrontal cortex and the integration of sensory, limbic and autonomic information. Prog Brain Res. 2000;126:3–28.

Hadeishi Y, Wood RI. Nitric oxide synthase in mating behaviour of male Syrian hamster brain. J Neurobiol. 1996;30(4):480–92.

Heaton JP, Varrin SJ. Effects of castration and exogenous testosterone supplementation in an animal model of penile erection. J Urol. 1994;151:797–800.

Heaton JPW, Morales A, Adams MA, Johnston B, El-Rashidy R. Recovery of erectile function by oral administration of apomorphine. Urology. 1995; 45(2): 200-6.

Horel JA, Keating EG, Misantone LJ. Partial Kluver Bucy syndrome produced by destroying temporal neocortex or amygdala. Brain Res. 1975;94(20):347–59.

Huber D, Veinante P, Stoop R. Vasopressin and oxytocin excite distinct neuronal populations in the central amygdala. Science. 2005;308(5719):245–8.

Hull EM, Rodríguez-Manzo G. Male Sexual Behaviour. In: Pfaff DW, Arnold AP, Etgen AM, Fahrbach SE, Rubin RT, editors. Hormones, Brain and Behavior. 2nd ed. San Diego: Academic Press; 2009. p. 5–65.

Hull EM, Eaton RC, Markowski VP, Moses J, Lumley LA, Loucks JA. Opposite influence of medial preoptic D1 and D2 receptors on genital reflexes: implications for copulation. Life Sci. 1992;51(22):1705–13.

Hull EM, Du J, Lorrain DS, Matuszevich L. Extracellular dopamine in medial preoptic area: Implication for sexual motivation and hormonal control of copulation. J Neurosci. 1995;15(11):7465–71.

Insel TR, Shapiro LE. Oxytocin receptor distribution reflects social organization in monogamous and polygamous voles. Proc Natl Acad Sci U S A. 1992;89:5981–5.

Insel TR, Wang ZX, Ferris CF. Patterns of brain vasopressin receptor distribution associated with social organization in microtine rodents. J Neurosci. 1994;14(9):5381–92.

Insel TR, Preston S, Winslow JT. Mating in the monogamous male: behavioral consequences. Physiol Behav. 1995;57:615–27.

Jarzab B, Kokocinska D, Kaminski M, Gubala E, Achtelik W, Wagiel J, et al. Influence of neurotransmitters of the brain: relationship between the volume of SDN-POA and functional characteristics. Compr Physiol. 1990;8:41–50.

Johnson SW, North RA. Opioid excite dopamine neurons by hyperpolarization of local interneurons. J Neurosci. 1992;12:483–8.

Kebabian JW, Calne DB. Multiple receptors for dopamine. Nature. 1979;277:93–6.

Kelley AE. Memory and addiction: shared neural circuitry and molecular mechanisms. Neuron. 2004;44:161–79.

Kendrick KM, Da Costa AP, Broad KD, Ohkura S, Guevara R, Lévy F, et al. Neural control of maternal behavior and olfactory recognition of offspring. Brain Res. 1997;44:383–95.

Kishi T, Aschkenasi CJ, Lee CE, Mountjoy KG, Saper CB, Elmquist JK. Expression of melanocortin 4 receptor Mrna in the central nervous system of rat. J Comp Neurol. 2003;457:213–35.

Kocsis K, Kiss J, Csáki A, Halász B. Location of putative glutamatergic neurons projecting to the medial preoptic area of the rat hypothalamus. Brain Res Bull. 2003;61(4):459–68.

Lim MM, Young LJ. Vasopressin dependent neural circuits underlying pair bond formation in monogamous prairie vole. Neuroscience. 2004;125(1):35–45.

Lim MM, Young LJ. Neuropeptidergic regulation of affiliative behavior and social bonding in animals. Horm Behav. 2006;50(4):506–17.

Lindblom J, Schioth HB, Larsson A, Wikeberg JE, Bergstrom L. Autoradiographic discrimination of melanocortin receptors indicates that the MC3 subtype dominates in the medial rat brain. Brain Res. 1998;810:161–71.

Lindvall O, Bjorklund A, Moore RY, Stenevi U. Mesencephalic dopamine neurons projecting to neocortex. Brain Res. 1974;81:325–31.

Liu Y, Wang ZX. Nucleus accumbens oxytocin and dopamine interact to regulate pair bond formation in female prairie voles. Neuroscience. 2003;121:537–44.

Lookingland KJ, Moore KE. Effects of estradiol and prolactin on incertohypothalamic dopaminergic neurons in male rat. Brain Res. 1984;323(1):83–91.

Lorrain DS, Matuszewich L, Howard RV, Du J, Hull EM. Nitric oxide promotes medial preoptic dopamine release during male rat copulation. Neuroreport. 1996;8(1):31–4.

Lu L, Grimm JW, Shaham Y, Hope BT. Molecular neuroadaptation in accumbens and VTA during first 90 days of forced abstinence from cocaine self administration in rat. J Neurochem. 2003;85:1604–13.

MacLusky NJ, Naftolin F. Sexual differentiation of the central nervous system. Science. 1981;211(4488):1294–302.

Maehr ML, Meyer H. Understanding motivation and schooling: where we have been, where we are and where we need to go. Educ Psychol Rev. 1997;9(44):371–409.

Manuck SB, Kaplan JR, Lotrich FE. Brain serotonin and aggressive disposition in humans and non human primates. In: Nelson RJ, editor. Biology of aggression. Oxford University Press: New York; 2006. p. 65–113.

Matochik JA, Sipos ML, Nyby JG, Barfield RJ. Intracranial androgenic activation of male typical behaviours in house mice: motivation versus performance. Behav Brain Res. 1994;60:141–9.

McDevitt RA, Tiran-Capello A, Shen H, Balderas I, Britt JP, Manno RAM, et al. Serotonergic versus nonserotonergic dorsal raphe projection neurons: differential participation in reward circuitry. Cell Rep. 2014;8(6):1857–69.

McDonald AJ. Glutamate and aspartate immunoreactive neurons of the rat basolateral amygdala: colocalization of excitatory amino acids and projections to the limbic circuit. J Comp Neurol. 1996;365:367–79.

Meisel RL, Sachs BD. The physiology of male sexual behavior. In: Knobil E, Neill JGS, Greenwald GS, Markert CL, Pfaff DW, editors. The physiology of reproduction, vol. 2. NewYork: Raven Press; 1994. p. 3–107.

Melis MR, Argiolas A. Nitric oxide synthase inhibitors prevent apomorphine-and oxytocin-induced penile erection and yawning in male rats. Brain Res Bull. 1993;32:71–4.

Melis MR, Argiolas A. Nitric oxide donors induce penile erection and yawning when injected in the central nervous system of male rats. Eur J Pharmacol. 1995;294:1–9.

Melis MR, Argiolas A. Role of central nitric oxide in the control of penile erection and yawning. Prog Neuropsychopharmacol Biol Psychiatry. 1997;21(6):899–922.

Melis MR, Argiolas A. Central oxytocinergic neurotransmission: a drug target for the therapy of psychogenic erectile dysfunction. Curr Drug Targets. 2003;4:55–66.

Melis MR, Mauri A, Argiolas A. Apomorphine- and oxytocin-induced penile erection and yawning in intact and castrated male rats: Effects of sexual steroids. Neuroendocrinology. 1994;59:349–54.

Melis MR, Succu S, Iannucci U, Argiolas A. Oxytocin increases nitric oxide production in the paraventricular nucleus of the hypothalamus of male rats: correlation with penile erection and yawning. Regul Pept. 1997a;69:105–11.

Melis MR, Succu S, Iannucci U, Argiolas A. Prevention by morphine of apomorphine and oxytocin induced penile erection and yawning: involvement of nitric oxide. Naunyn Schmiedebergs Arch Pharmacol. 1997b;355(5):595–600.

Melis MR, Succu S, Mauri A, Argiolas A. Nitric oxide production is increased in the paraventricular nucleus of the hypothalamus of male rats during non-contact penile erections and copulation. Eur J Neurosci. 1998;10:1968–74.

Melis MR, Melis T, Cocco C, Succu S, Sanna F, Pillolla G, et al. Oxytocin injected into the ventral tegmental area induces penile erection and increases extracellular dopamine in the nucleus accumbens and paraventricular nucleus of the hypothalamus of male rats. Eur J Neurosci. 2007;26:1026–35.

Melis MR, Succu S, Sanna F, Boi A, Argiolas A. Oxytocin injected into the ventral subiculum or the posteromedial cortical nucleus of the amygdala induces penile erection and increases extracellular dopamine levels in the nucleus accumbens of male rats. Eur J Neurosci. 2009;30:1349–57.

Michelsen KA, Schmitz C, Steinbusch HW. The dorsal raphe nucleus –from silver staining to a role in depression. Brain Res Rev. 2007;55:329–42.

Miczek KA, Hussain S, Faccidomo S. Alcohol-heightened aggression in mice: attenuation by 5-HT1A receptor agonists. Psychopharmacology (Berl). 1998;139:160–8.

Millan MJ, Peglion JL, Lavielle G. Perrin-Monneyron S. 5-HT2C receptors mediate penile erections in rats: actions of novel and selective agonists and antagonists. Eur J Pharmacol. 1997;325(1):9–12.

Mills TM, Reilly CM, Lewis RW. Androgens and penile erection: a review. J Androl. 1996;17(6):633–8.

Missale C, Nash SR, Robinson SW, Jaber M, Caron MG. Dopamine receptors:from structure to function. Physiol Rev. 1998;78(1):189–225.

Mogenson GJ, Yang CR. The contribution of basal forebrain to limbic-motor integration and the mediation of motivation to action. Adv Exp Med Biol. 1991;295:267–90.

Mogenson GJ, Jones DL, Yim CY. From motivation to action: functional interface between the limbic system and the motor system. Prog Neurobiol. 1980;14:69–97.

Mogenson GJ, Ciriello J, Garland J, Wu M. Ventral pallidum projections to mediodorsal nucleus of the thalamus: an anatomical and electrophysiological investigation in the rat. Brain Res. 1987;404:221–30.

Mora F, Segovia G, Del Arco A. Glutamate-dopamine-GABA interactions in the aging basal ganglia. Brain Res Rev. 2008;58:340–53.

Moreland RB, Nakane M, Donnelly-Roberts DL, Miller LN, Chang R, Uchic ME. Comparative pharmacology of human dopamine D(2)-like receptor stable cell lines coupled to calcium flux through Galpha(qo5). Biochem Pharmacol. 2004;68:761–72.

Mountjoy KG, Wild JM. Melanocortin 4 receptor mRNA expression in the developing autonomic and central nervous system. Brain Res Dev Brain Res. 1998;107:309–14.

Nestler EJ. Molecular basis of long term plasticity underlying addiction. Nat Rev Neurosci. 2001;2:119–28.

Nestler EJ, Malenka RC. The addicted brain. Sci Am. 2004;290:78–85.

Nestler EJ, Hyman SE, Malenka RC. Molecular basis of neuropharmacology: a foundation for clinical neuroscience. Columbus: McGraw-Hill; 2001.

Ogawa S, Lubahn DB, Korach KS, Pfaff DW. Behavioral effects of estrogen gene disruption in male mice. Proc Natl Acad Sci U S A. 1997;94:1476–81.

Olivier B. Serotonergic mechanisms in aggression. Novartis Found Symp. 2005;268:171–83.

Olivier B, Mos J. Serenics, serotonin and aggression. In: Meldrum BS, Williams M, editors. Current and future trends in anticonvulsant, anxiety, and stroke therapy. New York: Wiley; 1990. p. 203–30.

Olivier B, Mos J, van Oorschot R, Hen R. Serotonin receptors and animal models of aggressive behavior. Pharmacopsychiatry. 1995;28:80–90.

Parent A, Descarries L, Beaudet A. Organisation of ascending serotonin systems in adult rat brain. A radioautographic study after interventricular administration of [3H] 5-hydroxytryptamine. Neuroscience. 1981;6:115–38.

Pitkow LJ, Sharer CA, Ren X, Insel TR, Terwilliger EF, Young LJ. Facilitation of affiliation and pair bond formation by vasopressin receptor gene transfer into the ventral forebrain of a monogamous vole. J Neurosci. 2001;7392–6

Poggioli R, Benelli A, Arletti R, Cavazzuti E, Bertolini A. Nitric oxide is involved in the ACTH induced behavioral syndrome. Peptides. 1995;16:1263–8.

Pogun S, Baumann MH, Kuhar MJ. Nitric oxide inhibits [^{3}H] dopamine uptake. Brain Res. 1994;641:83–91.

Ra S, Aoki H, Fujioka T, Sato F, Kubo T, Yasuda N. In vitro contraction of canine corpus cavernosum penis by direct perfusion with prolactin or growth hormone. J Urol. 1996;156(2 Pt 1):522–5.

Rampin O, Jerome N, Suaudeau C. Proerectile effects of apomorphine in mice. Life Sci. 2003;72:2329–36.

Rehman J, Christ G, Melman A, Fleischmann J. Intracavernous pressure responses to physical and electrical stimulation of the cavernous nerve in rats. Urology. 1998;51(4):640–4.

Rehman J, Christ G, Alyskewycz M, Kerr E, Melman A. Experimental hyperprolactinemia in a rat model: alteration in centrally mediated neuroerectile mechanisms. Int J Impot Res. 2000;12(1):23–32.

Routtenberg A, Malsbury C. Brainstem pathways of reward. J Comp Physiol Psychol. 1969;68:22–30.

Sato F, Aoki H, Nakamura K, Taguchi M, Aoki T, Yasuda N. Suppressive effects of chronic hyperprolactinemia on penile erection and yawning following administration of apomorphine to pituitary transplanted rats. J Androl. 1997;18(1):21–5.

Sato Y, Horita H, Kurohata T, Adachi H, Tsukamoto T. Effect of nitric oxide level in the medial preoptic area on male copulatory behaviour in rats. Am J Physiol. 1998;274(1 Pt 2):R243–7.

Sato Y, Christ GJ, Horita H, Adachi H, Suzuki N, Tsukamoto T. The effects of alteration of nitric oxide levels in paraventricular nucleus on copulatory behavior and reflexive erections in male rats. J Urol. 1999;162(6):2182–5.

Schultz W. Subjective neuronal coding of reward: temporal value discounting and risk. Eur J Neurosci. 2010;31(12):2124–35.

Schuman EM, Madison DV. Nitric oxide and synaptic function. Ann Rev Neurosci. 1994;17:153–83.

Shaw-Lutchman SZ, Impey S, Storm D, Nestler EJ. Regulation of CRE mediated transcription in mouse brain by amphetamine. Synapse. 2003;48:10–7.

Simon NG, Cologer-Clifford A, Lu SF, McKenna SE, Hu S. Testosterone and its metabolites modulate 5HT1A and 5HT1B agonist effects on intermale aggression. Neurosci Biobehav Rev. 1998;23(2):325–36.

Skagerberg G, Lindvall O. Organisation of diencephalic dopamine neurons projecting to the spinal cord of the rat. Brain Res. 1985;342(2):340–1.

Smeltzer MD, Curtis JT, Aragona BJ, Wang ZX. Dopamine, oxytocin, and vasopressin receptor binding in the medial prefrontal cortex of monogamous and promiscuous voles. Neurosci Lett. 2006;394:146–51.

Sofroniew MV. Morphology of vasopressin and oxytocin neurons and their central and vascular projections. Prog Brain Res. 1983;60:101–14.

Tang Y, Rampin O, Calas A, Facchinetti P, Giuliano F. Oxytocinergic and serotonergic innervations of identified lumbosacral nuclei controlling penile erection in the male rat. Neuroscience. 1998;82(1):241–54.

Tarlaci S. The brain in love: has neuroscience stolen the secret of love? NeuroQuantology. 2012;10:744–53.

Tzschentke TM. Pharmacology and behavioural pharmacology of mesocortical dopamine system. Prog Neurobiol. 2001;63:241–320.

Tzschentke TM, Schmidt WJ. Functional relationship among medial prefrontal cortex, nucleus accumbens, and ventral tegmental area in locomotion and reward. Crit Rev Neurobiol. 2000;14:131–42.

Van Leeuwen FW, Caffe AR, De Vries GJ. Vasopressin cells in the bed nucleus of stria terminalis of the rat: sex differences and influence of androgens. Brain Res. 1985;325:391–4.

Verma S, Chhina GS, Kumar VM, Singh B. Inhibition of male sexual behavior by serotonin application in the medial preoptic area. Physiol Behav. 1989;46:327–30.

Véronneau-Longueville F, Rampin O, Freund-Mercier MJ, Tang Y, Calas A, Marson L, et al. Oxytocinergic innervations of autonomic nuclei controlling penile erection in the rat. Neuroscience. 1999;93(4):1437–47.

Wang ZX, Aragona BJ. Neurochemical regulation of pair bonding in male prairie voles. Physiol Behav. 2004;83:319–28.

Wang ZX, Insel TR. Parental behavior in voles. Adv Study Behav. 1996;25:361–84.

Wang C, Swerdloff RS, Iranmanesh A, Dobs A, Snyder PJ, Cunningham G. Transdermal testosterone gel improves sexual function, mood, muscle strength and body composition parameters in hypogonadal men. J Clin Endocrinol Metab. 2000;85:2839–53.

Warner RK, Thompson JT, Markowski VP, Loucks JA, Bazzett TJ, Eaton RC, et al. Microinjection of dopamine antagonists cis-flupenthixol into MPOA impairs copulation, penile reflexes and sexual motivation in male rats. Brain Res. 1991;540(1–2):177–82.

Watabe-Uchida M, Zhu L, Ogawa SK, Vamanrao A, Uchida N. Whole-brain mapping of direct inputs to midbrain opamine neurons. Neuron. 2012;74:858–73.

Wessells H, Hruby VJ, Hackett J, Han G, Balse-Srinivasan P, Vanderah TW. Ac-Nle-c[Asp-His-DPhe-Arg-Trp-Lys]-NH2 induces penile erection via brain and spinal melanocortin receptors. Neuroscience. 2003;118:755–62.

Whitton PS. Glutamatergic control over brain dopamine release in vivo and in vitro. Neurosci Biobehav Rev. 1997;21:481–8.

Wise RA. Brain reward circuitry: insights from unsensed incentives. Neuron. 2002;36(2):229–40.

Wise RA, Bozarth MA. Brain reward circuitry: four circuit elements "wired" in apparent series. Brain Res Bull. 1984;12:203–8.

Wu MV, Manoli DS, Fraser EJ, Coats JK, Tollkuhn J, Honda S, et al. Estrogen masculinizes neural pathways and sex specific behaviours. Cell. 2009;139:61–72.

Yonezawa A, Watanabe C, Ando R, Furuta S, Sakurada S, Yoshimura H. Characterization of p-chloroamphetamine- induced penile erection and ejaculation in anesthesized rats. Life Sci. 2000;67(25):3031–9.

Yoshimura R, Kiyama H, Kimura T, Araki T, Maeno H, Tanizawa O, Tohyama M. Localization of oxytocin receptor messenger-ribonucleic-acid in the rat-brain. Endocrinology. 1993;133:1239–46.

Young LJ. Oxcytocin and vasopressin receptors and species-typical social behaviors. Horm Behav. 1999;36:212–21.

Young LJ, Wang Z. The neurobiology of pair bonding. Nat Neurosci. 2004;7(10):1048–54.

Young LJ, Lim MM, Gingrich B, Insel TR. Cellular mechanisms of social attachment. Horm Behav. 2001;40(2):133–8.

Zeki S. The neurobiology of love. FEBS Lett. 2007;581(14):2575–9.

Zingg HH. Vasopressin and oxytocin receptors. Baillieres Clin Endocrinol Metab. 1996;10(1):75–96.

Zitzmann M, Nieschlag E. Testosterone levels in healthy men and the relation to behavioural and physical characteristics: facts and constructs. Eur J Endocrinol. 2001;144:183–97.

Male Behaviors III: Brain Polymorphism and Sexual Orientation

19

Rakesh Kumar Chadda, Mona Sharma, and Anand Kumar

According to Puranas, Shiv Puran in particular, Brahma first created five mental sons—Sanak, Sanandan, Vidvan Sanatan, Ribhu and Sanat Kumar—who were faultless yogis, who constantly meditated on the God, and were disinterested in anything else. They had no interest in increasing their number. Then, came emanations of Siva out of Brahma's mouth; they were 11 Rudras who ran and cried; they were beings of life, identical to Siva, beyond life and death. They asexually proliferated over the 14 mythical realms. But, still the world was not adequately populated. Then, Brahma requested Siva to create mortal, sexual beings so that the earth is populated. Benevolent Siva pulled out his transcendent power from the left part of his body in form of a resplendent Goddess, the universal Creatrix or the Mother. Till before her appearance, the feminine form was unknown. The Goddess, the Power of Siva, produced Shatroopa, the first woman. Brahma materialized a son, Virat Pususa, from the male half of his body. This son of Brahma was known as Manu, the first man. Shatroopa and Manu produced sexually reproducing human beings, the maithuni sristi, to populate the world, which otherwise was impossible despite every effort of gods.

R.K. Chadda, MD, FAMS, FRC (✉)
Department of Psychiatry, All India Institute of Medical Sciences, New Delhi, India
e-mail: drrakeshchadda@gmail.com

M. Sharma, MD, DNB, MAMS • A. Kumar, MD, FAMS
Department of Reproductive Biology, All India Institute of Medical Sciences, New Delhi, India

A. Kumar, M. Sharma (eds.), *Basics of Human Andrology*,
DOI 10.1007/978-981-10-3695-8_19

Learning Objectives

- Androgen-mediated brain sexual differentiation
- Differences in male and female brain
- Homosexual brain
- Transsexual brain

19.1 Introduction

Sexual dimorphisms are the sex-related differential expressions in the phenotype and genotype of individuals. Human behavior is influenced by the chromosomal sex, the phenotypic sex and the gender. Chromosomal sex includes XX for chromosomal females and XY for chromosomal males. Phenotypic sex is determined by internal and external genitals, secondary sexual characters and the behavior. Gender is defined as an individual's perception of his or her sex and sexual orientation.

Sexual dimorphisms in behavior have led researchers to study the neuroanatomical organization of the brain. Varying hormonal influences on brain development of men and women reflect their differential behavioral pattern. The potential factors governing the human sexual dimorphisms are the sex steroids and the distribution of their receptors in the developing brains; as already being confirmed by various animal studies (Pilgrim and Hutchison 1994; Gorski 2000).

19.2 Androgen and Phenotypic Sex

Androgens affect the phenotypic sex of an individual, and their presence lead to masculinization of the developing brain in phenotypic males (see Fig. 19.1).

The brain sexual differentiation in animals involves an essential process—aromatization—in which testosterone is converted to estrogen by the aromatase enzyme (Kawata 1995). In rodents, monkeys and ferrets, the activity of aromatase enzyme has been observed in the hypothalamus, amygdala (AMG) and hippocampus; these areas are also rich in sex steroid receptors (MacLusky et al. 1987; Miranda and Toran

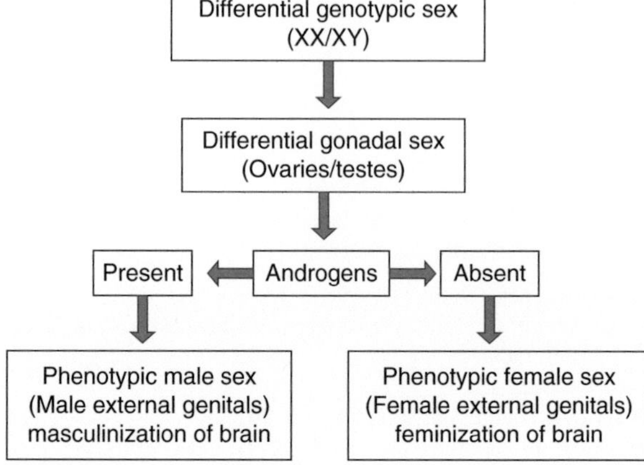

Fig. 19.1 Androgen-mediated development of phenotypic sex

Allarand 1992; Tobet et al. 1993). The animal studies have confirmed that the circulating androgens also maintain the volumes of sexually dimorphic brain areas such as medial AMG and preoptic nucleus (Cooke et al. 1999; Bloch and Gorski 1988).

The animal studies have also confirmed the role of estradiol in brain development (Miranda and Toran Allarand 1992; Toran Allarand 1996). Estradiol is one of the active compounds of estrogen family. The developing brain expresses high levels of estradiol receptors. There are a few studies reported in humans regarding the localization and sex differences in the sex steroid receptors in the brain (Rance et al. 1990; Sarrieau et al. 1990; Fernandez-Guasti et al. 2000). The distribution of estrogen receptors in the human brain is identical to the similar areas in various other mammals (Donahue et al. 2000). Sexual dimorphisms have been observed in the brain areas such as the hypothalamus, hippocampus, AMG, basal forebrain, frontal cortex and parahippocampus. The sexually dimorphic areas in humans are homologous to areas in animals with higher sex steroid receptors (Goldstein et al. 2001). See Table 19.1 for sexually dimorphic differences in male and female brain. The sexually dimorphic features of male brain are shown in Fig. 19.2.

Table 19.1 Sexually dimorphic differences in male and female brain

Brain dimorphic areas	Men	Women	Comments and references
Gross size	Larger (10%)	Smaller	Men's larger body size would require more number of neurons to control them (Cosgrove et al. 2007)
Composition	Higher white matter	Higher grey matter	Cosgrove et al. (2007)
Amygdala (AMG)	Larger	Smaller	AMG processes fear and aggression by the action of androgens (Clark et al. 1988)
Hippocampus	Smaller	Larger	Larger hippocampus helps in better emotional expression and remembering details of emotional events (Morse et al. 1986)
Inferior parietal lobule (IPL)	Larger (Left side larger than right side)	Smaller (Right side larger than left side)	IPL process information from stimuli and help in selective attention and perception Women have bigger right IPL. So, they have better spatial orientation, ability to have better sense of relationships between body parts and emotional perception. Men have bigger size of the left IPL that helps them in better perception of time and speed (Frederikse et al. 1999)
Third interstitial nucleus of anterior hypothalamus (INAH-3)	Larger	Smaller	Cosgrove et al. (2007)
Wernicke's area and Broca's area	Smaller	Larger	Women have stronger language perception as compared to men. This is because both sides of the brain function in woman during language processing. Whereas in man, the brain tends to process it in the left side only
Anterior commissure	Smaller	Larger	Allen and Gorski (1992)
Bed nucleus of stria terminalis (BNST)	Larger	Smaller	The role of BNST is implicated in sexually dimorphic functions such as aggression and sexual behavior in man (Allen and Gorski 1990)

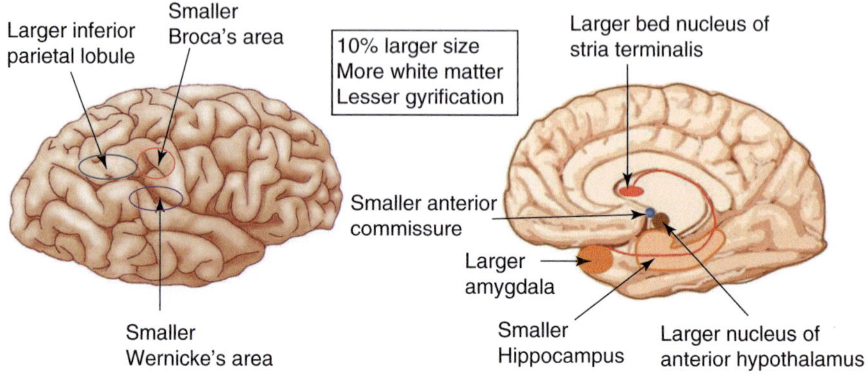

Fig. 19.2 Morphological characteristics of male brain

19.3 The Homosexual Brain

The homosexual brain is a distinct biological entity and not any kind of pathological disease as suggested by homophobic and religious men. Plato…..years back wrote:

'Long ago, our nature was not the same as it is now but quite different. For one thing, there were three human genders, not just the present two, male and female. There was also a third one, a combination of these two; now its name survives, although the gender has vanished. Then "androgynous" was a distinct gender as well as a name, combining male and female; now nothing is left but the name, which is used as an insult.'

'Zeus cut humans into two.'

Each of us is a matching half of human being, because we've been cut in half like flatfish, making two out of one, and each of us is looking for his own matching half. Those men who are cut from the combined gender (the androgynous, as it was called then) are attracted to women, and many adulterers are from this group. Similarly, the women who are attracted to men and become adulteresses come from this group. Those women who are cut from the female gender are not at all interested in men, but are drawn much more towards women; female homosexuals come from this group.

Those who are cut from the male gender go for males. While they are boys, because they are slices of the male gender, they are attracted to men and enjoy sleeping with men and being embraced by them. These are the best of their generation, both as boys and young men, because they are naturally the bravest. Some people say that they are shameless, but that isn't true. It's not out of shamelessness that they do this but because they are bold, brave and masculine, and welcomes the same qualities in others. Here is clear evidence of this: men like this are the only ones who, when grown up, end up as politicians. When they become men, they're sexually attracted by boys; they have no natural interest in getting married and having children, although they are forced to this by convention. They are quite satisfied by spending their lives together and not getting married. In short, such people become lovers of boys and boys who love their male lovers, always welcoming their shared natural character.

Aristophanes speech in Plato's Symposium, translated by Christopher John Gill 1999, *Penguin Classics*

Anatomical differences in the brain of homosexual men were first documented by Swaab. The sexual orientation in human beings is associated with anterior commissure. The anterior commissure is the bundle of axonal fibres that connects the inferior and middle parts of temporal lobes of left and right hemispheres. It is involved in exchanging visual, auditory and olfactory information between two hemispheres. The anterior commissure is widest in midsagittal sections in homosexual men than heterosexual men and women (Allen and Gorski 1992). Homosexual men also have smaller third interstitial nucleus of the anterior hypothalamus (INAH3) as compared to heterosexual men (LeVay 1991). Isthmus of the corpus callosum is significantly larger in homosexual men than in heterosexual men (Witelson et al. 2008). The size of suprachiasmatic nucleus (SCN) of hypothalamus in homosexual men is almost twice as large as that of heterosexual males (Swabb and Hofman 1990). It is concluded that the homosexual and heterosexual men differ in the central neuronal mechanisms that control sexual behavior and that this difference in anatomy is no product of upbringing or environment, but rather prenatal cerebral development and structural differentiation.

19.4 Transsexual Brain

Transsexuality is a gender identity disorder having a sense of belonging to opposite sex (Blanchard 1989). The true perception of own sex depends on sexual differentiation of the brain, and in transsexuals, this sense differs from phenotypic sex. The sexual differentiation of the brain occurs after the genital differentiation. The two brain regions implicated in sexually dimorphic functions are BNST and INAH3. Both the structures are smaller in male to female transsexuals as compared to heterosexual males (Zhou et al. 1995; Berglund et al. 2008). These neuroanatomical differences can be the basis of altered sexual orientation in transsexuals.

Key Questions

- How do androgens influence brain sexual differentiation?
- Describe the differences in language skills and spatial orientation between heterosexual men and women based on anatomical differences in the brain.
- How do transsexual and homosexual brains differ from the heterosexual brain?

References

Allen LS, Gorski RA. Sex differences in the bed nucleus of stria terminalis of the human brain. J Comp Neurol. 1990;302(4):697–706.

Allen LS, Gorski RA. Sexual orientation and size of anterior commissure in human brain. Proc Natl Acad Sci U S A. 1992;89(15):7199–202.

Aristophanes speech in Plato's Symposium, translated by Christopher John Gill, 1999, Penguin Classics.

Berglund H, Lindstrom P, Dhejne-Helmy C, Savic I. Male to female transsexuals show sex atypical hypothalamus activation when smelling odorous steroids. Cereb Cortex. 2008;18:1900–8.

Blanchard R. The classification and labeling of nonhomosexuals gender dysphorias. Arch Sex Behav. 1989;18:315–34.

Bloch GJ, Gorski RA. Cytoarchitectonic analysis of SDN-POA of the intact and gonadectomized rat. J Comp Neurol. 1988;275:604–12.

Clark AS, Maclusky NJ, Goldman PS. Androgen binding and metabolism in the cerebral cortex of the developing rhesus monkey. Endocrinology. 1988;123:932–40.

Cooke BM, Tabibnia G, Breedlove SM. Abrain sexual dimorphism controlled by adult circulating androgens. Proc Natl Acad Sci U S A. 1999;96:7538–40.

Cosgrove KP, Mazure CM, Staley JK. Evolving knowledge of sex differences in brain structure, function and chemistry. Biol Psychiatry. 2007;62(8):847–55.

Donahue JE, Stopa EG, Chorsky RL, King LC, Schipper HM, Tobet SA, et al. Cell containing immunoreactive estrogen receptor α in the human basal forebrain. Brain Res. 2000;856:142–51.

Fernandez-Guasti A, Kruijver FPM, Fodor M, Swaab DF. Sex differences in the distribution of androgen receptors in the human hypothalamus. J Comp Neurol. 2000;425:422–35.

Frederikse ME, Lu A, Aylward E, Barta P, Pearlson G. Sex differences in the inferior parietal lobule. Cereb Cortex. 1999;9(8):896–901.

Goldstein JM, Seidman LJ, Horton NJ, Makris N, Kennedy DN, Caviness VS, et al. Normal sexual dimorphism of the adult human brain assessed by in vivo magnetic resonance imaging. Cereb Cortex. 2001;11:490–7.

Gorski RA. Sexual differentiation of the nervous system. In: Kandel ER, Schwartz JH, Jessel TM, editors. Principles of neuroscience. New York: McGraw Hill; 2000. p. 1131–46.

Kawata M. Roles of steroid hormones and their receptors in structural organization in the nervous system. Neurosci Res. 1995;24:1–46.

LeVay S. A difference in hypothalamic structure between heterosexual and homosexual men. Science. 1991;253(5023):1034–7.

MacLusky NJ, Clark AS, Naftolin F, Goldman-Rakic PS. Estrogen formation in mammalian brain: possible role of aromatase in sexual differentiation of the hippocampus and neocortex. Steroids. 1987;50:459–74.

Miranda RC, Toran Allarand D. Developmental expression of estrogen receptor mRNA in the rat cerebral cortex: a non isotopic in situ hybridization histochemistry study. Cereb Cortex. 1992;2:1–15.

Morse JK, Scheff SW, Dekosky ST. Gonadal steroids influence axonal sprouting in the hippocampal dentate gyrus: a sexually dimorphic response. Exp Neurol. 1986;94:649–58.

Pilgrim C, Hutchison JB. Developmental regulation of sex differences in the brain: can the role of gonadal steroids be redefined? Neuroscience. 1994;60:843–55.

Rance NE, McMullen NT, Smialek JE, Price DL, Young III WS. Postmenopausal hypertrophy of neurons expressing the estrogen receptor gene in the human hypothalamus. J Clin Endocrinol Metab. 1990;71:79–85.

Sarrieau A, Mitchell JB, Lal S, Olivier A, Quirion R, Meaney MJ. Androgen binding sites in human temporal cortex. Neuroendocrinology. 1990;51:713–6.

Swabb DF, Hofman MA. An enlarged suprachiasmatic nucleus in homosexual men. Brain Res. 1990;537:141–8.

Tobet SA, Basham ME, Baum MJ. Estrogen receptor immunoreactive neurons in the fetal ferret forebrain. Brain Res Dev Brain Res. 1993;72:167–80.

Toran Allarand CD. The estrogen/neurotrophin connection during neural development: is co-localization of estrogen receptors with the neurotrophins and their receptors biologically relevant? Dev Neurosci. 1996;18:36–41.

Witelson SF, Kigar DL, Scamvougeras A, Kideckel DM, Buck B, Stanchev PL, et al. Corpus callosum anatomy in right- handed homosexual and heterosexual men. Arch Sex Behav. 2008;37:857–63.

Zhou JN, Hofman MA, Gooren LJ, Swaab DF. A sex difference in the human brain and its relation to transsexuality. Nature. 1995;378:68–70.

Male Behaviors IV: Ageing Brain 20

Mona Sharma and Anand Kumar

Learning Objectives

- Neuroanatomical and neurotransmitter changes in the ageing brain
- Ageing brain and androgens
- Behavioral deficits with brain ageing

20.1 Introduction

Brain ages with increasing chronological age. Age-related deficits in behaviors are related to brain ageing. The predominant markers of ageing process are the cognitive deficit and sexual dysfunction. Cognition is a mental process by which sensory inputs are transformed, elaborated, reduced, stored or used to acquire knowledge and to exhibit behaviors (Neisser 1967). Cognition is an integral part of behavior formation. Cognition has various components which are used during behavior formation such as attention, memory, learning, execution, judgement and decision making. Humans exhibit behaviors based on sensory integration, motivation generation, goal-directed motor output or movements and cognitive control by the brain.

M. Sharma, MD, DNB, MAMS (✉) • A. Kumar, MD, FAMS
Department of Reproductive Biology, All India Institute of Medical Sciences, New Delhi, India
e-mail: dr.mona18sharma@gmail.com

A. Kumar, M. Sharma (eds.), *Basics of Human Andrology*,
DOI 10.1007/978-981-10-3695-8_20

Brain ageing causes anatomical and neurotransmitter changes which affect hypothalamic-pituitary-gonadal axis and functioning of neural circuits. The chemical and anatomical disconnection between neural circuits leads to cognitive loss and other behavioral deficits of ageing.

20.2 Neuroanatomical Changes and Behavioral Deficits

At morphological level, ageing brain shows atrophic changes by neuronal cell death in grey matter especially in prefrontal cortex (PFC), medial temporal lobes, basal ganglia and white matter atrophy in frontal lobes (Ranga-Krishnan et al. 1990; Hof and Morrison 2004; Salat et al. 2005). Ageing male brain shows more atrophic changes as compared to ageing female brain (Xu et al. 2000). Ageing brain also shows ventricular dilation and decrease in synaptic density (Brody and Vijayashankar 1977; Miller et al. 1980; Hubbard and Anderson 1981; Anderson et al. 1983). Decreased neuronal pool is also observed in the hippocampus, cerebellum, amygdala (AMG) and substantia nigra (SN; Hall et al. 1975; Ball 1977; Mann and Yates 1979; Herzog and Kemper 1980).

With ageing, there is cortical thinning in primary visual sensory and motor cortices (Salat et al. 2004). Older subjects with thicker cortex show better cognition as compared to the ones with thinner cortex (Fjell et al. 2006). Age-related reduction in cerebral blood flow is observed in cortical areas such as frontal, temporal, parietal lobes and in limbic areas (Martin et al. 1991). There is also widespread cortical reduction in brain oxygen consumption with ageing (Marchal et al. 1992).

The characteristic histological changes in ageing brain are senile plaques (SPs), neurofibrillary tangles (NFTs) and granulovacuolar degeneration (GVD; Tomlinson et al. 1968; Mountjoy et al. 1983). The large number of SPs, NFTs and GVD becomes the hallmark of the Alzheimer's disease (AD), a memory deficit disorder.

Most common sites for SPs and NFTs are the hippocampus and AMG (Terry and Wisniewski 1970). SPs are the extracellular beta amyloid (Aβ) protein deposits. NFTs are the intracellular deposits, most commonly of microtubule-associated tau protein. Microtubules are important for transport of nutrients and cellular organelle. Tau proteins hold the microtubules together in parallel. With ageing, tau proteins undergo hyperphosphorylation and form aggregates. Due to the tau protein aggregates, normal functions of microtubules, essential for neuronal survival, are lost (Bancher et al. 1989). The other protein constituents of NFTs are neurofilament-associated proteins, microtubule-associated protein, vimentin, tropomyosin and ubiquitin (Anderton et al. 1982; Yen et al. 1983; Ksiezak and Yen 1987; Cole and Timiras 1987; Galloway et al. 1990).

The GVD is restricted to the hippocampal neurons and was first described in AD (Simchowicz 1911; Ball and Lo 1977). GVD consists of intraneuronal spherical vacuoles of diameter 3–5 μm, with centrally placed granules of diameter 0.5–1.5 μm (Tomlinson and Kitchener 1972). Tau protein is also one of the constituents of granules in GVD (Bondareff et al. 1991). The degree of GVD in hippocampal neurons increases with age and correlates with cognitive function (Woodard 1962; Ghoshal et al. 2002).

Ageing neurons also show decrease in nucleolar volume, RNA content, disruption of cell organelle such as endoplasmic reticulum and Golgi apparatus. There is also accumulation of neurodegenerative pigments such as lipofuscin and neuromelanin in ageing neurons (Brody and Vijayashankar 1977).

Brain areas such as the hypothalamus, hippocampus, AMG and frontal cortex show the presence of androgen receptors and are prone for age-related neurodegenerative changes (Finley and Kritzer 1999; Beyenburg et al. 2000; Roselli et al. 2001). Androgens are responsible for masculinizing the brain. Male rats have larger dentate gyrus, a part of hippocampus, as compared to females. The hippocampus shows both estrogen and testosterone receptors (Roof 1993). Testosterone exposure of the neonatal female rats produces a masculine hippocampus (Roof and Havens 1992; Roof 1993). Androgen deprivation in animal studies shows the neuropathological changes similar to age-related degenerative changes such as loss of synaptic density in the hippocampus that is restored with testosterone or dihydrotestosterone (DHT) administration (Leranth et al. 2003). Age-related neurodegenerative protein Aβ increases in rodent brain and human plasma after androgen deprivation and decreases on administration of testosterone or DHT (Ramsden et al. 2003; Almeida et al. 2004).

Various animal studies show that testosterone exerts neurotropic action by promoting neuronal growth, differentiation and synapse formation (Beyer and Hutchison 1997). Testosterone depletion induces neurodegenerative changes of ageing. Testosterone mediates two types of cellular effects – the slow genomic and the fast non-genomic effects. The genomic effects include upregulation of gene expression for various cellular events. The non-genomic effects include stimulation of signal transduction pathways and ion channels (Kousteni et al. 2001). Testosterone also exerts neuroprotective effects by decreasing neuronal death induced by Aβ deposition (Pike 2001). The process of neural protection by testosterone is androgen receptor dependent and not through aromatization to estrogen pathway (Hammond et al. 2001). It has also been shown that testosterone-mediated neuronal protection is through rapid non-genomic cell signalling pathways (Pike 2001). Mitogen-activated protein kinase/extracellular signal-regulated kinase pathway (MAPK/ERK) is linked to testosterone-mediated neuronal survival in the hippocampus (Zhu et al. 2002; Nguyen et al. 2005). Testosterone binds to androgen receptor and activates MAPK/ERK signalling cascade where ERK undergoes phosphorylation that leads to activation of ribosomal S6 kinases. Downstream pathway includes phosphorylation and inactivation of pro-apoptotic factor Bcl2 associated death protein or Bad that leads to inhibition of apoptotic cascade and neuronal survival (Zhu et al. 2002).

The neuroanatomical changes in ageing brain affect cognition. Brain ageing is characterized by decline in different domains of cognition – memory, attention, execution such as problem solving and abstract thinking (Lindenberger et al. 1993; Glisky 2007). Aβ deposition is an early event that further progresses and leads to neuronal death, cortical atrophy, decreased structural and functional connectivity and cognitive decline (Rodrigue et al. 2009). The loss of working memory and long-term memory is the cognitive hallmarks of brain ageing. Various studies show negative correlation of Aβ deposition and memory performance; association between Aβ deposits in frontal and parietal cortices and cognition (Pike et al. 2007; Braskie et al. 2010). In humans, working memory holds and updates information over seconds to minutes and is dependent on the function of PFC (Baddeley et al. 1986). Long term memory, a function of hippocampus, encodes and stores the events permanently (Squire and Zola 1997). Hippocampal and PFC atrophy in ageing is correlated with poor memory performance (Cardenas et al. 2011). Various experiments show that larger PFC is associated with better executive tasks, working memory and visuospatial mental performance in older subjects (Raz et al. 1999; Tapp et al. 2004; Head et al. 2009). Procedural learning and motor tasks are also better in subjects with larger dorsal striatum or caudate nucleus and PFC volume (Kennedy and Raz 2005).

20.3 Neurotransmitter Changes and Behavioral Deficits

Ageing reduces the efficacy of brain circuitries due to decline in neurotransmitter functions. The dopaminergic, serotonergic, noradrenergic and cholinergic neurons are responsive to the changing levels of gonadal hormones (McEwen et al. 1995; Bethea et al. 1999). Role of dopamine (DA) in behaviors such as motivation, pleasure and sex is also androgen dependent as described in the previous chapter. DA is important in pleasure-mediated behaviors, cognition and motor performance or movements. Dopaminergic system is vulnerable to age-induced changes (Joseph et al. 1990). There is decrease in DA receptors- D1, D2 and D3 in brain areas such as the hippocampus, cingulate cortex, frontal cortex and AMG with ageing (Suhara et al. 1991; Volkow et al. 1996b; Mukherjee et al. 2002). Age-related decline in DA transporters has also been reported (Volkow et al. 1996a). The decreasing levels of DA with ageing are responsible for poor cognitive function, decreased motor performance and decline in motivation (Bäckman et al. 2006).

Pleasure circuitry is vulnerable to ageing due to age-induced changes in the midbrain DA synthesis, nucleus accumbens (NAc) and PFC activity. Motivation to approach towards goals involves activation of NAc whereas motivation to avoid threats or fear involves activation of insula and deactivation of NAc (Knutson et al. 2014). Final execution of behavior involves activation of PFC (Knutson and Greer 2008). Human studies have shown that during pleasure-associated learning, there is decreased NAc and PFC activity in older subjects as compared to younger subjects (Knutson et al. 2001; Rajah and D'Esposito 2005; Marschner et al. 2005). DA concentration and receptor density decreases in PFC and striatum in older brains (Bäckman and Farde 2005). Decreased DA activity in PFC causes cognitive deficit in ageing (West 1996). Brain ageing also decreases concentration of tyrosine hydroxylase, a synthetic enzyme of DA, in NAc and striatum (McGeer and McGeer 1976).

Another neurotransmitter important for motivation and pleasure is serotonin (5-HT). Decrease in 5-HT and its receptors in hippocampus and frontal cortex has been found with ageing (Yamamoto et al. 2002; Mattson et al. 2004; Moller et al. 2007). 5-HT is also responsible for sleep wake cycle. Reduced 5-HT activity with ageing is responsible for sleep disorders in ageing males (Rehman and Masson 2001).

Noradrenaline (NA) concentration also decreases with ageing in the hypothalamus (Robinson et al. 1972). Reduced NA and DA action in hypothalamus reduces secretion of gonadotropin-releasing hormone (GnRH) and growth hormone-releasing hormone (GHRH; Meites 1990). The effects of hormonal deficits will be discussed in the next section. NA concentration is also found to be less in the brain of people with AD, suggesting its role in cognition (Palmer 1996).

The neurotransmitters such as DA, 5-HT and NA are metabolized by enzyme monoamine oxidase (MAO). Two forms of MAO enzyme are found – MAO-A and MAO-B. Ageing causes decrease or no change in the activity and concentration of MAO-A. With ageing, MAO-B activity increases in all the brain areas (Fowler et al. 1980). The increase in MAO-B activity reduces synaptic concentration of DA, 5-HT and NA which causes age-related decline in their activity. Depletion of 5-HT and NA is associated with depression in elderly people (Zubenko et al. 1990).

Brain cholinergic activity also decreases with ageing (McGeer et al. 1984). Decrease in muscarinic receptors has been shown in striatum, hippocampus and frontal cortex which is responsible for cognitive and motor deficit with ageing (Rinne 1987).

Glutamate is known to be involved in cognition, motivation, emotional responses and motor behavior. Age-related decrease in glutamate receptors in the hippocampus and behavioral deficits have been confirmed (Cotman et al. 1987; Mora and Cobo 1990; Schmidt et al. 1992; Cobo et al. 1992, 1993).

20.4 Ageing Brain, Androgens and Behaviors

The neural development throughout life is influenced by various neurotransmitters and hormones. Testosterone levels in plasma and brain decline decade by decade with ageing in males (Morley 2001; Rosario et al. 2004). Various age-related factors such as obesity, diabetes mellitus and cardiovascular diseases may act as confounders, thereby affecting testosterone levels in ageing males. The effect of confounders on testosterone levels has been described in the chapter Andropause.

There is also increase in sex hormone-binding globulin (SHBG) with ageing which causes less bioavailability of free testosterone (Hijazi and Cunningham 2005). The decline in testosterone levels is due to the changes in the hypothalamic-pituitary-testicular axis. Brain ageing causes decreased activity of neurotransmitters such as NA in the hypothalamus which causes fall in GnRH outflow to pituitary (Robinson et al. 1972; Keenan et al. 2006). Decrease GnRH outflow causes decrease in luteinizing hormone (LH) stimulated testosterone secretion (Winters and Troen 1982; Keenan et al. 2006). The number of Leydig cells producing testosterone also decreases with ageing (Kaler and Neaves 1978). The number of androgen receptors in the hypothalamus decreases in older men (Tohgi et al. 1995). The GnRH neurons do not contain steroid receptors (Fox et al. 1990; Lehman and Karsch 1993; Huang and Harlan 1993). Testicular steroids may act on GnRH neurons via steroid receptive neuronal systems such as dopaminergic and opioidergic (Tortonese 1999). The normal hypothalamus-mediated circadian variation with peak testosterone levels in morning is also blunted in older men as compared to younger men (Bremner et al. 1983).

Sleep patterns and efficiency change with age. Older individuals show decreased sleep efficiency and increased sleep latency as compared to younger ones (Ohayon 2004). The age and sleep efficiency association is stronger in males as compared to females (Redline et al. 2004). Testosterone levels in males vary with sleep pattern. Testosterone levels increase during first 3 h of uninterrupted sleep and remain high during sleep till morning time (Luboshitzky et al. 2001). Testosterone levels decrease with increase in awake time (Axelsson et al. 2005). The total night time sleep is correlated with morning testosterone levels (Penev 2007). The basis of such an association is suggested to be the effect of sleep deprivation on hypothalamic-pituitary-testicular axis. The effect of sleep deprivation on declining levels of testosterone is more in older as compared to younger men (Oh et al. 2012). Testosterone levels also affect sleep quality. Older males with low testosterone levels have decreased sleep efficiency and increased nocturnal awakenings (Barrett-Connor et al. 2008). The basis of such association between testosterone levels and effects on sleep pattern is not completely understood although association of obesity and decreased neurotransmitter activity with low testosterone levels can be proposed.

Androgens are responsible for development of secondary sexual characters, libido and erectile function in males. In hypogonadal males, there is decreased libido and erection which show improvement with androgen replacement therapy (Matsumoto 1991). The androgen deficiency in male reproductive ageing has been discussed in the chapter, Andropause. The age-related effects on other behaviors such as cognition, motivation and pleasure are described in this chapter.

Androgen loss in males parallels cognitive deficit in ageing. Testosterone or DHT supplementation improves cognition in older men as compared to younger men (Muller et al. 2005). The hippocampus, AMG and PFC are involved in memory formation. Testosterone levels in older men are correlated with verbal fluency and spatial memory (Morley et al. 1993). Spatial memory is an important function mediated by the hippocampus (Cammaleri et al. 1996). Younger individuals show stronger activation of hippocampus and frontal areas during memory encoding tasks (Grady et al. 1995). Testosterone replacement therapy in older men shows improved spatial memory and verbal fluency (Flood et al. 1995; Morley et al. 1993; Cherrier et al. 2003). Gonadectomy in rats causes impaired memory and learning tasks (Frye et al. 2001; Kritzer et al. 2001). The functions in these rats are restored with testosterone or DHT administration (Edinger and Frye 2004). Men with AD show low testosterone levels as compared to healthy men (Hogervorst et al. 2003). Low testosterone level in men is considered a risk factor for AD (Moffat et al. 2004). Few studies also show weak association between testosterone and cognition or AD (Almeida and Flicker 2003).

Apart from decline in sexual function and cognitive deficit, brain ageing also reduces motor performance. Androgens upregulate the anabolic effects of GH by increasing muscle mass and bone mineral density (Matsumoto 1991; Gibney et al. 2005). Testosterone stimulates expression of GHRH-RNA in hypothalamic neurons by directly acting on androgen receptors (Zeitler et al. 1990). GH deficiency with ageing is due to decreased hypothalamic GHRH output (Russell-Aulet et al. 1999). Decrease androgen level in ageing is associated with GH deficiency (Iranmanesh et al. 1998). Testosterone supplementation stimulates GH production in hypogonadal older men (Gentili et al. 2002). The rate of decline in GH production is 50% greater in males than female (Weltman et al. 1994).

Decrease in androgens with ageing leads to decrease in muscle mass. Testosterone replacement therapy in older males increases muscle mass and strength (Morley et al. 1993; Urban et al. 1995). Hypogonadism is a risk factor for fractures in elderly males. Testosterone replacement therapy in older males shows a decline in bone degradation parameters such as alkaline phosphatase, serum osteocalcin and rise in bone mineral density (Tenover 1994; Morley et al. 1993). Androgen deprivation therapy (ADT) in men with prostate cancer decreases concentration of both testosterone and GH. The side effects of ADT in these patients are bone and muscle weakness, multiple fractures, depression, erectile dysfunction and impaired cognition (Chen et al. 2002).

With the help of the above discussions, it can be inferred that brain structural changes, decreasing levels of more than one neurotransmitter, and declining levels and actions of androgens and other hormones in males lead to age-related cognitive decline and decrease in motivation, pleasure and motor behavior.

See Fig. 20.1 for structural and functional changes in the ageing brain.

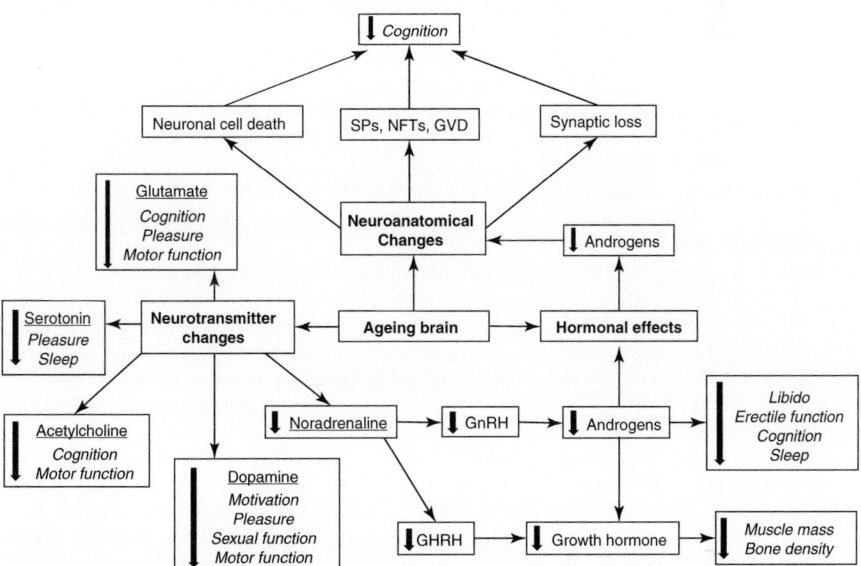

Fig. 20.1 Flow chart showing structural and functional changes of ageing brain. *SPs* senile plaques, *NFTs* neurofibrillary tangles, *GVD* granulovacuolar degeneration, *GnRH* gonadotropin-releasing hormone, *GHRH* growth hormone-releasing hormone; ↓: Decrease

Key Questions

- Describe the age-related structural changes in the brain.
- Discuss the androgen-dependent pathway of neural protection.
- What are the neurotransmitter systems prone to neurodegenerative changes with ageing?
- How do the changing levels of monoamine oxidase enzyme with ageing affect the neurotransmitter functions?
- Describe the age-induced changes in the pleasure circuitry.
- Describe the role of prefrontal cortex and hippocampus in cognition.
- What are the testosterone-mediated behavioral deficits in older men?

References

Almeida OP, Flicker L. Testosterone and dementia: too much ado about too little data. J Br Menopause Soc. 2003;9:107–10.

Almeida OP, Waterreus A, Spry N, Flicker L, Martins RN. One year follow-up study of the association between chemical castration, sex hormones, beta-amyloid, memory and depression in men. Psychoneuroendocrinology. 2004;29:1071–81.

Anderson JM, Hubbard BM, Coghill GR, Slidders W. The effect of advanced old age on the neurone content of the cerebral cortex. Observations with an automatic image analyser point counting method. J Neurol Sci. 1983;58:235–46.

Anderton BH, Breinberg D, Downes MJ, Green PJ, Tomlinson BE, Ulrich J, et al. Monoclonal antibodies show that neurofibrillary tangles and neurofilaments share antigenic determinants. Nature. 1982;298:84–6.

Axelsson J, Ingre M, Akerstedt T, Holmback U. Effects of acutely displaced sleep on testosterone. J Clin Endocrinol Metab. 2005;90:4530–5.

Bäckman L, Farde L. The role of dopamine systems in cognitive aging. In: Cabeza R, Nyberg L, Park DC, editors. Cognitive neuroscience of aging: linking cognitive and cerebral aging. New York: Oxford University Press; 2005. p. 58–84.

Bäckman L, Nyberg L, Lindenberger U, Li SC, Farde L. The correlative triad among aging, dopamine, and cognition: current status and future prospects. Neurosci Biobehav Rev. 2006;30:791–807.

Baddeley A, Logie R, Bressi S, Sala SD, Spinnler H. Dementia and working memory. QJEP Sec A. 1986;38(4):603–18.

Ball MJ, Lo P. Granulovacuolar degeneration in the ageing brain and in dementia. J Neuropathol Exp Neurol. 1977;36:474–87.

Ball MJ. Neuron loss, neurofibrillary tangles, and granulovacuolar degeneration in the hippocampus with ageing and dementia. A quantitative study. Acta Neuropathol (Berl). 1977;37:111–8.

Bancher C, Brunner C, Lassmann H, Budka H, Jellinger K, Wiche G, et al. Accumulation of abnormally phosphorylated x precedes the formation of neurofibrillary tangles in Alzheimer's disease. Brain Res. 1989;477(1–2):90–9.

Barrett-Connor E, Dam TT, Stone K, Harrison SL, Redline S, et al. The association of testosterone levels with overall sleep quality, sleep architecture, and sleep-disordered breathing. J Clin Endocrinol Metab. 2008;93:2602–9.

Bethea CL, Pecins-Thompson M, Schutzer WE, Gundlah CLZN. Ovarian steroids and serotonin neural function. Mol Neurobiol. 1999;18:87–123.

Beyenburg S, Watzka M, Clusmann H, Blumcke I, Bidlingmaier F, Elger CE, et al. Androgen receptor mRNA expression in the human hippocampus. Neurosci Lett. 2000;294:25–8.

Beyer C, Hutchison JB. Androgens stimulate the morphological maturation of embryonic hypothalamic aromatase-immunoreactive neurons in the mouse. Dev Brain Res. 1997;98:74–81.

Bondareff W, Wischik CM, Novak M, Roth M. Sequestration of Tau by granulovacuolar degeneration in Alzheimer's disease. AJP. 1991;139(3):641–7.

Braskie MN, Klunder AD, Hayashi KM, Protas H, Kepe V, Miller KJ, et al. Plaque and tangle imaging and cognition in normal aging and Alzheimer's disease. Neurobiol Aging. 2010;31(10):1669–78.

Bremner WJ, Vitiello MV, Prinz PN. Loss of circadian rhythmicity in blood testosterone levels with aging in normal men. J Clin Endocrinol Metab. 1983;56:1278–81.

Brody H, Vijayashankar N. Anatomical changes in the nervous system. In: Finch CE, Hayflick L, editors. Handbook of the biology of aging. New York: Van Nostrand; 1977. p. 241–61.

Cammaleri R, Gangitano M, D'Amelio M, Raieli V, Raimondo D, Camarda R. Transient topographical amnesia and cingulated cortex damage: a case report. Neuropsychologia. 1996;34(4):321–6.

Cardenas VA, Chao LL, Studholme C, Yaffe K, Miller BL, Madison C, et al. Brain atrophy associated with baseline and longitudinal measures of cognition. Neurobiol Aging. 2011;32(4):572–80.

Chen Z, Maricic M, Nguyen P, Ahmann FR, Bruhn R, Dalkin BL. Low bone density and high percentage of body fat among men who were treated with androgen deprivation therapy for prostate carcinoma. Cancer. 2002;95:2136–44.

Cherrier MM, Craft S, Matsumoto AH. Cognitive changes associated with supplementation of testosterone or dihydrotestosterone in mildly hypogonadal men: a preliminary report. J Androl. 2003;24568–76

Cobo M, Expo'sito I, Mora F. Ageing, prefrontal cortex, and amino acid neurotransmitters: differential effects produced by electrical stimulation. Neurobiol Ageing. 1993;14:187–90.

Cobo M, Expo'sito I, Porras A, Mora F. Release of amino acid neurotransmitters in different cortical areas of conscious adult and aged rats. Neurobiol Ageing. 1992;13:705–9.

Cole GM, Timiras PS. Ubiquitin-protein conjugated in Alzheimer's disease. Neurosci Lett. 1987;79:207–12.

Cotman CW, Monaghan DT, Ottersen OP, Storm-Mathisen J. Anatomical organization of excitatory amino acid receptors and their pathways. Trends Neurosci. 1987;10:273–80.

Edinger KL, Frye CA. Testosterone's analgesic, anxiolytic, and cognitive-enhancing effects may be due in part to actions of its 5alpha-reduced metabolites in the hippocampus. Behav Neurosci. 2004;118:1352–64.

Finley S, Kritzer M. Immunoreactivity for intracellular androgen receptors in identified subpopulations of neurons, astrocytes and oligodendrocytes in primate prefrontal cortex. J Neurobiol. 1999;40:446–57.

Fjell AM, Walhovd KB, Reinvang I, Lundervold A, Salat D, Quinn BT, et al. Selective increase of cortical thickness in high-performing elderly—structural indices of optimal cognitive aging. NeuroImage. 2006;29:984–94.

Flood JF, Farr SA, Kaiser FE, La Regina M, Morley JE. Age related decrease of plasma testosterone in SAMP8 mice: replacement improves age related impairment of learning and memory. Physiol Behav. 1995;57(4):669–73.

Fowler CJ, Wiberg A, Oreland L, Marcussen J, Winbald B. The effect of age on the activity and molecular properties of human brain monoamine oxidase. J Neural Transm. 1980;49:1–20.

Fox SR, Harlan RE, Shivers BD, Pfaff DW. Chemical characterization of neuroendocrine targets for progesterone in the female rat brain and pituitary. Neuroendocrinology. 1990;51:276–83.

Frye C, Park D, Tanaka M, Rosellini R, Svare B. The testosterone metabolite and neurosteroid 3alpha-androstanediol may mediate the effects of testosterone on conditioned place preference. Psychoneuroendocrinology. 2001;26:731–50.

Galloway PG, Mulvihill P, Siedlak S, Magdalena M, Kawai M, Padget H, et al. Immunochemical demonstration of tropomyosin in the neurofibrillary pathology of Alzheimer's lesions. Am J Pathol. 1990;137:291–300.

Gentili A, Mulligan T, Godschalk M, Clore J, Patrie J, Iranmanesh A, et al. Unequal impact of short-termtestosterone repletion on the somatotropic axis of young and older men. J Clin Endocrinol Metab 2002;87: 825–834.

Ghoshal N, Garcia-Sierra F, Wuu J, Leurgans S, Bennett DA, Berry RW, et al. Tau conformational changes correspond to impairments of episodic memory in mild cognitive impairment and Alzheimer's disease. Exp Neurol. 2002;177:475–93.

Gibney J, Wolthers T, Johannsson G, Umpleby AM, Ho KK. Growth hormone and testosterone interact positively to enhance protein and energy metabolism in hypopituitary men. Am J Physiol Endocrinol Metab. 2005;289:266–71.

Glisky EL. Changes in cognitive function in human aging. In: Riddle DR, editor. Brain aging: model, methods and mechanisms. Boca Raton: CRC Press/Taylor and Francis; 2007. p. 1–34.

Grady CL, McIntosh AR, Horwitz B, Maisog JM, Ungerleider LG, Mentis MJ, et al. Age-related reductions in human recognition memory due to impaired encoding. Science. 1995;269(5221):218–21.

Hall TC, Miller AKH, Corsellis JAN. Variations in the human Purkinje cell population according to age and sex. Neuropathol Appl Neurobiol. 1975;1:267–92.

Hammond J, Le Q, Goodyer C, Gelfand M, Trifiro M, LeBlanc A. Testosterone-mediated neuroprotection through the androgen receptor in human primary neurons. J Neurochem. 2001;77:1319–26.

Head D, Kennedy KM, Rodrigue KM, Raz N. Age differences in perseveration: Cognitive and neuroanatomical mediators of performance on the Wisconsin Card Sorting Test. Neuropsychologia. 2009;47(4):1200–3.

Herzog AG, Kemper TL. Amygdaloid changes in aging and dementia. Arch Neurol. 1980;37:625–9.

Hijazi RA, Cunningham GR. Andropause: is androgen replacement therapy indicated for the aging male? Annu Rev Med. 2005;56:117–37.

Hof PR, Morrison JH. The ageing brain: morphomolecular senescence of cortical circuits. Trends Neurosci 2004; 27:607–613.

Hogervorst E, Combrinck M, Smith AD. Testosterone and gonadotropin levels in men with dementia. Neuroendocrinol Lett. 2003;24:203–8.

Huang X, Harlan RE. Absence of androgen receptors in LHRH immunoreactive neurons. Brain Res. 1993;624:309–11.

Hubbard BM, Anderson JM. Age, senile dementia and ventricular enlargement. J Neurol Neurosurg Psychiatry. 1981;44:631–5.

Iranmanesh A, South S, Liem AY, Clemmons D, Thorner MO, Weltman A, et al. Unequal impact of age, percentage body fat, and serum testosterone concentrations on the somatotrophic, IGF-I, and IGF-binding protein responses to a three-day intravenous growth hormone-releasing hormone pulsatile infusion in men. Eur J Endocrinol. 1998;139:59–71.

Joseph JA, Roth GS, Strong R. The striatum, a microcosm for the examination of age-related alterations in the CNS: a selected review. Rev Biol Res Aging. 1990;4:181–99.

Kaler LW, Neaves WB. Attrition of the human Leydig cell population with advancing age. Anat Rec. 1978;192:513–8.

Keenan DM, Takahashi PY, Liu PY, Roebuck PD, Nehra AX, Iranmanesh A, et al. An ensemble model of the male gonadal axis: illustrative application in aging men. Endocrinology. 2006;147:2817–28.

Kennedy KM, Raz N. Age, sex, and regional brain volumes predict perceptual-motor skill acquisition in healthy adults. Cortex Special Issue Cognition Ageing Brain. 2005;41(4):560–9.

Knutson B, Fong GW, Adams CM, Varner JL, Hommer D. Dissociation of reward anticipation and outcome with event-related fMRI. Neuroreport. 2001;12:3683–7.

Knutson B, Greer SM. Anticipatory affect: neural correlates and consequences for choice. Philos Trans R Soc B. 2008;363:3771–86.

Knutson B, Katovich K, Suri G. Inferring affect from fMRI data. Trends Cogn Sci. 2014;18:422–8.

Kousteni S, Bellido T, Plotkin LI, O'Brien CA, Bodenner DL, Han L, et al. Nongenotropic, sex-nonspecific signaling through the estrogen or androgen receptors: dissociation from transcriptional activity. Cell. 2001;104:719–30.

Kritzer M, McLaughlin P, Smirlis T, Robinson J. Gonadectomy impairs T-maze acquisition in adult male rats. Horm Behav. 2001;39:167–74.

Ksiezak H, Yen SH. Two monoclonal antibodies recognize Alzheimer's neurofibrillary tangles, neurofilament and microtubule- associated proteins. J Neurochem. 1987;48:455–62.

Lehman MN, Karsch FJ. Do gonadotropin-releasing hormone, tyrosine hydroxylase-, and b-endorphin-immunoreactive neurons contain oestrogen receptors? A double-label immunocytochemical study in the Suffolk ewe. Endocrinology. 1993;133:887–95.

Leranth C, Petnehazy O, MacLusky NJ. Gonadal hormones affect spine synaptic density in the CA1 hippocampal subfield of male rats. J Neurosci. 2003;23:1588–92.

Lindenberger U, Mayr U, Kliegl R. Speed and intelligence in old age. Psychol Aging. 1993;8:207–20.

Luboshitzky R, Zabari Z, Shen-Orr Z, Herer P, Lavie P. Disruption of the nocturnal testosterone rhythm by sleep fragmentation in normal men. J Clin Endocrinol Metab. 2001;86:1134–9.

Mann DMA, Yates PO. The effects of ageing on the pigmented nerve cells of the human locus caeruleus and substantia nigra. Acta Neuropathol (Berl). 1979;47:93–7.

Marchal G, Rioux P, Petit-Taboué MC, Marchal G, Rioux P, Petit-Taboué MC, et al. Regional cerebral oxygenconsumption, blood flow, and blood volume in healthy human aging. Arch Neurol. 1992;49:1013–20.

Marschner A, Mell T, Wartenburger I, Villringer A, Reischies FM, Heekeren HR. Reward-based decision-making and aging. Brain Res Bull. 2005;67:382–90.

Martin AJ, Friston KJ, Colebatch JG, Frackowiak RS. Decreases in regional cerebral blood flow with normal aging. J Cereb Blood Flow Metab. 1991;11:684–9.

Matsumoto AM. Testis and male sexual function. In: Smith LH, Bennet JC, editors. Cecil textbook of medicine. Philadelphia: WB Saunders; 1991. p. 1333–50.

Mattson M, Maudsley S, Martin B. BDNF and 5-HT: a dynamic duo in age related neuronal plasticity and neurodegenerative disorders. Trends Neurosci. 2004;27:589–94.

McEwen BS, Gould E, Orchinik M, Weiland NG, Woolley CS. Oestrogens and the structural and functional plasticity of neurons: implications for memory, ageing and neurodegenerative processes. In: Bock GR, Goode JA, editors. Ciba Foundation Symposium 191- non-reproductive actions of sex steroids, vol. 191. Chichester: John Wiley & Sons, Ltd; 1995. p. 52–73.

McGeer E, McGeer PL. Neurotransmitter metabolism in the aging brain. In: Terry RD, Gershon S, editors. Neurobiology of aging, vol. 3. New York: Raven; 1976. p. 387–403.

McGeer PL, McGeer EG, Suzuki J, Dolman CE, Nagai T. Aging, Alzheimer's disease and the cholinergic system of the basal forebrain. Neurology. 1984;34:741–5.

Meites J. Aging: hypothalamic catecholamines, neuroendocrine immune interations and dietary restriction. Proc Soc Exp Biol Med. 1990;195:304–11.

Miller AKH, Alston RL, Corsellis JAN. Variation with age in the volumes of grey and white matter in the cerebral hemispheres of man: measurements with an image analyser. Neuropathol Appl Neurobiol. 1980;16:119–32.

Moffat SD, Zonderman AB, Metter EJ, Kawas C, Blackman MR, Harman SM, et al. Free testosterone and risk for Alzheimer disease in older men. Neurology. 2004;62:188–93.

Moller M, Jakobsen S, Gjedde A. Parametric and regional maps of free serotonin 5HT1A receptor sites in human brain as function of age in healthy humans. Neuropsychopharmacology. 2007;32:1707–14.

Mora F, Cobo M. The neurobiological basis of prefrontal cortex self-stimulation: a review and an integrative hypothesis. In: Uylings HBM, Van Eden CG, De Bruin JPC, Comer MA, Feenstra MGP, editors. The prefrontal cortex: its function, structure and pathology — progress in brain research, vol. 85. Amsterdam: Elsevier; 1990. p. 409–31.

Morley JE. Androgens and aging. Maturitas. 2001;38:61–71.

Morley JE, Perry HM, Kaiser FE, Kraenzel D, Jensen J, Houston K, et al. Effects of testosterone replacement therapy in old hypogonadal males: a preliminary study. J Am Geriatr Soc. 1993;41:149–52.

Mountjoy CQ, Roth M, Evans NJR, Evans HM. Cortical neuronal counts in normal elderly controls and demented patients. Neurobiol Aging. 1983;4:1–11.

Mukherjee J, Christian BT, Dunigan KA, Shi B, Narayanan TK, Satter M, et al. Brain imaging of 18F-fallypride in normal volunteers: blood analysis, distribution, test-retest studies, and preliminary assessment of sensitivity to aging effects on dopamine D-2/D-3 receptors. Synapse. 2002;46(3):170–88.

Muller M, Aleman A, Grobbee DE, de Haan EH, van der Schouw YT. Endogenous sex hormone levels and cognitive function in aging men: is there an optimal level? Neurology. 2005;64:866–71.

Neisser U. Cognitive psychology. In: Cognitive approach. New York/London: Psychology Press, Taylor and Francis group; 1967. p. 3–12.

Nguyen TV, Yao M, Pike CJ. Androgen activates mitogen-activated protein kinase signaling: role in neuroprotection. J Neurochem. 2005;94:1639–51.

Oh MM, Kim JW, Jin MH, Kim JJ, Moon du G. Influence of paradoxical sleep deprivation and sleep recovery on testosterone level in rats of different ages. Asian J Androl. 2012;14:330–4.

Ohayon MM. Sleep and the elderly. J Psychosom Res. 2004;56(5):463–4.

Palmer AM. Neurocheminal studies of Alzheimer's disease. Neurodegeneration. 1996;5:381–91.

Penev PD. Association between sleep and morning testosterone levels in older men. Sleep. 2007;30:427–32.

Pike CJ. Testosterone attenuates b-amyloid toxicity in cultured hippocampal neurons. Brain Res. 2001;919:160–5.

Pike KE, Savage G, Villemagne VL, Ng S, Moss SA, Maruff P, et al. Beta-amyloid imaging and memory in nondemented individuals: evidence for preclinical Alzheimer's disease. Brain. 2007;130:2837–44.

Rajah MN, D'Esposito M. Region-specific changes in prefrontal function with age: a review of PET and fMRI studies on working and episodic memory. Brain. 2005;128:1964–83.

Ramsden M, Nyborg AC, Murphy MP, Chang L, Stanczyk FZ, Golde TE, Pike CJ. Androgens modulate beta-amyloid levels in male rat brain. J Neurochem. 2003;87:1052–5.

Ranga-Krishnan K, Husain MM, McDonald WM, Doraiswamy PM, Figiel GS, Boyko OB, et al. In vivo stereological assessment of caudate volume in man: effect of normal ageing. Life Sci. 1990;47:1325–9.

Raz N, Briggs SD, Marks W, Acker JD. Age-related deficits in generation and manipulation of mental images: II. The role of dorsolateral prefrontal cortex. Psychol Aging. 1999;14:436–44.

Redline S, Kirchner HL, Quan SF, Gottlieb DJ, Kapur V, Newman A. The effects of age, sex, ethnicity, and sleep-disordered breathing on sleep architecture. Arch Intern Med. 2004;164(4):406–18.

Rehman HU, Masson EA. Neuroendocrinology of ageing. Age Ageing. 2001;30:279–87.

Rinne JO. Muscarininc and dopaminergic receptors in aging human brain. Brain Res. 1987;404:161–8.

Robinson DS, Nies A, Davis JN, Bunney WE, Davis JM, Colburn RW, et al. Aging monoamines and monoamines oxidase. Lancet. 1972;1:290–1.

Rodrigue KM, Kennedy KM, Park DC. Beta-amyloid deposition and the aging brain. Neuropsychol Rev. 2009;19:436–50.

Roof RL, Havens MD. Testosterone improves maze performance and induces development of a male hippocampus in females. Brain Res. 1992;572:310–3.

Roof RL. The dentate gyrus is sexually dimorphic in prepubescent rats: testosterone plays a significant role. Brain Res. 1993;610:148–51.

Rosario ER, Chang L, Stanczyk FZ, Pike CJ. Age related testosterone depletion and the development of Alzheimer disease. J Am Med Assoc. 2004;292:1431–2.

Roselli C, Klosterman S, Resko J. Anatomic relationships between aromatase and androgen receptor mRNA expression in the hypothalamus and amygdala of adult male ynomolgus monkeys. J Comp Neurol. 2001;439:208–23.

Russell-Aulet M, Jaffe CA, Mott-Friberg R, Barkan AL. In vivo semiquantification of hypothalamic growth hormone-releasing hormone (GHRH) output in humans: evidence for relative GHRH deficiency in aging. J Clin Endocrinol Metab. 1999;84(10):3490–7.

Salat DH, Buckner RL, Snyder AZ, Greve DN, Desikan RS, Busa E, et al. Thinning of the cerebral cortex in aging. Cereb Cortex. 2004;14:721–30.

Salat DH, Tuch DS, Greve DN, van der Kouwe AJ, Hevelone ND, Zaleta AK, et al. Age-related alterations in white matter microstructure measured by diffusion tensor imageing. Neurobiol Ageing. 2005;26:1215–27.

Schmidt WJ, Bubser M, Hauber W. Behavioural pharmacology of glutamate in the basal ganglia. J Neural Transm. 1992;38:65–89.

Simchowicz T. Histopathologische Studien, uber die Senile Demenz. In: Nissl F, Alzheimer A, editors. Histologie und Histopathologische Arbiten uber die Grosshirnrinde. Jena: Fisher; 1911. p. 267–444.

Squire L, Zola S. Amnesia memory and brain systems. Philos Trans R Soc Lond Ser B Biol Sci. 1997;352:1663–73.

Suhara T, Fukuda H, Inoue O, Itoh T, Suzuki K, Yamasaki T, et al. Age-related changes in human D1 dopamine receptors measured by positron emission tomography. Psychopharmacology (Berl). 1991;103:41–5.

Tapp PD, Siwak CT, Gao FQ, Chiou JY, Black SE, Head E, et al. Frontal lobe volume, function, and beta-amyloid pathology in a canine model of aging. J Neurosci. 2004;24:8205–13.

Tenover JS. Androgen administration to ageing men. Endocrinol Metab Clin N Am. 1994;23(4):877–92.

Terry RD, Wisniewski H. The ultrastructure of the neurofibrillary tangle and the senile plaque. In: Wolstenholme GEW, O'Connor M, editors. Alzheimer's Disease and Related Conditions, a Ciba Foundation Symposium. London: Churchill; 1970. p. 145–68.

Tohgi H, Utsugisawa K, Yamagata M, Yoshimura M. Effects of age on messenger RNA expression of glucocorticoid, thyroid hormone, androgen, and estrogen receptors in postmortem human hippocampus. Brain Res. 1995;700:245–53.

Tomlinson BE, Kitchener D. Granulovacuolar degeneration of hippocampal pyramidal cells. J Pathol. 1972;106:165–85.

Tomlinson BE, Blessed G, Roth M. Observations on the brains of nondemented old people. J Neurol Sci. 1968;7:331–56.

Tortonese DJ. Interaction between hypothalamic dopaminergic and opioidergic systems in the photoperiodic regulation of pulsatile luteinizing hormone secretion in sheep. Endocrinology. 1999;140:750–7.

Urban RJ, Bodenburg YH, Gilkison C, Foxworth J, Coggan AR, Wolfe RR, et al. Testosterone administration to elderly men increases skeletal muscle strength and protein synthesis. Am J Phys. 1995;269:E820–6.

Volkow ND, Ding YS, Fowler JS, Wang GJ, Logan J, Gatley SJ, et al. Dopamine transporters decrease with age in healthy subjects. J Nucl Med. 1996a;37:554–8.

Volkow ND, Wang GJ, Fowler JS, Logan J, Gatley SJ, MacGregor RR, et al. Measuring age-related changes in DA D2 receptors with [11C]raclopride and with [18F]N-methylspiroperidol. Psychiatry Res. 1996b;67:11–6.

Weltman A, Weltman JY, Hartman ML, Abbott RD, Rogol AD, Evans WS, et al. Relationship between age, percentage body fat, fitness, and 24-hour growth hormone release in healthy young adults: effects of gender. J Clin Endocrinol Metab. 1994;78:543–8.

West RL. An application of prefrontal cortex function theory to cognitive aging. Psychol Bull. 1996;120:272–92.

Winters SJ, Troen P. Episodic luteinizing hormone (LH) secretion and the response of LH and follicle-stimulating hormone to LH-releasing hormone in aged men: evidence for coexistent primary testicular insufficiency and an impairment in gonadotropin secretion. J Clin Endocrinol Metab. 1982;55:560–5.

Woodard JS. Clinicopathological significance of granulovacuolar degeneration in Alzheimer's disease. J Neuropathol Exp Neurol. 1962;21:85–91.

Xu J, Kobayashi S, Yamaguchi S, Iijima K, Okada K, Yamashita K. Gender effects on age related changes in brain structure. AJNR Am J Neuroradiol. 2000;21:112–8.

Yamamoto M, Suhara T, Okubo Y, Ichimiya T, Sudo Y, Inoue M, et al. Age-related decline of serotonin transporters in living human brain of healthy males. Life Sci. 2002;71:751–7.

Yen SH, Gaskin F, Fu SM. Neurofibrillary tangles in senile dementia of the Alzheimer type share an antigenic determinant with intermediate filaments of the vimentin class. Am J Pathol. 1983;113:373–81.

Zeitler P, Argente J, Chowen-Breed JA, Clifton DK, Steiner RA. Growth hormone releasing hormone messenger ribonucleic acid in the hypothalamus of the adult male rat is increased by testosterone. Endocrinology. 1990;127:1362–8.

Zhu Y, Yang GY, Ahlemeyer B, Pang L, Che XM, Culmsee C, et al. Transforming growth factor-b1 increases bad phosphorylation and protects neurons against damage. J Neurosci. 2002;22:3898–909.

Zubenko GS, Moossy J, Kopp U. Neurochemical correlates of major depression in primary dementia. Arch Neurol. 1990;47:209–14.

Part VI

Systemic Influences on Male Reproduction

Immune System

21

Surabhi Gupta and Anand Kumar

Learning Objectives

- Immune cells and cytokines in the testis
- Male reproduction in infections and autoimmune diseases
- Thymus and male reproduction

21.1 Introduction

Both immune and male reproductive systems influence each other. The time line of development for both is different. Spermatogenesis in human males starts only around the time of puberty. In contrast, the process of establishment of self-tolerance by the immune system is mostly completed during the fetal period. The germ cells in males are at the spermatogonial stage at that time, and hence, the unique antigens present on the mature sperm are unavailable to the immune system for establishing self-tolerance. Hence, the cell-specific molecules on the spermatogenic cells pose a challenge to the immune system as they are perceived as "foreign" antigens. In spite of this, continuous production of millions of sperm proceeds without any apparent problem in majority of males. Breakdown of the normal testicular physiology can lead to autoimmune infertility. Hence, a unique immunological privilege is maintained for the male reproductive tract. The main constituent for this is the

S. Gupta, PhD (✉) • A. Kumar, MD, FAMS
Department of Reproductive Biology, All India Institute of Medical Sciences, New Delhi, India
e-mail: surabhi72@rediffmail.com

A. Kumar, M. Sharma (eds.), *Basics of Human Andrology*,
DOI 10.1007/978-981-10-3695-8_21

blood-testis barrier that is made up physically by the tight junctions between the Sertoli cells of the testis. These tight junctions are impermeable to all cells, and hence, the developing and mature sperm towards the lumen are protected from any kind of immune attack. The details of the blood-testis barrier have been discussed in the chapter, The Testis. Testicular cells, such as Sertoli cells, Leydig cells, testicular macrophages, and regulatory T lymphocytes, also contribute to the immune privilege status of the testis by producing anti-inflammatory cytokines and other immunosuppressive factors.

This chapter will focus on how the components of immune system interact with the male reproductive system with emphasis on immunoregulation of the normal testicular functions of steroidogenesis and spermatogenesis.

21.2 Immune Cells in the Testis

The cellular components of the innate immune system include monocytes or macrophages, neutrophils, and natural killer (NK) cells, whereas the adaptive immune system uses T lymphocytes, B lymphocytes, and the antigen-presenting macrophages and dendritic cells. A simple histology of the testicular cross-section shows a large number of leukocytes in the interstitium. Majority of these leukocytes are macrophages followed by lymphocytes and granulocytes (see Table 21.1). Dendritic cells and mast cells are also present though in much lower numbers (Pollanen and Niemi 1987; el-Demiry and James 1988; Hedger 2015).

21.2.1 Testicular Macrophages and Dendritic Cells

The most abundant immune cell type in the testis are the macrophages. Macrophages in the testis have a hematopoietic origin as they are derived from the circulating blood monocytes which migrate into the testicular interstitium during the fetal period. In fetal rat testis, their presence has been reported as early as 16th day

Table 21.1 Immune cells in the human testis

Immune cells in testis	Abundance in adult testis
Phagocytic and Antigen-presenting cells (APCs)	
Macrophages	10–25 million/g tissue
Dendritic cells	Present
Lymphocytes	
T cells ($CD8^+$ and $CD4^+$)	1.4–2.4 million/g tissue
Natural Killer cells	1.0–2.8 million/g tissue
Granulocytes	
Mast cells	Present
Neutrophils	Present only if testicular inflammation or damage occurs

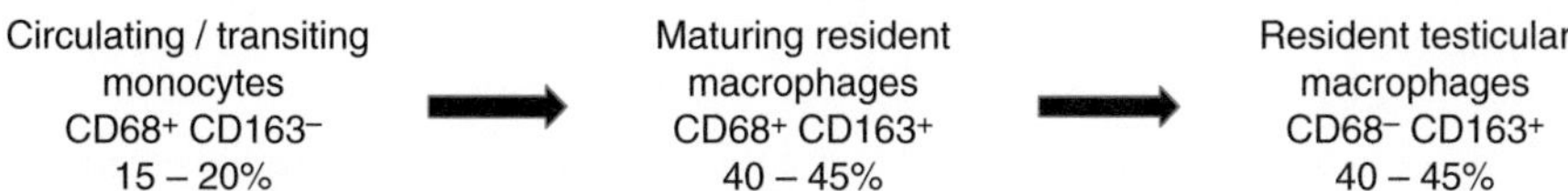

Fig. 21.1 Heterogeneity of testicular macrophages

post-conception (Livera et al. 2006). However, they gradually lose their monocyte phenotypic markers and acquire resident macrophage markers (Hedger 2002). Even in adult testis, a heterogeneous population of macrophages exists (see Fig. 21.1). Majority of them have the surface marker cluster of differentiation 163 (CD163) which belongs to the group B scavenger receptor cysteine-rich protein superfamily (van den Berg et al. 2001). CD163 is a marker of tissue-fixed or resident macrophages in other non-lymphoid tissues. In rat testis, it is recognized by the monoclonal antibody ED2 (Barbe et al. 1990). However, around 15–20% of the testicular macrophages express the monocytic marker, CD68. CD68 is a lysosomal antigen, which is recognized by the monoclonal antibody ED1. It is hypothesized that the $CD68^{+}$ population represents the recently arrived circulating macrophages, which gradually acquire CD163 expression. This change in surface marker is accompanied by changes in their functional properties. The $CD68^{+}$ testicular macrophages have a smaller nuclear diameter and produce the pro-inflammatory cytokines interleukin 1 beta (IL1β) and tumor necrosis factor alpha (TNF-α). In contrast, the $CD163^{+}$ testicular macrophages have increased nuclear volume with a loss of ability to produce pro-inflammatory cytokines and constitutively produce the anti-inflammatory IL-10.

The immunosuppressive properties of the resident testicular macrophages contribute to the immune privileged status of the testis. They express the MHC class II molecules required for antigen presentation to T helper cells, but lack the co-stimulatory molecules CD80 and CD86, which are required to activate T cells (Pollanen and Niemi 1987; Haas et al. 1988; Sainio-Pollanen et al. 1996). The immunosuppressive role of the testicular macrophages is also supported by the observation that the inflammatory response of the testis is enhanced upon depletion of resident macrophages (Bergh et al. 1993).

Several lines of evidence support the idea that development of testicular macrophages is dependent on Leydig cells (Hedger 2002). The number of testicular macrophages increases at the time of puberty, which coincides with proliferation of adult Leydig cells (Hardy et al. 1989; Raburn et al. 1993; Vergouwen et al. 1993). Selective destruction of Leydig cells by ethane dimethane sulfonate (EDS) leads to reduction in the macrophage number (Wang et al. 1994). Leydig cells are responsible for recruitment and proliferation of macrophages, while Sertoli cells induce maturation as evidenced by increase in the macrophage nuclear volume in response to follicle-stimulating hormone (FSH). Normal development of the testicular macrophages is dependent on the growth factor colony-stimulating factor 1 (CSF1). Leydig cells produce CSF1, and its absence in osteopetrotic mice ($csfm^{op}/csfm^{op}$) leads to reduced male fertility (Cohen et al. 1996).

Testicular macrophages have close physical interaction with Leydig cells in the form of inter-cytoplasmic digitations (Hutson 2006). The ratio of macrophages to

Leydig cells in rodent testis is fixed at 1:4 or 1:5 (Hutson 1994). This suggests functional association between the two cell types. The 25-hydroxycholesterol produced and secreted by the macrophages can be used by Leydig cells as precursor for pregnenolone which is ultimately converted to testosterone (Nes et al. 2000; Lukyanenko et al. 2001). This acts as an alternate pathway for steroidogenesis since it can bypass the steroidogenic acute regulatory (StAR) protein-dependent movement of cholesterol into the mitochondria (Lukyanenko et al. 1998). On the other hand, these macrophages also have a negative effect on steroidogenesis which is mediated by the production of nitric oxide (NO), TNF-α, and IL-1 (Hales 2002). A number of studies have shown that functional macrophages are required for the development of Leydig cells in prepubertal rats as well as for regeneration after EDS treatment (Gaytan et al. 1994a, b).

Dendritic cells are also seen throughout the interstitium but at much lower frequency than macrophages (Haas et al. 1988; Hoek et al. 1997; Rival et al. 2006). They can be distinguished by their distinctive morphology and express specific markers such as integrin αx (CD11c) and integrin αE2. Unlike macrophages, they function primarily as antigen-presenting cells other than functioning as phagocytes. They play a role in regulating immune response and tolerance in normal testis (Sainio-Pollanen et al. 1996; Rival et al. 2007; Guazzone et al. 2011).

21.2.2 Testicular Lymphocytes

In human testicular interstitium, around 10–20% of total leukocytes are lymphocytes (Pollanen and Niemi 1987; Hedger 2015). The intratesticular lymphocyte population has not been characterized intensively but most studies show the presence of cytotoxic $CD8^+$ T cells, $CD4^+$ T helper cells and NK cells but no B cells have been reported (Wang et al. 1994; Hedger and Meinhardt 2000). In circulation, there is a predominance of the $CD4^+$ T helper cells; however, the testicular lymphocytes are predominantly $CD8^+$ T cells and NK cells. The physiological relevance of this altered ratio of $CD8^+$:$CD4^+$ (testis vs. circulation) is not known. The presence of $FOXP3^+$ regulatory T cells has been reported recently (Duan et al. 2011). Their presence under physiological conditions suggests their involvement in testicular immune privilege.

21.2.3 Testicular Granulocytes

Mast cells are present in the normal human testis, but neutrophils are only detected if testis inflammation or damage occurs (Nistal et al. 1984, 1986; Anton et al. 1998; Welter et al. 2011). The mast cells are distributed throughout the interstitium in human testis, while in rodent testis, they associate with blood vessels in the testicular capsule region only. However, depletion of Leydig cells by EDS in rats leads to increased proliferation of mast cells throughout the interstitium (Gaytan et al. 1992). This effect may be mediated by the Stem Cell Factor (SCF) produced by the Sertoli cells which acts as a growth factor for both Leydig cells and mast cells (Maurer et al. 1998; Yan et al. 2000).

Table 21.2 Interplay of cytokines in the testis

Cytokine	Produced by	Acts on
Interleukin-1α (IL-1α)	Sertoli cells, germ cells, macrophages	Sertoli cells, Leydig cells
Interleukin-1β (IL-1β)	Macrophages, Leydig cells	Leydig cells
Interleukin-6 (IL-6)	Sertoli cells, Leydig cells, macrophages	Sertoli cells, preleptotene spermatocytes
Tumor necrosis factor-α (TNF- α)	Germ cells, macrophages	Leydig cells, Sertoli cells, peritubular cells
Tumor growth factor- β (TGF- β)	Peritubular cells, Sertoli cells, Leydig cells, germ cells	Leydig cells, germ cells
Interferon- γ (IFN-γ)	Sertoli cells, Leydig cells, peritubular cells	Leydig cells
Leukemia inhibitory factor (LIF)	Peritubular cells	Primordial germ cells, Sertoli cells, Leydig cells
Macrophage migration inhibitory factor (MIF)	Leydig cells, Sertoli cells (?)	Sertoli cells, peritubular cells
Colony-stimulating factor-1 (CSF-1)	Leydig cells	Macrophages

21.3 Cytokines in the Testis

The secreted components of the immune system include various cytokines and growth factors. In the testis, these are produced by the resident immune cells as well as the other cells of the testis like Leydig cells, Sertoli cells, as well as germ cells themselves. The cytokines influence the testicular functions in both autocrine and paracrine manner. Table 21.2 provides a list of the major cytokines found in the testis and the cells producing them. A brief description of the cytokine families expressed in the testis and how they influence the testicular functions follows.

21.3.1 Interleukin-1 Family

The two major forms of IL-1, namely IL-1α and IL-1β, have about 25% homology at the mature protein level and bind to the same receptor. Both are single-chain proteins which are synthesized as 31–33 kDa precursor proteins but get cleaved enzymatically to produce the 17 kDa active form. While both precursor and mature IL-1α are bioactive, only the mature form of IL-1β is active. Though both IL-1α and IL-1β are secreted by the cells producing them, the majority of IL-1α produced is retained within the cell and hence mostly acts as an autocrine growth factor. IL-1α is also found associated with the cell membrane and may mediate direct cell-to-cell communication.

Expression of IL-1α and IL-1β has been found in the testis of both rodents and humans (Abu Elheija et al. 2011). In the rat testis, IL-1α is produced by Sertoli cells, and its production starts around postnatal day 20, which coincides with progression of the first wave of spermatogenesis (Gerard et al. 1991). It is synthesized in two forms: the mature form of 17 kDa and a testis-specific form of 24 kDa. The testis-specific isoform lacks the calpain cleavage site, which is necessary for the processing of the precursor protein, and is produced by an altered mRNA molecule (Sultana et al. 2000, 2003). Although both of these forms are biologically active, the 17 kDa form is more potent (Sultana et al. 2000; Svechnikov et al. 2001). Constitutive expression of IL-1α is seen in rat spermatocytes and spermatids also (Haugen et al. 1994). Phagocytosis of the residual cytoplasm by the spermatids stimulates Sertoli cells to produce IL-1α (Gerard et al. 1992; Syed et al. 1995). Sertoli cells from adult rats can produce and secrete bioactive form of IL-1α, whereas Sertoli cells of the prepubertal rats produce but are not able to secrete IL-1α (Gerard et al. 1991; Winnall et al. 2009).

Under normal conditions, IL-1β is not produced in significant amounts, but during inflammation, it is upregulated by the Leydig cells and a subset of testicular macrophages (Gerard et al. 1991; O'Bryan et al. 2005). Interestingly, majority of the secreted testicular IL-1β is present as the precursor protein (O'Bryan et al. 2005).

IL-1 regulates Leydig cell steroidogenesis, spermatogenesis-related Sertoli cell functions, as well as development of spermatogonia and spermatocyte. Proliferation of Leydig cells during prepubertal development is regulated by macrophage-derived IL-1β (Khan et al. 1992). Due to differences in experimental conditions used, the in vitro effect of IL-1 on mature Leydig cell steroidogenesis shows stimulation, inhibition, or no effect. However, the consensus suggests that the basal testosterone production of IL-1 is either stimulated or not affected, whereas the luteinizing hormone (LH)/human chorionic gonadotropin (hCG) and/or cyclic adenosine monophosphate (cAMP)-stimulated testosterone production is inhibited (Calkins et al. 1988; Verhoeven et al. 1988; Moore and Moger 1991; Hales 2002). The major site of inhibition is the expression of CYP17A, while CYP11A expression is also regulated by IL-1 (Lin et al. 1991; Hales 1992; Xiong and Hales 1997). However, gene expression and protein synthesis of STAR are not affected (Lin et al. 1998b). The variant of IL-1α involved and the developmental stage of Leydig cell also determine the effect produced. While the precursor 32 kDa and mature 17 kDa IL-1α inhibit LH/cAMP-stimulated steroid production in mature Leydig cells, there is no effect of the testis-specific 24 kDa variant. However, in immature rat Leydig cells, testosterone production is stimulated by all the three forms of IL-1α (Svechnikov et al. 2001).

IL-1α stimulates mitotic DNA synthesis in intermediate and B-type spermatogonia and meiotic DNA synthesis in preleptotene spermatocytes (Pollanen et al. 1989; Parvinen et al. 1991; Soder et al. 1991). Proliferation of prepubertal Sertoli cells is stimulated by IL-1 in vitro (Petersen et al. 2005). It inhibits aromatase activity, but stimulates the production of lactate, transferrin, and glucose transporters in mature Sertoli cells, thereby controlling spermatogenesis (Khan and Nieschlag 1991; Hoeben et al. 1996; Nehar et al. 1998; Galardo et al. 2008). IL-1 also alters the Sertoli cell cytoskeleton and stimulates opening of the occluding junctions of the blood-testis barrier (Lie et al. 2011).

Interlukin-18 (IL-18) also belongs to the IL-1 family and is most closely related to IL-1α and IL-1β but acts through a distinct IL-18 receptor (Gracie et al. 2003).

IL-18 has been localized in human testis, but its expression is lower in impaired spermatogenesis (Komsky et al. 2012). IL-18 and its receptor are present in the seminiferous epithelium of the rat, while IL-18 is localized to spermatocytes and round spermatids (Strand et al. 2005).

21.3.2 Tumor Necrosis Factor Family (TNF)

TNF, a 17 kDa glycosylated polypeptide, can exert pro-inflammatory or cytotoxic effects depending upon the receptor subtypes present in the target cells. TNF is present in round spermatids, pachytene spermatocytes, and testicular macrophages in mice (De et al. 1993). Bioactive TNF is produced by round spermatids, and its receptor is located on Sertoli and Leydig cells in vitro.

TNF inhibits Leydig cell steroidogenesis by inhibiting LH binding, STAR expression, and cholesterol mobilization (Mauduit et al. 1998; Morales et al. 2003). In cultured mouse Leydig cells, TNF suppresses basal expression of HSD3b and cAMP-induced expression of CYP11A, CYP17A, and HSD3b (Xiong and Hales 1993, 1997). After puberty, the testis of TNF knockout mice shows significantly lower testosterone levels and mRNA levels of steroidogenesis-related genes compared to wild-type testis. However this result is attributed to increased levels of anti-Mullerian hormone in these mice (Suh et al. 2008).

Similar to IL-1, TNF also stimulates the spermatogenesis-supporting functions of the Sertoli cells including lactate production, aromatase activity, and inhibin production (Nehar et al. 1997; Riera et al. 2001; Kazutaka et al. 2011). It also stimulates androgen receptor expression in the Sertoli cells by upregulating NF-κB (Delfino et al. 2003). TNF inhibits the production of occludin, a tight junction protein, hence disrupting the inter-Sertoli cell tight junction assembly (Siu et al. 2003; Li et al. 2006).

Men having single nucleotide polymorphisms in the TNF gene cluster produce increased TNF levels which leads to decreased sperm count and sperm motility, resulting in infertility (Tronchon et al. 2008). In support of this, cytotoxic injury to the Sertoli cells leads to the secretion of TNF by germ cells, which disrupts spermatogenesis (Yao et al. 2007). Similarly, activation of NF-kB in the Sertoli cells or in the germ cells induces germ cell apoptosis in various testicular damage models (Pentikainen et al. 2002; Lysiak et al. 2005; Rasoulpour and Boekelheide 2005; Mizuno et al. 2009). In contrast, TNF inhibits germ cell apoptosis and promotes cell survival during spermatogenesis in cultured human and rat seminiferous tubules possibly by regulating FAS ligand or NF-kB levels in the Sertoli cells (Pentikainen et al. 2001; Suominen et al. 2004).

Another member of the TNF family is the death receptor, Fas, and its ligand (FasL). The binding of Fas-FasL on activated T cells is necessary for immune response (Ju et al. 1995). In the testis, Fas and FasL regulate germ cell apoptosis during testicular damage (Lee et al. 1997). A role of Fas/FasL in maintaining immune privilege has also been proposed. FasL is constitutively expressed on Sertoli cells and can interact with the Fas receptor on activated inflammatory T cells leading to apoptosis of the T cells (Bellgrau et al. 1995).

21.3.3 Interleukin-6

IL-6 can have both pro-inflammatory and anti-inflammatory actions. In vitro, IL-6 is constitutively secreted by Sertoli cells and Leydig cells (Boockfor et al. 1994; Syed et al. 1995; Cudicini et al. 1997). Production of IL-1α by Sertoli cells in response to phagocytosis of residual cytoplasm shed by spermatids, in turn, stimulates IL-6 secretion in an autocrine manner (Syed et al. 1995). Production of IL-6 in Sertoli cells is also stimulated by FSH, testosterone, TNF, and lipopolysaccharide (LPS), but it is inhibited by IFN-γ (Stephan et al. 1997; Riccioli et al. 2000). Leydig cells are major producers of IL-6 in the testis and can be stimulated by LH, LPS, or IL-1 in vitro (Boockfor et al. 1994; Cudicini et al. 1997). Human peritubular cells express IL-6 mRNA when stimulated by TNF (Schell et al. 2008). Germ cells isolated from adult mice also express IL-6 mRNA and protein (Wang et al. 2012).

IL-6 increases the basal and FSH-induced transferrin and cyclic guanosine monophosphate (cGMP) secretion by Sertoli cells (Boockfor and Schwarz 1991; Hoeben et al. 1997). It inhibits meiotic DNA synthesis in preleptotene spermatocytes (Hakovirta et al. 1995). Suppression of Leydig cell steroidogenesis has been shown in human volunteers injected subcutaneously with recombinant IL-6. A prolonged suppression of serum testosterone levels without appreciable change in gonadotropin levels suggests inhibition at testicular level only (Tsigos et al. 1999).

21.3.4 Transforming Growth Factor-β (TGF-β)

Similar to IL-6, TGF-β also shows both pro-inflammatory and anti-inflammatory properties. It exists in three forms – TGF-β1, TGF-β2, and TGF-β3 – which are disulfide-linked dimers of approximately 25 kDa. All the three forms are differentially produced by Sertoli cells, Leydig cells, and peritubular cells in the fetal and in the immature testis (Teerds and Dorrington 1993). However, the production reduces considerably during puberty (Mullaney and Skinner 1993; Konrad et al. 2000). It is also localized to spermatogenic cells in the postpubertal testis (Teerds and Dorrington 1993).

TGF-β controls the development and apoptosis of germ cells, differentiation of Leydig cells, and interactions of Sertoli cell-peritubular cell in the fetal and immature testis (Olaso et al. 1998; Konrad et al. 2000; Dickson et al. 2002; Gonzalez et al. 2010). In the adult testis also, stimulation of apoptosis and inhibition of meiosis are mediated by TGF-β during spermatogenesis (Maire et al. 2005; Perrard and Durand 2009; Moreno et al. 2010). It also has an inhibitory effect on Leydig cell steroidogenesis by reducing the number of LH receptor as well as lowering expression of CYP17A (Avallet et al. 1987; Chuzel et al. 1996). TGF-β also regulates the permeability of the blood-testis barrier by regulating Sertoli cell tight junction dynamics (Lui et al. 2003).

21.3.5 Interferons

Viral infections stimulate interferon-gamma (IFN-γ) production by Sertoli cells, peritubular cells, Leydig cells, and testicular macrophages (Dejucq et al. 1998).

Table 21.3 Modulation of steroidogenesis by various cytokines

Cytokine	Mode of action	References
Interleukin-1α (IL-1α)	Biphasic – inhibits LH-stimulated testosterone production but stimulates basal steroidogenesis; inhibits P450 enzymes	Verhoeven et al. (1988), Lin et al. (1991), Moore and Moger (1991), Hales (1992), Sun et al. (1993), Svechnikov et al. (2001)
Interleukin-6 (IL-6)	Inhibits steroidogenesis	Tsigos et al. (1999)
Tumor necrosis factor-α (TNF- α)	Inhibits steroidogenesis	Mealy et al. (1990), Xiong and Hales (1993)
Tumor growth factor- β (TGF- β)	Biphasic effect on testosterone production – stimulatory at low dose and inhibitory at higher doses; inhibits estradiol production by inhibiting expression and activity of Cyp19 / P450 aromatase.	Morera et al. (1988), Liu et al. (2013)
Interferon- γ (IFN-γ)	Inhibits steroidogenesis by inhibiting expression of StAR, CYP11A, and CYP17A	Orava et al. (1989), Lin et al. (1998a)
Leukemia inhibitory factor (LIF)	Inhibits steroidogenesis	Mauduit et al. (2001)

Production of IFN-γ in the testis has been suggested to regulate immune privilege as well as the onset of autoimmune orchitis (Itoh et al. 1998; Dal Secco et al. 2008). It stimulates the production of Fas and caspase 1 in Sertoli cells and also mediates germ cell apoptosis (Kanzaki and Morris 1998; Riccioli et al. 2000). Overexpression of IFN-β in the seminiferous tubule of a transgenic mice model causes spermatogenic cell loss and sterility (Satie et al. 2011).

In primary porcine Leydig cells, testosterone production is inhibited by both IFN-α and IFN-γ. The inhibition occurs due to the suppression of STAR protein, CYP11A and CYP17A, and decreased cholesterol transport into the mitochondria (Orava et al. 1989). Similar inhibition of steroidogenesis in rat Leydig cells is seen by IFN-α (Lin et al. 1998a). Treatment of healthy human males with IFN-α significantly decreased serum testosterone levels but not serum gonadotropin levels indicating inhibition at the Leydig cell level (Orava et al. 1986). This correlates with the overall reduction in steroidogenic function that is usually seen during viral infections (Dobs et al. 1988).

Table 21.3 summarizes the effect of individual cytokines on steroidogenesis in the Leydig cells.

21.4 Male Reproduction in Infectious and Autoimmune Diseases

Bi-directional relationships operate between the hypothalamus-pituitary-gonadal (HPG) axis and the immune system. Infection or other inflammatory stimuli lead to activation of the innate immune system, in particular the activation of macrophages.

This is in turn sets off a series of events and secretions which influence the HPG axis at all levels and has a detrimental effect on male reproductive functions. Gonadotropin secretion as well as testicular steroidogenesis and spermatogenesis is disrupted. As detailed above, cytokines produced during inflammation can suppress the activity of many enzymes of the steroidogenesis pathway like StAR, CYP11A1, HSD3B, and HSD17B in Leydig cells.

Many systemic diseases lead to reduced sperm production and lower androgen levels. Infections like tuberculosis (TB) lead to higher levels of pro-inflammatory cytokines, e.g., IFN-γ, IL-6, and TGF-β in the peripheral plasma. This increased cytokine level parallels a decreased testosterone level in male TB patients. The in vitro incubation of Leydig cells with these cytokines inhibits testosterone production (Bini et al. 2015).

Human immunodeficiency virus (HIV)-positive men show hypogonadism and impaired semen parameters including lower sperm number and motility, altered sperm morphology, and leukospermia (Kushnir and Lewis 2011). However, the underlying mechanisms are not clear and could possibly involve the infection itself, associated coinfections, effect of the cytokines on the testicular cells, or the side effect of medications (Poretsky et al. 1995; Kushnir and Lewis 2011).

Males with autoimmune inflammatory conditions like rheumatoid arthritis (RA) also show abnormal HPG axis function, and rheumatoid flares can cause prolonged suppression of testicular function (Gordon et al. 1988). Long-standing active RA male patients, not on glucocorticoid treatment, show a state of compensated partial gonadal failure where there is higher FSH and LH but normal testosterone. However, patients on low-dose glucocorticoid therapy had lower testosterone and gonadotropin levels suggesting that the medication may suppress the HPG axis (Martens et al. 1994). Men with other autoimmune disorders, like ankylosing spondylitis, systemic lupus erythematosus, or osteoarthritis, show normal testicular steroidogenesis, but treatment with the cytotoxic drugs may lead to spermatogenic damage (Sartorius and Handelsman 2010).

21.5 Thymus and Male Reproduction

Thymus is an immune-endocrine gland with a major role in the selection and maturation of T lymphocytes. A connection between thymus and reproductive system has been proposed because the thymus enlarges until late puberty and then undergoes involution. The lymphocytes and epithelial cells of thymus are almost completely replaced with fat. The fact that involution of thymus starts at around the time of puberty may be related to changes in the sex hormone production. Changes in androgen status lead to alteration of thymus weight and cellular composition with androgen deficiency in male animals leading to enlargement of the thymus (Sutherland et al. 2005). Thymus is a primary sex hormone-responsive organ. However, neonatal thymectomy of mice leads to developmental arrest of the ovary but not of the testis (Nishizuka and Sakakura 1969).

Diseases showing abnormal thymus include myasthenia gravis (MG)- a neuromuscular disease with autoimmune origin (Vincent et al. 2000), and DiGeorge syndrome – a genetic disorder with deletion at 22q 11.2 chromosome. In adults with MG, thymus remains large and shows lymphoid hyperplasia, possibly giving incorrect instructions to the developing T lymphocytes and thereby resulting in production of autoimmune acetylcholine receptor antibodies. Though a few cases have been reported of MG patients having hypergonadotropic hypogonadism, the association is probably coincidental (Lichiardopol et al. 2006). Patients with DiGeorge syndrome have hypoplastic or aplastic thymus, but fertility is generally unaffected (Kapadia and Bassett 2008; Chan et al. 2015).

Key Questions

- Why does the male reproductive tract require an immune privilege status?
- Describe the immune cells found in the testis and their location.
- Discuss the essential role of macrophages and Leydig cells for each other.
- Describe the effect of IL-1 and TNF on the functions of Leydig cells and Sertoli cells.
- Discuss the effect of HIV infection and rheumatoid arthritis on Leydig cell steroidogenesis.

References

Abu Elheija M, Dyomin V, Ganaiem M, Lunenfeld E, Vardy NS, Huleihel M. Distinct expression of interleukin-1alpha, interleukin-1beta, and interleukin-1 receptor antagonist in testicular tissues and cells from human biopsies with normal and abnormal histology. J Interf Cytokine Res. 2011;31:401–8.

Anton F, Morales C, Aguilar R, Bellido C, Aguilar E, Gaytan F. A comparative study of mast cells and eosinophil leukocytes in the mammalian testis. Zentralbl Veterinarmed A. 1998;45:209–18.

Avallet O, Vigier M, Perrard-Sapori MH, Saez JM. Transforming growth factor beta inhibits Leydig cell functions. Biochem Biophys Res Commun. 1987;146:575–81.

Barbe E, Damoiseaux JG, Dopp EA, Dijkstra CD. Characterization and expression of the antigen present on resident rat macrophages recognized by monoclonal antibody ED2. Immunobiology. 1990;182:88–99.

Bellgrau D, Gold D, Selawry H, Moore J, Franzusoff A, Duke RC. A role for CD95 ligand in preventing graft rejection. Nature. 1995;377:630–2.

Bergh A, Damber JE, van Rooijen N. The human chorionic gonadotrophin-induced inflammation-like response is enhanced in macrophage-depleted rat testes. J Endocrinol. 1993;136:415–20.

Bini EI, D'Attilio L, Marquina-Castillo B, Mata-Espinosa D, Diaz A, Marquez-Velasco R, et al. The implication of pro-inflammatory cytokines in the impaired production of gonadal androgens by patients with pulmonary tuberculosis. Tuberculosis (Edinb). 2015;95:701–6.

Boockfor FR, Schwarz LK. Effects of interleukin-6, interleukin-2, and tumor necrosis factor alpha on transferrin release from Sertoli cells in culture. Endocrinology. 1991;129:256–62.

Boockfor FR, Wang D, Lin T, Nagpal ML, Spangelo BL. Interleukin-6 secretion from rat Leydig cells in culture. Endocrinology. 1994;134:2150–5.

Calkins JH, Sigel MM, Nankin HR, Lin T. Interleukin-1 inhibits Leydig cell steroidogenesis in primary culture. Endocrinology. 1988;123:1605–10.

Chan C, Costain G, Ogura L, Silversides CK, Chow EW, Bassett AS. Reproductive health issues for adults with a common genomic disorder: 22q11.2 deletion syndrome. J Genet Couns. 2015;24:810–21.

Chuzel F, Clark AM, Avallet O, Saez JM. Transcriptional regulation of the lutropin/human choriogonadotropin receptor and three enzymes of steroidogenesis by growth factors in cultured pig Leydig cells. Eur J Biochem. 1996;239:8–16.

Cohen PE, Chisholm O, Arceci RJ, Stanley ER, Pollard JW. Absence of colony-stimulating factor-1 in osteopetrotic (csfmop/csfmop) mice results in male fertility defects. Biol Reprod. 1996;55:310–7.

Cudicini C, Lejeune H, Gomez E, Bosmans E, Ballet F, Saez J, et al. Human Leydig cells and Sertoli cells are producers of interleukins-1 and -6. J Clin Endocrinol Metab. 1997;82:1426–33.

Dal Secco V, Riccioli A, Padula F, Ziparo E, Filippini A. Mouse Sertoli cells display phenotypical and functional traits of antigen-presenting cells in response to interferon gamma. Biol Reprod. 2008;78:234–42.

De SK, Chen HL, Pace JL, Hunt JS, Terranova PF, Enders GC. Expression of tumor necrosis factor-alpha in mouse spermatogenic cells. Endocrinology. 1993;133:389–96.

Dejucq N, Lienard MO, Guillaume E, Dorval I, Jegou B. Expression of interferons-alpha and -gamma in testicular interstitial tissue and spermatogonia of the rat. Endocrinology. 1998;139:3081–7.

Delfino FJ, Boustead JN, Fix C, Walker WH. NF-kappaB and TNF-alpha stimulate androgen receptor expression in Sertoli cells. Mol Cell Endocrinol. 2003;201:1–12.

Dickson C, Webster DR, Johnson H, Cecilia Millena A, Khan SA. Transforming growth factor-beta effects on morphology of immature rat Leydig cells. Mol Cell Endocrinol. 2002;195:65–77.

Dobs AS, Dempsey MA, Ladenson PW, Polk BF. Endocrine disorders in men infected with human immunodeficiency virus. Am J Med. 1988;84:611–6.

Duan YG, Yu CF, Novak N, Bieber T, Zhu CH, Schuppe HC, et al. Immunodeviation towards a Th17 immune response associated with testicular damage in azoospermic men. Int J Androl. 2011;34:e536–45.

el-Demiry M, James K. Lymphocyte subsets and macrophages in the male genital tract in health and disease. A monoclonal antibody-based study. Eur Urol. 1988;14: 226–35.

Galardo MN, Riera MF, Pellizzari EH, Chemes HE, Venara MC, Cigorraga SB, et al. Regulation of expression of Sertoli cell glucose transporters 1 and 3 by FSH, IL1 beta, and bFGF at two different time-points in pubertal development. Cell Tissue Res. 2008;334:295–304.

Gaytan F, Aceitero J, Lucena C, Aguilar E, Pinilla L, Garnelo P, et al. Simultaneous proliferation and differentiation of mast cells and Leydig cells in the rat testis. Are common regulatory factors involved? J Androl. 1992;13:387–97.

Gaytan F, Bellido C, Aguilar E, van Rooijen N. Requirement for testicular macrophages in Leydig cell proliferation and differentiation during prepubertal development in rats. J Reprod Fertil. 1994a;102:393–9.

Gaytan F, Bellido C, Morales C, Reymundo C, Aguilar E, Van Rooijen N. Effects of macrophage depletion at different times after treatment with ethylene dimethane sulfonate (EDS) on the regeneration of Leydig cells in the adult rat. J Androl. 1994b;15:558–64.

Gerard N, Syed V, Bardin W, Genetet N, Jegou B. Sertoli cells are the site of interleukin-1 alpha synthesis in rat testis. Mol Cell Endocrinol. 1991;82:R13–6.

Gerard N, Syed V, Jegou B. Lipopolysaccharide, latex beads and residual bodies are potent activators of Sertoli cell interleukin-1 alpha production. Biochem Biophys Res Commun. 1992;185:154–61.

Gonzalez CR, Gonzalez B, Rulli SB, Dos Santos ML, Mattos Jardim Costa G, Franca LR, et al. TGF-beta1 system in Leydig cells. Part II: TGF-beta1 and progesterone, through Smad1/5, are involved in the hyperplasia/hypertrophy of Leydig cells. J Reprod Dev. 2010;56:400–4.

Gordon D, Beastall GH, Thomson JA, Sturrock RD. Prolonged hypogonadism in male patients with rheumatoid arthritis during flares in disease activity. Br J Rheumatol. 1988;27:440–4.

Gracie JA, Robertson SE, McInnes IB. Interleukin-18. J Leukoc Biol. 2003;73:213–24.

Guazzone VA, Hollwegs S, Mardirosian M, Jacobo P, Hackstein H, Wygrecka M, et al. Characterization of dendritic cells in testicular draining lymph nodes in a rat model of experimental autoimmune orchitis. Int J Androl. 2011;34:276–89.

Haas Jr GG, D'Cruz OJ, De Bault LE. Distribution of human leukocyte antigen-ABC and -D/DR antigens in the unfixed human testis. Am J Reprod Immunol Microbiol. 1988;18:47–51.

Hakovirta H, Syed V, Jegou B, Parvinen M. Function of interleukin-6 as an inhibitor of meiotic DNA synthesis in the rat seminiferous epithelium. Mol Cell Endocrinol. 1995;108:193–8.

Hales DB. Interleukin-1 inhibits Leydig cell steroidogenesis primarily by decreasing 17 alpha-hydroxylase/C17-20 lyase cytochrome P450 expression. Endocrinology. 1992;131:2165–72.

Hales DB. Testicular macrophage modulation of Leydig cell steroidogenesis. J Reprod Immunol. 2002;57:3–18.

Hardy MP, Zirkin BR, Ewing LL. Kinetic studies on the development of the adult population of Leydig cells in testes of the pubertal rat. Endocrinology. 1989;124:762–70.

Haugen TB, Landmark BF, Josefsen GM, Hansson V, Hogset A. The mature form of interleukin-1 alpha is constitutively expressed in immature male germ cells from rat. Mol Cell Endocrinol. 1994;105:R19–23.

Hedger MP. Macrophages and the immune responsiveness of the testis. J Reprod Immunol. 2002;57:19–34.

Hedger MP. The immunophysiology of male reproduction. In: Plant TM, Zeleznik AJ, editors. Knobil and Neill's physiology of reproduction. 4th ed. Boston: Academic Press; 2015. p. 805–92.

Hedger MP, Meinhardt A. Local regulation of T cell numbers and lymphocyte-inhibiting activity in the interstitial tissue of the adult rat testis. J Reprod Immunol. 2000;48:69–80.

Hoeben E, Van Damme J, Put W, Swinnen JV, Verhoeven G. Cytokines derived from activated human mononuclear cells markedly stimulate transferrin secretion by cultured Sertoli cells. Endocrinology. 1996;137:514–21.

Hoeben E, Wuyts A, Proost P, Van Damme J, Verhoeven G. Identification of IL-6 as one of the important cytokines responsible for the ability of mononuclear cells to stimulate Sertoli cell functions. Mol Cell Endocrinol. 1997;132:149–60.

Hoek A, Allaerts W, Leenen PJ, Schoemaker J, Drexhage HA. Dendritic cells and macrophages in the pituitary and the gonads. Evidence for their role in the fine regulation of the reproductive endocrine response. Eur J Endocrinol. 1997;136:8–24.

Hutson JC. Testicular macrophages. Int Rev Cytol. 1994;149:99–143.

Hutson JC. Physiologic interactions between macrophages and Leydig cells. Exp Biol Med (Maywood). 2006;231:1–7.

Itoh M, Yano A, Xie Q, Iwahashi K, Takeuchi Y, Meroni PL, et al. Essential pathogenic role for endogenous interferon-gamma (IFN-gamma) during disease onset phase of murine experimental autoimmune orchitis. I. In vivo studies. Clin Exp Immunol. 1998;111:513–20.

Ju ST, Panka DJ, Cui H, Ettinger R, el-Khatib M, Sherr DH, et al. Fas(CD95)/FasL interactions required for programmed cell death after T-cell activation. Nature. 1995;373:444–8.

Kanzaki M, Morris PL. Identification and regulation of testicular interferon-gamma (IFNgamma) receptor subunits: IFNgamma enhances interferon regulatory factor-1 and interleukin-1beta converting enzyme expression. Endocrinology. 1998;139:2636–44.

Kapadia RK, Bassett AS. Recognizing a common genetic syndrome: 22q11.2 deletion syndrome. CMAJ. 2008;178:391–3.

Kazutaka S, Winnall WR, Muir JA, Hedger MP. Regulation of Sertoli cell activin A and inhibin B by tumour necrosis factor alpha and interleukin 1alpha: interaction with follicle-stimulating hormone/adenosine 3′,5′-cyclic phosphate signalling. Mol Cell Endocrinol. 2011;335:195–203.

Khan SA, Khan SJ, Dorrington JH. Interleukin-1 stimulates deoxyribonucleic acid synthesis in immature rat Leydig cells in vitro. Endocrinology. 1992;131:1853–7.

Khan SA, Nieschlag E. Interleukin-1 inhibits follitropin-induced aromatase activity in immature rat Sertoli cells in vitro. Mol Cell Endocrinol. 1991;75:1–7.

Komsky A, Huleihel M, Ganaiem M, Kasterstein E, Komorovsky D, Bern O, et al. Presence of IL-18 in testicular tissue of fertile and infertile men. Andrologia. 2012;44:1–8.

Konrad L, Albrecht M, Renneberg H, Aumuller G. Transforming growth factor-beta2 mediates mesenchymal-epithelial interactions of testicular somatic cells. Endocrinology. 2000;141:3679–86.

Kushnir VA, Lewis W. Human immunodeficiency virus/acquired immunodeficiency syndrome and infertility: emerging problems in the era of highly active antiretrovirals. Fertil Steril. 2011;96:546–53.

Lee J, Richburg JH, Younkin SC, Boekelheide K. The Fas system is a key regulator of germ cell apoptosis in the testis. Endocrinology. 1997;138:2081–8.

Li MW, Xia W, Mruk DD, Wang CQ, Yan HH, Siu MK, et al. Tumor necrosis factor {alpha} reversibly disrupts the blood-testis barrier and impairs Sertoli-germ cell adhesion in the seminiferous epithelium of adult rat testes. J Endocrinol. 2006;190:313–29.

Lichiardopol C, Herlea V, Ioan V, Tomulescu V, Mixich F. 46,XY hypergonadotropic hypogonadism and myasthenia gravis. Romanian J Morphol Embryol. 2006;47:295–9.

Lie PP, Cheng CY, Mruk DD. Interleukin-1alpha is a regulator of the blood-testis barrier. FASEB J. 2011;25:1244–53.

Lin T, Hu J, Wang D, Stocco DM. Interferon-gamma inhibits the steroidogenic acute regulatory protein messenger ribonucleic acid expression and protein levels in primary cultures of rat Leydig cells. Endocrinology. 1998a;139:2217–22.

Lin T, Wang D, Stocco DM. Interleukin-1 inhibits Leydig cell steroidogenesis without affecting steroidogenic acute regulatory protein messenger ribonucleic acid or protein levels. J Endocrinol. 1998b;156:461–7.

Lin T, Wang TL, Nagpal ML, Calkins JH, Chang WW, Chi R. Interleukin-1 inhibits cholesterol side-chain cleavage cytochrome P450 expression in primary cultures of Leydig cells. Endocrinology. 1991;129:1305–11.

Liu ML, Wang H, Wang ZR, Zhang YF, Chen YQ, Zhu FH, et al. TGF-beta1 regulation of estrogen production in mature rat Leydig cells. PLoS One. 2013;8:e60197.

Livera G, Delbes G, Pairault C, Rouiller-Fabre V, Habert R. Organotypic culture, a powerful model for studying rat and mouse fetal testis development. Cell Tissue Res. 2006;324:507–21.

Lui WY, Wong CH, Mruk DD, Cheng CY. TGF-beta3 regulates the blood-testis barrier dynamics via the p38 mitogen activated protein (MAP) kinase pathway: an in vivo study. Endocrinology. 2003;144:1139–42.

Lukyanenko YO, Carpenter AM, Brigham DE, Stocco DM, Hutson JC. Regulation of Leydig cells through a steroidogenic acute regulatory protein-independent pathway by a lipophilic factor from macrophages. J Endocrinol. 1998;158:267–75.

Lukyanenko YO, Chen JJ, Hutson JC. Production of 25-hydroxycholesterol by testicular macrophages and its effects on Leydig cells. Biol Reprod. 2001;64:790–6.

Lysiak JJ, Bang HJ, Nguyen QA, Turner TT. Activation of the nuclear factor kappa B pathway following ischemia-reperfusion of the murine testis. J Androl. 2005;26:129–35.

Maire M, Florin A, Kaszas K, Regnier D, Contard P, Tabone E, et al. Alteration of transforming growth factor-beta signaling system expression in adult rat germ cells with a chronic apoptotic cell death process after fetal androgen disruption. Endocrinology. 2005;146:5135–43.

Martens HF, Sheets PK, Tenover JS, Dugowson CE, Bremner WJ, Starkebaum G. Decreased testosterone levels in men with rheumatoid arthritis: effect of low dose prednisone therapy. J Rheumatol. 1994;21:1427–31.

Mauduit C, Gasnier F, Rey C, Chauvin MA, Stocco DM, Louisot P, et al. Tumor necrosis factor-alpha inhibits leydig cell steroidogenesis through a decrease in steroidogenic acute regulatory protein expression. Endocrinology. 1998;139:2863–8.

Mauduit C, Goddard I, Besset V, Tabone E, Rey C, Gasnier F, et al. Leukemia inhibitory factor antagonizes gonadotropin induced-testosterone synthesis in cultured porcine leydig cells: sites of action. Endocrinology. 2001;142:2509–20.

Maurer M, Echtenacher B, Hultner L, Kollias G, Mannel DN, Langley KE, et al. The c-kit ligand, stem cell factor, can enhance innate immunity through effects on mast cells. J Exp Med. 1998;188:2343–8.

Mealy K, Robinson B, Millette CF, Majzoub J, Wilmore DW. The testicular effects of tumor necrosis factor. Ann Surg. 1990;211:470–5.

Mizuno K, Hayashi Y, Kojima Y, Nakane A, Tozawa K, Kohri K. Activation of NF-kappaB associated with germ cell apoptosis in testes of experimentally induced cryptorchid rat model. Urology. 2009;73:389–93.

Moore C, Moger WH. Interleukin-1 alpha-induced changes in androgen and cyclic adenosine 3′,5′-monophosphate release in adult rat Leydig cells in culture. J Endocrinol. 1991;129:381–90.

Morales V, Santana P, Diaz R, Tabraue C, Gallardo G, Lopez Blanco F, et al. Intratesticular delivery of tumor necrosis factor-alpha and ceramide directly abrogates steroidogenic acute regulatory protein expression and Leydig cell steroidogenesis in adult rats. Endocrinology. 2003;144:4763–72.

Moreno SG, Attali M, Allemand I, Messiaen S, Fouchet P, Coffigny H, et al. TGFbeta signaling in male germ cells regulates gonocyte quiescence and fertility in mice. Dev Biol. 2010;342:74–84.

Morera AM, Cochet C, Keramidas M, Chauvin MA, de Peretti E, Benahmed M. Direct regulating effects of transforming growth factor beta on the Leydig cell steroidogenesis in primary culture. J Steroid Biochem. 1988;30:443–7.

Mullaney BP, Skinner MK. Transforming growth factor-beta (beta 1, beta 2, and beta 3) gene expression and action during pubertal development of the seminiferous tubule: potential role at the onset of spermatogenesis. Mol Endocrinol. 1993;7:67–76.

Nehar D, Mauduit C, Boussouar F, Benahmed M. Tumor necrosis factor-alpha-stimulated lactate production is linked to lactate dehydrogenase A expression and activity increase in porcine cultured Sertoli cells. Endocrinology. 1997;138:1964–71.

Nehar D, Mauduit C, Boussouar F, Benahmed M. Interleukin 1alpha stimulates lactate dehydrogenase A expression and lactate production in cultured porcine sertoli cells. Biol Reprod. 1998;59:1425–32.

Nes WD, Lukyanenko YO, Jia ZH, Quideau S, Howald WN, Pratum TK, et al. Identification of the lipophilic factor produced by macrophages that stimulates steroidogenesis. Endocrinology. 2000;141:953–8.

Nishizuka Y, Sakakura T. Thymus and reproduction: sex-linked dysgenesia of the gonad after neonatal thymectomy in mice. Science. 1969;166:753–5.

Nistal M, Santamaria L, Paniagua R. Mast cells in the human testis and epididymis from birth to adulthood. Acta Anat (Basel). 1984;119:155–60.

Nistal M, Santana A, Paniaqua R, Palacios J. Testicular toxoplasmosis in two men with the acquired immunodeficiency syndrome (AIDS). Arch Pathol Lab Med. 1986;110:744–6.

O'Bryan MK, Gerdprasert O, Nikolic-Paterson DJ, Meinhardt A, Muir JA, Foulds LM, et al. Cytokine profiles in the testes of rats treated with lipopolysaccharide reveal localized suppression of inflammatory responses. Am J Physiol Regul Integr Comp Physiol. 2005;288:R1744–55.

Olaso R, Pairault C, Boulogne B, Durand P, Habert R. Transforming growth factor beta1 and beta2 reduce the number of gonocytes by increasing apoptosis. Endocrinology. 1998;139:733–40.

Orava M, Cantell K, Vihko R. Treatment with preparations of human leukocyte interferon decreases serum testosterone concentrations in men. Int J Cancer. 1986;38:295–6.

Orava M, Voutilainen R, Vihko R. Interferon-gamma inhibits steroidogenesis and accumulation of mRNA of the steroidogenic enzymes P450scc and P450c17 in cultured porcine Leydig cells. Mol Endocrinol. 1989;3:887–94.

Parvinen M, Soder O, Mali P, Froysa B, Ritzen EM. In vitro stimulation of stage-specific deoxyribonucleic acid synthesis in rat seminiferous tubule segments by interleukin-1 alpha. Endocrinology. 1991;129:1614–20.

Pentikainen V, Erkkila K, Suomalainen L, Otala M, Pentikainen MO, Parvinen M, et al. TNFalpha down-regulates the Fas ligand and inhibits germ cell apoptosis in the human testis. J Clin Endocrinol Metab. 2001;86:4480–8.

Pentikainen V, Suomalainen L, Erkkila K, Martelin E, Parvinen M, Pentikainen MO, et al. Nuclear factor-kappa B activation in human testicular apoptosis. Am J Pathol. 2002;160:205–18.

Perrard MH, Durand P. Redundancy of the effect of TGFbeta1 and beta-NGF on the second meiotic division of rat spermatocytes. Microsc Res Tech. 2009;72:596–602.

Petersen C, Svechnikov K, Froysa B, Soder O. The p38 MAPK pathway mediates interleukin-1-induced Sertoli cell proliferation. Cytokine. 2005;32:51–9.

Pollanen P, Niemi M. Immunohistochemical identification of macrophages, lymphoid cells and HLA antigens in the human testis. Int J Androl. 1987;10:37–42.

Pollanen P, Soder O, Parvinen M. Interleukin-1 alpha stimulation of spermatogonial proliferation in vivo. Reprod Fertil Dev. 1989;1:85–7.

Poretsky L, Can S, Zumoff B. Testicular dysfunction in human immunodeficiency virus-infected men. Metabolism. 1995;44:946–53.

Raburn DJ, Coquelin A, Reinhart AJ, Hutson JC. Regulation of the macrophage population in postnatal rat testis. J Reprod Immunol. 1993;24:139–51.

Rasoulpour RJ, Boekelheide K. NF-kappaB is activated in the rat testis following exposure to mono-(2-ethylhexyl) phthalate. Biol Reprod. 2005;72:479–86.

Riccioli A, Starace D, D'Alessio A, Starace G, Padula F, De Cesaris P, et al. TNF-alpha and IFN-gamma regulate expression and function of the Fas system in the seminiferous epithelium. J Immunol. 2000;165:743–9.

Riera MF, Meroni SB, Gomez GE, Schteingart HF, Pellizzari EH, Cigorraga SB. Regulation of lactate production by FSH, iL1beta, and TNFalpha in rat Sertoli cells. Gen Comp Endocrinol. 2001;122:88–97.

Rival C, Guazzone VA, von Wulffen W, Hackstein H, Schneider E, Lustig L, et al. Expression of co-stimulatory molecules, chemokine receptors and proinflammatory cytokines in dendritic cells from normal and chronically inflamed rat testis. Mol Hum Reprod. 2007;13:853–61.

Rival C, Lustig L, Iosub R, Guazzone VA, Schneider E, Meinhardt A, et al. Identification of a dendritic cell population in normal testis and in chronically inflamed testis of rats with autoimmune orchitis. Cell Tissue Res. 2006;324:311–8.

Sainio-Pollanen S, Saari T, Simell O, Pollanen P. CD28-CD80/CD86 interactions in testicular immunoregulation. J Reprod Immunol. 1996;31:145–63.

Sartorius GA, Handelsman DJ. Testicular dysfunction in systemic diseases. In: Nieschlag E, Behre HM, Nieschlag S, editors. Male reproductive health and dysfunction. 3rd ed. Berlin: Springer.

Satie AP, Mazaud-Guittot S, Seif I, Mahe D, He Z, Jouve G, et al. Excess type I interferon signaling in the mouse seminiferous tubules leads to germ cell loss and sterility. J Biol Chem. 2011;286:23280–95.

Schell C, Albrecht M, Mayer C, Schwarzer JU, Frungieri MB, Mayerhofer A. Exploring human testicular peritubular cells: identification of secretory products and regulation by tumor necrosis factor-alpha. Endocrinology. 2008;149:1678–86.

Siu MK, Lee WM, Cheng CY. The interplay of collagen IV, tumor necrosis factor-alpha, gelatinase B (matrix metalloprotease-9), and tissue inhibitor of metalloproteases-1 in the basal lamina regulates Sertoli cell-tight junction dynamics in the rat testis. Endocrinology. 2003;144:371–87.

Soder O, Syed V, Callard GV, Toppari J, Pollanen P, Parvinen M, et al. Production and secretion of an interleukin-1-like factor is stage-dependent and correlates with spermatogonial DNA synthesis in the rat seminiferous epithelium. Int J Androl. 1991;14:223–31.

Stephan JP, Syed V, Jegou B. Regulation of Sertoli cell IL-1 and IL-6 production in vitro. Mol Cell Endocrinol. 1997;134:109–18.

Strand ML, Wahlgren A, Svechnikov K, Zetterstrom C, Setchell BP, Soder O. Interleukin-18 is expressed in rat testis and may promote germ cell growth. Mol Cell Endocrinol. 2005;240:64–73.

Suh JH, Gong EY, Hong CY, Park E, Ahn RS, Park KS, et al. Reduced testicular steroidogenesis in tumor necrosis factor-alpha knockout mice. J Steroid Biochem Mol Biol. 2008;112:117–21.

Sultana T, Svechnikov K, Weber G, Soder O. Molecular cloning and expression of a functionally different alternative splice variant of prointerleukin-1alpha from the rat testis. Endocrinology. 2000;141:4413–8.

Sultana T, Wahab-Wahlgren A, Assmus M, Parvinen M, Weber G, Soder O. Expression and regulation of the prointerleukin-1alpha processing enzymes calpain I and II in the rat testis. Int J Androl. 2003;26:37–45.

Sun XR, Hedger MP, Risbridger GP. The effect of testicular macrophages and interleukin-1 on testosterone production by purified adult rat Leydig cells cultured under in vitro maintenance conditions. Endocrinology. 1993;132:186–92.

Suominen JS, Wang Y, Kaipia A, Toppari J. Tumor necrosis factor-alpha (TNF-alpha) promotes cell survival during spermatogenesis, and this effect can be blocked by infliximab, a TNF-alpha antagonist. Eur J Endocrinol. 2004;151:629–40.

Sutherland JS, Goldberg GL, Hammett MV, Uldrich AP, Berzins SP, Heng TS, et al. Activation of thymic regeneration in mice and humans following androgen blockade. J Immunol. 2005;175:2741–53.

Svechnikov KV, Sultana T, Soder O. Age-dependent stimulation of Leydig cell steroidogenesis by interleukin-1 isoforms. Mol Cell Endocrinol. 2001;182:193–201.

Syed V, Stephan JP, Gerard N, Legrand A, Parvinen M, Bardin CW, et al. Residual bodies activate Sertoli cell interleukin-1 alpha (IL-1 alpha) release, which triggers IL-6 production by an autocrine mechanism, through the lipoxygenase pathway. Endocrinology. 1995;136:3070–8.

Teerds KJ, Dorrington JH. Localization of transforming growth factor beta 1 and beta 2 during testicular development in the rat. Biol Reprod. 1993;48:40–5.

Tronchon V, Vialard F, El Sirkasi M, Dechaud H, Rollet J, Albert M, et al. Tumor necrosis factor-alpha −308 polymorphism in infertile men with altered sperm production or motility. Hum Reprod. 2008;23:2858–66.

Tsigos C, Papanicolaou DA, Kyrou I, Raptis SA, Chrousos GP. Dose-dependent effects of recombinant human interleukin-6 on the pituitary-testicular axis. J Interf Cytokine Res. 1999;19:1271–6.

van den Berg TK, Dopp EA, Dijkstra CD. Rat macrophages: membrane glycoproteins in differentiation and function. Immunol Rev. 2001;184:45–57.

Vergouwen RP, Huiskamp R, Bas RJ, Roepers-Gajadien HL, Davids JA, de Rooij DG. Postnatal development of testicular cell populations in mice. J Reprod Fertil. 1993;99:479–85.

Verhoeven G, Cailleau J, Van Damme J, Billiau A. Interleukin-1 stimulates steroidogenesis in cultured rat Leydig cells. Mol Cell Endocrinol. 1988;57:51–60.

Vincent A, Beeson D, Lang B. Molecular targets for autoimmune and genetic disorders of neuromuscular transmission. Eur J Biochem. 2000;267:6717–28.

Wang J, Wreford NG, Lan HY, Atkins R, Hedger MP. Leukocyte populations of the adult rat testis following removal of the Leydig cells by treatment with ethane dimethane sulfonate and subcutaneous testosterone implants. Biol Reprod. 1994;51:551–61.

Wang T, Zhang X, Chen Q, Deng T, Zhang Y, Li N, et al. Toll-like receptor 3-initiated antiviral responses in mouse male germ cells in vitro. Biol Reprod. 2012;86:106.

Welter H, Kohn FM, Mayerhofer A. Mast cells in human testicular biopsies from patients with mixed atrophy: increased numbers, heterogeneity, and expression of cyclooxygenase 2 and prostaglandin D2 synthase. Fertil Steril. 2011;96:309–13.

Winnall WR, Okuma Y, Saito K, Muir JA, Hedger MP. Regulation of interleukin 1alpha, activin and inhibin by lipopolysaccharide in Sertoli cells from prepubertal rats. Mol Cell Endocrinol. 2009;307:169–75.

Xiong Y, Hales DB. The role of tumor necrosis factor-alpha in the regulation of mouse Leydig cell steroidogenesis. Endocrinology. 1993;132:2438–44.

Xiong Y, Hales DB. Differential effects of tumor necrosis factor-alpha and interleukin-1 on 3 beta-hydroxysteroid dehydrogenase/delta 5-- > delta 4 isomerase expression in mouse Leydig cells. Endocrine. 1997;7:295–301.

Yan W, Kero J, Huhtaniemi I, Toppari J. Stem cell factor functions as a survival factor for mature Leydig cells and a growth factor for precursor Leydig cells after ethylene dimethane sulfonate treatment: implication of a role of the stem cell factor/c-Kit system in Leydig cell development. Dev Biol. 2000;227:169–82.

Yao PL, Lin YC, Sawhney P, Richburg JH. Transcriptional regulation of FasL expression and participation of sTNF-alpha in response to sertoli cell injury. J Biol Chem. 2007;282:5420–31.

Thyroid

22

Skand Shekhar, Bodhana Dhole, and Anand Kumar

Learning Objectives
- Thyroid hormone receptors on male reproductive organs
- Action of thyroid hormones on testicular cells
- Action of thyroid hormones on accessory sex glands
- Thyroid hormones and oxidative stress
- Reproductive dysfunction in thyroid disorders

22.1 Introduction

The development and function of the testis and male accessory sex glands are regulated by the hypothalamic-pituitary-testicular (HPT) axis. The details of the regulating HPT axis have already been discussed in the chapter, Hypothalamic-Pituitary-Testicular Axis. In the recent years, thyroid hormones, especially 3, 5, 3′-L-triiodothyronine (T_3), have emerged as an important regulator of the male reproduction. The structure of T_3 is given in Fig. 22.1. In this chapter, we will focus on the specific functions of T_3 in the testis and the male accessory sex glands along with reproductive dysfunction in thyroid disorders.

S. Shekhar, MD
Saint Peter's University Hospital, Rutgers Robert Wood Johnson Medical School, New Jersey, USA

B. Dhole, PhD • A. Kumar, MD, FAMS (✉)
Department of Reproductive Biology, All India Institute of Medical Sciences, New Delhi, India
e-mail: anandkumarrepbiol@hotmail.com

A. Kumar, M. Sharma (eds.), *Basics of Human Andrology*,
DOI 10.1007/978-981-10-3695-8_22

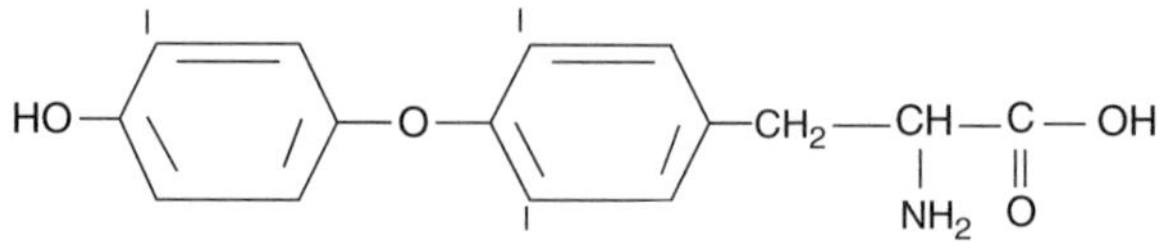

Fig. 22.1 Structure of 3, 5, 3′-L-triiodothyronine (T_3)

22.2 Thyroid Hormone Receptors on Testis and Accessory Sex Glands

Thyroid hormone receptors (TR) are nuclear receptors which belong to a family of ligand-dependent transcription factors (Wrutniak-Cabello et al. 2001). TRs are formed from a single peptide that is folded into three modular functional domains—an amino-terminal domain (A–B domain), a central DNA-binding domain (DBD) and a carboxyl-terminal ligand-binding domain (LBD; see Fig. 22.2). The LBD regulates the specificity of the receptor for its ligand and modulates its capacity to form either homodimers or heterodimers with other members of the nuclear receptor superfamily. The human TR proteins are encoded by two separate genes- TRα and TRβ. Each of which has a homologue in mice, known as Trα and Trβ, respectively. The α gene is located on chromosome 17 in humans and chromosome 11 in mouse, while the β gene is located on chromosome 3 in humans and chromosome 14 in mouse. In humans, three splice variants of TRα—α1, α2 and α3—are identified. TRα1 binds T_3 while TRα2 and TRα3 have a truncated LBD and do not bind T_3. The TRβ is expressed as either TRβ1 or TRβ2; a third isoform, Trβ3, is found only in rats. All the β isoforms bind to T_3 (Ortiga-Carvalho et al. 2014).

T_3 enters the cell via thyroid hormone transporters or by passive diffusion. In the nucleus, T_3 binds to TR which usually exists as a heterodimer with retinoid X receptor (RXR). The hormone-bound receptor then binds to specific sequences on target genes called thyroid response elements (TREs). The carboxyl-terminal portion of the LBD contains an activation function 2 (AF-2) region that binds to other proteins called cofactors which includes coactivators and corepressors. In the absence of T_3, TR bound to a target TRE recruits corepressor proteins. The corepressor proteins form a complex with histone deacetylases, which condense the chromatin structure resulting in a subsequent inhibition in gene transcription. In the presence of T_3, AF-2 region undergoes a conformational change, corepressors are released, and coactivators are recruited. Coactivators induce remodelling of chromatin by acetylating or methylating histones, which promote the interactions among RNA polymerase and other transcriptional factors and thereby induce gene transcription (Ortiga-Carvalho et al. 2014; see Fig. 22.3).

Thyroid-stimulating hormone (TSH) is synthesized and secreted from the anterior pituitary. TSH stimulates the secretion of thyroid hormones. TSH receptors belong to the seven-transmembrane, G protein-coupled glycoprotein receptor

N C

N-Terminal domain	DNA binding domain	Ligand binding domain

Fig. 22.2 Schematic diagram showing structure of thyroid hormone receptor

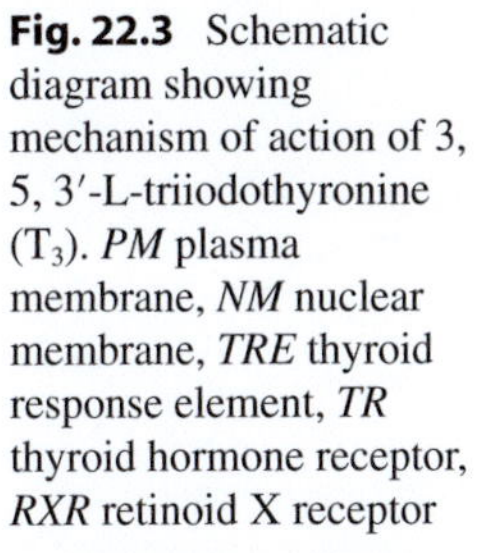

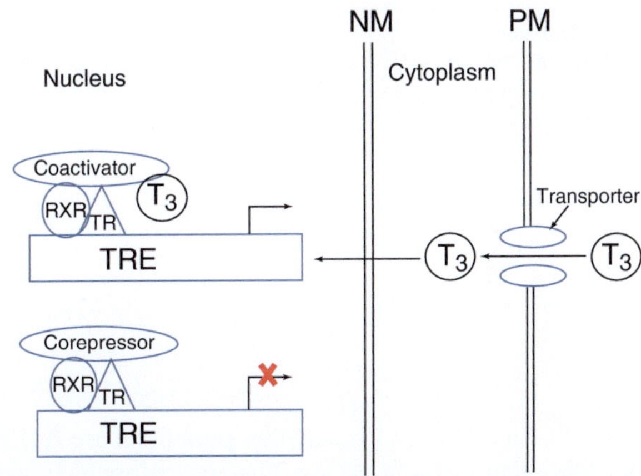

Fig. 22.3 Schematic diagram showing mechanism of action of 3, 5, 3′-L-triiodothyronine (T_3). *PM* plasma membrane, *NM* nuclear membrane, *TRE* thyroid response element, *TR* thyroid hormone receptor, *RXR* retinoid X receptor

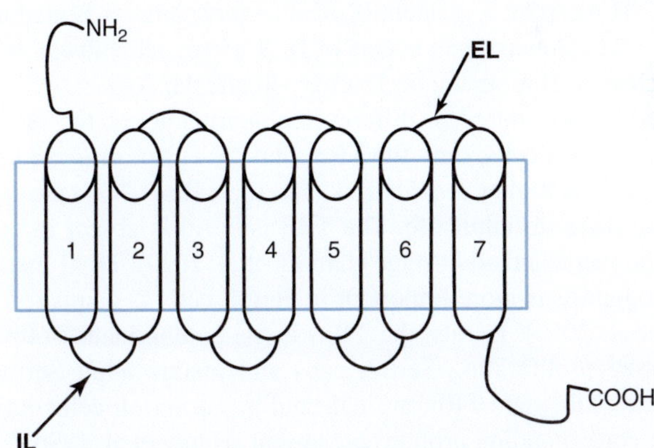

Fig. 22.4 Structure of TSH receptor. *EL* extracellular loop, *IL* intracellular loop

superfamily (see Fig. 22.4). In humans, the TSH gene is located on chromosome 14, and in mouse, it is located on chromosome 12 (Low 2011).

When TSH binds to its receptor, it induces two signalling pathways—the adenyl cyclase/cyclic adenosine monophosphate (AC/cAMP) pathway and the inositol triphosphate/calcium (IP_3/Ca^{2+}) responses. The cAMP signal is mediated by the stimulatory G protein (G_S). The activated Gs binds to and activates AC, which, in turn, catalyses the conversion of adenosine triphosphate (ATP) into cAMP. Increase in

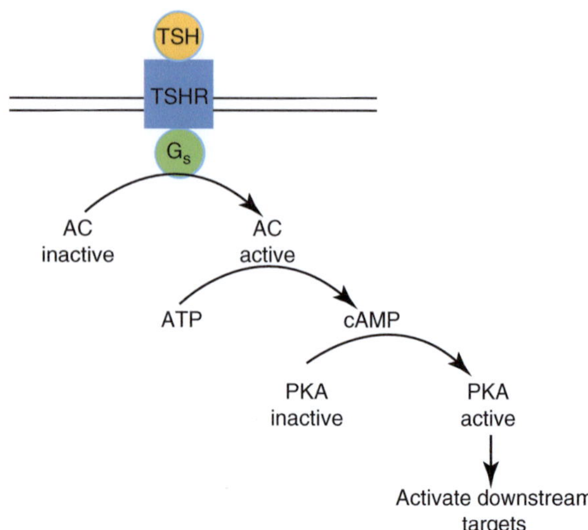

Fig. 22.5 The cAMP-PKA pathway activated by TSH-TSHR interaction. *TSH* thyroid-stimulating hormone, *TSHR* thyroid-stimulating hormone receptor, *Gs* stimulatory G protein, *AC* adenyl cyclase, *ATP* adenosine triphosphate, *cAMP* cyclic adenosine monophosphate, *PKA* protein kinase A

intracellular cAMP levels activates protein kinase A (PKA). Activated PKA in turn phosphorylates and activates other downstream targets (see Fig. 22.5; Allgeier et al. 1994).

The TSH-TSH receptor-dependent IP_3/Ca^{2+} signal pathway is mediated by the G protein, $Gq/G_{11.}$ The downstream events of IP_3/Ca^{2+} signal pathway have been discussed in the chapter, Hypothalamic-Pituitary-Testicular Axis.

TRs are widely distributed in different compartments of the testis. In human foetal and adult Sertoli cells, only the $TR\alpha_1$ and $TR\alpha_2$ isoforms are expressed; the expression of $TR\alpha_2$ is higher at all stages, and the $TR\alpha_2/TR\alpha_1$ ratio increases progressively from foetal to adult life. The TRβ isoform is absent in human Sertoli cells, both in the foetal and adult stage (Jannini et al. 2000). TRα1 mediated regulation of steroidogenesis is more important in Sertoli cells as compared to in Leydig cells (Fumel et al. 2015). In rats, the $TR\alpha_1$ is the predominant isoform expressed both in immature proliferating Sertoli cells and mature adult Sertoli cells. The mRNAs of $TR\alpha_2$, $TR\alpha_3$ and $TR\beta_1$ are detected in rat Sertoli cells during development, but their corresponding proteins are absent (Hardy et al. 1996).

Rat mesenchymal stem cells and immature and adult Leydig cells express the TRα isoform, their expression being maximal in the postnatal age and decreases to almost negligible levels in adulthood. T_3 binds specifically to nuclei of goat Leydig cells and consequently stimulates androgen production from these cells (Jana and Bhattacharya 1994).

The presence of TRs on germ cells suggests a role of thyroid hormones in sustaining different population of germ cells. TRs are identified on different stages of developing rat germ cells such as gonocyte, spermatogonia, preleptotene, leptotene, pachytene, zygotene and round and elongating spermatids. Both TRα and $TR\beta_1$ are expressed during different stages of germ cell development. $TR\beta_1$ first appears in intermediate-type spermatogonia, while TRα first appears in type B spermatogonia

(Buzzard et al. 2000). Thyroid hormone is also required in proliferation and differentiation of Sertoli cells in the developmental phase, which in turn influences spermatogenesis (Chatonnet et al. 2014).

The epithelial cells from different segments of rat epididymis—caput, corpus and cauda—express TRs. However, unlike classical TRs, the TRs in the epithelial cells of the epididymis are predominantly located in the cytoplasm. Both the TRα_1 and TRβ_1 isoforms are expressed in all three segments of the epididymis. Both the protein and mRNA levels of TR isoforms increase significantly in hypothyroid rats (De Paul et al. 2008). TRβ_1 is also identified on the nuclear membrane of PZ-HPV-10 cell line derived from prostatic tissue (Hsieh and Juang 2005).

It has been shown that TSH binds to human cryopreserved testicular slices with high affinity (Trokoudes et al. 1979). We have recently identified the presence of TSH receptor mRNA and protein in mouse Leydig cell-derived line, MLTC-1 (unpublished data from our laboratory).

22.3 Testicular Actions of Thyroid Hormones

22.3.1 Sertoli Cells

Sertoli cells provide support and sustain the developing germ cells. Each Sertoli cell supports a limited number of developing germ cells. The ratio of Sertoli cells to germ cells is 1:11 in humans and 1:50 in the adult rat testis (Bendsen et al. 2003; Mruk and Cheng 2004). Therefore, the number of Sertoli cells is one of the indicators of daily sperm production (DSP) of the testis. There are two distinct populations of Sertoli cells—the immature proliferating Sertoli cell and the mature differentiated Sertoli cell (Sharpe et al. 2003). In Sertoli cells, expressions of certain genes and proteins are associated with its state of maturation. Anti-Mullerian hormones (AMH), aromatase and neural cell adhesion molecule (NCAM) are expressed exclusively in the immature Sertoli cells. The expressions of cell cycle inhibitors—p27^{Kip1}, p21^{Cip1}—and androgen receptor are characteristic markers of mature Sertoli cells (Sharpe et al. 2003).

T_3 suppresses the expression of immature Sertoli cell markers, including AMH, aromatase and NCAM. Hypothyroidism induced in neonatal rats delays the fall in AMH-mRNA levels, and T_3 treatment decreases AMH-mRNA in Sertoli cells (Bunick et al. 1994). In mouse Sertoli cell line, TM4, T_3 decreases transcription of aromatase gene (Catalano et al. 2003). T_3 downregulates NCAM production in Sertoli cell-germ cell co-cultures (Laslett et al. 2000). T_3 increases the levels of cell cycle inhibitory proteins p27^{Kip1} and p21^{Cip1} in Sprague-Dawley rats. The inhibitory proteins are negative regulators of cyclin-dependent kinases Cdk2 and Cdk4 required for G1 to S transition in the cell cycle (Buzzard et al. 2003). T_3 increases androgen receptor-mRNA levels in 5- and 20-day-old cultured Sertoli cells (Arambepola et al. 1998). In contrast, T_3 maintains inter-Sertoli cell junctions by regulating the levels of gap junction protein, connexin. T_3 increases the levels of connexin 43 (Cx43). Cx43 is the most abundant gap junction protein in the Sertoli cell cultures (Gilleron et al. 2006). Two specific inhibitors of Cx43-18α-glycyrrhetinic acid (AGA) and oleamide significantly lower the T_3-induced Cx43 levels in cultured Sertoli cells (Gilleron et al. 2006). Some studies also suggest that thyroid hormone may inhibit Sertoli cell production via upregulation of Cx43 levels (St-Pierre et al. 2003).

22.3.2 Leydig Cells

The stem cells for both foetal and adult Leydig cell populations are the mesenchymal cells in the testis interstitium. The mesenchymal stem cells are spindle shaped and non-steroidogenic. While some of the mesenchymal cells differentiate into foetal Leydig cells, others retain their undifferentiated characteristics and serve as precursor cells for the adult Leydig cells in the postnatal testis (Mendis-Handagama and Ariyaratne 2001).

Thyroid hormones regulate Leydig cell development and steroidogenesis. It has been shown that propylthiouracil (PTU) treatment increases Leydig cell number in rats (Hardy et al. 1993; Ariyaratne et al. 2000). However, Leydig cells from the hypothyroid rats have a 50% lower human chorionic gonadotropin (hCG) binding sites and exhibit lower steroid-producing potential (Hardy et al. 1993). The increase in adult Leydig cell population in hypothyroid rats is mainly due to the proliferation of immature Leydig cells (Hardy et al. 1993). Thyroid hormone is essential for differentiation of mesenchymal stem cells into Leydig cells in adult rats (Ariyaratne et al. 2000). Ethane dimethanesulfonate, a specific inhibitor of Leydig cells, completely eliminates Leydig cell population in adult Sprague-Dawley rats by 2 days of treatment. The Leydig cells reappear by 14 and 21 days in euthyroid and T_3-treated rats, respectively. The number of new Leydig cells doubles in the T_3-treated rats. In PTU-treated hypothyroid rats, Leydig cells do not reappear. The above results confirm the role of T_3 in the differentiation of mesenchymal stem cells into adult Leydig cells (Ariyaratne et al. 2000).

Thyroid hormones also stimulate Leydig cell steroidogenesis. Leydig cells synthesize steroid hormones from cholesterol. In the Leydig cells, cholesterol is esterified by cholesterol acyl transferase and stored in Leydig cell cytoplasmic lipid droplets. Upon luteinizing hormone (LH) stimulation, cholesteryl ester is hydrolysed and is released from the lipid droplets. Steroidogenic acute regulatory (StAR) protein, a de novo synthesized labile protein, catalyses the translocation of cholesterol from outer to the inner mitochondrial membrane. Steroidogenic factor 1 (SF-1), a 52 KDa orphan nuclear receptor transcription factor, stimulates the transcription of StAR gene (Sugawara et al. 1996). In the inner mitochondrial membrane, conversion of cholesterol to pregnenolone is catalysed by cytochrome P450 side-chain cleavage enzyme (CYP11A1) using nicotinamide adenine dinucleotide phosphate (NADPH) oxidase as a cofactor. Pregnenolone then diffuses out to the cytoplasm where remaining steps of testosterone biosynthesis are carried out. T_3 induces de novo synthesis of a 52 KDa soluble protein, which augments androgen synthesis in the goat Leydig cells (Jana et al. 1996). Similarly, T_3 increases steroid secretion by primary mouse Leydig cells and their derived cell lines (Manna et al. 1999, 2001; Maran et al. 2000).

T_3 treatment for 8 h increases StAR- mRNA in MLTC-1 and mouse primary Leydig cells (Manna et al. 1999). However, chronic stimulation of MLTC-1 cells with T_3 beyond 8 h and up to 30 h decreases StAR mRNA and protein levels and subsequently the steroid production (Manna et al. 2001). T_3 does not alter StAR mRNA stability but decreases its transcription. The reduction in progesterone production upon chronic T_3 exposure is partially restored by 22R hydroxycholesterol.

Hydroxycholesterol is a lipid-soluble side-chain oxygenated sterol which can directly diffuse from the outer mitochondrial to the inner mitochondrial membrane without the need for StAR or pregnenolone (Manna et al. 2001). Chronic T_3 treatment also decreases 3 beta-hydroxysteroid dehydrogenase (3β-HSD) mRNA levels required for catalysing the conversion of pregnenolone to progesterone (Manna et al. 2001). T_3 also increases hCG binding to the MLTC-1 cells. The hCG binding peaks at 16 h and thereafter, binding decreases in a time-dependent manner. Hypothyroid mice express higher amount of receptor for LH-mRNA (LHR-mRNA), whereas hyperthyroid mice express lower amount of LHR-mRNA (Manna et al. 2001). The action of T_3 is mediated via a 173-bp fragment on the LHR gene promoter region. Mutation studies show that the SF-1 binding region on the mouse StAR promoter region is involved in T_3 response (Manna et al. 2001).

22.3.3 Sperm Count

Hypothyroidism induced in rats by PTU treatment during the first postnatal week results in a significant increase in testis weight, DSP and efficiency of sperm production (Cooke et al. 1991). The rise in sperm production could be due to (1) a rise in gonadotropins, LH and follicle-stimulating hormone (FSH), which are important for spermatogenesis and germ cell survival, (2) increase in Sertoli cell number and (3) decrease in germ cell apoptosis. Serum gonadotropins are reduced in PTU-treated hypothyroid rats eliminating the fact that the rise in sperm production was due to increase in LH and FSH levels (Kirby et al. 1992). The hypothyroid rats have significantly higher number of Sertoli cells which in turn support the survival of greater number of germ cells (Hess et al. 1993). One study also showed a positive correlation between sperm concentration and free thyroxine (T_4) levels (Meeker et al. 2007). In hypothyroid rats, the differentiation of immature Sertoli cells to the mature state is delayed as shown by a delay in the expression of clusterin, a marker of differentiated Sertoli cell. In the hypothyroid rats, the delay in Sertoli cell differentiation results in delay in differentiation of spermatocyte to the more matured germ cell stage (Silva et al. 2011). In rats, the large number of germ cells undergoes apoptosis during the third postnatal week. The third week is characterized by a significant rise in pro-apoptotic proteins of both the intrinsic and extrinsic pathways. In hypothyroid rats, the apoptosis of germ cells is delayed. Maximum apoptosis is seen on day 45 as compared to day 25. Thus, thyroid hormones do not act directly on germ cell apoptosis, but these delay Sertoli cell maturation which results in delayed germ cell apoptosis (Silva et al. 2011).

In addition, there is growing evidence that thyroid hormone regulators such as deiodinases affect the bioactive T_3 levels by regulating its production from the inactive T_4 form. This is primarily achieved via deiodinase types 1 (D1), 2 (D2) and 3 (D3; Bates et al. 1999). D1 and D2 are responsible for producing the activated form, whereas D3 has the opposite effect (Bianco et al. 2002). A mutant D3 has been shown to have a more profound impact on spermiogenesis compared to D1 or D2 (Martinez et al. 2016). However, the clinical significance of this mediation is still under study (Hernandez et al. 2006).

22.3.4 Sperm Morphology

Morphological abnormalities during the development of sperm may result in deformed sperm head and tail. Both hyperthyroid and hypothyroid men have a lower proportion of morphologically normal sperm (Krassas et al. 2002, 2008). Thyroid hormones exert their effect on the cell cytoskeleton. In hypothyroid rats, there is increase in the phosphorylation and immunoreactivity of vimentin protein which is a cytoskeleton-associated protein expressed in Sertoli cell. As a result, there is loss of Sertoli cell cytoskeleton integrity (Zamoner et al. 2008). The high proportion of morphologically abnormal sperm observed in altered thyroid state can be due to effect of thyroid hormones on sperm cytoskeleton.

22.3.5 Motility

Hypothyroid and hyperthyroid patients show a decrease in progressive motility of the sperm (Corrales-Hernandez et al. 1990; Hudson and Edwards 1992). Thyroid hormones increase basal metabolic rate and stimulate oxygen consumption in metabolically active cells by promoting the action of sodium potassium adenosine triphosphatases (Na+/K+ATPase), increasing mitochondrial number and mitochondrial gene expression (Wrutniak-Cabello et al. 2001; Lei et al. 2003). The role of T_3 on ATP generation in males is speculated. However, specific studies on the modulatory effect of T_3 on sperm ATPase and energy production need to be studied.

22.4 Fertility

Gestational exposure of rats to methimazole (MMI) results in a reduction in testicular interstitial fluid levels of testosterone and estradiol. The hypothyroid rats show reduced pregnancy induction capacity on mating with normal females and show reduction in litter size. Additionally, the ratio of male to female pups is also reduced significantly (Anbalagan et al. 2010). Furthermore, there is some evidence of fertility in hypothyroid and hyperthyroid states being influenced via the differential expression of thyroid hormone receptor, deiodinase and transporter in both these states (Romano et al. 2017).

22.5 Thyroid Hormones and Accessory Sex Glands

Thyroid hormones influence both epididymal structure and its secretary activity. Gestational exposure of rats to MMI results in reduced epididymal weight, but sperm content in cauda and corpus epididymis remains unchanged. The caudal sperm forward motility decreases significantly. The epididymal secretions such as sialic acid, glycerophosphorylcholine and carnitine are also reduced in the hypothyroid rats. Gestational MMI exposure also decreases 5α-reductase and androgen receptor levels (Anbalagan et al. 2010).

Thyroid hormones regulate the contractile activity of vas deferens in response to prostaglandin E2 (PGE2). Removal of thyroid glands in albino rats completely inhibits the contractile activity of the vas deferens in response to PGE2. But treatment with T_4 further increases the contractibility of the vas deferens in response to PGE2 (Amadi et al. 1999).

In PTU-induced hypothyroid rats, there is a significant fall in their seminal vesicle and prostate gland weight (Choudhury et al. 2003). Gestational MMI exposure to rats decreases androgen receptor-mRNA levels in the dorsolateral prostate lobe but increases androgen receptor-mRNA levels in the ventral lobe in pups (Aruldhas et al. 2010). Thyroid hormones stimulate rat prostatic glycosidase activities such as β-glucosidase, β-galactosidase, β-N-acetylglucosaminidase and β-N-acetylgalactosaminidase, while opposite effects are reported in thyroidectomized rats (Maran et al. 1998). T_3 decreases hexosamine and sialic acid concentrations in the ventral prostatic tissues of 30- and 60-day-old thyroidectomized rats. Fucose concentrations in the ventral prostatic tissues increase at 30 days but decrease at 60 days in the hypothyroid rats (Maran et al. 1998). In the dorsolateral prostatic lobe, hypothyroidism enhances the concentration of hexosamines but results in a decrease in fucose, sialic acid and fructose levels irrespective of the duration of hypothyroidism. In the anterior prostate, hypothyroidism decreases fucose, fructose and hexosamine levels and increases sialic acid concentration in 60-day-old rats (Maran et al. 1998). Thyroid hormones influence the risk of prostate cancer through their function in cell differentiation, growth and metabolism. Hypothyroid men show a decreased risk of prostate cancer when compared to euthyroid men, although no association between hyperthyroidism and risk of prostate cancer has been established (Mondul et al. 2012).

22.6 Thyroid Hormones and Oxidative Stress

Reactive oxygen species (ROS) are chemically highly reactive reduced forms of oxygen and their products. All ROS including superoxide radical, hydroxyl radical and hydrogen peroxide contain one or more unpaired electrons. Mitochondria are the primary biological source of ROS. Under physiological conditions, ROS can oxidize a number of biological molecules such as unsaturated fatty acids, sulphhydryl proteins and nucleic acids (Tremellen 2008). To counteract the effects of ROS, cells produce many antioxidant molecules such as superoxide dismutase (SOD), glutathione peroxidase (GSHPx), catalase (CAT), tocopherols and carotenoids.

Oxidative stress results when ROS levels are much higher than antioxidant levels in the cells (Tremellen 2008). In the semen, sperm and some leukocytes contaminating the seminal plasma are the principal sources of ROS (Ochsendorf 1999). The plasma membrane of sperm contains high amount of polyunsaturated fatty acid such as docosahexaenoic acid, which is rapidly oxidized by ROS, thereby decreasing the flexibility and motility of the sperm tail (De Lamirande and Gagnoc 1992). ROS also decrease ATP generated by sperm mitochondrial sheath which provides energy for sperm motility (De Lamirande and Gagnoc 1992). Teratozoospermic sperm in most cases retain excess residual body which is normally lost during sperm maturation. The residual bodies have large amount of glucose-6-phosphate dehydrogenase which generates NADPH. NADPH produces ROS catalysed by the NADP oxidase enzyme present in sperm plasma membrane (Gomez et al. 1996). Thyroid hormones are important in maintaining the balance between ROS and antioxidant molecules in the testis. The oxygen consumption of the testis in the hypothyroid rats is significantly lower, but there is no difference in their lipid peroxide levels as compared to euthyroid controls (Zamoner et al. 2008). The antioxidant SOD activity is significantly lower in hypothyroid group in comparison to euthyroid controls (Zamoner et al. 2008). In the hypothyroid state, the ROS levels do not increase, but reduced antioxidant activity leads to oxidative stress (Zamoner et al. 2008). In PTU-treated rats, hydrogen peroxide and protein carbonyl content levels in the crude homogenate of the testis increase, but the endogenous lipid peroxide levels remain unaltered. SOD and catalase activity decreases in hypothyroid rats, but on addition of T_3, only the catalase activity enhances without any change in the SOD activity (Choudhury et al. 2003).

22.7 Reproductive Dysfunctions in Thyroid Disorders

Thyroid disorders are mainly categorized into two groups—hypothyroidism and hyperthyroidism. Any alteration in serum thyroid hormone levels has profound effects on male reproduction.

22.7.1 Hypothyroidism

A rise in serum LH and FSH levels is reported in hypothyroid men (Jaya Kumar 1990). However, other studies have shown a rise in only FSH levels without any change in LH levels (Velazquez and Arata 1997). On the other hand, there are yet some studies which report no change in serum LH and FSH levels in hypothyroid men (Donnelly and White 2000). Thyroid hormones also alter pituitary response to gonadotropin-releasing hormone (GnRH). Administration of GnRH to hypothyroid patients results in an attenuated LH response (Prasannakumar et al. 1987).

Studies have shown a fall in circulating testosterone levels in hypothyroid patients (Donnelly and White 2000; Jaya Kumar et al. 1990). But few studies show no change in free testosterone levels in hypothyroid men (Velazquez and Arata

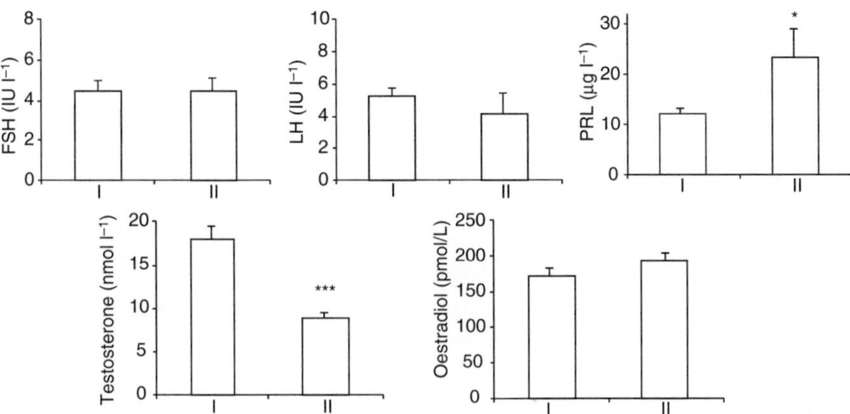

Fig. 22.6 Serum hormone levels in hypothyroid men. Bars I and II represent mean ± SEM of normal ($n = 22$) and hypothyroid patients ($n = 12$); *FSH* follicle-stimulating hormone, *LH* luteinizing hormone, *PRL* prolactin (Adapted from Kumar et al. 2007)

1997). The fall in testosterone levels could be due to (1) low cholesterol uptake by Leydig cells since serum cholesterol levels are high, (2) lower conversion of progesterone to testosterone as testosterone to progesterone ratio is low, (3) higher conversion of estradiol to testosterone as estradiol to testosterone ratio is high and (4) hyperprolactinemia (Kumar et al. 2007; see Fig. 22.6).

High prolactin suppresses 17α-hydroxylase enzyme in rat testicular cells which catalyses the conversion of progesterone to 17α-hydroxyprogesterone (Welsh et al. 1986). Our study has shown that gonadotropin levels remain unchanged in hypothyroid men. The same study also reports no change in the serum estradiol levels in hypothyroid men, suggesting the primary role of estradiol in the feedback regulation of pituitary gonadotropin secretion (Kumar et al. 2007). In aromatase-deficient men, there is an increase in basal and GnRH-stimulated LH and FSH secretion even with normal testosterone levels (Rochira et al. 2002). Administration of estrogen to the aromatase-deficient men decreases the amplitude and frequency of LH and the amplitude of FSH. The results prove estradiol as the prime regulator of negative feedback on pituitary levels of gonadotropins (Rochira et al. 2002, 2006).

Clinically, Chinese cretins show symptoms of both myxoedema and neurological cretinism. The majority of cretins show normal TSH levels. However, the hypothyroid cretin men have very high TSH levels. The hypothyroid cretins show a significant rise in serum LH, FSH and prolactin levels. Hypothyroid cretins also have testicular volume lower than 10 ml suggesting hypogonadism (Boyages and Halpem 1993).

Hypothyroid men have high TSH levels and low thyroid hormone levels confounding the exact mechanism of inhibition of testicular steroidogenesis, whether it is low levels of T_3 or a rise in TSH. The subclinical hypothyroid men show an endocrine profile similar to frankly hypothyroid males. Subclinical hypothyroid men show a significant fall in the levels of serum testosterone and progesterone and rise in serum prolactin levels but no significant change in pituitary gonadotropin and estradiol levels (Kumar et al. 2006). As only TSH is raised in subclinical hypothyroid patients without any change in thyroid hormones, this suggests a possible role of TSH in regulating gonadal steroidogenesis in males.

In the MLTC-1 cells, TSH directly inhibits progesterone and testosterone synthesis by inhibiting the 3β-HSD. The levels of gonadal steroids and gonadotropins in subclinical hyperthyroid men, with low TSH levels and normal thyroid hormone levels, are no different from euthyroid controls (unpublished data from our laboratory).

22.7.1.1 Sperm Function and Semen

Men with hypothyroidism show abnormal sperm morphology. Treatment with levothyroxine shows significant improvement in morphology (Krassas et al. 2008). Sperm motility is also reduced in hypothyroid men.

22.7.1.2 Sexual Function

Hypothyroid males also show altered sexual behavior. In adult hypothyroid males, impaired sexual behavior including hypoactive sexual desire, erectile dysfunction (ED) and ejaculatory disorders are prevalent. In hypothyroid patients, the sexual behavior improves with restoration of euthyroid status (Krassas et al. 2010). The exact cause of sexual dysfunction in hypothyroidism is not clear. It can either be attributed to persistent mild hyperprolactinemia as reported in the hypothyroid patients or to a rise in estrogen to testosterone ratio as reported by us.

22.7.2 Hyperthyroidism

In hyperthyroid men, there is an increase in the serum levels of testosterone and estradiol, but their progesterone levels are similar to euthyroid controls (Abalovich et al. 1999; Kumar et al. 2012). Testosterone to progesterone ratio in the hyperthyroid group is significantly higher, suggesting a higher conversion rate of progesterone to testosterone. However, the cause of normal levels of progesterone in spite of increased conversion to testosterone is due to an increase in synthesis of progesterone (Kumar et al. 2012). A fall in serum total cholesterol and low-density lipoprotein-C (LDL-C) levels in hyperthyroid men suggests an increase in utilization of serum cholesterol which may further be utilized for increased progesterone synthesis (Kumar et al. 2012). The increase in serum estradiol levels indicates an increase in peripheral aromatization of testosterone to estradiol. Further, hyperthyroidism being a state of hyperdynamic circulation is expected to increase the tissue blood flow along with increase in availability of testosterone to the peripheral tissues for aromatization (Longcope 1982). In hyperthyroid men, there is an increase in dialysable free estradiol without any change in dialysable free testosterone (Hudson and Edwards 1992). Also, in hyperthyroid men, there is a decrease in free testosterone to free estradiol ratio suggesting an increase in aromatization of testosterone. The doubling of estradiol levels suggests a rise in aromatization (Hudson and Edwards 1992). Hyperthyroid men also show an increase in serum sex hormone-binding globulin (SHBG). SHBG decreases the metabolic clearance rate of testosterone

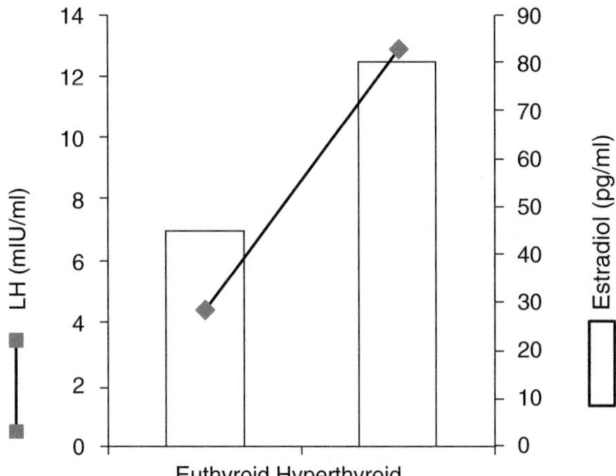

Fig. 22.7 Serum LH and estradiol levels in hyperthyroid and euthyroid males. *LH* luteinizing hormone (Adapted from Kumar et al. 2012)

which partly attributes to increase in serum total testosterone levels observed in hyperthyroid men. In spite of raised SHBG levels, there is sufficient amount of circulating free testosterone available as evident by a rise in calculated bioavailable testosterone (Kumar et al. 2012).

During the preovulatory stage, the positive feedback of estradiol is characterized by 3–12-fold rise in LH levels in cycling females. Studies show that maintenance of serum estradiol concentration, similar to that present in women at mid-cycle, for a period of 96–122 h, leads to a surge of LH in both intact and castrated men (Barbarino et al. 1983). However, the magnitude of LH surge is not as huge as found in females. Our studies, too, report a similar positive feedback effect of raised estradiol levels on LH secretion suggesting a key role of estrogen in regulating serum gonadotropin levels. The rise in LH is not as high as seen in cycling females which may be due to high testosterone levels (see Fig. 22.7, Kumar et al. 2012). Injections of estradiol to patients with Klinefelter syndrome for 5 days induce an initial decline in the serum levels of FSH and LH, followed by a 6.6-fold rise in estradiol levels on day 3 and 1.7-fold increase in LH levels on day 4 of injection. However, there is a suppression of FSH levels till day 7. A positive feedback effect of estradiol on LH secretion is also seen in patients with Klinefelter syndrome (Boucekkine and Semrouni 1990).

22.7.2.1 Sperm Function and Semen

In hyperthyroid men, there is a significant decrease in semen volume, sperm count, sperm motility and number of morphologically normal sperm (Krassas et al. 2002; Zamoner et al. 2008). There is also evidence to suggest the role of thyroid hormone in a permissive capacity on the ejaculatory mechanism (Lotti et al. 2016).

22.7.2.2 Breasts

Hyperthyroidism is often associated with gynaecomastia. A rise in serum SHBG and a subsequent fall in free testosterone are probable causes of development of gynaecomastia in hyperthyroid men (Karnath 2008). A fall in free testosterone to free estradiol ratio suggests that the hormonal imbalance may lead to gynaecomastia (Hudson and Edwards 1992; Abalovich et al. 1999).

22.7.2.3 Sexual Functions

Hyperthyroid men show impairment in sexual behavior. There is a higher prevalence of ED and ejaculatory disorders in hyperthyroid men. A return to euthyroid state reverses the abnormality in sexual functions (Krassas et al. 2010). A large survey-based study on sample size of 3369 adult males suggested a strong association between overt hyperthyroidism and severe ED (Corona et al. 2012).

Key Questions

1. Describe the types of thyroid hormone receptors present on the reproductive tissues.
2. Make a flow diagram of the mechanism of action of T_3 and TSH on target tissues.
3. How do thyroid hormones regulate maturation of Sertoli cells?
4. Explain briefly the role of thyroid hormones in regulating oxidative balance in the testis.

Acknowledgement We are thankful to the *Indian Journal of Endocrinology and Metabolism* for giving us permission to publish this article in a modified form (Kumar A, Shekhar S, Dhole B. Thyroid and male reproduction. IJEM. 2014;18:23–31).

References

Abalovich M, Levalle O, Hermes R. Hypothalamic–pituitary–testicular axis and seminal parameters in hyperthyroid males. Thyroid. 1999;9:857–63.

Allgeier A, Offermanns S, Van Sande J, Spicher K, Schultz G, Dumon JE. The human thyrotropin receptor activates G-proteins Gs and Gq/11. J Biol Chem. 1994;269:13733–5.

Amadi K, Nwana EJ, Otubu JA. Effect of thyroxine on the contractile responses of the vas deferens to prostaglandin E2. Arch Androl. 1999;42:55–62.

Anbalagan J, Sashi AM, Vengatesh G, Stanley JA, Neelamohan R, Aruldhas MM. Mechanism underlying transient gestational-onset hypothyroidism-induced impairment of posttesticular sperm maturation in adult rats. Fertil Steril. 2010;93:2491–7.

Arambepola NK, Bunick D, Cooke PS. Thyroid hormone effects on androgen receptor messenger RNA expression in rat Sertoli and peritubular cells. J Endocrinol. 1998;156:43–50.

Ariyaratne HB, Mills N, Mason IJ, Mendis-Handagama SM. Effects of thyroid hormone on Leydig cell regeneration in the adult rat following ethane dimethane sulphonate treatment. Biol Reprod. 2000;63:1115–23.

Aruldhas MM, Ramalingam N, Jaganathan A, John S, Arokya M, Stanley JA, et al. Gestational and neonatal-onset hypothyroidism alters androgen receptor status in rat prostate glands at adulthood. Prostate. 2010;70:689–700.

Barbarino A, Marinis LD, Mancini A. Estradiol modulation of basal and gonadotropin-releasing-hormone-induced gonadotropin release in intact and castrated men. Neuroendocrinology. 1983;36:105–11.

Bates JM, St Germain DL, Galton VA. Expression profiles of the three iodothyronine deiodinases, D1, D2, and D3, in the developing rat. Endocrinology. 1999;140:844–51.

Bendsen E, Byskov AG, Laursen SB, Larsen HE, Andersen CY, Westergaard LG. Number of germ cells and somatic cells in human fetal testes during the first weeks after sex differentiation. Hum Reprod. 2003; 18:13–8.

Bianco AC, Salvatore D, Gereben B, Berry MJ, Larsen PR. Biochemistry, cellular and molecular biology, and physiological roles of the iodothyronine selenodeiodinases. Endocr Rev. 2002;23:38–89.

Boucekkine C, Semrouni M. Presence of positive feedback in males with Klinefelter's syndrome. Horm Res. 1990;33:244–7.

Boyages SC, Halpern JP. Endemic cretinism: toward a unifying hypothesis. Thyroid. 1993;3:59–69.

Bunick D, Kirby J, Hess RA, Cooke PS. Developmental expression of testis messenger ribonucleic acids in the rat following propylthiouracil-induced neonatal hypothyroidism. Biol Reprod. 1994;51:706–13.

Buzzard JJ, Morrison JR, O'Bryan MK, Song Q, Wreford NG. Developmental expression of thyroid hormone receptors in the rat testis. Biol Reprod. 2000;62:664–9.

Buzzard JJ, Wreford NG, Morrison JR. Thyroid hormone, retinoic acid, and testosterone suppress proliferation and induce markers of differentiation in cultured rat Sertoli cells. Endocrinology. 2003;144:3722–31.

Catalano S, Pezzi V, Chimento A, Giordano C, Carpino A, Young M, et al. Triiodothyronine decreases the activity of the proximal promoter (PII) of the aromatase gene in the mouse Sertoli cell line, TM4. Mol Endocrinol. 2003;17:923–34.

Chatonnet F, Livera G, Fumel B, Fouch ÉCourt S, Flamant F. Direct and indirect consequences on gene expression of a thyroid hormone receptor alpha 1 mutation restricted to Sertoli cells. Mol Reprod Dev. 2014;81:1159–66.

Choudhury S, Chainy GB, Mishro MM. Experimentally induced hypo- and hyper-thyroidism influence on the antioxidant defence system in adult rat testis. Andrologia. 2003;35:131–40.

Cooke PS, Hess RA, Porcelli J, Meisami E. Increased sperm production in adult rats after transient neonatal hypothyroidism. Endocrinology. 1991;29:244–8.

Corona G, Wu FC, Forti G, Lee DM, O'Connor DB, O'Neill TW, et al. Thyroid hormones and male sexual function. Int J Androl. 2012;35:668–79.

Corrales-Hernandez JJ, Miralles-Garcia JM, Garcia-Diez LC. Primary hypothyroidism and human spermatogenesis. Arch Androl. 1990;25:21–7.

De Lamirande E, Gagnoc C. Reactive oxygen species and human spermatozoa. II Depletion of adenosine triphosphate plays an important role in the inhibition of sperm motility. J Androl. 1992;13:379–86.

De Paul AL, Mukdsi JH, Pellizas CG, Montesinos M, Gutierrez S, Susperreguy S, et al. Thyroid hormone receptor alpha 1-beta 1 expression in epididymal epithelium from euthyroid and hypothyroid rats. Histochem Cell Biol. 2008;129:631–42.

Donnelly P, White C. Testicular dysfunction in men with primary hypothyroidism; reversal of hypogonadotrophic hypogonadism with replacement thyroxine. Clin Endocrinol (Oxf). 2000;52:197–201.

Fumel B, Froment P, Holzenberger M, Livera G, Monget P, Fouchecourt S. Expression of dominant-negative thyroid hormone receptor alpha 1 in Leydig and Sertoli cell demonstrates no additional defect compared with expression in Sertoli cells only. PLoS One. 2015;10:e0119392.

Gilleron J, Nebout M, Scarabelli L, Senegas-Balas F, Palmero S, Segretain D, et al. A potential novel mechanism involving connexion 43 gap junction for control of Sertoli cell proliferation by thyroid hormones. J Cell Physiol. 2006;209:153–61.

Gomez E, Buckingham DW, Brindle J, Lanzafame F, Irvine DS, Aitken RJ. Development of an image analysis system to monitor the retention of residual cytoplasm by human spermatozoa: correlation with biochemical markers of the cytoplasmic space, oxidative stress, and sperm function. J Andol. 1996;17:276–87.

Hardy MP, Kirby JD, Hess RA, Cooke PS. Leydig cell increase their numbers but decline in steroidogenic function in the adult after neonatal hypothyroidism. Endocrinology. 1993;132:2417–20.

Hardy MP, Sharma RS, Arambepola NK, Sottas CM, Russell LD, Bunick D, et al. Increased proliferation of Leydig cells induced by neonatal hypothyroidism in the rat. J Androl. 1996;17:231–8.

Hernandez A, Martinez ME, Fiering S, Galton VA, St Germain D. Type 3 deiodinase is critical for the maturation and function of the thyroid axis. J Clin Invest. 2006;116:476–84.

Hess RA, Cooke PS, Bunick D, Kirby JD. Adult testicular enlargement induced by neonatal hypothyroidism is accompanied by increased Sertoli and germ cell numbers. Endocrinology. 1993;132:2607–13.

Hsieh ML, Juang HH. Cell growth effects of triiodothyronine and expression of thyroid hormone receptor in prostate carcinoma cells. J Androl. 2005;26:422–8.

Hudson R, Edwards A. Testicular function in hyperthyroidism. J Androl. 1992;13:117–24.

Jana NR, Bhattacharya S. Binding of thyroid hormone to goat testicular cell induces generation of a proteinaceous factor which stimulates androgen release. J Endocrinol. 1994;143:549–56.

Jana NR, Halder S, Bhattacharya S. Thyroid hormone induces a 52 kDa soluble protein in goat testis Leydig cell which stimulates androgen release. Biochim Biophys Acta. 1996;1292:209–14.

Jannini EA, Crescenzi A, Rucci N, Screponi E, Carosa E, De Matteis A, et al. Ontogenetic pattern of thyroid hormone receptor expression in the human testis. J Clin Endocrinol Metab. 2000;85:3453–7.

Jaya Kumar B, Khurana ML, Ammini AC, Karmarkar MG, Ahuja MM. Reproductive endocrine functions in men with primary hypothyroidism: effect of thyroxine replacement. Horm Res. 1990;34:215–8.

Karnath BM. Gynecomastia. Hosp Physician. 2008;44:45–51.

Kirby JD, Jetton AE, Cooke PS, Hess RA, Bunick D, Ackland JF, et al. Developmental hormonal profiles accompanying the neonatal hypothyroidism-induced increase in adult testicular size and sperm production in the rat. Endocrinology. 1992;131:559–65.

Krassas GE, Papadopoulou F, Tziomalos K, Zeginiadou T, Pontikides N. Hypothyroidism has an adverse effect on human spermatogenesis: a prospective, controlled study. Thyroid. 2008;18:1255–9.

Krassas GE, Pontikides N, Deligianni V, Miras K. A prospective controlled study of the impact of hyperthyroidism on reproductive function in males. J Clin Endocrinol Metab. 2002;87:3667–71.

Krassas GE, Poppe K, Glinoer D. Thyroid function and human reproductive health. Endocr Rev. 2010;31:702–55.

Kumar A, Chaturvedi PK, Mohanty BP. Hypoandrogenaemia is associated with subclinical hypothyroidism in men. Int J Androl. 2006;30:14–20.

Kumar A, Dewan R, Suri J, Kohli S, Shekhar S, Dhole B, et al. Abolition of endocrine dimorphism in hyperthyroid males? An argument for the positive feedback effect of hyperestrogenaemia on LH secretion. Andrologia. 2012;44:217–25.

Kumar A, Mohanty BP, Rani L. Secretion of testicular steroids and gonadotrophins in hypothyroidism. Andrologia. 2007;39:253–60.

Laslett AL, Li LH, Jester Jr WF, Orth JM. Thyroid hormone downregulates neural cell adhesion molecule expression and affects attachment of gonocytes in Sertoli cell-gonocyte cocultures. Endocrinology. 2000;141:1633–41.

Lei J, Nowbar S, Mariash CN, Ingbar DH. Thyroid hormone stimulates Na-K-ATPase activity and its plasma membrane insertion in rat alveolar epithelial cells. Am J Physiol Lung Cell Mol Physiol. 2003;285:762–72.

Longcope C. Methods and results of aromatization studies in vivo. Cancer Res. 1982;42:3307s–11.

Lotti F, Maseroli E, Fralassi N, Degl'Innocenti S, Boni L, Baldi E, et al. Is thyroid hormones evaluation of clinical value in the work-up of males of infertile couples? Hum Reprod. 2016;31:518–29.

Low MJ. Neuroendocrinology. In: Melmed S, Polonsky KS, Larsen PR, Kronenberg HM, editors. Williams textbook of endocrinology. 12th ed. Philadelphia: Elsevier; 2011. p. 103–74.

Manna PR, Kero J, Tena-Sempere M, Pakarinen P, Stocco DM, Huhtaniemi IT. Assessment of mechanisms of thyroid hormone action in mouse Leydig cells: regulation of the steroidogenic acute regulatory protein, steroidogenesis and luteinizing hormone receptor function. Endocrinology. 2001;142:319–31.

Manna PR, Tena-Sempere M, Huhtaniemi IT. Molecular mechanisms of thyroid hormone-stimulated steroidogenesis in mouse leydig tumor cells. J Biol Chem. 1999;274:5909–18.

Maran RR, Aruldhas MM, Udhayakumar RC, Subramanian S, Rajendiren G, Antony FF, et al. Impact of altered thyroid hormone status on prostatic glycosidases. Intl J Androl. 1998;21:121–8.

Maran RR, Arunakaran J, Aruldhas MM. T_3 directly stimulates basal and modulates LH induced testosterone and estradiol production by rat Leydig cells in vitro. Endocr J. 2000;47:417–28.

Martinez ME, Karaczyn A, Stohn JP, Donnelly WT, Croteau W, Peeters RP, et al. The Type 3 Deiodinase is a critical determinant of appropriate thyroid hormone action in the developing testis. Endocrinology. 2016;157:1276–88.

Meeker JD, Godfrey-Bailey L, Hauser R. Relationships between serum hormone levels and semen quality among men from an infertility clinic. J Androl. 2007;28:397–406.

Mendis-Handagama SM, Ariyaratne HB. Differentiation of the adult leydig cell population in the postnatal testis. Biol Reprod. 2001;65:660–71.

Mondul AM, Weinstein SJ, Bosworth T, Remaley AT, Virtamo J, Albanes D. Circulating thyroxine, thyroid stimulating hormone, and hypothyroid status and the risk of prostate cancer. PLoS One. 2012;7:1–7.

Mruk DD, Cheng Y. Sertoli-Sertoli and Sertoli-Germ cell interactions and their significance in germ cell movement in the seminiferous epithelium during spermatogenesis. Endocr Rev. 2004;25:747–806.

Ochsendorf FR. Infections in the male genital tract and reactive oxygen species. Hum Reprod Update. 1999;5:399–420.

Ortiga-Carvalho TM, Sidhaye AR, Wondisford FE. Thyroid hormone receptors and resistance to thyroid hormone disorders. Nature Rev Endocrinol. 2014;10:582–91.

Prasannakumar KM, Khurana ML, Ammini AC, Godbole MM, Ahuja MM. Pituitary response to luteinizing hormone releasing hormone in hypothyroidism. In: Vichanrat A, Nitiyanant W, Eastmen C, editors. Recent progress in thyroidology. New Delhi: Crystal; 1987. p. 511–5.

Rochira V, Balestrieri A, Faustini-Fustini M, Borgato S, Beck-Peccoz P, Carani C. Pituitary function in a man with congenital aromatase deficiency: Effect of different doses of transdermal E2 on basal and stimulated pituitary hormones. J Clin Endocrinol Metab. 2002;87:2857–62.

Rochira V, Zirilli L, Genazzani AD, Balestrieri A, Aranda C, Fabre B, et al. Hypothalamic-pituitary-gonadal axis in two men with aromatase deficiency: evidence that circulating estrogens are required at the hypothalamic level for the integrity of gonadotropin negative feedback. Eur J Endocrinol. 2006;155:513–22.

Romano RM, Gomes SN, Cardoso NC, Schiessl L, Romano MA, Oliveira CA. New insights for male infertility revealed by alterations in spermatic function and differential testicular expression of thyroid-related genes. Endocrine. 2017;55:607–17.

Sharpe RM, McKinnell C, Kivlin C, Fisher JS. Proliferation and functional maturation of Sertoli cells, and their relevance to disorders of testis function in childhood. Reproduction. 2003; 125:769–84.

Silva D, Lizama C, Tapia V, Moreno RD. Propylthiouracil-induced hypothyroidism delays apoptosis during the first wave of spermatogenesis. Biol Res. 2011;44:181–8.

St-Pierre N, Dufresne J, Rooney AA, Cyr DG. Neonatal hypothyroidism alters the localization of gap junctional protein connexin 43 in the testis and messenger RNA levels in the epididymis of the rat. Biol Reprod. 2003;68:1232–40.

Sugawara T, Holt JA, Kiriakidou M, Strauss JF. Steroidogenic factor 1-dependent promoter activity of the human steroidogenic acute regulatory protein (StAR) gene. Biochemistry. 1996;35: 9052–9.

Tremellen K. Oxidative stress and male infertility-A clinical perspective. Hum Reprod Update. 2008;14:243–58.

Trokoudes KM, Sugenoya A, Hazani E, Row VV, Volpe R. Thyroid-stimulating hormone (TSH) binding to extrathyroidal human tissues: TSH and thyroid-stimulating immunoglobulin effects on adenosine 3′,5′-monophosphate in testicular and adrenal tissues. J Clin Endocrinol Metab. 1979;48:919–23.

Velazquez EM, Arata GB. Effects of thyroid status on pituitary gonadotropin and testicular reserve in men. Arch Androl. 1997;38:85–92.

Welsh TH, Kasson BG, Hsueh AW. Direct biphasic modulation of gonadotropin-stimulated testicular androgen biosynthesis by prolactin. Biol Reprod. 1986;34:796–804.

Wrutniak-Cabello C, Casas F, Cabello G. Thyroid hormone action in mitochondria. J Mol Endocrinol. 2001;26:67–77.

Zamoner A, Barreto KP, Filho DW, Sell F, Woehl VM, Guma FC, et al. Propylthiouracil-induced congenital hypothyroidism upregulates vimentin phosphorylation and depletes antioxidant defences in immature rat testis. J Mol Endocrinol. 2008;40:125–35.

Metabolism and Endocrines

23

Mohd Ashraf Ganie, Sk Hammadur Rahaman, and Nishant Raizada

Learning Objectives
- Obesity and gonadal axis in men
- Diabetes, hypogonadism and fertility
- Role of insulin in spermatogenesis
- Endocrine disorders and reproduction

23.1 Introduction

The hypothalamic-pituitary-gonadal (HPG) axis in men has two main functions: androgen synthesis and spermatogenesis. Apart from the conditions directly affecting the HPG axis, several other endocrine disorders also cause alterations in the functioning of the testis or the pituitary gonadotrophs (Horiba and Rinsho 1997; Maneesh et al. 2006; Kyrou and Tsigos 2008). The various hormonal pathways show variable degrees of interdependence because of which non-gonadal endocrine disorders also affect testosterone levels and spermatogenesis in men. This chapter summarizes the interaction and alterations of male gonadostat by metabolic syndrome (MS), diabetes mellitus (DM), Addison's disease, Cushing's syndrome,

M.A. Ganie, MD, DM (✉) • SK. H. Rahaman, MD • N. Raizada, MD, DM
Department of Endocrinology and Metabolism,
All India Institute of Medical Sciences, New Delhi, India
e-mail: ashraf.endo@gmail.com

A. Kumar, M. Sharma (eds.), *Basics of Human Andrology*,
DOI 10.1007/978-981-10-3695-8_23

acromegaly and prolactinoma. Before proceeding to discussion about the role of individual endocrine disorders and their contribution in male hypogonadism, one should be aware that any critical illness, whether endocrine or non-endocrine, can cause a state of functional hypopituitarism (Karagiannis and Harsoulis 2005). The examples of these disorders are diabetic ketoacidosis, myxoedema coma, thyrotoxic crises and adrenal insufficiency. Among these, hypothyroidism and DM contribute to a maximum number of cases as these are common conditions. The impact of thyroid dysfunction on HPG axis has been described in the previous chapter, Thyroid. This chapter also addresses the issues of poor fertility in diabetic men despite apparently normal semen analysis and the recently discovered role of insulin in spermatogenesis.

23.2 Metabolic Syndrome/Obesity and Gonadal Axis in Men

Several clinical conditions may result in secondary hypogonadism in obese subjects including obesity itself. Bidirectional relationships exist between serum testosterone and obesity as well as between testosterone and MS. Total testosterone levels are often lower in obese men and levels decrease as body mass index (BMI) increases. MS is the existence of various metabolic risk factors such as hypertension, abdominal obesity, dyslipidaemia and impaired glucose tolerance. These risk factors are associated with severalfold increased risk of cardiovascular disease and type 2 DM (Ford 2005). The definitions of MS according to the National Cholesterol Education Program-Adult Treatment Panel III (NCEP ATP III), the World Health Organization (WHO) and the International Diabetes Federation (IDF) are widely accepted. Most of these criteria include waist circumference as a marker of obesity along with elevated blood pressure, elevated blood glucose and dyslipidaemia. The prevalence of MS is more common in women and increases with age. There is enough data which suggests that MS and its different components are more frequent in women with hyperandrogenism, such as polycystic ovary syndrome (PCOS), and hypoandrogenimic men (Glueck et al. 2003). To support this idea, studies show that prostate cancer patients undergoing androgen deprivation therapy and hypogonadal men have been associated with MS (Chubb et al. 2008).

The impact of MS on reproductive functions is not clearly understood; one can consider it to be mediated by one or both the components – obesity and insulin resistance. Obesity is a state of excess free fatty acids (FFAs), proinflammatory cytokines and adipokines. Visceral fat is the main source of inflammatory cytokines, adipokines, biochemical modulators and proinflammatory factors like interleukin (IL) 6, IL-1β, tumor necrosis factor alpha (TNFα), vascular endothelial growth factor, angiotensinogen, plasminogen activator inhibitor-1 and serum amyloid A (see Fig 23.1). These risk factors predispose significantly to the development of type 2 DM, androgen deficiency and MS (Traish et al. 2009). Besides, they may lead to systemic and peripheral vascular inflammation and endothelial dysfunction (Guzik et al. 2006). The pathways of nuclear factor-kB (NFkB) are activated by free fatty acids resulting in excess synthesis of TNF-α. TNF-α by

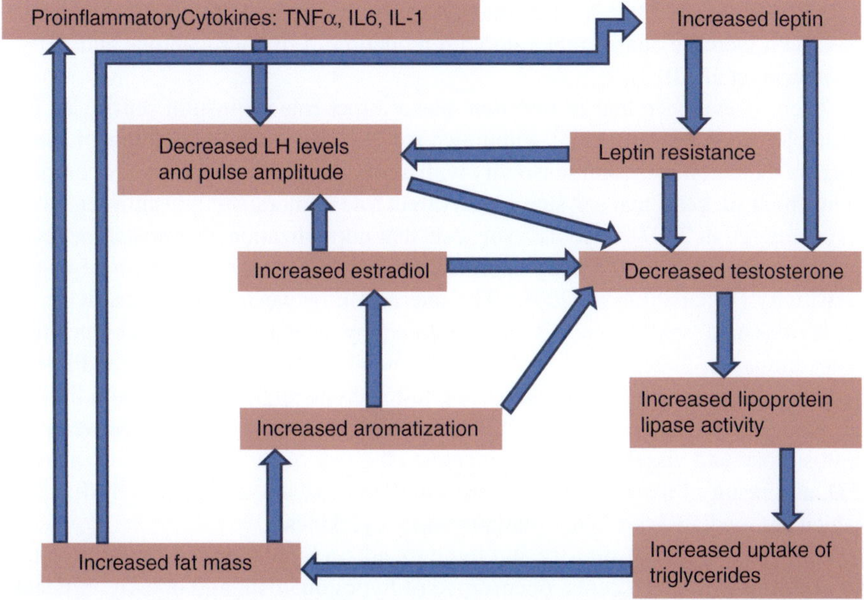

Fig. 23.1 Flow chart showing the effects of obesity and adipocytokines on gonadal axis. *TNFα* tumor necrosis factor alpha, *IL-6* interleukin 6, *IL-1* interleukin 1, *LH* luteinizing hormone

itself increases the production of free fatty acids, IL-6 and macrophage chemoattractant protein-1 that in turn causes recruitment of more macrophages. TNF-α acts on vascular endothelium and smooth muscle cells to increase expression of adhesion molecules. IL-6 increases levels of C-reactive protein, a marker of vascular inflammation, and TNF-α promotes vasoconstriction by altering endothelin-1-mediated vasoconstriction and nitric oxide-mediated vasodilatation. This is the other mechanism by which release of FFAs by visceral fat alters response to insulin sensitivity. Obesity has also been shown to be associated with elevated estrogen in men due to high aromatase activity, the enzyme that converts testosterone to estradiol. The estrogens activate hypothalamic estrogen receptors leading to inhibition of the HPG axis. That is why the hypogonadotropic hypogonadism associated with obesity can be reversed by treatment with aromatase inhibitor in some patients. The low testosterone and high estradiol levels due to aromatase excess are partially reversed by aromatase inhibitors like letrozole (Loves et al. 2008). Low serum testosterone is associated with the development of central obesity, and the likely explanation for this is peripheral insulin resistance and resultant hyperinsulinaemia causing decrease in sex hormone-binding globulin (SHBG). Data indicates that there is attenuated Leydig cell responsiveness to exogenous gonadotropin stimulation in men with obesity and insulin resistance (Pitteloud et al. 2005). In the European Male Ageing Study, three sexual symptoms – poor morning erections, low libido and erectile dysfunction (ED) – were related with low testosterone levels (Wu et al. 2010).

In the European Male Ageing Study, it has been shown that weight loss is associated with increase and weight gain with reduction in free testosterone and SHBG (Camacho et al. 2013).

There is evidence that testosterone has a direct role on insulin sensitivity. For example, when testosterone is withdrawn for 2 weeks in men with hypogonadism, there is reduction in insulin sensitivity with no significant changes in body composition which suggests that sex steroid has direct role in modulating insulin sensitivity (Yialamas et al. 2007). Also data suggests that normalization of testosterone levels in older men changes the body composition favourably that may improve insulin sensitivity (Agledahl et al. 2008). The interaction between insulin sensitivity, triglycerides and sex steroids is not modified by changes in body composition. Concomitantly, there are favourable postprandial triglyceride disposal and insulin sensitivity in young men with increasing testosterone and decreasing estradiol levels for 1 week (Lapauw et al. 2010). These studies reflect that interplay between low testosterone and visceral adiposity increases the risk of cardiovascular disease and ED, as a result of increased insulin resistance and vascular endothelial dysfunction which are mediated through proinflammatory cytokines.

The coexistence of obesity and hypogonadism is not limited to adult males. In children and adolescents, occurrence of hypogonadism and obesity indicates syndromic causes. Obese adolescents show significantly lower total testosterone levels compared with nonobese ones (Mogri et al. 2013). Obese boys show delayed pubertal development unlike girls who show early onset of puberty. Testing of anterior pituitary functions and magnetic resonance imaging (MRI) may be needed to rule out mass lesion in pituitary or hypothalamus. Lesions in the satiety centre of the hypothalamus may be the underlying aetiology in obesity; however, this is uncommon. During clinical assessment, it should be remembered that the penis buried under the suprapubic pad of fat may give a false appearance of micropenis or buried penis, thereby erroneously linking obesity and hypogonadism. In addition to buried penis, obese adolescents can have lipomastia. These two findings should not be misinterpreted as micropenis and gynaecomastia in obese adolescents to avoid unnecessary investigations directed towards hypogonadism.

The testosterone link with obesity is supported by relatively stronger evidence, but the relationship between obesity and sperm count is weaker. Most of the studies have found no relationship between obesity and total sperm count, whereas sperm morphology has been shown to be affected by obesity in some of the studies (Jensen et al. 2004; Aggerholm et al. 2008). Therefore, it can be suggested that hormonal regulation of spermatogenesis is maintained even in obese men. Despite of these evidences, obesity remains an important risk factor for infertility in men which has been shown by at least one study where erectile function improves with weight loss (Sallmen et al. 2006; Passquali et al. 2007). Thus, it is interesting that although BMI does not alter semen parameters, it is strongly related to hypogonadism in males. This discrepancy can be explained by two findings: firstly, free testosterone is less affected in obese than total testosterone, and secondly, FSH is not related to BMI.

23.3 Diabetes Mellitus and Hypogonadism

DM represents a group of disorders which share the common phenotype of hyperglycaemia. The most common type of diabetes is type 2 DM which is mostly associated with insulin resistance. As discussed above, MS and related insulin resistance are the risk factors which frequently predispose to type 2 DM. Hypogonadism and type 2 DM frequently coexist. The impact of type 2 DM on gonadal axis is due to the effects of MS, but these effects are more pronounced in overt type 2 DM.

Androgen deficiency has a role in various pathologic processes of DM and insulin resistance, but the relationship of DM and low testosterone is interesting. It is known that adipose tissue particularly visceral adipose tissue is involved in alterations in glucose and lipid metabolism as well as oxidative stress via free radical injury. There are several proinflammatory substances and adipokines which modulate these effects (Gustafson et al. 2007). These include interleukin (IL) 6, TNFα IL-1b, plasminogen activator inhibitor 1, vascular endothelial growth factor, angiotensin-converting enzymes, angiotensinogen and serum amyloid A proteins. IL-6 and TNF-α are associated with obesity and insulin resistance by interfering with insulin signalling in mature adipocytes. Testosterone has been found to cause a reduction in production of proinflammatory cytokines and chemokines, whereas androgen deprivation therapy leads to increased levels of proinflammatory factors and decreased anti-inflammatory cytokines (Malkin et al. 2004; Maggio et al. 2006). Conversely, treatment with testosterone prevents increase of visceral adipose tissue in nonobese ageing men (Schroeder et al. 2004). It has been shown that steroidogenesis in the Leydig cells is affected by circulating hormones and local cytokines and thus provides a link of insulin resistance and suppression of testosterone synthesis (Saez 1994).

The effects of diabetes and insulin resistance on the HPG axis are more complicated. DM subjects often have lower luteinizing hormone (LH) and follicle-stimulating hormone (FSH) levels representing a form of secondary hypogonadism (Ballester et al. 2004). Clinical studies have shown that subjects with low testosterone are at a greater risk of developing type 2 DM as one of the meta-analyses has suggested that higher plasma testosterone levels are associated with lower risk of diabetes and vice versa (Ding et al. 2006). Several observational studies have reported an inverse relationship between testosterone and insulin resistance in men as free testosterone shows stronger relationship than total testosterone (Rhoden et al. 2005). The role of testosterone deficiency in development of insulin resistance has been confirmed (Stellato et al. 2000). One of the studies has found no correlation between insulin sensitivity and LH secretion or LH response to exogenous gonadotropin-releasing hormone (GnRH), suggesting that low testosterone levels are not as a result of major fall in hypothalamic or pituitary hormone secretion (Pitteloud et al. 2008).

Recent studies in rats have shown that insulin is expressed in testicular tissue. Male HPG axis is regulated by pancreatic insulin and it is essential for fertility. Exogenous insulin replacement in insulin-deficient rats restores the lost fertility, while hyperglycaemia alone does not impact fertility. The effects are probably not mediated at the testicular level as insulin does not cross blood-testis barrier and because levels of LH and testosterone levels are lower during insulin deficiency.

These levels, however, returned to normal after insulin therapy (Schoeller et al. 2012). Further questions raised by this research include the role of insulin produced by the Sertoli cells within the testis; insulin could act as a growth factor or survival signal and additionally affects glucose homeostasis in the testis. Sertoli cells also possess insulin receptors and thus may respond to circulating insulin in an unidentified manner. Additionally, the fact that Sertoli cells span the blood-testis barrier places them in a unique position in which they may deliver either insulin or insulin-mediated signals to the developing spermatocytes and spermatids.

Leptin is an adipokine secreted from fat cells and has effects on HPG axis. Increased serum leptin levels are inversely correlated with serum testosterone levels, and hence, it is postulated that leptin acts directly to inhibit steroidogenesis in the Leydig cells (Caprio et al. 1999). Similarly, TNFα also suppresses testicular steroidogenesis by several mechanisms (Hong et al. 2004). A significant negative correlation among BMI, insulin and homeostasis model assessment of insulin resistance (HOMA-IR) to testosterone levels has also been demonstrated. Data indicates that diabetic men having higher BMI are more likely to have hypogonadism and visceral fat may have stronger correlation with hypogonadism than DM (Corona et al. 2007). Insulin effect can stimulate testosterone production in vivo and reduce SHBG concentration in both obese and nonobese men (Pasquali et al. 1995). The body fat distribution may have a complex relationship with low total testosterone and elevated insulin levels (Blouin et al. 2005). Hypogonadism is perhaps a stronger risk factor for hyperinsulinaemia and hyperglycaemia compared with overweight/obesity that is substantiated by lower total testosterone and free testosterone levels in lean hypogonadal men than in overweight or obese men consequently having higher fasting insulin, glucose and HOMA-IR values (Pagotto et al. 2003).

The following are the proposed mechanisms of hypogonadism in obesity and DM:

1. Aromatization of testosterone by adipose tissue
2. Reduced Leydig cell steroidogenesis due to proinflammatory cytokines
3. Reduced SHBG
4. Leptin-mediated decrease in testosterone synthesis
5. Reduced LH release by unidentified mechanisms
6. Insulin deficiency-mediated impaired HPG function

In men with obesity or MS, low total testosterone and SHBG predict the risk of development of type 2 DM. Conversely, men with type 2 DM frequently have lower total testosterone levels than those of euglycaemic men. About 25% of type 2 diabetic men have subnormal testosterone concentrations with inappropriately low gonadotropin levels. The Endocrine Society, therefore, now recommends testosterone to be measured on a routine basis in patients with type 2 DM. Although some studies have shown that testosterone replacement improves glycaemic control and reduces insulin resistance, testosterone replacement in such cases is not recommended at present.

Erectile and ejaculatory dysfunctions are common in men with type 2 DM and not always due to androgen deficiency. Neuropathy, vascular, psychological factors and drugs usually contribute. Men with type 1 DM usually have normal testosterone and gonadotropin levels at the time of diagnosis; however, basal and stimulated LH secretion may be compromised in poorly controlled DM. In men with DM, spermatogenesis and fertility are often intact if there is no erectile or ejaculatory dysfunction. No association exists between subnormal testosterone and glycosylated haemoglobin or duration of diabetes, but association between obesity, very high C-reactive protein concentrations and mild anaemia is found to be important (Dandona and Dhindsa 2011). Subnormal testosterone levels in these men predispose to an elevated risk of cardiovascular events and death (Wang et al. 2011). In hypogonadal men with type 2 DM, testosterone therapy has shown improvement in insulin sensitivity and waist circumference. Improvement in libido has been shown after testosterone therapy in hypogonadal men and either MS or type 2 DM (Jones et al. 2011).

23.4 Diabetes and Fertility

Literature available on effect of diabetes on spermatogenesis is confusing and inconclusive. Nevertheless DM has been found to influence fertility in numerous ways. Only conventional sperm parameters like semen volume, sperm count, motility and morphology are available in studies regarding sperm quality in DM. In the evaluation of fertility status, conventional semen analysis is of limited value unless there are major problems like severe oligoasthenoteratozoospermia or azoospermia. There is no consensus in current literature on how DM affects male reproductive function, because of paucity of data. Animal studies suggest strongly that DM affects male reproductive functions. If similar effects exist in the context of human reproduction, the human fertility is at risk because of rising rates of DM. Despite this, the potential impact on reproductive health among young diabetics is less important till date. There is evidence to suggest a higher prevalence of bad obstetric outcomes such as spontaneous abortions in their partners and infertility in diabetic men. A retrospective review has shown that out of 3000 infertile couples, 80 couples have diabetic male partners. Out of these, 18 couples have gone for assisted reproduction treatment. The fertilization rates and embryo quality are similar to nondiabetic patients (Mulholland et al. 2011).

Both animal and clinical studies have shown the deleterious effect of diabetes on HPG axis. BioBreeding (BB) rat, a genetic model of type 1 diabetes, is associated with diminished testicular weight, disruption of seminiferous tubular morphology, Sertoli cell vacuolization and severe germ cell depletion. Consequently, it has low serum testosterone levels, decreased sperm production and less fertility (Murray et al. 1985; Seethalakshmi et al. 1987). Similarly, Goto-Kakizaki (GK) rat, an example of genetic model of type 2 diabetes, is associated with low sperm production (Amaral et al. 2006). Clinically induced models of type 1 diabetes (streptozotocin rat) have low testicular weight, altered morphology of epididymis, decreased sperm production and motility and low testosterone along with low gonadotropins. Clinically, they are less fertile, having decreased mating behavior and erectile as well as ejaculatory dysfunction

(Seethalakshmi et al. 1987; Soudamani et al. 2005; Amaral et al. 2006; Scarano et al. 2006). Alloxan (ALX) rat, a clinically induced model of type 2 diabetes, has diminished levels of gonadotropins as well as testosterone. It also has decreased number of spermatogonia (Jelodar et al. 2009). Similarly, in clinical studies also, both type 1 and 2 diabetics have deranged gonadal microscopic structure and function. Subjects with type 1 diabetes have depleted germ cells and disruption in seminiferous tubule structure and blood-testis barrier (Cameron et al. 1985). In concordance with type 1, type 2 diabetics were also found to have low semen volume, decreased sperm count and abnormal sperm DNA fragmentation (Ali et al. 1993). There is increased incidence of ED in both types 1 and 2 (Schoeffling et al. 1963; Bartak 1979). In addition, type 1 diabetics have ejaculatory dysfunction (Ellenberg and Weber 1966).

Dyslipidaemia is common in DM. For a normal spermatogenesis to occur, normal lipid metabolism is essential. Enormous metabolic plasticity of Sertoli cells helps them to use a variety of substrates including palmitate, ketone bodies, glutamine, alanine, leucine, glycine and valine. In addition, these cells can utilize free fatty acids and generate adenosine triphosphate (ATP) by its beta-oxidation and even through recycled lipids from apoptotic sperm. In testicular cells, lactate is produced from glucose metabolism. DM predisposes cell to face oxidative stress as a result of overproduction of reactive oxygen species (ROS) and reduced antioxidant efficiency. There is increasing levels of oxidative stress which is potentially toxic for sperm quality and function. Hyperglycaemia-induced superoxide overproduction inhibits glucose-6-phosphate dehydrogenase leading to a decrease in antioxidant defences, ROS overproduction and mitochondrial dysfunction owing to sensitivity of mitochondrial fission/fusion machinery. In fact, hyperglycaemia, by altering energy production and free radical damage, induces important alteration in sperm concentration and motility. Advanced glycation end products (AGEs) and ROS are reported to mediate sperm DNA damage in young diabetic men. ROS can also cause sperm plasma membrane and nuclear or mitochondrial DNA fragmentation, leading to compromised fertility potential of the individual. Recent data suggests that ROS in the pathogenesis of insulin resistance and hyperglycaemia of DM may go far beyond the glucose transport and metabolism.

To investigate male infertility, light microscopic assessment of sperm parameters proved inadequate, and assessment of sperm nuclear DNA (nDNA) or mitochondrial DNA (mtDNA) is an alternative approach. A study on diabetic men has shown seminal volume reduction but normal sperm count, motility and morphology (Agbaje et al. 2007). The other parameters examined in this study are nDNA fragmentation and mtDNA deletion analysis. Diabetic subjects show significantly higher nDNA fragmentation and mtDNA deletions as compared to nondiabetic subjects. Increased sperm nDNA damage may lead to poor embryo quality, lower implantation rates and some childhood onset cancers. There are accumulating evidence that the sperm nDNA damage occurs while transit through and storage in epididymis because of oxidative stress.

Sperm are rich in unsaturated fatty acid, and they also lack DNA repair mechanisms which make them susceptible to ROS that causes oxidative damage and generation of AGEs. AGEs form complexes with proteins, lipids and nucleic acids that induce DNA damage through activation of the receptor for AGEs (RAGE). RAGE belongs to IgG superfamily. AGE binding to RAGE activates downstream activation of nuclear factor (NFkB) and other signalling pathways leading to ROS generation and certain proinflammatory responses.

In DM, expression of RAGE is associated with tissue damage and altered homeostasis (Schmidt et al. 2001). The presence of RAGE has been shown in the testis, in caput epididymis and on sperm acrosomes (Mallidis et al. 2007). The overall protein content of sperm and seminal plasma and the number of sperm with positive RAGE expression are significantly higher in subjects with diabetes. In diabetic men, RAGE may affect sperm nDNA integrity. The significantly greater number of RAGE found in sperm of diabetic men in comparison to nondiabetic subjects is probably secondary to excess rate of AGE production and accumulation found in diabetics. However, specific site of RAGE expression in acrosomal cap and the equatorial region of the sperm head are particularly noteworthy, regardless of diabetic status of the man.

23.5 Role of Insulin in Spermatogenesis: Newer Concepts

Male HPG axis is regulated by insulin and it is essential for fertility (Schoeller et al. 2012). Rat testis has been found to express insulin. While primates have only one insulin gene, rodents have two such genes, namely, Ins1 and Ins2. It is the duplication of the ancestral Ins2 gene that resulted in Ins1 gene about 20 million years ago (Shiao et al. 2008). It is likely to be expressed in mouse as well as in human testes.

The pancreas shows the presence of both Ins1 and Ins2. Only the Ins2 gene is transcribed in the testis. Abnormal insulin signalling in the testis, systemic effects of insulin or both are responsible for the adverse effects of DM. Mutation in the Ins2 allele in Akita mouse model results in misfolded protein product that leads to endoplasmic reticulum stress and pancreatic beta cell death (Izumi et al. 2003). The homozygous Akita males when treated with subcutaneous insulin implants had restoration of spermatogenesis and fertility.

Insulin cannot cross the blood-testis barrier; therefore, the above effects cannot be due to restoration of insulin within the testis. Leydig cells contain insulin receptors and are located outside the blood-testis barrier and might be affected by exogenous insulin and not be responsive to insulin from Sertoli cells (Kimura et al. 1992). Akita homozygotes with significant reduction in LH and testosterone levels show improvement in levels by insulin treatment. LH production and spermatogenesis are impaired when insulin receptor gene is knocked out (Brüning et al. 2000). Insulin acts on the hypothalamus directly or it determines the effect of hypothalamic GnRH on pituitary gonadotropes, as FSH levels are not significantly different between wild-type, heterozygous and homozygous Akita males.

23.6 Cushing's Syndrome

In men with Cushing's syndrome, studies have demonstrated decreased libido, ED and infertility. Serum total testosterone, basal LH and GnRH-stimulated LH concentrations are frequently suppressed in these subjects. Adrenalectomy is known to restore gonadal function in subjects with Cushing's syndrome. This is attributed to the fact that glucocorticoids primarily act to suppress gonadotropins via inhibition of hypothalamic GnRH secretion. However, glucocorticoids may also have direct suppressive effects on testicular function leading, therefore, to both primary and secondary hypogonadism. Exogenous glucocorticoids as low as 7.5 mg of prednisolone have been shown to cause hypogonadism by interfering with HPG axis. Glucocorticoids also contribute to hypogonadism by inducing aromatase activity and decreasing SHBG levels. Therefore, serum-free testosterone levels may be needed in these subjects to confirm a biochemical diagnosis of hypogonadism for the same reason. Hypogonadotropic hypogonadism disappears when hypercortisolism is controlled (Luton et al. 1977).

23.7 Congenital Adrenal Hyperplasia

Congenital adrenal hyperplasia (CAH) due to deficiency of 21-hydroxylase or 11β-hydroxylase results in excessive secretion of adrenal androgens that are converted to testosterone which suppresses gonadotropin secretion by negative feedback and decreases endogenous testosterone secretion and, thereby, sperm production. Men with CAH, however, do not manifest with androgen deficiency due to excess adrenal androgens; instead they manifest with infertility due to isolated impairment of sperm production and function. The mechanisms described above apply to inadequately treated or untreated CAH men, and overtreatment with steroids also results in hypogonadism as described in Cushing's syndrome. It has been shown that optimal steroid therapy in these cases restores gonadotropin secretion and spermatogenesis. Testicular adrenal rest tumors arise from cells with mixed adrenal and Leydig cell function producing all major adrenal steroids resulting in oligo-/azoospermia. Testicular adrenal rest tumors can occur during childhood and adolescence and impair both Leydig and Sertoli cell function. In severe cases, these tumors may lead to azoospermia and testicular damage due to seminiferous tubule obstruction. Testis-sparing conservative surgery may not restore sperm count in these cases.

23.8 Growth Hormone Deficiency and Acromegaly

Growth hormone stimulates the synthesis of insulin-like growth factor-1 which may mediate the actions of growth hormone in testes. Therefore, children with growth hormone deficiency often have delayed puberty which suggests a role of growth hormone on Leydig cell function. That is why if growth hormone stimulation testing is done without priming with sex steroids, it may falsely mimic growth hormone deficiency. However, delayed puberty with growth hormone deficiency may be a manifestation of coexistent gonadotropin deficiency. There is no negative impact of excess growth hormone on HPG axis in men unlike growth hormone deficiency which is associated with delayed puberty. The mechanism related to hypogonadism in acromegaly includes hyperprolactinaemia or hypogonadotropic hypogonadism.

23.9 Prolactinoma

Prolactinoma presents in women with galactorrhoea or irregular menstrual cycles, but in males it often presents late with features of raised intracranial tension due to macroadenoma, visual disturbances or hypogonadism mainly ED. Hyperprolactinaemia causes hypogonadism by gonadotropin deficiency primarily by suppressing pulsatile hypothalamic GnRH secretion. Pulsatile secretion of LH is lost and decline in LH secretion is responsible for the low testosterone production and GnRH administration restores testicular function to normal. During clinical assessment, the presence of eunuchoid proportions in adult men indicates the onset in childhood and adolescence before epiphyseal fusion. The symptoms apart from hypogonadism include headache, disturbed vision, symptoms of coexistent hypopituitarism and galactorrhoea. However, only 20% present with galactorrhoea because estrogen and progesterone levels are too low to cause mammary ductal hyperplasia and glandular formation. Whether prolactin has direct effects on the human testis remains unclear. Dopamine agonists normalize gonadal functions, though serum levels usually take months to reach normal. If, despite normalization of prolactin and shrinking of tumor, testosterone levels do not normalize, then testosterone supplementation may be required. One should be aware that testosterone may get aromatized to estrogen and serum prolactin and tumor size can increase. Aromatase inhibitors may be helpful in this uncommon situation.

Key Questions

- How does metabolic syndrome affect gonadal axis?
- Describe the role of cytokines in the pathogenesis of hypoandrogenism in metabolic syndrome.
- What is the association between low testosterone levels and obesity?
- What are the different causes of hypogonadism in diabetes mellitus?
- Discuss the role of insulin in spermatogenesis.

References

Agbaje IM, Rogers DA, McVicar CM, McClure AB, Atkinson C, Mallidis C, et al. Insulin dependent diabetes mellitus: implications for male reproductive function. Hum Reprod. 2007;22:1871–7.

Aggerholm AS, Thulstrup AM, Toft G, Ramlau-Hansen CH, Bonde JP. Is overweight a risk factor for reduced semen quality and altered serum sex hormone profile? Fertil Steril. 2008;90:619–26.

Agledahl I, Hansen JB, Svartberg J. Impact of testosterone treatment on postprandial triglyceride metabolism in elderly men with subnormal testosterone levels. Scand J Clin Lab Invest. 2008;68:641–8.

Ali ST, Shaikh RN, Siddiqi NA, Siddiqi PQ. Semen analysis in insulin-dependent/ non-insulin-dependent diabetic men with/without neuropathy. Arch Androl. 1993;30:47–54.

Amaral S, Moreno AJ, Santos MS, Seica R, Ramalho-Santos J. Effects of hyperglycemia on sperm and testicular cells of Goto–Kakizaki and streptozotocin-treated rat models for diabetes. Theriogenology. 2006;66:2056–67.

Ballester J, Muñoz MC, Domínguez J, Rigau T, Guinovart JJ, Rodríguez-Gil JE. Insulin-dependent diabetes affects testicular function by FSH- and LH-linked mechanisms. J Androl. 2004;25:706–19.

Bartak V. Sperm quality in adult diabetic men. Int J Fertil. 1979;24:226–32.

Blouin K, Despres JP, Couillard C, Tremblay A, Prud'homme D, Bouchard C, et al. Contribution of age and declining androgen levels to features of the metabolic syndrome in men. Metabolism. 2005;54:1034–40.

Brüning JC, Gautam D, Burks DJ, Gillette J, Schubert M, Orban PC, et al. Role of brain insulin receptor in control of body weight and reproduction. Science. 2000;289:2122–5.

Camacho EM, Huhtaniemi IT, O'Neill TW, Finn JD, Pye SR, Lee DM, et al. Age-associated changes in hypothalamic-pituitary-testicular function in middle-aged and older men are modified by weight change and lifestyle factors: longitudinal results from the European Male Ageing Study. Eur J Endocrinol. 2013;168(3):445–55.

Cameron DF, Murray FT, Drylie DD. Interstitial compartment pathology and spermatogenic disruption in testes from impotent diabetic men. Anat Rec. 1985;213:53–62.

Caprio M, Isidori AM, Carta AR, Moretti C, Dufau ML, Fabbri A. Expression of functional leptin receptors in rodent Leydig cells. Endocrinology. 1999;140:4939–47.

Chubb SA, Hyde Z, Almeida OP, Flicker L, Norman PE, Jamrozik K, et al. Lower sex hormone-binding globulin is more strongly associated with metabolic syndrome than lower total testosterone in older men: the health in men study. Eur J Endocrinol. 2008;158:785–92.

Corona G, Mannucci E, Petrone L, Balercia G, Paggi F, Fisher AD, et al. NCEP-ATPIII-defined metabolic syndrome, type 2 diabetes mellitus, and prevalence of hypogonadism in male patients with sexual dysfunction. J Sex Med. 2007;4:1038–45.

Dandona P, Dhindsa S. Update: hypogonadotropic hypogonadism in type 2 diabetes and obesity. J Clin Endocrinol Metab. 2011;96(9):2643–51.

Ding EL, Song Y, Malik VS, Liu S. Sex differences of endogenous sex hormones and risk of type 2 diabetes: a systematic review and meta-analysis. JAMA. 2006;295:1288–99.

Ellenberg M, Weber H. Retrograde ejaculation in diabetic neuropathy. Ann Intern Med. 1966;65:1237–46.

Ford ES. Risks for all-cause mortality, cardiovascular disease, and diabetes associated with the metabolic syndrome: a summary of the evidence. Diabetes Care. 2005;28:1769–78.

Glueck CJ, Papanna R, Wang P, Goldenberg N, Sieve-Smith L. Incidence and treatment of metabolic syndrome in newly referred women with confirmed polycystic ovarian syndrome. Metabolism. 2003;52:908–15.

Gustafson B, Hammarstedt A, Andersson CX, Smith U. Inflamed adipose tissue: a culprit underlying the metabolic syndrome and atherosclerosis. Arterioscler Thromb Vasc Biol. 2007;27:2276–83.

Guzik TJ, Mangalat D, Korbut R. Adipocytokines: novel link between inflammation and vascular function? J Physiol Pharmacol. 2006;57:505–28.

Hong CY, Park JH, Ahn RS, Im SY, Choi HS, Soh J, et al. Molecular mechanism of suppression of testicular steroidogenesis by proinflammatory cytokine tumor necrosis factor alpha. Mol Cell Biol. 2004;24:2593–604.

Horiba N, Rinsho N. Sexual and gonadal dysfunction in adrenal disorders. Nihon Rinsho. 1997;55(11):2979–84.

Izumi T, Yokota-Hashimoto H, Zhao S, Wang J, Halban PA, Takeuchi T. Dominant negative pathogenesis by mutant proinsulin in the Akita diabetic mouse. Diabetes. 2003;52:409–16.

Jelodar G, Khaksar Z, Pourahmadi M. Endocrine profile and testicular histomorphometry in adult rat offspring of diabetic mothers. J Physiol Sci. 2009;59:377–82.

Jensen TK, Andersson A, Jørgensen N, Andersen A, Carlsen E, Petersen JH, et al. Body mass index in relation to semen quality and reproductive hormones among 1,558 Danish men. Fertil Steril. 2004;82:863–70.

Jones TH, Arver S, Behre HM, Buvat J, Meuleman E, Moncada I, et al. Testosterone replacement in hypogonadal men with type 2 diabetes and/or metabolic syndrome (The TIMES2 Study). Diabetes Care. 2011;34:828–37.

Karagiannis A, Harsoulis F. Gonadal dysfunction in systemic diseases. Eur J Endocrinol. 2005;152:501–13.

Kimura M, Lukinius A, Ericsson JL, Grimelius L. Distribution of insulin binding sites on Leydig cells of rat testes using insulin-coated gold particles. Histochemistry. 1992;97:213–20.

Kyrou I, Tsigos C. Chronic stress, visceral obesity and gonadal dysfunction. Hormones. 2008;7(4):287–93.

Lapauw B, Ouwens M, 'tHart LM, Wuyts B, Holst JJ, T'Sjoen G, et al. Sex steroids affect triglyceride handling, glucose-dependent insulinotropic polypeptide, and insulin sensitivity: a 1-week randomized clinical trial in healthy young men. Diabetes Care. 2010;33:1831–3.

Loves S, Ruinemans-Koerts J, de Boer H. Letrozole once a week normalizes serum testosterone in obesity-related male hypogonadism. Eur J Endocrinol. 2008;158:741–7.

Luton JP, Thieblot P, Valcke JC, Mahoudeau JA, Bricaire H. Reversible gonadotropin deficiency in male Cushing's disease. J Clin Endocrinol Metab. 1977;45(3):488–95.

Maggio M, Blackford A, Taub D, Carducci M, Ble A, Metter EJ, et al. Circulating inflammatory cytokine expression in men with prostate cancer undergoing androgen deprivation therapy. J Androl. 2006;27:725–8.

Malkin CJ, Pugh PJ, Jones RD, Kapoor D, Channer KS, Jones TH. The effect of testosterone replacement on endogenous inflammatory cytokines and lipid profiles in hypogonadal men. J Clin Endocrinol Metab. 2004;89:3313–8.

Mallidis C, Agbaje I, Rogers D, Glenn J, McCullough S, Atkinson AB, et al. Distribution of the receptor for advanced glycation end products in the human male reproductive tract: prevalence in men with diabetes mellitus. Hum Reprod. 2007;22(8):2169–77.

Maneesh M, Jayalakshmi H, Singh TA, Chakrabarti A. Impaired hypothalamic-pituitary-gonadal axis function in men with diabetes mellitus. Indian J Clin Biochem. 2006;21(1):165–8.

Mogri M, Dhindsa S, Quattrin T, Ghanim H, Dandona P. Testosterone concentrations in young pubertal and post-pubertal obese males. Clin Endocrinol (Oxf). 2013;78(4):593–9.

Mulholland J, Mallidis C, Agbaje I, McClure N. Male diabetes mellitus and assisted reproduction treatment outcome. Reprod Biomed Online. 2011;22:215–9.

Murray FT, Cameron DF, Orth JM, Katovich MJ. Gonadal dysfunction in the spontaneously diabetic BB rat: alterations of testes morphology, serum testosterone and LH. Horm Metab Res. 1985;17:495–501.

Pagotto U, Gambineri A, Pelusi C, Genghini S, Cacciari M, Otto B, et al. Testosterone replacement therapy restores normal ghrelin in hypogonadal men. J Clin Endocrinol Metab. 2003;88:4139–43.

Pasquali R, Casimirri F, De Iasio R, Mesini P, Boschi S, Chierici R, et al. Insulin regulates testosterone and sex hormone-binding globulin concentrations in adult normal weight and obese men. J Clin Endocrinol Metab. 1995;80:654–8.

Pasquali R, Patton L, Gambineri A. Obesity and infertility. Curr Opin Endocrinol Diabetes Obes. 2007;14:482–7.

Pitteloud N, Dwyer AA, DeCruz S, Lee H, Boepple PA, Crowley Jr WF, et al. Inhibition of luteinizing hormone secretion by testosterone in men requires aromatization for its pituitary but not its hypothalamic effects: evidence from the tandem study of normal and gonadotropin-releasing hormone-deficient men. J Clin Endocrinol Metab. 2008;93:784–91.

Pitteloud N, Hardin M, Dwyer AA, Valassi E, Yialamas M, Elahi D, et al. Increasing insulin resistance is associated with a decrease in Leydig cell testosterone secretion in men. J Clin Endocrinol Metab. 2005;90:2636–41.

Rhoden EL, Ribeiro EP, Teloken C, Souto CA. Diabetes mellitus is associated with subnormal serum levels of free testosterone in men. BJU Int. 2005;96:867–70.

Saez JM. Leydig cells: endocrine, paracrine, and autocrine regulation. Endocr Rev. 1994;15:574–626.

Sallmen M, Sandler DP, Hoppin JA, Blair A, Baird DD. Reduced fertility among overweight and obese men. Epidemiology. 2006;17:520–3.

Scarano WR, Messias AG, Oliva SU, Klinefelter GR, Kempinas WG. Sexual behaviour, sperm quantity and quality after short-term streptozotocin-induced hyperglycaemia in rats. Int J Androl. 2006;29:482–8.

Schmidt AM, Yan SD, Yan SF, Stern DM. The multiligand receptor RAGE as a progression factor amplifying immune and inflammatory responses. J Clin Invest. 2001;108:949–55.

Schoeffling K, Federlin K, Ditschuneit H, Pfeiffer EF. Disorders of sexual function in male diabetics. Diabetes. 1963;12:519–27.

Schoeller EL, Albanna G, Frolova AI, Moley KH. Insulin rescues impaired spermatogenesis via the hypothalamic-pituitary-gonadal axis in Akita diabetic mice and restores male fertility. Diabetes. 2012;61:1869–78.

Schroeder ET, Zheng L, Ong MD, Martinez C, Flores C, Stewart Y, et al. Effects of androgen therapy on adipose tissue and metabolism in older men. J Clin Endocrinol Metab. 2004;89:4863–72.

Seethalakshmi L, Menon M, Diamond D. The effect of streptozotocin-induced diabetes on the neuroendocrine-male reproductive tract axis of the adult rat. J Urol. 1987;138:190–4.

Shiao MS, Liao BY, Long M, Yu HT. Adaptive evolution of the insulin two gene system in mouse. Genetics. 2008;178:1683–91.

Soudamani S, Malini T, Balasubramanian K. Effects of streptozotocin-diabetes and insulin replacement on the epididymis of prepubertal rats: histological and histomorphometric studies. Endocr Res. 2005;31:81–98.

Stellato RK, Feldman HA, Hamdy O, Horton ES, McKinlay JB. Testosterone, sex hormone-binding globulin, and the development of type 2 diabetes in middle-aged men: prospective results from the Massachusetts male aging study. Diabetes Care. 2000;23:490–4.

Traish AM, Feeley RJ, Guay A. Mechanisms of obesity and related pathologies: androgen deficiency and endothelial dysfunction may be the link between obesity and erectile dysfunction. FEBS J. 2009;276:5755–67.

Wang C, Jackson G, Jones TH, Matsumoto AM, Nehra A, Perelman MA, et al. Low testosterone associated with obesity and the metabolic syndrome contributes to sexual dysfunction and cardiovascular disease risk in men with type 2 diabetes. Diabetes Care. 2011;34(7):1669–75.

Wu FC, Tajar A, Beynon JM, Pye SR, Silman AJ, Finn JD, et al. Identification of late-onset hypogonadism in middle-aged and elderly men. N Engl J Med. 2010;363:123–35.

Yialamas MA, Dwyer AA, Hanley E, Lee H, Pitteloud N, Hayes FJ. Acute sex steroid withdrawal reduces insulin sensitivity in healthy men with idiopathic hypogonadotropic hypogonadism. J Clin Endocrinol Metab. 2007;92:4254–9.

Part VII

Disorders and Diseases of Male Reproductive System

Primary Testicular Failure

24

Manish Jain, Prashant Kumar, Venkateswaran Iyer, and Ashutosh Halder

Learning Objectives

- Primary testicular failure and its subtypes
- Etiology
- Biomarkers
- Principles of diagnostic evaluation and management

24.1 Introduction

Primary testicular failure (PTF) or hypergonadotropic hypogonadism refers to a condition where testes fail to produce sperm despite of adequate hormonal support. PTF is a major cause of nonobstructive azoospermia (NOA) and oligospermia. It affects approximately 1% of all men and 10% of those seeking fertility evaluations (Schlegel 1991). NOA can be primary or secondary. In primary NOA, pathology is present within testes, whereas in secondary NOA, pathology is present outside testes, mostly hypothalamic-pituitary axis or pre-testicular.

M. Jain, PhD • P. Kumar, MSc, MPhil • A. Halder, MD, DNB, DM, FAMS, MRCOG-I (✉)
Department of Reproductive Biology, All India Institute of Medical Sciences,
New Delhi, India
e-mail: om_704@yahoo.co.in; prashant03micro@gmail.com; ashutoshhalder@gmail.com

V. Iyer, MD
Department of Pathology, All India Institute of Medical Sciences, New Delhi, India
e-mail: iyer_venkat@hotmail.com

A. Kumar, M. Sharma (eds.), *Basics of Human Andrology*,
DOI 10.1007/978-981-10-3695-8_24

The hypogonadotropic hypogonadism or pre-testicular causes of azoospermia may be defined as extra-gonadal endocrine disorders such as those originating in the pineal, hypothalamic, pituitary, or adrenal glands, which have an adverse effect on spermatogenesis through aberrant hormonal stimulation or suppression.

24.2 Types

PTF is classified according to histopathology or cytology findings into four distinct subtypes—Sertoli cell-only syndrome (SCOS), maturation arrest (MA), hypospermatogenesis (HS), and tubular fibrosis (TF). However, accurate categorization is possible only by multiple testicular biopsies. Further, one may find differences in subtypes between testes, e.g., SCOS in one testis and TF in other testis or any other combinations. Such cases are usually labeled as mixed group (MG). In addition, many patients may present as complete germinal cell aplasia or SCOS in some seminiferous tubules (STs) whereas complete spermatogenesis in adjacent ST. This condition is called focal germinal cell aplasia. The categorization is based on the description of histopathologic phenotypes of spermatogenetic failure and not on manifestations of disease entities (Cerilli et al. 2010).

24.2.1 Sertoli Cell-Only Syndrome (SCOS)/Germ Cell Aplasia

SCOS is a common cause of NOA. SCOS is a histopathologic phenotype of spermatogenic failure that was first described by Del Castillo et al. in 1947. In complete germ cell aplasia, STs are reduced in diameter and contain only Sertoli cells. There is absence of any other cell involved in spermatogenesis (see Fig. 24.1a). Fine-needle aspiration cytology (FNAC) smear shows only Sertoli cells (see Fig. 24.1b). In congenital germ cell aplasia, the primordial germ cells either do not migrate from

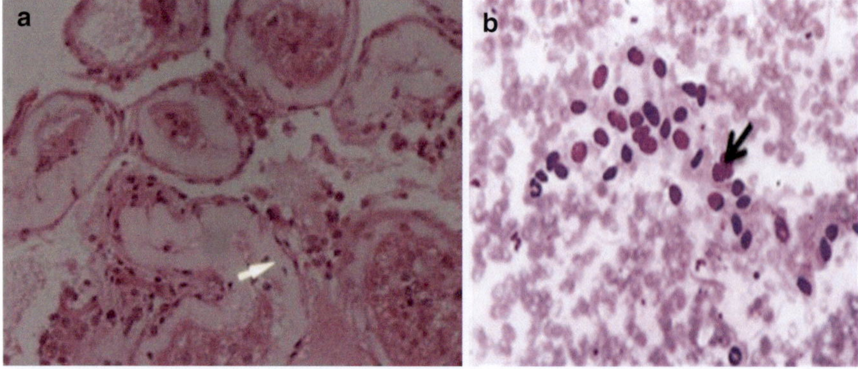

Fig. 24.1 Sertoli cell-only syndrome: (**a**) Testicular biopsy showing seminiferous tubules containing Sertoli cells (*arrow*) only. (**b**) Fine-needle aspiration cytology (FNAC) showing Sertoli cells aggregates on background of leucocytes and red blood cells. There is no evidence of germ cell lineages (spermatogonia, spermatocytes, spermatids, and mature sperm)

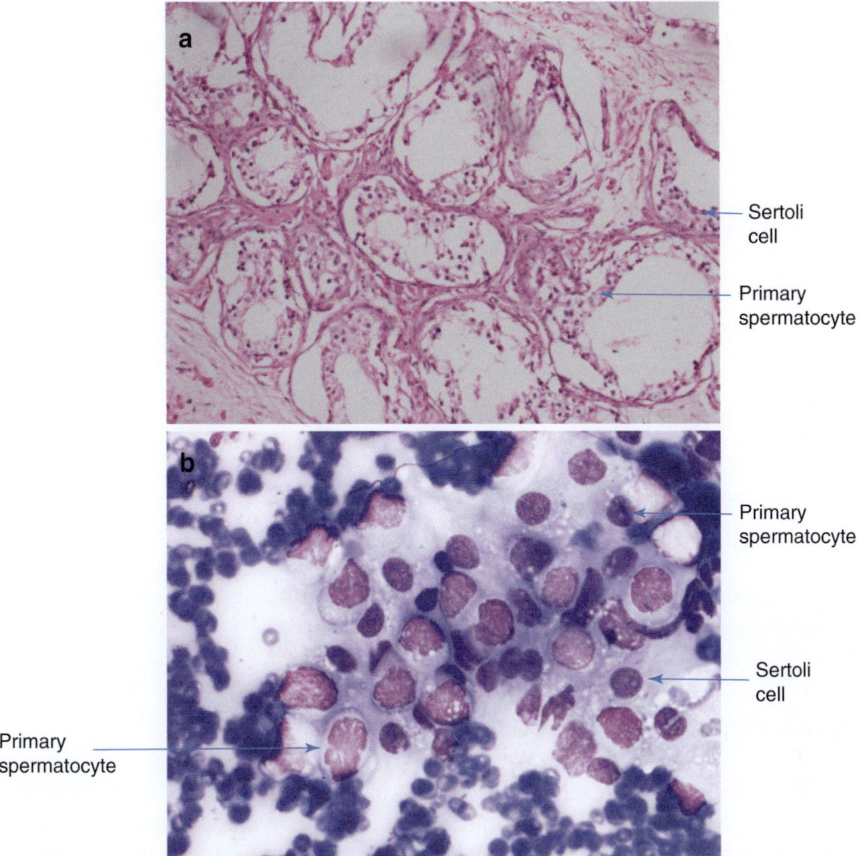

Fig. 24.2 Early maturation arrest: (**a**) Testicular biopsy showing seminiferous tubules containing Sertoli cells and primary spermatocytes. (**b**) FNAC showing primary spermatocytes and Sertoli cells

the yolk sac into the future gonads or do not survive in the STs. SCOS can also be focal with a variable percentage of STs containing germ cells. Such STs show limited activity of spermatogenesis and, hence, should be categorized as hypospermatogenesis (Silber et al. 1995; Cerilli et al. 2010).

24.2.2 Maturation Arrest (MA)

MA refers to conditions where testes fail to produce mature sperm due to interruption of germ cell development and differentiation despite presence of germ cells in adequate numbers. MA is further classified into two distinct subtypes—early MA and late MA. In early MA, germ cells do not mature beyond secondary spermatocyte stage and arrest mostly at primary spermatocyte stage (see Fig. 24.2).

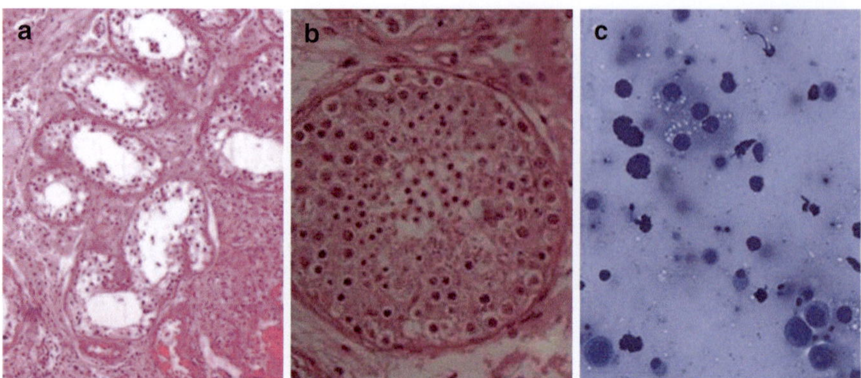

Fig. 24.3 Hypospermatogenesis: (**a**) Testicular biopsy (10×); (**b**) at 100× showing seminiferous tubules containing all stages of spermatogenesis including mature sperm but reduced to a varying degree. (**c**) Fine-needle aspiration cytology (FNAC) showing all germ cell lineages (spermatogonia, spermatocytes, spermatids, and mature sperm)

Late maturation arrest is post-meiotic. In late MA, germ cells do not mature beyond round spermatid stage (Cerilli et al. 2010).

24.2.3 Hypospermatogenesis

In hypospermatogenesis, all stages of spermatogenesis such as spermatogonia, spermatocytes, spermatids, and mature sperm are present in some or all STs but are reduced in number (see Fig. 24.3). Leydig cell number and functions are normal. Sertoli cell number is usually normal (Cerilli et al. 2010).

24.2.4 Tubular Fibrosis (TF)

Seminiferous TF presents as thickening of peritubular membrane and deposition of collagen on basement membrane with absence of intratubular germ cells and Sertoli cells (see Fig. 24.4). Leydig cells are usually decreased or lost (Cerilli et al. 2010).

24.3 Etiology

PTF can result from a variety of congenital or acquired disorders, including chromosomal defects such as Klinefelter syndrome (KS), Y chromosome microdeletions, single-gene mutations, cryptorchidism, varicocele, etc.

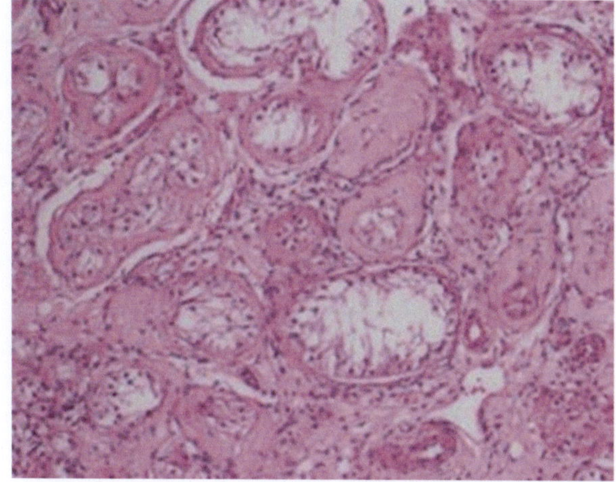

Fig. 24.4 Tubular fibrosis: Testicular biopsy showing seminiferous tubules containing thickening of the peritubular membranes and deposition of collagen on basement membrane

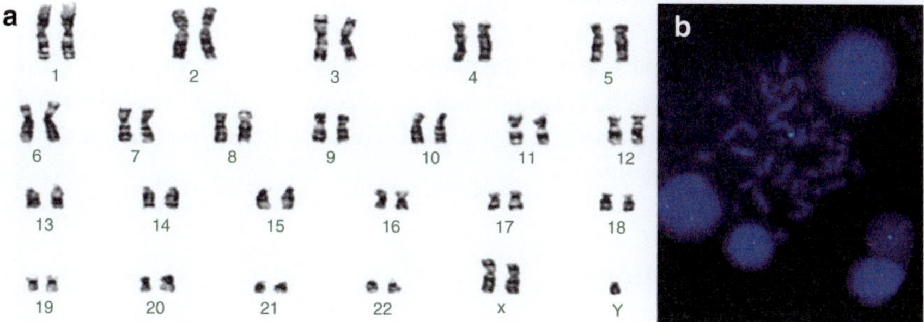

Fig. 24.5 (**a**) Karyotype showing 47,XXY state (Klinefelter syndrome) and (**b**) XY fluorescent in situ hybridization (FISH) showing some cells (metaphase and two interphase cells) with *one green* (X centromere) and *one red* (Y centromere) signal suggesting XY (male) state and other cells with *two green* (XX) and *one red* (*Y*) signals suggesting XXY state, i.e., mosaic XY/XXY state

24.3.1 Chromosomal Abnormalities

KS is the most common known cause of PTF. KS affects approximately 1 in 1000 males (Schwartz and Root 1991; Bojesen et al. 2003). KS is characterized by sex chromosome aneuploidy in which there is one extra X chromosome, 47,XXY. Although an extra X chromosome is the most common form (see Fig. 24.5a). Some patients with KS have a greater or lesser number of X chromosomes or mosaicism such as 48,XXXY and 46,XY/47,XXY (see Fig. 24.5b; Paulsen et al. 1968). Rarely, 46,XX males resulting from translocation of the testis-determining gene or sex-determining region Y (SRY) to an X chromosome may have KS phenotype. The phenotype varies with the number of extra X chromosomes and possibly also with the number of trinucleotide CAG repeats on the androgen receptor gene polymorphism. A longer

CAG repeat sequence has been associated with taller stature, lower bone mineral density, gynecomastia, and shorter penile length (Zitzmann et al. 2004; Zinn et al. 2005). Men with KS generally have small testes, resulting from damage to both ST and Leydig cells. Serum concentrations of follicle-stimulating hormone (FSH) and luteinizing hormone (LH) are elevated, and testosterone levels are decreased to varying extent. Affected men have severely reduced sperm counts and are undervirilized (Paulsen et al. 1968; Wang et al. 1975). Cryptorchidism is more common in men with KS and causes more severe testicular damage (Ferlin et al. 2008). Later in life, they have an increased risk for breast cancer, mediastinal germ cell tumor, varicose veins with leg ulcers, and diabetes mellitus (Campbell et al. 1980; Weiss et al. 2005; Bojesen et al. 2006; Volkl et al. 2006). Other chromosomal abnormalities associated with primary gonadal failure include the 46,XY/45,X karyotype causing a syndrome characterized by short stature and other features of Turner syndrome (Simpson 1978). Because the testes may be streaked, dysgenetic, or normal, the phenotype varies from female to male. In those with a streak and a dysgenetic testis or mixed gonadal dysgenesis, the risk of gonadoblastoma is increased to approximately 20%, and therefore, gonadectomy is indicated.

The XX male syndrome is a rare genetic disorder. The phenotype is variable; it ranges from a severe impairment of the external genitalia to a normal male phenotype with infertility. It generally results from an unequal crossing over between the short arms of the sex chromosomes X and Y. The XX syndrome has been described in the chapter Y Chromosome.

24.3.2 Yq Microdeletions

Microdeletions of the long arm of the Y chromosome are now recognized as one of the common causes of PTF. The deletions affect up to 20% of men with infertility (Ferlin et al. 2007). Most microdeletions are mapped to the Yq11 region which is named azoospermia factor (AZF). The Yq11 region contains three regions, AZFa, AZFb, and AZFc. Deletions of the AZFa or AZFb typically result in azoospermia (see Fig. 24.6). Mutations in the AZFc region lead to infertility of

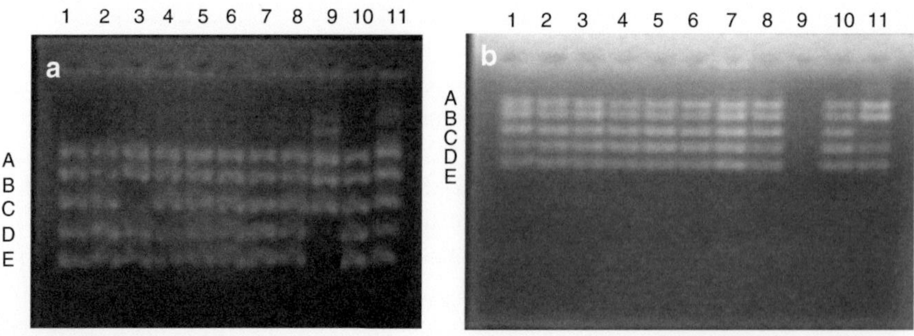

Fig. 24.6 Multiplex PCR results of Y chromosome microdeletion of AZFa (**a** lane 9 bands *D* and *E*) and AZFb (**b** lane 11 band *C*)

varying severity, ranging from oligospermia to azoospermia, and are the commonest microdeletions in humans (Kuroda-Kawaguchi et al. 2001; Krausz and Degl'Innocenti 2006). The DDX3Y and USP9Y genes, both located in the AZFa region, have important role in spermatogenesis. Deletions of these genes are consistently observed with azoospermia (Luddi et al. 2009; Tyler-Smith and Krausz 2009). Y chromosome microdeletions also have been identified in men with cryptorchidism, varicocele, and obstructions of the vas deferens (Foresta et al. 1999; Krausz et al. 1999).

24.3.3 Copy Number Variations

Chromosomal rearrangements can result in deletions or duplications which are termed as copy number variations (CNVs) or unbalanced translocations. CNVs are submicroscopic DNA (1 kb or more) segments of duplications and deletions of the genome (Lee et al. 2007). These chromosomal rearrangements may subsequently result in imbalance in gene dosage. Meiotic non- disjunction events in normal carriers of balanced translocations may also lead to a disturbance of gene dosage in offspring. Various rearrangements occur in specific regions of the genome, suggesting specific mechanisms. CNVs can be pathogenic where they predispose to diseases or benign CNVs where they apparently have no effect. CNVs sit at the interface between microscopically visible rearrangements and point mutations and are increasingly being assessed by using microarray methods. Recent studies have shown that CNVs may affect about 20% of the human genome. CNVs have been implicated in premature ovarian failure, XY gonadal dysgenesis, and Mayer-Rokitansky-Kuster-Hauser syndrome. These diseases are linked to genital development and function (Ledig et al. 2010a, b, 2011). Up to 1500 genes are thought to be involved in spermatogenesis, out of which 300–600 are specifically expressed in the male germline and, therefore, are candidate genes implicated for male infertility (Schultz et al. 2003; Matzuk and Lamb 2008; Yan 2009). Tuttelmann et al. in 2011 reported number of candidate genes as risk factors of spermatogenic failure (see Table 24.1). CATSPER2, PLSCR2, and SYT6 genes are involved in the last stage of spermatogenesis (Stouffs et al. 2012). SLC25A24, a mitochondrial carrier, has testis-enriched expression pattern (Fiermonte et al. 2004; Stouffs et al. 2012). Currently, we are working on genotype phenotype correlation by using array comparative genomic hybridization (CGH) on MA. Our initial findings indicate association between CNVs of pseudoautosomal regions (PAR) 1, 2, and 3 with testicular MA in approximately 30% of cases (Halder et al. 2015, 2017).

24.3.4 Gene Mutations

Normal male sexual differentiation and spermatogenesis require both normal androgen production and normal androgen receptors. The androgen receptors play an important role in the differentiation of spermatids and their release from the seminiferous epithelium. The number of trinucleotide CAG repeats in exon 1 of the androgen receptors gene is inversely correlated with its transcriptional activity (Zitzmann et al. 2004; Zinn et al. 2005). Men with spermatogenic disorders frequently have significantly longer CAG repeat of androgen receptor (Davis-Dao et al. 2007). Similarly, disorders of estrogen synthesis or action are also associated with spermatogenesis defect. Impaired spermatogenesis has been observed in mice and in men lacking a functional estrogen receptor alpha (Smith et al. 1994; Hess et al. 1997). In mice, inactivating mutation in the aromatase enzyme causes spermatogenic defect (Robertson et al. 1999). Polymorphisms involving variations in the number of TA tandem repeats in the promoter region of the estrogen receptor gene have been related to sperm production (Nuti and Krausz 2008). FSH receptor gene mutation affects spermatogenesis also (Tapanainen et al. 1997; Simoni et al.

Table 24.1 Candidate genes with proposed function (Tuttelmann et al. 2011)

Group	Phenotype	Region	Gene symbol	OMIM	Function (references)
Autosomal CNVs	Normal/oligoz/SCOS	4q13.2	UGT2B17	601903	Glucuronidase essential for urinary testosterone excretion (Jakobsson et al. 2006; Juul et al. 2009)
	4x oligoz/3x SCOS	3p11.1	EPHA3	179611	Anks proteins involved in modulating degradation of EphA receptors (Kim et al. 2010)
	2x oligoz	6p21.31	ANKS1A	608994	
	2x SCOS	12q23.1	ANKS1B	607815	
	2x SCOS	8q24.3	PLEC	601282	Plectin in Sertoli cells concentrated at intracellular junctions and nuclear surface (Guttman et al. 1999)
	2x oligoz	16q22.1	PRMT7	610087	Protein methyltransferase, cooperates with the testis-specific factor CTCFL (Jelinic et al. 2006)
Sex chromosome CNVs	1x oligoz	Xp11.4	TSPAN7	300096	Interaction with SPAG11B (serum-associated antigen) isoform D, associated with sperm (Radhakrishnan et al. 2009)

Table 24.1 (continued)

Group	Phenotype	Region	Gene symbol	OMIM	Function (references)
	1x oligoz	Xp11.22	SSX7	300542	Cancer-testis antigen, expressed in normal testis tissue (Gure et al. 2002; Chen et al. 2003)
	1x oligoz	Xp11.22	SPANXN 5	300668	Cancer-testis antigen, expressed in post-meiotic spermatid (Kouprina et al. 2007)
	1x oligoz	Xq22.1	BEX1	300690	In mice, expression in pachytene spermatocytes and spermatids (Yang et al. 2002)
	1x oligoz	Xq22.1	NXF3	300316	Belongs to a family of nuclear RNA export factors (NXF), Nxf2 plays a role in spermatogenesis (Sasaki et al. 2005; Pan et al. 2009)
	1x oligoz	Xq22.2	H2BFWT	300507	Testis-specific histone, SNP in 5′ untranslated region associated with oligozoospermia (Churikov et al. 2004; Boulard et al. 2006; Lee et al. 2009)
	1x oligoz/1x SCOS	Xq22.3	PAK3	300142	One isoform specifically expressed in the testis (Kohn et al. 2004)
	1x SCOS	Xq23	TRPC5	300334	Interacts and co-localizes with Enkurin in sperm (Sutton et al. 2004)

OMIM Online Mendelian Inheritance in Man (http://www.ncbi.nlm.nih.gov/omim), *Oligoz* oligozoospermia, *SCOS* Sertoli cell-only syndrome, *CNVs* copy number variations

1999). Men with myotonic dystrophy, an autosomal disorder, also show abnormal spermatogenesis (Takeda and Ueda 1977). Mutations in the SYCP3 gene, which is involved in regulation of the synapse between homologous chromosomes during meiosis, have been implicated as a potential cause of male infertility (Nuti and Krausz 2008). DAZL gene, an autosomal homolog of the DAZ, is deleted in azoospermia (Teng et al. 2002, 2006; Becherini et al. 2004; Tung et al. 2006a, b). Genes, such as PRM1 and PRM2 which are protamines involved in chromatin compaction, TNP1 and TNP2 which are transition nuclear proteins, and USP26 which is deubiquitinating enzyme family, are also associated with spermatogenic defects (Nuti and Krausz 2008).

24.3.5 Cryptorchidism

Undescended testes or cryptorchidism is an androgen-dependent process and is one of the major causes of spermatogenic failure in humans. It is commonly seen with Kallmann syndrome, androgen resistance, and defects in testosterone synthesis. Cryptorchidism can be unilateral or bilateral and, in either case, is associated with impaired spermatogenesis. This also increases risk for developing testicular tumors. In men with cryptorchidism, serum FSH levels are often elevated, but LH concentrations are generally normal reflecting normal Leydig cell function.

24.3.6 Varicocele

Varicocele results from dilatation of the pampiniform plexus of spermatic veins. Varicocele is 30% more prevalent in infertile men and ten times commonly involves left side venous system, probably because the left spermatic vein is longer and joins the left renal vein at a right angle (Romeo and Santoro 2009). Although increased testicular temperature, delayed removal of local toxins, hypoxia, and stasis are viewed as the mechanisms likely to be responsible for the association between varicocele and infertility, yet, no causal relationship has been established (Saypol et al. 1981; Pryor and Howards 1987; Mieusset et al. 1987).

24.3.7 Mumps Orchitis

Mumps orchitis is recognized as a cause of male infertility. Although rare in prepubertal males, it occurs in up to 25% of adult men with mumps. Some even become azoospermic. The mechanisms underlying the infertility are damage to germinal epithelium, ischemia, or immune dysfunction (Beard et al. 1977; Adamopoulos et al. 1978).

24.3.8 Gonadotoxins

Drugs that adversely affect spermatogenesis or Leydig cell function are alkylating agents such as cyclophosphamide, chlorambucil, antiandrogens such as flutamide, cyproterone, spironolactone, ketoconazole, cimetidine, and anabolic steroids (Gazvani et al. 1997). Doses of radiations as low as 0.015 Gy or 15 rads can suppress spermatogenesis, and doses above 6 Gy generally lead to permanent azoospermia and infertility (Rowley et al. 1974). Environmental exposures that may act as gonadotoxins include heat, smoking, heavy metals, organic solvents, and pesticides. Men with NOA with dark semen and without genital tract bleeding may rarely be linked with heavy metal toxicity as the underlying etiology (Halder et al. 2014).

24.3.9 Chronic Illness

Chronic renal or liver insufficiency as in cirrhosis of liver can cause PTF (Smith et al. 1975; Handelsman and Dong 1993).

24.4 Biomarkers of Primary Testicular Failure (PTF)

24.4.1 Follicle-Stimulating Hormone (FSH)

FSH is the classical endocrine parameter to differentiate testicular impairment of spermatogenesis from obstructive azoospermia (Behre et al. 1997). FSH is a valuable predictive marker of the histological status of the testis, but a few studies show wide overlap between values in normal and reduced spermatogenesis that limits its diagnostic accuracy (Bergmann et al. 1994; Halder et al. 2005).

24.4.2 Inhibin B

Inhibin B is a heterodimeric polypeptide hormone. It selectively suppresses the secretion of the FSH and has local paracrine action on the gonads. It is produced by the Sertoli cells of the testis and granulosa cells of the ovary. Its primary role appears to be in the regulation of gametogenesis via negative feedback on the production of FSH. Inhibin B is the major circulating inhibin in men (Bhasin and de Krestser 1996; Groome et al. 1996; Illingworth et al. 1996). In the prepubertal testis, it is predominantly produced by the Sertoli cells, but the site of production in the adult testis is still controversial (Illingworth et al. 1996; Meachem et al. 2001; Frydelund-Larsen et al. 2002).

Inhibin B production varies with age. There is an inhibin B peak shortly after birth which coincides with increased serum FSH due to proliferating activity of the Sertoli cells. Afterwards, inhibin B level decreases and remains low until puberty. The levels rise again, first as a consequence of FSH stimulation and later due to combined regulation of FSH and spermatogenesis (Meachem et al. 2001). In adult male, levels of inhibin B remain stable throughout life (Bohring and Krause 2003). Inhibin B expression and secretion in men are positively correlated with Sertoli cell function, sperm count, and spermatogenic status and negatively correlated with FSH levels (Luisi et al. 2005). Men with hypospermatogenesis and spermatogenic arrest have significantly lower levels of inhibin B. Men with SCOS syndrome have extremely low levels of inhibin B (Meachem et al. 2001). Inhibin B concentration is closely correlated with sperm concentration in the ejaculate and testicular volume (Jensen et al. 1997; Pierik et al. 1998). Suppression of spermatogenesis by exogenous testosterone or chemotherapy decreases serum inhibin B levels (Anderson et al. 1997; Wallace et al. 1997). Inhibin B is considered as a marker of spermatogenesis and may offer an improved diagnosis of testicular dysfunction (Hipler et al. 2001; Hu and Huang 2002). Inhibin B controls FSH secretion through negative feedback mechanism. Based on these observations, inhibin B has been considered as a reliable marker for spermatogenesis (Leifke et al. 1997). Our published data also indicates the utility of inhibin B as marker for male and female gonadal functions (Halder et al. 2005, 2007).

24.4.3 Anti-Mullerian Hormone

Anti-Mullerian hormone (AMH) is a glycoprotein. AMH belongs to the transforming growth factor-β (TGF-β) superfamily. It is involved in regulation of tissue growth and differentiation. AMH is produced by the Sertoli cells of the testis and by ovarian granulosa cells. During embryonic development, secretion of AMH from testicular Sertoli cells is essential for the regression of the Mullerian ducts, and thus, it is essential for normal development of the male reproductive tract. Secretion of AMH by the Sertoli cells begins during embryogenesis and continues throughout life. AMH levels decrease during puberty and remain low in adults (Josso et al. 1990; Lee et al. 1996). Functions of AMH in postnatal life are not clearly known. It has been shown that AMH controls Leydig cell proliferation and steroidogenic function and it may be related to germinal cell proliferation (Cazorla et al. 1998; Racine et al. 1998).

AMH levels are expected to be high in Sertoli cell immaturity and Sertoli cell hyperplasia or tumor. Low AMH levels are most often observed with NOA as in SCOS and TF (Muttukrishna et al. 2007; Jain and Halder 2012). Low AMH levels are also observed in cases of oligospermia (Al-Qahtani et al. 2005). Low AMH levels reflect an altered Sertoli cell function or number that may lead to spermatogenic arrest. Findings of normal as well as low AMH levels are also possible at different stages of PTF.

24.4.4 Lactate

Lactate is produced by Sertoli cells and used by germ cells as an energy substrate (Jutte et al. 1981a, b; Mita and Hall 1982; Robinson and Fritz 1979). Two to nine percent of the glucose utilized by Sertoli cells is converted to carbon dioxide, and 95–98% is converted to anionic compounds, commonly to lactate (Robinson and Fritz 1979). Round spermatids and pachytene spermatocytes are dependent on lactate, whereas ejaculated sperm can use glucose or fructose as energy source (Voglmayr 1975; Jutte et al. 1981a, b). Lactate is an important intermediate for the regulation of survival of pachytene spermatocytes and round spermatids. Low levels of lactate are expected in late MA as in post-meiotic or round stage. High lactate levels indicate either non-utilization of lactate by germ cells as in SCOS or Sertoli cell dysfunction.

The classical predictors of spermatogenesis are testicular size, semen analysis, FSH level, and testicular histology. However, we have observed cases having small testis with better seminal parameters. SCOS with occasional sperm has also been observed during semen analysis. We have observed lower predictive value of FSH with MA as well as hypospermatogenesis. FSH levels are often normal in such subgroups (Bergmann et al. 1994). Inhibin B seems to be a better predictor in such situations. It can discriminate between complete absence of germ cells and less severe disturbances of sperm production in the testis. Serum inhibin B levels are more accurate than serum FSH in predicting the presence of testicular sperm in NOA (Anawalt et al. 1996; Illingworth et al. 1996; Jensen et al. 1997; Klingmuller and Haidl 1997; Pierik et al. 1998; von Eckardstein et al. 1999; Ballesca et al. 2000; Brugo-Olmedo et al. 2001).

Variable effects of FSH and inhibin B on spermatogenesis have been observed. In infertile patients with primary gonadal failure, inhibin B levels decrease and FSH levels increase. Spermatogenic arrest at the earlier stages shows lowest inhibin B levels. However, there are studies showing FSH levels with a higher predictive value than inhibin B (Bohring and Krause 1999; Andersson et al. 2004). Combination of inhibin B and FSH or inhibin B/FSH ratio correlates better with spermatogenesis (von Eckardstein et al. 1999; Andersson et al. 2004). Serum AMH levels are not of diagnostic significance in men with impaired spermatogenesis as shown in few studies (Tuttelmann et al. 2009). However, others have shown lower serum AMH levels in NOA than in obstructive azoospermia (Muttukrishna et al. 2007). Therefore, it can be suggested that some unrecognized factor influences the complex interplay between FSH, inhibin B, AMH, lactate, and spermatogenesis.

24.5 Principles of Diagnostic Evaluation and Management

The diagnosis of a patient with azoospermia is done to determine the etiology. The numerous causes of azoospermia fall into three categories: pre-testicular or hypogonadotropic hypogonadism, testicular or hypergonadotropic hypogonadism, and post-testicular or normogonadotropic hypogonadism. Testicular etiology or PTF involves disorders of spermatogenesis intrinsic to testes.

Azoospermia is diagnosed when no sperm can be detected on high-power microscopic examination of seminal deposits after 15 min centrifugation at 3000 g or greater, on at least two occasions at an interval of 3 months.

The patient information should include fertility history, mumps, undescended testis with or without surgical correction, genital trauma or history of inguinal surgery, genital infections such as filariasis/tuberculosis, gonadotoxin exposures such as radiation therapy/chemotherapy or heat exposure, and current medications. All these conditions may cause testicular damage. Family history of cystic fibrosis should also be considered as it is usually associated with congenital absence of the vas deferens. Physical examination should be done to look for secondary sexual characteristics which include body habitus, hair distribution, beard, moustache, sideburn, gynecomastia, etc. BMI should be measured. Site, size, and consistency of the testis should be examined. The presence or absence for nodule/cyst/varicocele in the epididymis and vas deferens should be evaluated as they can cause obstructive azoospermia. The initial hormonal evaluation should include measurement of serum testosterone, prolactin, and FSH levels. One ultrasound with Doppler study should be advised to exclude varicocele.

Management of PTF is based on avoiding risk factors. Universal measles, mumps, and rubella (MMR) vaccination minimizes mumps orchitis that may lead to PTF. The risks arising from chemotherapy, radiation, or surgery can be prevented by taking care during surgery, prior cryopreservation of gamete or gonadal biopsy, and protective measures during radiotherapy. Genetic abnormalities associated with the disorder are usually inherited and transmitted to offspring; hence, genetic counseling should be provided to the patient once genetic etiology is detected.

24.6 Future Directions

Recent studies suggest that many cases of idiopathic PTF have underlying genetic etiology. It is important to investigate idiopathic PTF cases at genomic and epigenomic levels to find out the underlying etiology.

In azoospermic men with focal spermatogenesis or hypospermatogenesis or late maturation arrest, pregnancies can be achieved with testicular sperm extraction and injecting sperm into mature oocyte by intracytoplasmic sperm injection (ICSI). In vitro spermatogenesis from testicular biopsy in these cases may result in pregnancy and live births. However, currently there is no therapy for PTF with complete germ cell aplasia. Most cases of genetic etiology present with normal spermatogenesis in peripubertal age group. However, spermatogenesis decreases rapidly in few years indicating accelerated programmed cell death. The predictive genomic medicine will help in identifying individuals who are at risk of PTF and will help in planning gonadal cryopreservation and gamete cryobanking. Recent progress in stem cell research, in particular induced pluripotent stem cell as well as germ line stem cell isolation and culture, may provide a platform for in vitro gamete development. This may open a new era of gametogenesis in a dish and personalized infertility treatments in coming years.

Key Questions

- What are the different subtypes of primary testicular failure?
- Describe the histopathological findings in Sertoli cell-only syndrome and maturation arrest.
- Discuss the role of azoospermic factor region in primary testicular failure.
- What are copy number variations?
- Write a short note on biomarkers of primary testicular failure.

References

Adamopoulos DA, Lawrence DM, Vassilopoulos P, Contoyiannis PA, Swyer GI. Pituitary-testicular interrelationship in mumps orchitis and other viral infections. Br Med J. 1978;1: 1177–80.

Al-Qahtani A, Muttukrishna S, Appasamy M, Johns J, Cranfield M, Visser JA, et al. Development of a sensitive enzyme immunoassay for anti-Mullerian hormone and the evaluation of potential clinical applications in males and females. Clin Endocrinol (Oxf). 2005;63:267–73.

Anawalt BD, Bebb RA, Matsumoto AM, Groome NP, Illingworth PJ, McNeilly AS, et al. Serum inhibin B levels reflect Sertoli cell function in normal men and men with testicular dysfunction. J Clin Endocrinol Metab. 1996;81:3341–5.

Anderson RA, Wallace EM, Groome NP, Bellis AJ, Wu FC. Physiological relationships between inhibin B, follicle stimulating hormone secretion and spermatogenesis in normal men and response to gonadotrophin suppression by exogenous testosterone. Hum Reprod. 1997;12:746–51.

Andersson AM, Petersen JH, Jørgensen N, Jensen TK, Skakkebaek NE. Serum inhibin B and follicle stimulating hormone levels as tools in the evaluation of infertile men: significance of adequate reference values from proven fertile men. J Clin Endocrinol Metab. 2004;89: 2873–9.

Ballescá JL, Balasch J, Calafell JM, Alvarez R, Fábregues F, de Osaba MJ, et al. Serum inhibin B determination is predictive of successful testicular sperm extraction in men with non- obstructive azoospermia. Hum Reprod. 2000;15:1734–8.

Beard CM, Benson Jr RC, Kelalis PP, Elveback LR, Kurland LT. The incidence and outcome of mumps orchitis. In: Rochester Minnesota, 1935 to 1974. Mayo Clin Proc. 1977;52:3–7.

Becherini L, Guarducci E, Degl'Innocenti S, Rotondi M, Forti G, Krausz C. DAZL polymorphisms and susceptibility to spermatogenic failure: an example of remarkable ethnic differences. Int J Androl. 2004;27:375–81.

Behre HM, Yeung CH, Nieschlag E. Diagnosis of male infertility and hypogonadism. In: Nieschlag E, Behre HM, editors. Andrology: male reproductive health and dysfunction. Springer: Heidelberg; 1997. p. 87–114.

Bergmann M, Behre HM, Nieschlag E. Serum FSH and testicular morphology in male infertility. Clin Endocrinol (Oxf). 1994;40:133–6.

Bhasin S, de Krestser DD. Measurement of circulating inhibin levels: revisiting the inhibin hypothesis. J Clin Endocrinol Metab. 1996;81:1318–20.

Bohring C, Krause W. Serum levels of inhibin B in men of different age groups. Aging Male. 2003;6:73–8.

Bohring C, Krause W. Serum levels of inhibin B in men with different causes of spermatogenic failure. Andrologia. 1999;31:137–41.

Bojesen A, Juul S, Birkebaek NH, Gravholt CH. Morbidity in Klinefelter syndrome: a Danish register study based on hospital discharge diagnoses. J Clin Endocrinol Metab. 2006;91:1254–60.

Bojesen A, Juul S, Gravholt CH. Prenatal and postnatal prevalence of Klinefelter syndrome: a national registry study. J Clin Endocrinol Metab. 2003;88:622–6.

Boulard M, Gautier T, Mbele GO, Gerson V, Hamiche A, Angelov D, et al. The NH2 tail of the novel histone variant H2BFWT exhibits properties distinct from conventional H2B with respect to the assembly of mitotic chromosomes. Mol Cell Biol. 2006;26:1518–26.

Brugo-Olmedo S, De Vicentiis S, Calamera JC, Urrutia F, Nodar F, Acosta AA. Serum inhibin B may be a reliable marker of the presence of testicular spermatozoa in patients with nonobstructive azoospermia. Fertil Steril. 2001;76:1124–9.

Campbell WA, Newton MS, Price WH. Hypostatic leg ulceration and Klinefelter's syndrome. J Ment Defic Res. 1980;24:115–7.

Cazorla O, Seck M, Pisselet C, Perreau C, Saumande J, Fontaine J, et al. Anti-Mullerian hormone (AMH) secretion in prepubertal and adult rams. J Reprod Fertil. 1998;112:259–66.

Cerilli LA, Kuang W, Rogers D. A practical approach to testicular biopsy interpretation for male infertility. Arch Pathol Lab Med. 2010;134:1197–204.

Chen YT, Alpen B, Ono T, Gure AO, Scanlan MA, Biggs 3rd WH, et al. Identification and characterization of mouse SSX genes: a multigene family on the X chromosome with restricted cancer/testis expression. Genomics. 2003;82:628–36.

Churikov D, Siino J, Svetlova M, Zhang K, Gineitis A, Morton Bradbury E, et al. Novel human testis-specific histone H2B encoded by the interrupted gene on the X chromosome. Genomics. 2004;84:745–56.

Davis-Dao CA, Tuazon ED, Sokol RZ, Cortessis VK. Male infertility and variation in CAG repeat length in the androgen receptor gene: a meta-analysis. J Clin Endocrinol Metab. 2007;92:4319–26.

Del Castillo EB, Trabucco A, de la Balze A. Syndrome produced by absence of the germinal epithelium without impairment of the Sertoli or Leydig cells. J Clin Endocr Metab. 1947;7:493–502.

Ferlin A, Arredi B, Speltra E, Cazzadore C, Selice R, Garolla A, et al. Molecular and clinical characterization of Y chromosome microdeletions in infertile men: a 10-year experience in Italy. J Clin Endocrinol Metab. 2007;92:762–70.

Ferlin A, Zuccarello D, Zuccarello B, Chirico MR, Zanon GF, Foresta C. Genetic alterations associated with cryptorchidism. JAMA. 2008;300:2271–6.

Fiermonte G, De Leonardis F, Todisco S, Palmieri L, Lasorsa FM, Palmieri F. Identification of the mitochondrial ATP-Mg/Pi transporter. Bacterial expression, reconstitution, functional characterization, and tissue distribution. J Biol Chem. 2004;279:30722–30.

Foresta C, Moro E, Garolla A, Onisto M, Ferlin A. Y chromosome microdeletions in cryptorchidism and idiopathic infertility. J Clin Endocrinol Metab. 1999;84:3660–5.

Frydelund-Larsen L, Krause C, Leffers H, Andersson AM, Carlsen E, Bangsboell S, et al. Inhibin B: a marker for the functional state of the seminiferous epithelium in patients with azoospermia factor C microdeletions. J Clin Endocrinol Metab. 2002;87:5618–24.

Gazvani MR, Buckett W, Luckas MJM, Aird IA, Hipkin LJ, Lewis-Jones DI. Conservative management of azoospermia following steroid abuse. Hum Reprod. 1997;12:1706–8.

Groome NP, Illingworth PJ, O'Brien M, Pai R, Rodger FE, Mather JP, et al. Measurement of dimeric inhibin B throughout the human menstrual cycle. J Clin Endocrinol Metab. 1996;81:1401–5.

Gure AO, Wei IJ, Old LJ, Chen YT. The SSX gene family: characterization of 9 complete genes. Int J Cancer. 2002;101:448–53.

Guttman JA, Mulholland DJ, Vogl AW. Plectin is concentrated at intercellular junctions and at the nuclear surface in morphologically differentiated rat Sertoli cells. Anat Rec. 1999;254:418–28.

Halder A, Fauzdar A, Kumar A. Serum inhibin B and follicle stimulating hormone levels as markers in the evaluation of azoospermic men: a comparison. Andrologia. 2005;37:173–9.

Halder A, Fauzdar A, Ghosh M, Kumar A. Serum Inhibin B: A direct, precise & noninvasive marker of ovarian function. J Clin Diagn Res 2007;1:131–37.

Halder A, Jain M, Chaudhary I, Kumar G, Das T, Gupta YK. Dark colored semen in non-obstructive azoospermia: a report of four cases. Andrologia. 2014;46:316–21.

Halde A, Kumar P, Jain M, Iyer VK. Copy number variations in testicular maturation arrest. Andrology 2017 (in press; s1-ln2470135395844769).

Halder A, Jain M, Kumar P. Primary testicular failure: an overview. J Clin Diagn Res. 2015;3:1–5.

Handelsman DJ, Dong Q. Hypothalamo-pituitary gonadal axis in chronic renal failure. Endocrinol Metab Clin North Am. 1993;22:145–61.

Hess RA, Bunick D, Lee KH, Bahr J, Taylor JA, Korach KS, et al. A role for oestrogens in the male reproductive system. Nature. 1997;390:509–12.

Hipler UC, Hochheim B, Knoll B, Tittelbach J, Schreiber G. Serum inhibin B as a marker for spermatogenesis. Arch Androl. 2001;46:217–22.

Hu YA, Huang YF. A serum marker of spermatogenesis-inhibin B. Zhonghua Nan Ke Xue. 2002;8:57–60.

Illingworth PJ, Groome NP, Byrd W, Rainey WE, McNeilly AS, Mather JP, et al. Inhibin-B: a likely candidate for the physiologically important form of inhibin in man. J Clin Endocrinol Metab. 1996;81:1321–5.

Jain M, Halder A. Sertoli cell only syndrome: Status of Sertoli cell maturation and function. Indian J Endocrinol Metab. 2012;16:S514–5.

Jakobsson J, Ekström L, Inotsume N, Garle M, Lorentzon M, Ohlsson C, et al. Large differences in testosterone excretion in korean and swedish men are strongly associated with a UDP-glucuronosyl transferase 2B17 polymorphism. J Clin Endocrinol Metab. 2006;91:687–93.

Jelinic P, Stehle JC, Shaw P. The testis-specific factor CTCFL cooperates with the protein methyltransferase PRMT7 in H19 imprinting control region methylation. PLoS Biol. 2006;4:e355.

Jensen TK, Andersson AM, Hjollund NH, Scheike T, Kolstad H, Giwercman A, et al. Inhibin B as a serum marker of spermatogenesis: correlation to differences in sperm concentration and follicle-stimulating hormone levels. A study of 349 Danish men. J Clin Endocrinol Metab. 1997;82:4059–63.

Josso N, Legeai L, Forest MG, Chaussain JL, Brauner R. An enzyme linked immunoassay for anti-Mullerian hormone: new tool for the evaluation of testicular function in infants and children. J Clin Endocrinol Metab. 1990;70:23–7.

Jutte NH, Grootegoed JA, Rommerts FF, van der Molen HJ. Exogenous lactate is essential for metabolic activities in isolated rat spermatocytes and spermatids. J Reprod Fertil. 1981a;62:399–405.

Jutte NH, Koolen LM, Jansen R, Grootegoed JA, Rommerts FFG, van der Molen HJ. Lactate essential for biochemical activities in isolated germinal cells. Int J Androl (Suppl). 1981b;3:59.

Juul A, Sørensen K, Aksglaede L, Garn I, Rajpert-De Meyts E, Hullstein I, et al. A common deletion in the uridine diphosphate glucuronyltransferase (UGT) 2B17 gene is a strong determinant of androgen excretion in healthy pubertal boys. J Clin Endocrinol Metab. 2009;94:1005–11.

Kim J, Lee H, Kim Y, Yoo S, Park E, Park S. The SAM domains of anks family proteins are critically involved in modulating the degradation of EphA receptors. Mol Cell Biol. 2010;30:1582–92.

Klingmuller D, Haidl G. Inhibin B in men with normal and disturbed spermatogenesis. Hum Reprod. 1997;12:2376–8.

Kohn M, Steinbach P, Hameister H, Kehrer-Sawatzki H. A comparative expression analysis of four MRX genes regulating intracellular signalling via small GTPases. Eur J Hum Genet. 2004;12:29–37.

Kouprina N, Noskov VN, Pavlicek A, Collins NK, Schoppee Bortz PD, Ottolenghi C, et al. Evolutionary diversification of SPANX-N sperm protein gene structure and expression. PLoS One. 2007;2:e359.

Krausz C, Degl'Innocenti S. Y chromosome and male infertility: update, 2006. Front Biosci. 2006;11:3049–61.

Krausz C, Quintana-Murci L, Barbaux S, Siffroi JP, Rouba H, Delafontaine D, et al. A high frequency of Y chromosome deletions in males with nonidiopathic infertility. J Clin Endocrinol Metab. 1999;84:3606–12.

Kuroda-Kawaguchi T, Skaletsky H, Brown LG, Minx PJ, Cordum HS, Waterston RH, et al. The AZFc region of the Y chromosome features massive palindromes and uniform recurrent deletions in infertile men. Nat Genet. 2001;29:279–86.

Ledig S, Hiort O, Scherer G, Hoffmann M, Wolff G, Morlot S, et al. Array-CGH analysis in patients with syndromic and non-syndromic XY gonadal dysgenesis: Evaluation of array CGH as diagnostic tool and search for new candidate loci. Hum Reprod. 2010a;25:2637–46.

Ledig S, Ropke A, Wieacker P. Copy number variants in premature ovarian failure and ovarian dysgenesis. Sex Dev. 2010b;4:225–32.

Ledig S, Schippert C, Strick R, Beckmann MW, Oppelt PG, Wieacker P. Recurrent aberrations identified by array-CGH in patients with Mayer Rokitansky Kuster Hauser syndrome. Fertil Steril. 2011;95:1589–94.

Lee C, Iafrate AJ, Brothman AR. Copy number variations and clinical cytogenetic diagnosis of constitutional disorders. Nat Genet. 2007;39:S48–54.

Lee J, Park HS, Kim HH, Yun YJ, Lee DR, Lee S. Functional polymorphism in H2BFWT-59UTR is associated with susceptibility to male infertility. J Cell Mol Med. 2009;13:1942–51.

Lee MM, Donahoe PK, Hasegawa T, Silverman B, Crist GB, Best S, et al. Mullerian inhibiting substance in human: normal levels from infancy to adulthood. J Clin Endocrinol Metab. 1996;81:571–6.

Leifke E, Simoni M, Kamischke A, Gromoll J, Bergmann M, Nieschlag E. Does the gonadotrophic axis play a role in the pathogenesis of Sertoli-cell-only syndrome? Int J Androl. 1997;20:29–36.

Luddi A, Margollicci M, Gambera L, Serafini F, Cioni M, De Leo V, et al. Spermatogenesis in a man with complete deletion of USP9Y. New Engl J Med. 2009;360:881–5.

Luisi S, Florio P, Reis FM, Petraglia F. Inhibins in female and male reproductive physiology: role in gametogenesis, conception, implantation and early pregnancy. Hum Reprod Update. 2005;11:123–35.

Matzuk MM, Lamb DJ. The biology of infertility: Research advances and clinical challenges. Nat Med. 2008;14:1197–213.

Meachem SJ, Nieschlag E, Simoni M. Inhibin B in male reproduction: pathophysiology and clinical relevance. Eur J Endocrinol. 2001;145:561–71.

Mieusset R, Bujan L, Mondinat C, Mansat A, Pontonnier F, Grandjean H. Association of scrotal hyperthermia with impaired spermatogenesis in infertile men. Fertil Steril. 1987;48:1006–11.

Mita M, Hall PF. Metabolism of round spermatids from rats: lactate as the preferred substrate. Biol Reprod. 1982;26:445–55.

Muttukrishna S, Yussoff H, Naidu M, Barua J, Arambage K, Suharjono H, et al. Serum anti- Mullerian hormone and inhibin B in disorders of spermatogenesis. Fertil Steril. 2007;88:516–8.

Nuti F, Krausz C. Gene polymorphisms/mutations relevant to abnormal spermatogenesis. Reprod Biomed Online. 2008;16:504–13.

Pan J, Eckardt S, Leu NA, Buffone MG, Zhou J, Gerton GL, et al. Inactivation of Nxf2 causes defects in male meiosis and age-dependent depletion of spermatogonia. Dev Biol. 2009;330:167–74.

Paulsen CA, Gordon DL, Carpenter RW, Gandy HM, Drucker WD. Klinefelter's syndrome and its variants: a hormonal and chromosomal study. Recent Prog Horm Res. 1968;24:321–63.

Pierik FH, Vreeburg JT, Stijnen T, De Jong FH, Weber RF. Serum inhibin B as a marker of spermatogenesis. J Clin Endocrinol Metab. 1998;83:3110–4.

Pryor JL, Howards SS. Varicocele. Urol Clin North Am. 1987;14:499–513.

Racine C, Rey R, Forest MG, Louis F, Ferré A, Huhtaniemi I, et al. Receptors for anti-Mullerian hormone on Leydig cells are responsible for its effects on steroidogenesis and cell differentiation. Proc Natl Acad Sci U S A. 1998;95:594–9.

Radhakrishnan Y, Hamil KG, Tan JA, Grossman G, Petrusz P, Hall SH, et al. Novel partners of SPAG11B isoform D in the human male reproductive tract. Biol Reprod. 2009;81:647–56.

Robertson KM, O'Donnell L, Jones ME, Meachem SJ, Boon WC, Fisher CR, et al. Impairment of spermatogenesis in mice lacking a functional aromatase (cyp 19) gene. Proc Natl Acad Sci U S A. 1999;96:7986–91.

Robinson R, Fritz LB. Myoinositol biosynthesis by Sertoli cells, and levels of myoinositol biosynthetic enzymes in testis and epididymis. Can J Biochem. 1979;57:962–7.

Romeo C, Santoro G. Varicocele and infertility: why a prevention? J Endocrinol Invest. 2009;32:559–61.

Rowley MJ, Leach DR, Warner GA, Heller CG. Effect of graded doses of ionizing radiation on the human testis. Radiat Res. 1974;59:665–78.

Sasaki M, Takeda E, Takano K, Yomogida K, Katahira J, Yoneda Y. Molecular cloning and functional characterization of mouse nxf family gene products. Genomics. 2005;85:641–53.

Saypol DC, Howards SS, Turner TT, Miller Jr ED. Influence of surgically induced varicocele on testicular blood flow, temperature, and histology in adult rats and dogs. J Clin Invest. 1981;68:39–45.

Schlegel PN. How to do a workup for male infertility. Med Aspects Hum Sex. 1991:28–34.

Schultz N, Hamra FK, Garbers DL. A multitude of genes expressed solely in meiotic or post-meiotic spermatogenic cells offers a myriad of contraceptive targets. Proc Natl Acad Sci U S A. 2003;100:12201–6.

Schwartz ID, Root AW. The Klinefelter syndrome of testicular dysgenesis. Endocrinol Metab Clin North Am. 1991;20:153–63.

Silber SJ, Van Steirteghem AC, Devroiy P. Sertoli cell only revisited (editorial). Hum Reprod. 1995;10:1031–2.

Simoni M, Gromoll J, Hoppner W, Kamischke A, Krafft T, Stahle D, et al. Mutational analysis of the follicle-stimulating hormone (FSH) receptor in normal and infertile men: identification and characterization of two discrete FSH receptor isoforms. J Clin Endocrinol Metab. 1999;84:751–5.

Simpson JL. Male pseudohermaphroditism: genetics and clinical delineation. Hum Genet. 1978;44:1–49.

Smith EP, Boyd J, Frank GR, Takahashi H, Cohen RM, Specker B, et al. Estrogen resistance caused by a mutation in the estrogen-receptor gene in a man. New Engl J Med. 1994;331:1056–61.

Smith SR, Chhetri MK, Johanson J, Radfar N, Migeon CJ. The pituitary-gonadal axis in men with protein-calorie malnutrition. J Clin Endocrinol Metab. 1975;41:60–9.

Stouffs K, Vandermaelen D, Massart A, Menten B, Vergult S, Tournaye H, et al. Array comparative genomic hybridization in male infertility. Hum Reprod. 2012;27:921–9.

Sutton KA, Jungnickel MK, Wang Y, Cullen K, Lambert S, Florman HM. Enkurin is a novel calmodulin and TRPC channel binding protein in sperm. Dev Biol. 2004;274:426–35.

Takeda R, Ueda M. Pituitary-gonadal function in male patients with myotonic dystrophy-serum luteinizing hormone, follicle stimulating hormone and testosterone levels and histological damage of the testis. Acta Endocrinol. 1977;84:382–9.

Tapanainen JS, Aittomäki K, Min J, Vaskivuo T, Huhtaniemi I. Men homozygous for an inactivating mutation of the follicle stimulating hormone (FSH) receptor gene present variable suppression of spermatogenesis and fertility. Nat Genet. 1997;15:205–6.

Teng YN, Lin YM, Lin YH, Tsao SY, Hsu CC, Lin SJ, et al. Association of a single- nucleotide polymorphism of the deleted-in-azoospermia-like gene with susceptibility to spermatogenic failure. J Clin Endocrinol Metab. 2002;87:5258–64.

Teng YN, Lin YM, Sun HF, Hsu PY, Chung CL, Kuo PL. Association of DAZL haplotypes with spermatogenic failure in infertile men. Fertil Steril. 2006;86:129–35.

Tung JY, Rosen MP, Nelson LM, Turek PJ, Witte JS, Cramer DW, et al. Variants in Deleted in Azoospermia-Like (DAZL) are correlated with reproductive parameters in men and women. Hum Genet. 2006a;118:730–40.

Tung JY, Rosen MP, Nelson LM, Turek PJ, Witte JS, Cramer DW, et al. Novel missense mutations of the Deleted-in-Azoospermia-Like (DAZL) gene in infertile women and men. Reprod Biol Endocrinol. 2006b;4:40.

Tuttelmann F, Dykstra N, Themmen AP, Visser JA, Nieschlag E, Simoni M. Anti-Mullerian hormone in men with normal and reduced sperm concentration and men with maldescended testes. Fertil Steril. 2009;91:1812–9.

Tuttelmann F, Simoni M, Kliesch S, Ledig S, Dworniczak B, Wieacker P, et al. Copy number variants in patients with severe oligozoospermia and Sertoli-cell-only syndrome. PLoS One. 2011;6:e19426.

Tyler-Smith C, Krausz C. The will-o'-the-wisp of genetics–hunting for the azoospermia factor gene. N Engl J Med. 2009;360:925–7.

Voglmayr JK. Metabolic changes in spermatozoa during epididymal transit. In: Hamilton DW, Greep RD, editors. Handbook of physiology. Section 7, vol. V,. Washington D.C: American Physiological Society; 1975. p.437–51.

Volkl TM, Simm D, Beier C, Dorr HG. Obesity among children and adolescents with classic congenital adrenal hyperplasia due to 21-hydroxylase deficiency. Pediatrics. 2006;117:e98–105.

Von Eckardstein S, Simoni M, Bergmann M, Weinbauer GF, Gassner P, Schepers AG, et al. Serum inhibin B in combination with serum follicle-stimulating hormone (FSH) is a more sensitive marker than serum FSH alone for impaired spermatogenesis in men, but cannot predict the presence of sperm in testicular tissue samples. J Clin Endocrinol Metab. 1999;84:2496–501.

Wallace EM, Groome NP, Riley SC, Parker AC, Wu FCW. Effects of chemotherapy-induced testicular damage on inhibin, gonadotropin and testosterone secretion: a prospective longitudinal study. J Clin Endocrinol Metab. 1997;83:675–81.

Wang C, Baker HW, Burger HG, De Kretser DM, Hudson B. Hormonal studies in Klinefelter's syndrome. Clin Endocrinol (Oxf). 1975;4:399–411.

Weiss JR, Moysich KB, Swede H. Epidemiology of male breast cancer. Cancer Epidemiol Biomarkers Prev. 2005;14:20–6.

Yan W. Male infertility caused by spermiogenic defects: Lessons from gene knockouts. Mol Cell Endocrinol. 2009;306:24–32.

Yang QS, Xia F, Gu SH, Yuan HL, Chen JZ, Yang QS, et al. Cloning and expression pattern of a spermatogenesis- related gene, BEX1, mapped to chromosome Xq22. Biochem Genet. 2002;40:1–12.

Zinn AR, Ramos P, Elder FF, Kowal K, Samango-Sprouse C, Ross JL. Androgen receptor CAGn repeat length influences phenotype of 47, XXY (Klinefelter) syndrome. J Clin Endocrinol Metab. 2005;90:5041–6.

Zitzmann M, Depenbusch M, Gromoll J, Nieschlag E. X-chromosome inactivation patterns and androgen receptor functionality influence phenotype and social characteristics as well as pharmacogenetics of testosterone therapy in Klinefelter patients. J Clin Endocrinol Metab. 2004;89:6208–17.

Biology of Male Hermaphrodite and Intersex

25

Shilpa Sharma, Anand Kumar, and Devendra K. Gupta

Learning Objectives

- Gender
- Abnormal sexual differentiation
- Aetiology and classification of disorders of sexual differentiation
- Principles of clinical evaluation

25.1 Introduction

Around 1% of live births may have some form of gonadal dysgenesis or ambiguity. The incidence has increased with growing awareness, more investigative modalities and psychological help to diagnose the borderline cases. The determination of the biology of sex of an individual is indeed multifactorial. The most important

S. Sharma, MS, MCh, PhD (✉) • D.K. Gupta, MS, MCh, FAMS, FCS, FRCS, DSc (✉)
Department of Pediatric Surgery, All India Institute of Medical Sciences, New Delhi, India
e-mail: drshilpas@gmail.com; profdkgupta@gmail.com

A. Kumar, MD, FAMS (✉)
Department of Reproductive Biology, All India Institute of Medical Sciences, New Delhi, India
e-mail: anandkumarrepbiol@hotmail.com

A. Kumar, M. Sharma (eds.), *Basics of Human Andrology*,
DOI 10.1007/978-981-10-3695-8_25

determining factors are the genetic and hormonal influences on the internal and external anatomical development of the sexual organs and the sexual behavioral component of the individual. The process of sexual differentiation is complex, and science has yet to elucidate the complete network of factors that influence the biological features which determine the developmental path of an individual to fall in the grey area of intersex.

25.2 Gender

The most important feature of an individual's sex is their behavioral sex or gender. The term gender encompasses how an individual opines himself, as a man or woman, which is gender identity; how he or she is viewed by others, as masculine or feminine, which is gender role; and their erotic behavior (Money and Ehrhardt 1972).

25.3 Abnormal Sex Differentiation

Abnormal sex differentiation due to over-masculinization in subjects with a 46,XX karyotype is female pseudohermaphroditism or under-masculinization in subjects with a 46,XY karyotype is male pseudohermaphroditism.

Abnormal sex differentiation may also occur in patients with sex chromosome abnormalities such as Turner syndrome (45,XO), Klinefelter's syndrome (47,XXY), XY females (Swyer syndrome) or XX males (sex reversed).

Turner syndrome (45,XO): There is normal autosomal complement, with only 1-X chromosome and ovary. Activity of only one X in Turner syndrome causes degeneration of the oocytes, loss of the follicle cells and ovarian dysgenesis. As a result, the ovary gets highly regressed or becomes streak ovary.

Klinefelter's syndrome (47,XXY, male): Normal autosomal complement of chromosomes with 47,XXY karyotype. Testes are formed normally; all germ cells die in this condition.

Swyer syndrome: The condition is due to complete gonadal dysgenesis as a result of mutations in the SRY genes.

Sex-reversed male: The condition involves translocation of SRY gene to autosomes or X chromosome.

25.4 Disorders of Sexual Differentiation

Understanding the aetiology of disorders of sexual differentiation (DSD) can be based on the karyotype and the location of the gonads – palpable or impalpable. Thus, DSD can be divided broadly into those with karyotype 46,XX, 46,XY or 45,XO or 45,XO/46,XY (mosaicism).

Table 25.1 Aetiology of female pseudohermaphroditism

A. *Congenital adrenal hyperplasia* (*CAH*)
21-alfa-hydroxylase (21-OH) deficiency
11-beta-hydroxylase (11-beta-OH) deficiency
3-beta-hydroxy-steroido-dehydrogenase (3-beta-HSD) deficiency
B. *Maternal tumors*
C. *Iatrogenic fetal virilization*
D. *Idiopathic forms*

25.4.1 46,XX Chromosome (Female Karyotype)

46,XX chromosome (female karyotype) can be associated with:

1. Female pseudohermaphroditism (more frequent): Aetiological factors are given in Table 25.1.
2. True hermaphroditism (rare), with female karyotype in about 80% cases.
3. Mixed gonadal dysgenesis (very rare).
4. Female embryo exposure to androgens or progestins.

Characteristics of 46,XX

(a) If the gonads are impalpable in the presence of elevated 17-hydroxyprogesterone (17-OHP), the most common cause is 21-hydroxylase deficiency.
(b) If the gonads are palpable, with normal 17-OHP levels, the most common cause is true hermaphroditism.
(c) If the gonads are impalpable, with normal 17-OHP, the aetiology may be normal prenatal exposure to androgens or progestins. For example, maternal virilizing tumors, gonadal dysgenesis, true hermaphroditism or idiopathic female pseudohermaphroditism

25.4.2 46,XY (Male Karyotype)

The most common causes are (Migeon 1980):

1. Male pseudohermaphroditism: Aetiological factors are given in Table 25.2.
2. Pure gonadal dysgenesis in 40% of cases.
3. Male pseudohermaphroditism due to several aetiologies like testosterone biosynthesis deficiency, 5α-reductase deficiency, etc. in 10% of cases.

Characteristics of 46,XY

Both palpable gonads: Male pseudohermaphroditism in the presence of bilateral palpable gonads can be related to a defect in testosterone biosynthesis,

Table 25.2 Aetiology of male pseudohermaphroditism

A. *Adrenal steroidogenesis deficiency* (*congenital adrenal hyperplasia*)
20–22 desmolase deficiency
17-alfa-hydroxylase (17-OH) deficiency
3 beta-hydroxy-steroido-dehydrogenase (3-beta-HSD) deficiency
B. *Sexual hormones only synthesis deficiency*
17,20 desmolase deficiency
17 beta-hydroxysteroido-dehydrogenase (17-beta-HSD) deficiency
C. *Leydig cell aplasia*
D. *5α-reductase deficiency* (*lack of conversion of testosterone into DHT*)
E. *Defect of androgen receptor*
Complete androgen insensitivity (complete testicular feminization)
Incomplete androgen insensitivity (incomplete testicular feminization)
F. *Isolated AMH activity deficiency*
Mullerian duct persistent syndrome ("hernia uteri inguinalis")
G. *Idiopathic forms*

5α-reductase deficiency or partial androgen insensitivity. The complete form has no ambiguous genitalia which are totally feminine. Undetectable testosterone level after human chorionic gonadotropin (hCG) stimulation test can be related to a defect in its biosynthesis or, more rarely, to Leydig cell aplasia.

One gonad is palpable: The two possible diagnoses are true hermaphroditism which is more likely and exhibits 46,XY karyotype in 10% of cases and mixed gonadal dysgenesis with male karyotype. Pelvic ultrasonography and/or genitogram to detect Mullerian derivatives and gonadal biopsy will allow a correct diagnosis.

Impalpable gonads: The most likely aetiology is pure gonadal dysgenesis. The diagnosis must be confirmed by gonadal histology.

25.4.3 Mosaic Karyotype

The two most likely possibilities are:

1. True hermaphroditism, which is associated to mosaicism in one third of cases.
2. Gonadal dysgenesis. Generally its mixed form is associated to mosaicism.

In both conditions, there may be only one palpable gonad (see Fig. 25.1). However, in true hermaphroditism, there can be bilaterally palpable gonads with a dual consistency in either or both gonads. Gonadal biopsy is mandatory to establish the diagnosis (Gupta 1999).

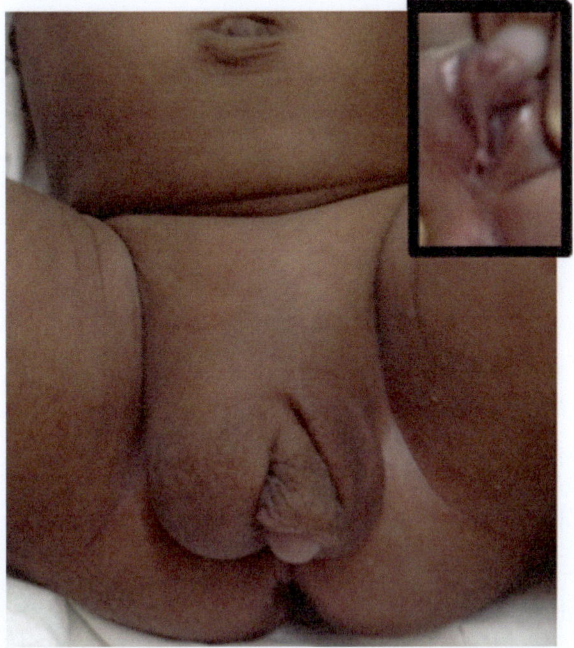

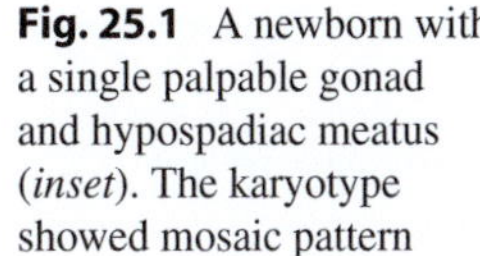

Fig. 25.1 A newborn with a single palpable gonad and hypospadiac meatus (*inset*). The karyotype showed mosaic pattern

25.5 Classification of Disorders of Sexual Differentiation (DSD)

There is no uniformly accepted classification of sex differentiation disorders. The most widely accepted classification is based on histology of the bilateral gonads present, with subclassification according to the aetiology (Allen 1976). In the Allen gonadal classification, there are five major categories:

I. Ovary only (female pseudohermaphrodite)
II. Testis only (male pseudohermaphrodite)
III. Ovary plus testis (true hermaphrodite)
IV. Testis plus streak (mixed gonadal dysgenesis)
V. Streak plus streak (pure gonadal dysgenesis)

The structural abnormalities affecting external genital development are usually associated with chromosomal anomalies and result in asymmetry in the phenotype.

The Lawson Wilkins Pediatric Endocrine Society (LWPES) and the European Society for Paediatric Endocrinology (ESPE) have proposed changes to the nomenclature and definitions of disorders in which the development of chromosomal, gonadal or phenotypic sex is atypical (see Table 25.3). The terminology mainly reflects the chromosomal sex or the gonadal tissue associated with the disorder (Gupta and Sharma 2011).

Table 25.3 Old and revised nomenclature of disorders of sexual development and their chromosomal pattern

Old nomenclature	Revised nomenclature	Chromosomal pattern
Female pseudohermaphrodite	46,XX DSD	46,XX
Male pseudohermaphrodite	46,XY DSD	46,XY
True hermaphrodite	Ovotesticular DSD	46,XX – 80%, 46,XY or mosaicism (45,X/46,XY) -10–20%
Mixed gonadal dysgenesis/XX male	46,XX testicular DSD, 46,XY DSD or sex chromosome DSD if there is mosaicism (45,X/46,XY)	Mosaicism (45,XO/46,XY) mostly, 46,XX, 46,XY
Pure gonadal dysgenesis/XY sex reversal	46,XY complete gonadal dysgenesis	46,XX, 46,XY, 46,XO

25.6 Clinical Evaluation

The clinical evaluation is based on the knowledge of the five main types of intersex disorders.

25.6.1 46,XX DSD or Congenital Adrenal Hyperplasia (CAH)

The condition is the result of a genetic female having 46,XX karyotype, being exposed to exogenous or endogenous androgens in utero. CAH constitutes the most common form of DSD presenting in the neonatal period.

In over 90% cases, CAH is due to 21-hydroxylase deficiency resulting in excess of 17-hydroxyprogesterone in a female child with 46,XX. There is clitoral enlargement and a large urogenital sinus. The vaginal orifice may be hidden under a skinfold.

The next common form is due to deficiency of 11-beta desmolase enzyme resulting in excessive vomiting and a serious salt loss that may even prove fatal shortly after birth.

The Mullerian system develops into the fallopian tubes, uterus and upper portion of the vagina, and the Wolffian system regresses. The degree of virilization that may vary from minimal phallic enlargement to almost complete masculinization depends on the stage of differentiation at the time of exposure and the androgenic effect of the hormone to which the fetus is exposed.

The affected child is born with a hypertrophy of the clitoris with severe chordae and variable degree of labioscrotal fold fusion. The labioscrotal folds may be bulbous and rugated, giving the general appearance of a male with underdeveloped scrotum and hypospadias. Hyperpigmentation of the genitalia may be seen. A completely virilized phallus with a glandular urethra may also be present in the more virilized cases (Sharma and Gupta 2012).

The degree of masculinization depends upon the time of exposure. If the fetal exposure occurs before 12 weeks of gestation, the external genitalia may have a

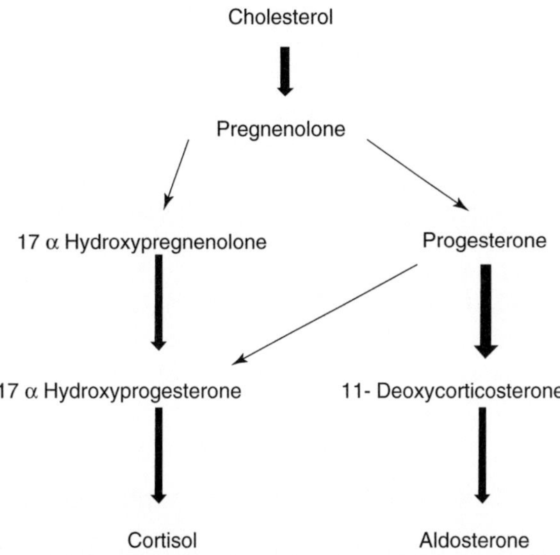

Fig. 25.2 Pathway of cortisol synthesis

well-developed phallus. If the exposure is after 12 weeks of gestation, when the vagina has already separated from the urethra, clitoromegaly occurs.

Cortisol biosynthesis requires five enzymes: 20-22 desmolase, 17-∝-hydroxylase, 3-ß–OH-dehydrogenase, 17–20 desmolase and 17-ß-OH-dehydrogenase (see Fig. 25.2). The enzymes 20–22 desmolase and 17-∝-hydroxylase do not affect the sexual development of the female. 3-ß–OH-dehydrogenase may produce female pseudohermaphroditism. Two other enzymes, 21 and 11-ß-hydroxylase required for the biosynthesis of cortisol, are found predominantly in the adrenal cortex. The deficiency of these causes hypersecretion of adrenocorticotropic hormone (ACTH), subsequent adrenocortical hyperplasia and accumulation of steroid's precursors, which are then metabolized into androgens. These androgens are converted in peripheral tissues to testosterone and dihydrotestosterone (DHT), with subsequent virilization of the external genitalia.

The salt-wasting form of 21-OHD is more common and involves complete enzyme deficiency, and adrenal insufficiency produces adrenal crisis. The simple virilizing form has a partial enzyme deficiency with compensation which produces adrenal insufficiency only in stressful circumstances. The third, late-onset form only affects adults, and a fourth, or cryptic, form has also been described.

The simple virilizing forms are less affected than females with the salt-wasting type. If untreated, excessive androgens are synthesized postnatally, accelerating growth and skeletal maturation and producing androgen-induced sexual changes, including abnormal genital development late in infancy (pseudo-precocious puberty), early pubic hair and penile or clitoral enlargement. Estrogenic breakthrough occurs in some girls at puberty, accompanied by menses and breast development.

Salt-wasting form is characterized by a significant impairment of cortisol and aldosterone synthesis. The adrenal insufficiency occurs few days after birth; 50% of the untreated neonates have the onset of acute adrenal crisis and accompanying sodium wasting between 7 and 14 days of age and the remaining 50% by 1 month of age.

11 β-hydroxylase deficiency is seen in about 5% of all CAH. The deficiency results in impairment of both the cortisol and aldosterone biosynthesis.

A nonclassic form is also described which is milder, has a later onset and manifests with hirsutism, amenorrhea, tall stature and advance bone age. Genetic males may have larger penis and hyperpigmentation of the genitalia and areolae; the nonclassic form has pseudo-precocious puberty, tall stature and advance bone age. CAH patients must be correctly diagnosed because they should be raised as females. The prognosis of these patients when raised as females is excellent for pubertal development and the attainment of normal female characteristics, sexual activity and reproduction.

25.6.2 46,XY DSD or Male Pseudohermaphroditism (MPH)

MPH accounts for about half of all intersex disorders that require surgical intervention. The gonads are testes, usually descended, but may be undescended unilaterally or bilaterally. There is proximal hypospadias with a large single urogenital sinus. The vaginal orifice is not visible in the perineum. The vaginal pouch is always present. The uterus is absent. The phallus may be small or normal in size. The chromosomes are 46,XY (Wisniewski 2012).

There are three subtypes of MPH.

25.6.2.1 Testicular Feminization Syndrome (TFS) or Androgen Insensitivity Syndrome (AIS Complete or Incomplete)

In TFS, the testes develop normally, the hormonal secretion such as androgens and AMH occurs normally, but the urogenital sinus and external genitalia do not develop along the male line because of the lack of response of all target organs including the pubic and axillary hair, larynx and hypothalamic-pituitary axis to the normally secreted androgen. The genitalia are genetically insensitive to the action of androgens. Complete regression of the androgen-dependent Wolffian ducts results in the development of female external genitalia. AMH action causes regression of the Mullerian ducts.

The defect is thought to be localized at the level of the cytoplasmic androgen receptor of the 5α-reductase enzyme. Two major types of defects have been described: androgen receptor negative form and androgen receptor positive form. In the androgen receptor negative type, the specific intracellular androgen receptors are undetectable in all target organs, whereas in the androgen receptor positive variant, there is a structural abnormality of the receptor (Brown et al. 1982).

Individuals affected by complete AIS present in the neonatal period as normal females. Androgen receptor DNA is present on the X chromosome, suggesting that the inheritance is X-linked recessive or sex-linked autosomal dominant.

All patients have exclusively testicular tissue. In AIS, seminiferous tubules show no spermatogenesis and increased numbers of Leydig cells. The internal ducts develop normally as Wolffian ducts, and there is regression of all Mullerian structures as AMH is normally secreted. They are phenotypically normal, tall and hairless females with non-ambiguous feminine external genitalia and symmetrically undescended gonads. Patients usually have a shallow vaginal cavity. At puberty, normal breasts develop, and axillary and pubic hair are absent or scanty. There is slight vulvar hair development, and amenorrhoea is the rule. Clinically, patients usually present in the postpubertal period for evaluation of primary amenorrhoea.

Gonadectomy is recommended after puberty in cases reared as female as there is a 5–10% risk for malignancy in the undescended gonads that rarely develop malignancy before puberty. Estrogen therapy will be required after pubertal gonadectomy to ensure normal growth and breast development.

Incomplete AIS forms present with ambiguity in the neonatal period. The individuals have exclusively testicular tissue, Wolffian ducts develop normally (sometimes there is hypoplastic or absent vas), and there is regression of the Mullerian structures. There is a minor defect of virilization of the external genitalia consisting of partial fusion of the labioscrotal folds and some degree of clitoromegaly present at birth leading to ambiguity. The gonads are usually symmetrically undescended, and there is a short blind-ending vagina. At puberty, some degree of virilization and feminization occurs, including the development of normal pubic hair and gynaecomastia. Sex assignment should be individualized. Since some virilization is expected at the time of puberty, in those assigned female sex, gonadectomy should be performed during the prepubertal period. The AIS patients remain infertile.

25.6.2.2 5-Alfa Reductase Deficiency

It was originally described in 1974 (Imperato-McGinley et al. 1974). Phenotypic appearance is generally represented by a severe hypospadias. The deficit is found in the genital sinus and hepatic tissues; the diagnosis in adults is made by measuring the ratio of testosterone to DHT, which is increased. However, in affected children past infancy, the basal plasma testosterone and DHT levels are too low for accurate determination of the testosterone/DHT ratio. The measurement can be done after hCG administration. The demonstration that the genital skin is unable to convert the testosterone to DHT in tissue culture is the ultimate test for diagnosis (Lorge et al. 1989). The baby is phenotypically a female with both the gonads in the labial folds.

The vaginal orifice may be visible in the vestibule. The other siblings may also be affected. The authors have seen that it runs in families quite often with 4–5 brothers of a single family affected. There is either androgen receptor failure or testosterone is not converted into DHT due to enzyme deficiency. The serum DHT level is low. The serum testosterone levels remain normal though part of it is converted to estradiol resulting in breast development at puberty.

Normal testosterone increase after hCG stimulation is related either to 5α-reductase deficiency or to partial androgen insensitivity. Testosterone/DHT ratio after hCG will allow the exact diagnosis: it is four- to fivefold increased in case of 5α-reductase deficiency, while it results to be normal in case of androgen receptor insensitivity. Diagnosis will be confirmed by evaluating the scrotal skin fibroblasts enzymatic activity. In 5α-reductase deficiency reduced 5α-reductase activity is detectable in vitro on genital skin fibroblasts (Allen 1985). In case of androgen insensitivity syndromes due to receptor deficiency, both testosterone and DHT increase after hCG stimulation, but the genital skin fibroblast test shows an abnormal binding of androgens to their receptors (Sultan and Migeon 1980).

25.6.2.3 Persistent Mullerian Duct Syndrome

The chromosomal pattern is 46-XY. The aetiology is that AMH production or the responsiveness to it is inadequate due to either a mutation in the AMH gene on 19p13 or its type II receptor, 12q13. The testicular androgens stimulate the external genitalia and Wolffian ducts. The Mullerian duct structures are retained. Genetically and gonadally, these individuals are male but possess the internal genitalia of both sexes (Gupta 1999).

25.6.3 Ovotesticular DSD or True Hermaphroditism (TH)

In TH individuals, the ambiguous genitalia are associated with an ovarian as well as a testicular tissue. There is a urogenital sinus. The vaginal orifice is usually covered with the skin flap. The scrotum develops only if the gonad has reached it. The phallus is usually triangular and well developed but could be small. The single meatus is wide. The presence of a dumbbell-shaped gonad with double consistency in case if palpable and descended, unilateral or bilateral, is diagnostic. The ovarian component is firmer and the testis is softer. 60–80% of TH cases have 46,XX karyotype, 10–20% cases are 46,XY, and only <10% have mosaicism (Gupta 1999).

In ovotesticular DSD, there could be various forms of combinations of the internal gonads resulting in a pair of ovary and testis, testis/ovotestis, ovary/ovotestis and bilateral ovotestis. Usually the ovarian gonads are abdominal in location and are accompanied by a fallopian tube, while the testis descends into the scrotum depending on the testicular volume and maturity and has a vas deferens as the draining duct. The descent and draining duct of the ovotestis depends on the amount of testicular component in it. The ovotestis may fully descend in the scrotum if the testicular component is significant and is palpable anywhere along the course of the descent of the testis, with the vas deferens as the draining duct. However, there is never a vas deferens as well as fallopian tube on the same side of the ovotestis.

Ovotesticular DSD comprises less than 10% of the DSD (Allen 1985). Most cases are sporadic, but a minority of cases has an autosomal recessive pattern of inheritance. Ovotesticular DSD is a disorder of gonadal differentiation at the level of the primitive germ cell probably caused by the chromosomal abnormality. The most frequent associations are an ovary and a testis in 30%, an ovary and an ovotestis in 30% and an ovotestis bilaterally in 20% (Van Niekerk 1981).

Internal duct differentiation follows the appropriate ipsilateral gonad; a fallopian tube develops when an ovary is present and a vas with the epididymis when a testis is present. When an ovotestis is present, the differentiation of the ducts is variable. In ovotesticular DSD, the uterus is present in about 90% cases, but usually it is hypoplastic, cordlike or unicornuate (Lorge et al. 1989; Allen 1985). The uterus may be bulky only in 10–15% cases. In such children, at puberty, bilateral breast development occurs, due to the estrogen effect from the ovary. Detection of high values of estradiol due to the presence of ovarian tissue on hormonal evaluation in a male child would strongly favour the diagnosis of ovotesticular DSD. Testosterone values may be normal or low depending on the amount of normally functioning testicular tissue.

Although the appearance of the external genitalia varies from feminine with slight clitoral prominence to full masculinization, a tendency to maleness with asymmetrically descended gonads and hypospadias predominates in 75% (Lorge et al. 1989).

The diagnosis of ovotesticular DSD should be considered in all patients with ambiguous genitalia. The finding of a 46,XX /46,XY karyotype or a bilobed gonad compatible with an ovotestis in the inguinal region or labioscrotal folds suggests the diagnosis. If all other forms of male and female pseudohermaphroditism have been excluded, laparotomy and histological confirmation of both ovarian and testicular tissue establish the diagnosis.

The changes at puberty are variable but generally correlate with the type of gonadal tissue present. Gynaecomastia is the rule. At puberty almost half of the ovotesticular DSD will have normal menstruation in the presence of ovarian tissue (Van Niekerk 1981). In the west, most authors advocate a female sex assignment as the reconstructive surgery for sex-assigned female gives by far the better result. Assignment to the male gender exposes the individual to several reconstructive procedures with sometimes poor functional and cosmetic result and inevitably primary hypogonadism since most of the testes are dysgenetic when adulthood is reached. However, the authors have not regretted assigning a male gender to these patients taking into consideration the sociocultural values in the region (Sharma and Gupta 2008). Gonadal malignancies have been reported; thus, all discordant and dysgenetic tissue should be removed after sex assignment. Anecdotal cases of pregnancy in true hermaphroditism have been reported. There is also a report of spermatogenesis in the testis of true hermaphrodites (Van Niekerk 1981).

25.6.4 46,XX Testicular DSD, 46,XY DSD or Sex Chromosome DSD or Mixed Gonadal Dysgenesis (MGD)

There is ambiguity of external genitalia. There is usually a tiny phallus. The vaginal and the urethral openings are separated in the vestibule. The scrotum is very poorly developed. The internal gonads include a testis on one side and a small, flat, shiny streak gonad on the other side, usually on left side but may be on right side in 20%

cases. The testis is dysgenetic and usually abdominal in location and does not descend unless better developed. The uterus is always present and is usually hypoplastic. The chromosomal pattern may be 46,XY/XO/XXO and the mosaicism is the most common finding.

The gonads of these subjects are classically asymmetric as there is functioning gonadal tissue which can be normal or dysgenetic on one side and a streak or absent on the other. A mosaic 45,X/46,XY karyotype is found in most of these patients (McCann-Crosby et al. 2014).

The testis in these patients is composed of Sertoli cells and Leydig cells, but no germinal cells are present. The streak gonad appears similar to the ovarian stroma but without oocytes. The Mullerian duct usually persists unilaterally or bilaterally; the testis is often provided with a fallopian tube rather than a vas and an epididymis. The streak gonad is usually drained by a Mullerian duct. A bicornuate or unicornuate uterus is generally present.

Malignancy will develop in about 9–40% of the mixed gonadal dysgenesis, usually in the form of gonadoblastomas, dysgerminomas or both (Lorge et al. 1989). Individuals with a 46,XY karyotype are at a much higher risk than those with a mosaic pattern. Individuals tend to virilize and have gynaecomastia at the time of puberty.

No fertility has been reported so far. Sex assignment should be individualized. The response of the phallus to androgenic stimulation should be assessed in borderline cases, but the female gender is usually preferred, as the development of the phallus is usually inadequate for functional and cosmetic surgical results.

Patients assigned as female require gonadectomy, clitoroplasty and vaginoplasty. Those assigned to the male gender most probably should undergo removal of any dysgenetic gonadal tissue and repair of the usually severe hypospadias.

25.6.5 46,XY Complete Gonadal Dysgenesis or Dysgenetic Male Pseudohermaphrodite (DMP)

The clinical picture resembles that of MGD. However, as both the gonads are always dysgenetic in DMP, they fail to descend and remain abdominal. The scrotum is hypoplastic. The uterus is also present. Phallic size is variable. The abdominal gonads are rounded and soft. The draining ducts resemble either the fallopian tubes or look like the epididymis.

The clinical diagnosis is based on the physical appearance, karyotype, genitogram and hormonal profile. Ultrasonography is routinely performed for the evaluation of the renal tract, internal gonads, uterus and adrenal glands.

The most common pattern of chromosome is 46,XY. Various other karyotype patterns may be present except 46,XX DSD. If the karyotype is 46,XX pattern, the case is either 46,XX DSD or ovotesticular DSD. The presence of a Y line with a mosaicism suggests a 46,XX testicular DSD, 46,XY DSD or sex chromosome DSD or 46,XY complete gonadal dysgenesis in most of the cases.

A genitogram establishes not only a vaginal pouch but may also reveal a cervical indentation, suggestive of the presence of the uterus. Rarely, the dye outlines the uterine cavity and fallopian tube and may even spill into the peritoneal cavity suggestive of the existence of Mullerian structures. The presence of the uterus can also be made on per rectal examination and on ultrasonography. An advanced bone age is suggestive of 46,XX DSD.

Hormonal evaluation is of particular importance in 46,XX DSD, ovotesticular DSD and few cases of 46,XY DSD. A 4–5 times higher level of 17-OHP is suggestive of 46,XX DSD. A normal value of estradiol in the presence of normal testosterone level is suggestive of the presence of an ovarian source as in ovotesticular DSD. Similarly, testosterone /DHT ratio is changed in 46,XY DSD with 5α-reductase deficiency, resulting in low levels of DHT, though testosterone levels remain normal. In testicular feminization syndrome, part of testosterone may be converted into estrogens and results in breast development.

Key Questions

- Describe the disorders of sexual differentiation.
- What are the conditions associated with a mosaic karyotype?
- Name the causes of male pseudohermaphroditism.
- Describe the aetiology of testicular feminization syndrome.

References

Allen T. Disorder of sexual differentiation. In: Kelalis PP, King LR, Belman AB, editors. Clinical pediatric urology. 2nd ed. Philadelphia: WB Saunders Co; 1985. p. 904–21.

Allen T. Disorder of sexual differentiation. Urology. 1976;7(4 suppl):1–32.

Brown TR, Maes M, Rothwell SW, Migeon CJ. Human complete androgen insensitivity with normal dihydrotestosterone receptor binding capacity in cultured genital skin fibroblasts: evidence for a qualitative abnormality of the receptor. J Clin Endocrinol Metab. 1982;55:61–8.

Gupta DK, Sharma S. Pediatric urology. In: Goel KM, Gupta DK, editors. Hutchison's pediatrics. 2nd ed. Jaypee: New Delhi; 2011. p. 244–57.

Gupta DK. Pediatric intersex disorders. In: Gupta DK, editor. Ambiguous genitalia and hypospadiology. New Delhi; Printext publishers: 1999. p. 57–62.

Imperato-McGinley J, Guerrero L, Gauthier T, Peterson RE. Steroid Sa-reductase deficiency in men: an inherited form of male pseudohermaphroditism. Science. 1974;186(4170):1213–5.

Lorge F, Wese FX, Sluysmans T, Hennebert PN, Malvaux P, Maes M, et al. L'ambiguite sexuelle: aspects urologiques. Acta Urol Belg. 1989;57:647–62.

McCann-Crosby B, Mansouri R, Dietrich JE, McCullough LB, Sutton VR, Austin EG, et al. State of the art review in gonadal dysgenesis: challenges in diagnosis and management. Int J Pediatr Endocrinol. 2014;4:1–17.

Migeon CJ. Male pseudohermaphroditism. Annee Endocrinol. 1980;41:311–43.

Money J, Ehrhardt AA. Man and woman, boy and girl. Baltimore: Johns Hopkins University Press; 1972.

Sharma S, Gupta DK. Gender assignment and hormonal treatment for disorders of sexual differentiation. Pediatr Surg Int. 2008;24:1131–5.

Sharma S, Gupta DK. Male genitoplasty for 46 XX congenital adrenal hyperplasia patients presenting late and reared as males. Indian J Endocrinol Metab. 2012;16:935–8.
Sultan CH, Migeon CJ. La culture de fibroblastes: moyen d'etude des anomalies de l'action cellulaire des androgenes au cours du pseuhermaphroditisme male. Annee Endocrinol. 1980;41(4):305–10.
Van Niekerk WA. True hermaphroditism. Pediatr Adolesc Endocrinol. 1981;8:80.
Wisniewski AB. Gender development in 46,XYDSD: infuences of chromosomes, hormones, and interactions with parents and health care professionals. Scientifica. 2012;Volume 2012, Article ID 834967:15 pages.

Gynaecomastia

26

Sweety Agrawal, Mohd Ashraf Ganie, and Sobia Nisar

Learning Objectives

- Aetiopathogenesis and histopathology of gynaecomastia
- Physiological gynaecomastia
- Gynaecomastia and breast cancer
- Principles of diagnostic evaluation and management

26.1 Introduction

Gynaecomastia is defined as an enlargement of the male breast glandular tissue. Gynaecomastia is common and may be seen in up to two-thirds of all adult men (Nuttall 1979; Carlson 1980; Georgiadis et al. 1994). There are no well-defined clinical thresholds to describe gynaecomastia. Various authors have used varying definitions ranging from 0.5 to 3 cm of palpable glandular tissue. Gynaecomastia must be differentiated from lipomastia or pseudogynaecomastia, which is characterized by an increase in subareolar fat alone in the absence of glandular enlargement.

S. Agrawal, MD • M.A. Ganie, MD, DM (✉)
Department of Endocrinology and Metabolism, All India Institute of Medical Sciences, New Delhi, India
e-mail: agrawalsweety81@gmail.com; ashraf.endo@gmail.com

S. Nisar, MD
Department of Geriatric Medicine, All India Institute of Medical Sciences, New Delhi, India
e-mail: sobianisar78@gmail.com

A. Kumar, M. Sharma (eds.), *Basics of Human Andrology*,
DOI 10.1007/978-981-10-3695-8_26

To differentiate gynaecomastia from pseudogynaecomastia, the subareolar tissue is compared with adjacent subcutaneous fat in the anterior axillary fold. In true gynaecomastia, a firm mound of tissue is palpable which is concentric with the areola (Braunstein 2007). Gynaecomastia may be unilateral, bilateral or asymmetric. Although most cases of gynaecomastia are benign, it can be a sign of underlying systemic disease.

26.2 Aetiopathogenesis

Gynaecomastia is caused by an imbalance between the free androgen and free estrogen effects (Mathur and Braunstein 1997). Gynaecomastia occurs when there is deficiency of androgen levels or their action or increase in estrogen levels.

Peripheral aromatization of androgens into estrogen is the primary source of estrogen in man (Braunstein 1999). Estrogen acts through the estrogen receptor alpha (ERα) to cause epithelial cell proliferation and differentiation. In estrogen-secreting tumors such as Leydig cell tumors, there is also suppression of luteinizing hormone (LH) secretion along with decreased androgen levels (Bercovici et al. 1984). Moreover, estrogen increases sex hormone-binding globulin (SHBG) levels, leading to a decrease in free testosterone levels. Estrogen also inhibits 17-alpha-hydroxylase, thereby exerting an inhibitory effect on intratesticular steroidogenesis. Androgens exert anti-proliferative and apoptotic effects on the breast. Hence, decreased testosterone levels lead to decreased androgen-mediated inhibition of growth of breast tissue. Also, a compensatory rise in LH occurs stimulating the Leydig cells to secrete estradiol preferentially. As a result, there is increased SHBG and decreased free testosterone leading to a further decrease in free androgen to free estrogen ratio. Human chorionic gonadotropin (hCG) hormone has similar functions as of LH. hCG stimulates Leydig cells to secrete estrogen (Forest et al. 1979).

Hyperprolactinaemia causes gynaecomastia primarily by suppression of gonadotropin-releasing hormone (GnRH) secretion. Progesterone plays a role in lobulo-alveolar differentiation in the female breast, but its physiological effect on male breast tissue is not clear (Humphreys et al. 1997).

Estrogen enhances receptor expression for insulin like growth factor-1 (IGF-1) in breast cancer cell cultures (Stewart et al. 1990). IGF-1 is responsible for terminal end bud formation and ductal morphogenesis.

Leptin receptors have been discovered on epithelial cells of the breast and may directly stimulate the breast tissue. Further, leptin has been proposed to enhance aromatase activity in adipose tissue and breast tissue. Leptin also activates estrogen receptors in breast tissue (Catalano et al. 2003, 2004).

Growth factors like transforming growth factor alpha and beta (TGF-α, β), epidermal growth factor (EGF), platelet derived growth factor (PDGF) and tumor necrosis factor alpha (TNFα) may contribute to gynaecomastia by autocrine or paracrine pathways (Peres et al. 1987; Nahta et al. 2003) Most of the information has been derived from studies of female breast cancer tissue. Sex differences as well as differences between normal and malignant breast cells may exist. All observations may not be generalizable to male breast tissue.

The aetiological factors associated with gynaecomastia can be categorized into four types (Burgess and Shousha 1993; Sasano et al. 1996; Kanhai et al. 2000; Dimitrikakis et al. 2002):

1. Estrogen excess due to exogenous estrogen intake as in the treatment of prostate cancer; excess production of estrogen in Leydig cell tumors and Sertoli cell tumors; increased aromatization as seen in obesity, hyperthyroidism, etc.
2. Androgen deficiency seen in primary hypogonadism such as Klinefelter syndrome, chemotherapy, radiotherapy. 50–70% cases of Klinefelter syndrome are associated with gynaecomastia (Smyth and Bremner 1998; Lanfranco et al. 2004).
3. Altered androgen/estrogen ratio as seen in puberty, ageing, chronic renal and hepatic disorders, hyperthyroidism, etc.
4. Decreased androgen action in case of use of antiandrogen drugs such as spironolactone, cimetidine, etc.; the presence of androgen receptor defects.

26.3 Histological Findings

Histologically, there is proliferation of breast ductal system or mammary ductules in fibroconnective tissue stroma. True acinar lobules, which are characteristic of the female breast tissue, do not form. Early disease is characterized by extensive ductal hyperplasia and proliferation, while at later stages, stromal fibrosis becomes predominant (Nicolis et al. 1971; Bannayan and Hajdu 1972). See Figs. 26.1, 26.2 and 26.3 for histological changes in gynaecomastia.

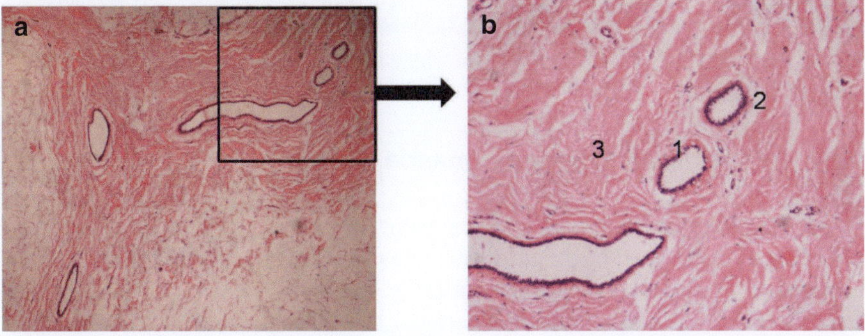

Fig. 26.1 (**a**) 4×, (**b**) 10× magnification – haematoxylin and eosin staining of breast tissue showing epithelial hyperplasia (*1*) with periductal oedema (*2*) and cellular stroma (*3*) (Courtesy: Dr. Sandeep Mathur, Additional Professor, Department of Pathology, All India Institute of Medical Sciences, New Delhi)

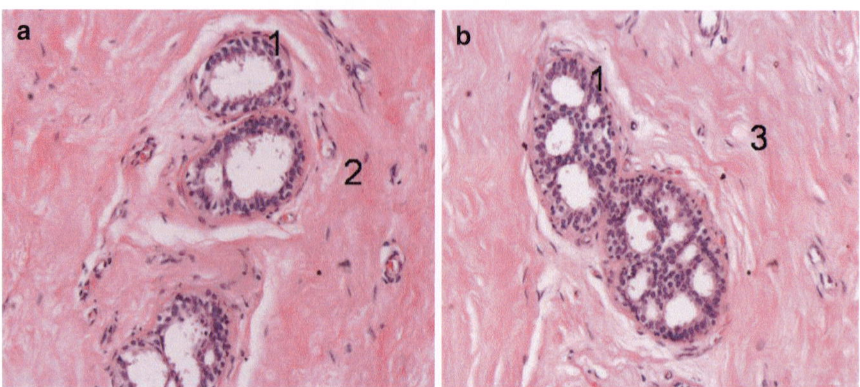

Fig. 26.2 (**a**, **b**) 20× magnification – haematoxylin and eosin staining of breast tissue showing epithelial cell hyperplasia (*1*) surrounding myxoid (*2*) and fibrotic stroma (*3*) (Courtesy: Dr. Sandeep Mathur, Additional Professor, Department of Pathology, All India Institute of Medical Sciences, New Delhi)

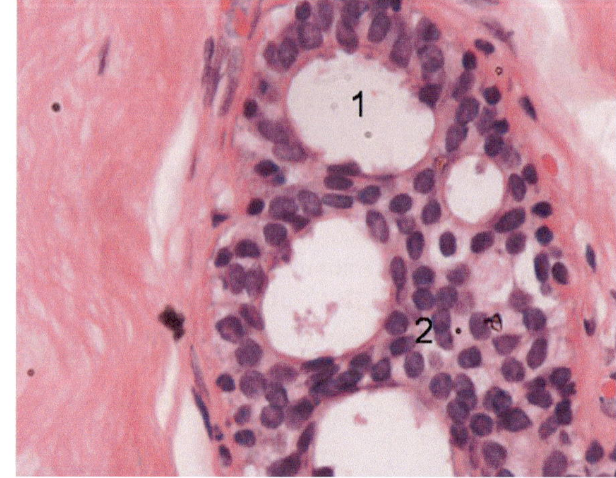

Fig. 26.3 60× magnification – haematoxylin and eosin staining of breast tissue showing dilated duct (*1*) with epithelial proliferation (*2*) (Courtesy: Dr. Sandeep Mathur, Additional Professor, Department of Pathology, All India Institute of Medical Sciences, New Delhi)

26.4 Physiological Gynaecomastia

Physiological gynaecomastia is seen in neonates, pubertal boys and elderly men. Transient gynaecomastia is seen in 60–90% of infants as a result of transfer of estrogens transplacentally. Gynaecomastia is seen in 48–64% of boys at puberty (Nydick et al. 1961). Age of onset is between 13 and 14 years, with peak incidence corresponding to Tanner development stage 3 or 4. Spontaneous resolution usually occurs in a year or two, with persistent gynaecomastia in 20%. Physiological gynaecomastia at puberty is due to a transient imbalance between estrogen and androgens. Estrogens rise by threefold and androgens by 30-fold during puberty. The peak estrogen levels may be achieved prior to testosterone, leading to an altered estrogen/testosterone ratio. Other mechanisms may be increased sensitivity of the breast tissue to normal male estrogen concentrations, gene polymorphisms involving leptin receptor or estrogen receptor beta, increased IGF-1 levels and high leptin levels in puberty (Kleinberg 1998). Physiological gynaecomastia is also seen in elderly men due to age-related decrease in gonadotropins and testosterone (Allan and Mclachlan 2004).

26.5 Gynaecomastia and Breast Cancer

Male breast cancer is distinctly uncommon. Gynaecomastia is not a risk factor for breast cancer in men except Klinefelter syndrome which has 20 times more risk of developing breast cancer (Swerdlow et al. 2005; Fentiman et al. 2006). Breast cancer in a male is a rapidly growing breast mass, which is hard, eccentrically located and usually unilateral. Overlying skin may be dimpled and nipple retraction may be present. Nipple bleeding or discharge may be present but tenderness is unusual.

26.6 Principles of Diagnostic Evaluation

Evaluation of gynaecomastia begins with a thorough history including duration of symptoms, the presence of pain or nipple discharge. History of liver or renal disease should be asked for as they are associated with poor metabolism of androgens. Changes in libido or sexual function or regression of secondary sexual characters point towards hypogonadism and should be enquired. Abnormal weight gain also indicates a possibility of lipomastia or pseudogynaecomastia. Finally, detailed drug history and history of dietary or occupational exposure to estrogen are to be taken to rule out drug-induced gynaecomastia.

Physical examination is first directed towards differentiation between true and pseudogynaecomastia as described earlier in this chapter. Subsequently, examination of genitalia is done to look for signs of hypogonadism. Systemic examination may also reveal signs of chronic liver disease or hyperthyroidism.

Finally, investigations are directed to localize aetiology of gynaecomastia. Chronic liver and kidney disease leads to altered androgen/estrogen ratio and may lead to gynaecomastia (Kley et al. 1985). Hence, liver and kidney function tests are to be done in all patients. Hyperthyroidism is characterized by increased levels of SHBG and aromatization of androgen to estrogen, thereby leading to gynaecomastia (Ford et al. 1992; Ridgway et al. 1982). Hence, thyroid function test is advised. Serum levels of testosterone, LH, follicle-stimulating hormone (FSH), estradiol and prolactin are measured to look for hypogonadism. Tumor markers like β-hCG are done to look for underlying germ cell tumors secreting estrogen. Mammography and breast ultrasonography are indicated only if breast cancer is suspected clinically (Hanavadi et al. 2006; Chen et al. 2006). If an estrogen-secreting testicular or adrenal mass is suspected, an ultrasound of the scrotum or abdomen should be ordered.

26.7 Principles of Management

In case of pseudogynaecomastia or lipomastia, weight loss should be encouraged. If this does not work, plastic surgery should be offered. For asymptomatic pubertal gynaecomastia and prolonged asymptomatic gynaecomastia, no treatment is necessary. Patients with recent onset or progressive gynaecomastia or associated breast pain require evaluation to find the cause of gynaecomastia. If a reversible cause is obtained, it should be treated. Medical therapy for gynaecomastia is indicated in cases of symptomatic patients with breast pain or tenderness or for cosmetic reasons.

26.7.1 Medical Management

There is lack of well-designed prospective studies on the treatment of gynaecomastia. Most often, empirical therapy is used. Only patients with recent onset, within 2 years, breast enlargement show benefit with medical therapy. After 2 years, treatment is ineffective as the stroma becomes fibrotic. Tamoxifen has been used as a first-line therapy for gynaecomastia. Tamoxifen and raloxifene produce antiestrogenic effects by acting as estrogen receptor antagonists and have shown an efficacy of up to 90% for resolution of gynaecomastia (Ting et al. 2000; Khan et al. 2004; Lawrence et al. 2004). In case of recurrence of gynaecomastia, workup should be done for any reversible cause, and a second course of therapy may be used. Raloxifene and clomiphene have also been used and have been efficacious in 90% and 42% cases, respectively. Moreover, tamoxifen prophylaxis may be considered in men with prostate cancer who are to be started on antiandrogen therapy to prevent the development of gynaecomastia. Aromatase inhibitors like anastrozole have been used but found to be less effective than antiestrogens (Perdona et al. 2005). Testosterone therapy is used to treat men with hypogonadism. However, it may cause gynaecomastia in some men due to aromatization to estrogens. Nonaromatizable androgens like dihydrotestosterone have been used topically to treat gynaecomastia. Radiation therapy had been tried in the past, but long-term risk of breast cancer is a significant concern; hence it is no more advocated (Boccardo et al. 2005; Perdona et al. 2005; Van Poppel et al. 2005).

26.7.2 Surgery for Gynaecomastia

As already discussed, medical therapy is ineffective in long-standing gynaecomastia. In such cases, surgery can be advised. Either suction lipectomy or removal of glandular breast tissue via a periareolar incision can be done (Fruhstorfer and Malata 2003).

Key Questions

1. How gynaecomastia is different from pseudogynaecomastia?
2. Describe the hormonal basis of physiological gynaecomastia.
3. What are the aetiological factors causing gynaecomastia?
4. What is the risk of breast cancer in patients with gynaecomastia?
5. Describe the histopathological findings in gynaecomastia.
6. Write the principles of diagnostic evaluation and management of gynaecomastia.

References

Allan CA, McLachlan RI. Age-related changes in testosterone and the role of replacement therapy in older men. Clin Endocrinol (Oxf). 2004;60(6):653–70.

Bannayan GA, Hajdu SI. Gynecomastia: clinicopathologic study of 351 cases. Am J Clin Pathol. 1972;57:431–7.

Bercovici JP, Nahoul K, Tater D, Charles JF, Scholler R. Hormonal profile of Leydig cell tumors with gynecomastia. J Clin Endocrinol Metab. 1984;59:625–30.

Boccardo F, Rubagotti A, Battaglia M, Di Tonno P, Selvaggi FP, Conti G, et al. Evaluation of tamoxifen and anastrozole in the prevention of gynecomastia and breast pain induced by bicalutamide monotherapy of prostate cancer. J Clin Oncol. 2005;23(4):808–15.

Braunstein GD. Aromatase and gynecomastia. Endocr Relat Cancer. 1999;6:315–24.

Braunstein GD. Clinical practice. Gynecomastia N Engl J Med. 2007;357:1229–37.

Burgess HE, Shousha S. An immunohistochemical study of the long-term effects of androgen administration on female-to-male transsexual breast: a comparison with normal female breast and male breast showing gynecomastia. J Pathol. 1993;170:37–43.

Carlson HE. Gynecomastia. N Engl J Med. 1980;303:795–9.

Catalano S, Marsico S, Giordano C, Mauro L, Rizza P, Panno ML, et al. Leptin enhances, via AP-1, expression of aromatase in the MCF-7 cell line. J Biol Chem. 2003;278:28668–76.

Catalano S, Mauro L, Marsico S, Giordano C, Rizza P, Rago V, et al. Leptin induces, via ERK1/ERK2 signal, functional activation of estrogen receptor α in MCF-7 cells. J Biol Chem. 2004;279:19908–15.

Chen L, Chantra PK, Larsen LH, Barton P, Rohitopakarn M, Zhu EQ, et al. Imaging characteristics of malignant lesions of the male breast. Radiographics. 2006;26:993–1006.

Dimitrikakis C, Zhou J, Bondy CA. Androgens and mammary growth and neoplasia. Fertil Steril. 2002;77(4):S26–33.

Fentiman IS, Fourquet A, Hortobagyi GN. Male breast cancer. Lancet. 2006;367:595–604.

Ford HC, Cooke RR, Keightley EA, Feek CM. Serum levels of free and bound testosterone in hyperthyroidism. Clin Endocrinol (Oxf). 1992;36:187–92.

Forest MG, Lecoq A, Saez JM. Kinetics of human chorionic gonadotropin-induced steroidogenic response of the human testis. II. Plasma 17 alpha-hydroxy-progesterone, delta4-androstenedione, estrone, and 17 beta-estradiol: evidence for the action of human chorionic gonadotropin on intermediate enzymes implicated in steroid biosynthesis. J Clin Endocrinol Metab. 1979;49:284–91.

Fruhstorfer BH, Malata CM. A systematic approach to the surgical management of gynaecomastia. Br J Plast Surg. 2003;56:237–46.

Georgiadis E, Papandreou L, Evangelopoulou C, Aliferis C, Lymberis C, Panitsa C, et al. Incidence of gynaecomastia in 954 young males and its relationship to somatometric parameters. Ann Hum Biol. 1994;21:579–87.

Hanavadi S, Monypenny IJ, Mansel RE. Is mammography overused in male patients? Breast. 2006;15(1):123–6.
Humphreys RC, Lydon JP, O'Malley BW, Rosen JM. Use of PRKO mice to study the role of progesterone in mammary gland development. J Mammary Gland Biol Neoplasia. 1997;2:343–54.
Kanhai RCJ, Hage JJ, VanDiest PJ, Bleomena E, Mulder JW. Short-term and long-term histologic effects of castration and estrogen treatment on breast tissue of 14 male-to-female transsexuals in comparison with two chemically castrated men. Am J Surg Pathol. 2000;24:74–80.
Khan HN, Rampaul R, Blamey RW. Management of physiological gynecomastia with tamoxifen. Breast. 2004;13:61–5.
Kleinberg DL. Role of IGF-I in normal mammary development. Breast Cancer Res Treat. 1998;47(3):201–8.
Kley HK, Niederau C, Stremmel W, Lax R, Strohmeyer G, Kruskemper HL. Conversion of androgens to estrogens in idiopathic hemochromatosis: comparison with alcoholic liver cirrhosis. J Clin Endocrinol Metab. 1985;61(1):1–6.
Lanfranco F, Kamischke A, Zitzmann M, Nieschlag E. Klinefelter's syndrome. Lancet. 2004;364:273–83.
Lawrence SE, Faught KA, Vethamuthu J, Lawson ML. Beneficial effects of raloxifene and tamoxifen in the treatment of pubertal gynecomastia. J Pediatr. 2004;145(1):71–6.
Mathur R, Braunstein GD. Gynecomastia: pathomechanisms and treatment strategies. Horm Res. 1997;48:95–102.
Nahta R, Hortobagyi GN, Esteva FJ. Growth factor receptors in breast cancer: potential for therapeutic intervention. Oncologist. 2003;8(1):5–17.
Nicolis GL, Modlinger RS, Gabrilove JL. A study of the histopathology of human gynecomastia. J Clin Endocrinol Metab. 1971;32:173–8.
Nuttall FQ. Gynecomastia as a physical finding in normal men. J Clin Endocrinol Metab. 1979;48:338–40.
Nydick M, Bustos J, Dale Jr JH, Rawson RW. Gynecomastia in adolescent boys. JAMA. 1961;178:449–54.
Perdona S, Autorino R, De Placido S, D'Armiento M, Gallo A, Damiano R, et al. Efficacy of tamoxifen and radiotherapy for prevention and treatment of gynaecomastia and breast pain caused by bicalutamide in prostate cancer: a randomised controlled trial. Lancet Oncol. 2005;6(5):295–300.
Peres R, Betsholtz C, Westermark B, Heldin CH. Frequent expression of growth factors for mesenchymal cells in human mammary carcinoma cell lines. Cancer Res. 1987;47:3425–9.
Ridgway EC, Maloof F, Longcope C. Androgen and estrogen dynamics in hyperthyroidism. J Endocrinol. 1982;95:105–15.
Sasano H, Kimura M, Shizawa S, Kimura N, Nagura H. Aromatase and steroid receptors in gynecomastia and male breast carcinoma: an immunohistochemical study. J Clin Endocrinol Metab. 1996;81:3063–7.
Smyth CM, Bremner WJ. Klinefelter syndrome. Arch Intern Med. 1998;12:1309–14.
Stewart AJ, Johnson MD, May FEB, Westley BR. Role of insulin-like growth factors and the type I insulin-like growth factor receptor in the estrogen-stimulated proliferation of human breast cancer cells. J Biolumin Chemilumin. 1990;265:21172–8.
Swerdlow AJ, Schoemaker MJ, Higgins CD, Write AF, Jacobs PA. Cancer incidence and mortality in men with Klinefelter syndrome: a cohort study. J Natl Cancer Inst. 2005;97(16):1204–10.
Ting ACW, Chow LWC, Leung YF. Comparison of tamoxifen with danazol in the management of idiopathic gynecomastia. Am Surg. 2000;66:38–40.
Van Poppel H, Tyrrell CJ, Haustermans K, Van Cangh P, Keuppens F, Colombeau P, et al. Efficacy and tolerability of radiotherapy as treatment for bicalutamide-induced gynaecomastia and breast pain in prostate cancer. Eur Urol. 2005;47:587–92.

Sexually Transmitted Infections in Males 27

Mohit Agrawal and Ashok Rattan

Learning Objectives

- Causative organisms, pathogenesis and clinical features of sexually transmitted infections
- Principles of diagnostic evaluation and management

27.1 Introduction

The older terminology of 'venereal disease' has over the past few decades been superseded by 'sexually transmitted diseases' (STDs). The term sexually transmitted infections (STIs) has replaced the STDs because many sexually transmitted infections remain asymptomatic and do not cause any disease.

There are approximately 30 sexually transmissible bacterial, viral and parasitic pathogens which cause STIs. The STIs discussed in this chapter are divided into groups based on their clinical presentations such as genital ulcers, urethritis, genital warts and opportunistic infections. The details of the pathogens and diseases caused by them are given in Table 27.1. See also table 27.2 for various cutaneous manifestations of STIs.

M. Agrawal, MD • A. Rattan, MD, MAMS (✉)
Department of Microbiology, Mahatma Gandhi University of Medical Sciences and Technology, Jaipur, India
e-mail: mohit_wal@yahoo.com; drashokrattan@gmail.com

A. Kumar, M. Sharma (eds.), *Basics of Human Andrology*,
DOI 10.1007/978-981-10-3695-8_27

Table 27.1 Sexually transmitted infections in males

Presentation	Organism	Disease	Sign and symptoms
Ulcer	Herpes simplex virus (HSV)	Genital herpes	Clusters of tender penile ulcers of uniform size, clean base with inflamed borders; tender inguinal nodes
	Treponema pallidum	Syphilis	One or two painless penile lesions with clean base and inflamed border ; rubbery non-tender nodes
	Haemophilus ducreyi	Chancroid	One or two large painful penile ulcers with ragged and necrotic base and undermined edges; very tender and fluctuant inguinal nodes
	Chlamydia trachomatis	Lymphogranuloma venereum (LGV)	One or two penile ulcers, regional lymphadenopathy, inflamed rectum and anal canal
	Klebsiella granulomatis	Granuloma inguinale	Multiple painless, red, beefy and rough penile ulcers
Urethritis	*Neisseria gonorrhoeae*	Gonococcal urethritis	Profuse purulent urethral discharge and painful micturition
	Chlamydia trachomatis, *Ureaplasma urealyticum*, *Mycoplasma genitalium*, *Trichomonas vaginalis*	Nongonococcal urethritis	Serous and clear discharge with painful micturition
Genital warts	Human papilloma virus (HPV)		Condyloma acuminatum, carcinoma of penis
Asymptomatic or opportunistic infections	Human immunodeficiency virus (HIV)	Acquired immunodeficiency syndrome (AIDS)	Asymptomatic or clinical features of opportunistic infections

27.2 Genital Herpes

27.2.1 Causative Organism

Genital herpes is caused by human herpes virus 1 and 2, also known as herpes simplex virus type 1 (HSV-1) and herpes simplex virus type 2 (HSV-2), respectively. HSV type 1 and type 2 are among the most ubiquitous human infections. HSV-1 and

Table 27.2 Cutaneous manifestations of STIs

Flat lesions	Raised/elevated solid lesions	Raised fluid-filled lesions	Lesions causing skin loss
Macule is a < 1 cm flat, discoloured area *Patch* is a > 1 cm flat, discoloured area	*Papule* is a < 1 cm solid raised area with defined borders *Plaque* is a > 1 cm solid raised area with ill-defined borders *Nodule* is a 1–2 cm solid raised area which is easily palpable and movable *Tumor* is a > 2 cm solid raised area which is easily palpable above or beneath the skin	*Vesicle* is a < 1 cm clear fluid-filled area *Bulla* is a > 1 cm clear fluid-filled area *Pustule* is a < 1 cm pus-filled area	*Ulcer* is caused by loss of superficial epidermis, dermis and sometimes subcutaneous fat; ulcer sometimes gets indurated or becomes hard as a result of increased fibrous tissue due to inflammation *Crust* is formed as a result of drying of pus containing exudates of an infected ulcer

HSV-2 belong to genus simplex virus. Both viruses are morphologically similar, possessing an internal core containing linear double-stranded DNA, an icosahedral capsid, a tegument which is amorphous material surrounding the capsid and a lipid envelope containing viral glycoproteins on its surface (Grünewald et al. 2003).

27.2.2 Pathogenesis

The word herpes, in Greek, means to creep. The viruses produce a wide variety of illnesses, including mucocutaneous infections. During the primary or initial infection, virus attaches to the cutaneous receptors of local sensory nerves. Most commonly, lumbosacral nerves are affected in case of genital HSV infection. The virus induces a latent infection in the local ganglia. HSV-1 affects trigeminal ganglia, whereas HSV-2 affects sacral ganglia (Corey et al. 1983). Once the virus gets reactivated by the triggers, it produces recurrent infection. The triggering stimuli can be trauma to the tissue which is innervated by the infected neurons, physical stress, emotional stress, hormonal imbalance and radiations (Lafferty et al. 1987).

Humans are the natural reservoir of HSV. Almost every case of genital herpes is sporadic, acquired via person-to-person transmission. HSV-1 infection is common in early life, while HSV-2 infection occurs after the onset of sexual activity.

27.2.3 Clinical Features

Genital herpes is an infection of long duration. The initial infection is followed by latency and frequent recurrences. The clinical manifestations depend upon the affected anatomical site, age and immune status of the host. Both viral subtypes can cause genital and orofacial infections (Reeves et al. 1981).

The incubation period of primary genital infection is variable, averaging 4 or 5 days. The clinically apparent disease may be heralded by fever, headache, local genital pain and burning sensation. Males have less severe disease than females. After 1 or 2 days, erythematous macules appear which progress to vesicles, pustules, ulcers and finally to crusts (see Table 27.2).

The mean time from the onset of a primary genital HSV lesion to complete healing is around 15–16 days for men. In males, lesions typically appear on glans penis and penile shaft. Perianal involvement is common among homosexual men. Tender bilateral inguinal lymphadenopathy is generally present (Corey et al. 1983).

The clinical manifestations of recurrent genital herpes are less severe and resolve rapidly. HSV-2 genital infection reactivates much more readily than HSV-1 genital herpes (Benedetti et al. 1994).

27.2.4 Diagnostic Evaluation

Herpes virus can be demonstrated by electron microscopy of vesicle fluid or scrapings. Scrapings are taken from the base of the lesions and stained with Wright stain, Giemsa stain, or Papanicolaou stain. The characteristic giant cells or intranuclear inclusions are diagnostic of HSV infection.

Viral culture is the gold standard technique for diagnosis in a patient presenting with ulcers. Both HSV1 and HSV2 grow rapidly in cell cultures and produce typical cytopathic effects in host cells such as shape changes, cell fusion and formation of inclusion bodies within 48–72 h (Ashley 1993).

Herpes simplex antibodies are readily detectable following infection. Serologic testing is done for antibody to G1 glycoprotein produced by HSV-1 and G2 glycoprotein produced by HSV-2. Assays based on HSV DNA polymerase chain reaction (PCR) are rapid and sensitive (Singh et al. 2005).

27.2.5 Principles of Management

Acyclic guanine derivatives are the mainstay of the treatment (Workowski and Bolan 2015). The drug group inhibits the DNA polymerases thereby inhibiting viral DNA replication.

27.3 Syphilis

27.3.1 Causative Organism

Syphilis is caused by the spirochete *Treponema pallidum* (TP). TP belongs to the order *Spirochaetales*, family *Spirochaetaceae* and genus *Treponema*. The organism is spiral in shape, having 6–14 spiral turns with pointed ends. The bacterium shows typical corkscrew motility (Norris and Larsen 1995).

27.3.2 Pathogenesis

Syphilis is a chronic infection with diverse clinical manifestations. The early mucocutaneous lesions harbour the organism from where these are transmitted, most commonly via sexual contact. The other modes of transfer are contact with active lesion which can be chancre or condyloma, blood transfusion and through placenta. Primary infection initiates an inflammatory response after which the infection is disseminated to deeper tissues. Syphilis is characterized by vascular endarteritis and periarteritis as well as the granulomatous inflammation (Engelkens et al. 1991).

27.3.3 Clinical Features

Syphilis present in three clinical stages: primary, secondary and tertiary. The primary stage takes 3–90 days to develop from the day of transmission of organism.

Primary syphilis – The classic skin lesion in primary syphilis is the chancre. Chancre is a single, painless and indurated ulcer with a clean base (Stokes et al. 1944). Coronal sulcus and glans penis are the most common sites affected in men. Homosexual men show anorectal chancres. Non-tender inguinal lymphadenopathy is common (Mindel et al. 1989). The chancre heals on its own within 3–6 weeks.

Secondary syphilis – The manifestations of secondary syphilis may vary. The most common are the skin lesions. Classically, the skin lesions are copper coloured. The palms and soles are commonly involved. Condylomata lata are moist lesions that can enlarge and coalesce to produce painless grey-white to erythematous plaques. The highly infectious lesions may affect mucous membranes where they are referred as mucous patches. The mucous patches coalesce to form 'snail-track' ulcers. The epitrochlear lymphadenopathy is diagnostic of syphilis (Stokes et al. 1944).

Latent syphilis – Latent syphilis shows absence of clinical features but positive-specific treponemal antibody test (Howles 1943).

Tertiary syphilis – The tertiary syphilis stage has three types: (1) cardiovascular syphilis, (2) neurosyphilis and (3) gummatous syphilis (Stokes et al. 1944). See Table 27.3 for different types of tertiary syphilis.

Table 27.3 Different types of tertiary syphilis

Cardiovascular syphilis	Neurosyphilis	Gummatous syphilis
Syphilitic aortitis is most common, with complications such as aortic regurgitation and aortic aneurysms	Asymptomatic with positive laboratory findings or meningeal involvement present as fever, headache, stiff neck or meningovascular involvement present with brain vascular ischemia	Granuloma-like lesions of the skin, bone and liver

Table 27.4 Laboratory tests for syphilis

Microscopy	Treponemal test	Nontreponemal tests
Dark field microscopy based on morphology and motility	(a) Serum fluorescent treponemal antibody absorption test (FTA-ABS) (b) Microhemagglutination test for TP (MHA-TP) Tests detect IgG antibodies to antigenic determinants of treponemes	(a) Venereal Disease Research Laboratory (VDRL) (b) Rapid plasma reagin (RPR) card test (c) Toluidine red unheated serum test (TRUST) All the tests detect IgM and IgG antibodies against the lipoidal-like antigens of TP and infected host cell-derived lipoidal material

27.3.4 Diagnostic Evaluation

Laboratory tests for syphilis are basically three types: (1) direct microscopic examination, used when skin lesions are present; (2) nontreponemal tests are used as screening test; and (3) treponemal tests are the confirmatory tests (Larsen et al. 1995). See Table 27.4 for various types of diagnostic tests of syphilis.

27.3.5 Principles of Management

Penicillin is the drug of choice to treat syphilis (Workowski and Bolan 2015). The drug kills bacteria by inhibiting cell wall synthesis.

27.4 Chancroid

27.4.1 Causative Organism

Chancroid or soft chancre is caused by *Haemophilus ducreyi* (HD). HD is a Gram-negative coccobacillus. The genus *Haemophilus* is a member of the family *Pasteurellaceae*. The organisms stain with Gram's stain and are arranged in 'school of fish' or 'railway tracks' patterns (Albritton 1989).

27.4.2 Pathogenesis

HD spreads only through sexual transmission. Chancroid is characterized by a single painful genital ulcer with associated inguinal lymphadenopathy. Risk of disease is more with low socioeconomic status, poor hygiene, drug abuse and prostitution.

27.4.3 Clinical Features

The incubation period of infection is 3–10 days followed by the development of typical, soft sore lesion. The inflammatory response starts as an erythematous lesion which eventually undergoes central necrosis, creating the typical painful, non-indurated ulcer. The base of the ulcer is covered with necrotic material. As the ulcer edge is not indurated, it is known as a 'soft chancre'. In men, the ulcer is usually located on the prepuce, urethral meatus, glans, coronal sulcus or penile shaft. Unilateral, painful lymph node enlargement is common (Lewis 2003).

27.4.4 Diagnostic Evaluation

Culture remains the gold standard for diagnosis. But, the most promising approach is a PCR-based assay (Lewis 2003). Multiplex PCR assay is also available for the detection of HSV, TP and HD. The multiplex assay simplifies the diagnosis and management of genital ulcer disease (Orle et al. 1996).

27.4.5 Principles of Management

Azithromycin is the drug of choice to treat chancroid (Workowski and Bolan 2015). The drugs kill bacteria by inhibiting bacterial protein synthesis.

27.5 Lymphogranuloma Venereum and Nongonococcal Urethritis

27.5.1 Causative Organism

Chlamydia trachomatis (CT) is responsible for causing lymphogranuloma venereum (LGV) and 30–50% of cases of nongonococcal urethritis (NGU) (Holmes et al. 1975). Chlamydiae are the obligate intracellular parasites as these are unable to synthesize energy compounds which make them dependent upon the host cell. Genital tract infections with CT are quite prevalent. NGU is the most common chlamydial infection in men.

27.5.2 Pathogenesis

The chlamydiae have two specialized morphologic forms: elementary bodies (EBs) and reticulate bodies (RBs). EBs are optimized for extracellular survival and are the infective form, while RBs are optimized for intracellular reproduction. EB is phagocytosed by epithelial cells to form a phagosome. EB survives within the phagosome.

EB transforms into RB by unfolding its dense nucleoid into diffuse form of DNA. RB begins to divide by binary fission followed by condensation into EBs. Some other RBs continue to divide. Rupture of the host cell membrane causes release of the EBs (Hatch 1999).

27.5.3 Clinical Feature

The L1–L3 serovars of CT cause lymphogranuloma venereum (LGV). There are three different stages in LGV. The first stage is primary lesion which, in men, is often painless ulcer or vesicle on the penis. Once infected with CT, the primary lesion takes 3–30 days to appear (Schachter and Dawson 1978). Primary lesion heals rapidly without leaving a scar. The secondary stage occurs 1–6 weeks after the primary infection. Secondary stage is characterized by systemic symptoms and regional lymphadenopathy. Systemic symptoms include fever, headache and muscular pain. The third stage is reached if the second stage remains untreated and is identified by the presence of hypertrophic chronic granulomatous enlargement with ulceration of the external genitalia (D'Aunoy and von Haam 1939).

The incubation period of chlamydial urethritis or NGU varies from 7–21 days. Clinical symptoms include painful micturition, clear or white urethral discharge and frequency of micturition (Stamm et al. 1984).

CT is also a cause of postgonococcal urethritis (PGU) which follows treatment of mixed gonococcal-chlamydial infections with an antibiotic that is active against *N. gonorrhoeae* but not against *C. trachomatis* (Stamm et al. 1984).

NGU may lead to reactive arthritis in 1% men due to immune-mediated inflammatory response (Rahman et al. 1992). Arthritis usually affects the large joints, particularly of the legs, or the sacroiliac joints. Skin involvement may cause the penile inflammation which is termed as circinate balanitis. NGU also causes keratoderma of the palms and soles.

27.5.4 Diagnostic Evaluation

CT-caused urethral infections are diagnosed by urethral swab or urine specimens in men. Chlamydia is intracellular parasite, so it can't be cultured in artificial media. Cell culture is a useful method of diagnosis. Direct immunofluorescent test (DIF) identifies EBs in clinical material (Uyeda et al. 1984). Nucleic acid amplification technology (NAAT) is another useful technique for the diagnosis of chlamydial infections (Cherneskey 2002).

27.5.5 Principles of Management

Patients with LGV or NGU can be treated with doxycycline or azithromycin (Workowski and Bolan 2015). The drugs kill bacteria by inhibiting bacterial protein synthesis.

27.6 Granuloma Inguinale (Donovanosis)

27.6.1 Causative Organism

Donovanosis is caused by the Gram-negative pleomorphic bacterium, *Klebsiella granulomatis* (KG). Large mononuclear cells containing inclusion bodies are characteristic (Pund and Greenblatt 1937). The bacteria have bipolar densities that give inclusion bodies or Donovan bodies, the appearance of closed safety pins.

27.6.2 Pathogenesis

Donovanosis is a progressive ulcerative disease usually of the genital region. KG invades epidermal and dermal tissues and causes local inflammation. KG resides in vacuoles in the cytoplasm of large mononuclear cells. Donovanosis generally has low infectivity.

27.6.3 Clinical Features

The donovanosis begins as small painless papule or indurated nodule that usually manifests after an incubation period of 8–80 days. The primary lesion subsequently ulcerates which can later form non-tender ulcerogranulomatous lesion. The other forms of progressive disease can be hypertrophic form with raised irregular edges, necrotic ulcer form and sclerotic lesion with fibrous and scar tissue (Hart 1997).

The large ulcerative lesions can remain painless unless there is severe secondary infection. The primary lesions can be accompanied by complications such as development of lymphoedema which later may lead to genital elephantiasis. Sometimes, the lesions may lead to genital deformity including penile autoamputation (Watsford and Alderman 1953).

27.6.4 Diagnostic Evaluation

Donovan bodies are diagnostic which are identified in the lesion specimen by staining with Giemsa, Wright or Leishman stains (Richens 1991). The other diagnostic test is PCR which is based on colorimetric detection system (Carter and Kemp 2000).

27.6.5 Principles of Management

Azithromycin is the drug of choice to treat granuloma inguinale (Workowski and Bolan 2015). Drug inhibits bacteria by inhibiting protein synthesis.

27.7 Gonorrhoea

27.7.1 Causative Organism

Gonorrhoea is caused by organism *Neisseria gonorrhoeae* (NG) which belongs to the family *Neisseriaceae*. NG is a Gram-negative aerobic, nonmotile bacteria.

27.7.2 Pathogenesis

NG affects mucosal surfaces. Primarily, the lower genital tract is involved. NG has specialized structure pili on its surface which help in adhesion and gene transfer (Merz and So 2000). The binding of NG pili to host cells involve the complement regulatory protein CD46, Opa proteins, lipooligosaccharide and the gonococcal porin, Por (Kallstrom et al. 1997). Further survival and colonization requires iron. Once bacteria colonizes, it elicits an inflammatory host response.

27.7.3 Clinical Features

Male urethra is the primary site of colonization that leads to inflammatory response and purulent urethral discharge which is the most common symptom (Jacobs and Kraus 1975). Dysuria or painful micturition is usually present without urinary frequency or urgency. In few cases, the upper genital tract infections such as prostatitis or epididymitis are also possible. Disseminated gonococcal infections are rarely encountered.

Homosexual men practising anal intercourse may present with rectal symptoms such as pruritus and painless rectal discharge, severe rectal pain and mucopurulent discharge.

Acute epididymitis is the most common complication of urethral gonorrhoea (Berger 1981). Penile oedema is occasionally seen in gonococcal urethritis.

27.7.4 Diagnostic Evaluation

Gonorrhoea is diagnosed by preparing a smear from the urethral discharge that shows the presence of Gram-negative intracellular diplococcus bacteria. The culture is the gold standard for detection of NG. Direct detection of gonococci involves rapid diagnostic nucleic acid amplification tests (Bignell and Unemo 2013).

27.7.5 Principles of Management

Gonorrhoea is treated with regimen of dual treatment with ceftriaxone and azithromycin (Workowski and Bolan 2015). Ceftriaxone is a bacterial cell wall synthesis inhibitor, while azithromycin is protein synthesis inhibitor.

27.8 Genital Warts

27.8.1 Causative Organism

Genital warts are caused by papilloma viruses (HPVs). Papilloma viruses are classified into five super groups: A–E. Genital human papilloma viruses belong to super group A, and human cutaneous papilloma viruses to super group E. HPVs are non-enveloped viruses with a single molecule of circular double-stranded DNA.

27.8.2 Pathogenesis

The most common sexually transmitted viral infection in humans is caused by HPVs (Satterwhite et al. 2013). HPVs produce epithelial tumors of the skin and mucous membranes.

HPV genome produces early and late proteins out of which two proteins, E6 and E7, are most important. E6 binds to the human tumor suppressor gene, p53 protein, and leads to its rapid degradation (Scheffner et al. 1990). The E7 protein binds to another tumor suppressor gene product, the retinoblastoma protein (pRb), and inhibits its functions (Dyson et al. 1989). Both p53 and pRb are negative regulators of the cell cycle. Failure of functions of p53 and pRb leads to uncontrolled cell division. Expression of E6 and E7 divides the viruses into low risk HPVs and high risk HPVs. Based on the oncogenic potential, HPV-6 and HPV-11 belong to low-risk group, while HPV-16 and HPV-18 belong to high-risk group (zur Hausen 2002).

HPVs replicate in squamous epithelium. Direct contact through sexual intercourse is the most common way to transmit infection.

27.8.3 Clinical Features

HPV infection usually affects skin and mucosa of the anogenital area. HPV infection usually causes anogenital papules or warts which are termed as condylomata acuminata (CA) and squamous intraepithelial lesions (SILs). Low-risk HPVs cause CA, whereas both low- and high-risk genital HPV types cause SILs.

Anogenital warts are hyperkeratotic, exophytic lesions either sessile on the skin or attached by a short peduncle on the penis and anus in men (Oriel 1971). Most lesions are caused by HPV-6, HPV- 11 and HPV-16.

Squamous intraepithelial lesions (SILs) are classified histologically and are some variants of intraepithelial neoplasias. The lesions include penile and anal regions (Okagaki 1992).

27.8.4 Diagnostic Evaluation

The diagnosis of genital warts is most commonly made clinically with physical examination. But, the gold standard for diagnosis of HPV infection is the nucleic acid detection. HPV genotyping can be performed using PCR-based methods (Burd 2016).

27.8.5 Principles of Management

Podophyllin, a resin extract, is the principal mode of therapy. Podophyllin is a mitotic poison and inhibits cell proliferation and growth of warts. Imiquimod is an imidazoquinolineamine that induces production of interferon-α and other cytokines and is used for the treatment of condyloma acuminatum (Hurwitz et al. 2003). Cryotherapy is used with liquid nitrogen or cryoprobe. Cryotherapy leads to cytolysis and inhibits growth of warts (Workowski and Bolan 2015).

27.9 Human Immunodeficiency Virus (HIV) Infections

27.9.1 Causative Organism

HIV infection contributes to the major global burden of the disease. HIV belongs to the family *Retroviridae* and subfamily *Lentiviruses*. The virus has a reverse transcriptase with two copies of genomic RNA. The major component of the virus is capsid protein p24 and lipid envelope containing surface glycoproteins, gp41 and gp120. The surface glycoproteins bind to cluster of differentiation 4 (CD4)-positive T lymphocytes and initiate the infection process. HIV is highly divergent organism with different subtypes. The two major HIV subtypes are HIV-1 and HIV-2 (Parekh et al. 2013).

27.9.2 Pathogenesis

The infection is most commonly transmitted by sexual contact. The other modes of transfer are through HIV-infected women to the infant during antenatal period, during delivery or after birth through breast feeding. HIV transmission is also possible through contaminated blood or blood products. HIV-positive donor can also transmit infection to the organ transplant recipient.

Through gp120 on the surface of the virus, HIV binds the receptor CD4 molecule, which is expressed on the macrophages and T lymphocytes. A second coreceptor, CCR5 or CXR4 in addition, is necessary for HIV to gain entry to cells. The coreceptor helps in fusion of the virus with the cells (Maartens 2014).

HIV infection leads to depletion of T helper lymphocytes. The consequences of CD4 cell dysfunction caused by HIV infection are devastating because CD4 cells play a critical role in human immune response. There is also immune activation which includes activation of macrophages, induction of natural killer cells and production of pro-inflammatory cytokines such as tumor necrosis factor alpha (TNFα) and interleukin 6 (IL6). Immune activation in HIV infection has been associated with mortality, cardiovascular diseases, cancers and neurological disorders (Maartens 2014).

27.9.3 Clinical Features

HIV infection has been classified based on the CD4 counts and clinical presentations. In clinical settings where CD4 count estimation facility is not available, clinical staging given by World Health Organization (WHO) is useful.

Primary HIV infection usually remains asymptomatic.

Patients in stage 1 may present with unexplained weight loss, persistent generalized lymphadenopathy or recurrent respiratory infections.

Patients in stage 2 present with recurrent oral infections, skin eruptions, herpes zoster infection, unexplained chronic diarrhoea, and unexplained persistent fever.

Stage 3 patients may present with pulmonary tuberculosis, decreased blood counts, unexplained anaemia, severe bacterial infections, chronic herpes infections, oral candidiasis, encephalopathy, etc.

Stage 4 patients present with disseminated non-tubercular mycobacterial infections, severe bacterial and fungal infections, lymphomas, nephropathy, cardiomyopathy, etc. (WHO 2007).

27.9.4 Diagnostic Evaluation

The diagnosis is made by detecting HIV antibodies in the body fluids and blood. Detection is possible usually 4–6 weeks after infection, although detectable antibody level may take up to 3–6 months.

During the initial period of virus replication, HIV diagnosis is not usually made as the HIV antibodies are absent. The window period of antibody absence can be shortened by directly detecting p24 antigen or RNA of HIV (Parekh et al. 2013).

Enzyme immunoassays (EIAs) detect HIV-specific antibodies by using virus lysate as antigens, synthetic peptides and enzyme-labelled antigens. The enzyme-labelled antigens detect early IgM responses reducing the window period. The newer EIAs detect p24 and HIV antibodies. Molecular assays are also developed which detect viral nucleic acid earlier than the p24 antigen (Parekh et al. 2013).

Rapid tests take 1–15 min to detect HIV antibodies, whereas EIAs usually take up to 2–4 h. The tests use high concentration of antigen and detect antigen-antibody complexes at faster rate.

Apart from detecting HIV antibodies, newer rapid tests also detect p24 antigen. Detection of acute infection is difficult. So, a positive antigen test must be confirmed by repeating it at 4 weeks or later (Parekh et al. 2013).

Use of oral fluid specimens for faster diagnosis is also suggested. These specimens are easily accessible and are biohazard-free.

Confirmatory assays are immunofluorescence assays (IFAs) and Western blot (WB) which detect only HIV-specific IgG. The presence of HIV-specific IgM or virus antigens detected by EIAs cannot be detected by IFAs and WB. Detection of virus or viral components such as p24 antigen, RNA or proviral DNA can also be used for the diagnosis of HIV infection (Parekh et al. 2013).

27.9.5 Principles of Management

Multiple antiretroviral drug regimens are used for HIV infection management. Antiretroviral agents target different stages of the HIV life cycle. The use of multiple drug regimens for HIV infection treatment is known as highly active antiretroviral therapy (HAART). The major goals of HAART are to maintain body immune function and to prevent occurrence of opportunistic infections which are life-threatening.

Treatment with HAART has shown successful results and has made HIV infection no more a chronic disease (Deeks 2013).

Drugs used in HAART target different phases of retroviral life cycle.

The drugs inhibiting binding and entry of HIV into the host cells belong to entry inhibitors. Maraviroc and enfuvirtide are the drugs used in this class. Maraviroc targets CCR5 which is a coreceptor located on human helper T cells (Lieberman-Blum et al. 2008). Enfuvirtide blocks fusion by inhibiting gp41 (Bai et al. 2013).

Another group of drugs targets the HIV reverse transcriptase enzyme and causes DNA chain termination. The drugs can be nucleoside reverse transcriptase inhibitors (NRTI) or nucleotide reverse transcriptase inhibitors (NtRTI). Commonly used NRTIs are zidovudine, lamivudine, tenofovir, etc. (Das and Arnold 2013).

Reverse transcriptase enzyme is also inhibited by non-nucleoside reverse transcriptase inhibitors (NNRTI). NNRTIs binds to an allosteric site of the enzyme thereby inhibiting its action. NNRTI group of drugs are nevirapine, etravirine, etc. (Das and Arnold 2013).

Drugs inhibiting viral integration into host DNA are integrase inhibitors which are also known as integrase nuclear strand transfer inhibitors or INSTIs. INSTIs inhibit viral enzyme integrase. Raltegravir, elvitegravir and dolutegravir are the drugs used in this group (Metifiot et al. 2013).

Drugs used in treatment of HIV infection also target protease enzyme that leads to production of defective and non-infective viral particles. HIV protease inhibitors are lopinavir, indinavir, nelfinavir, etc. (Wensing et al. 2010).

27.10 Men Who Have Sex with Men

AIDS was first diagnosed in men who have sex with men (MSM) and was called ‘the gay plague’. The term MSM includes men who have sex with men, regardless of their bisexual or gay identity.

MSM are at higher risk for HIV because of indulgence in anal sex, multiple sex partners and substance abuse. Increased incidence of STIs such as HIV, syphilis, gonorrhoea, HPV and chlamydial infections in MSM has been reported. Incidence of rectal gonorrhoea is also more in MSM which also predisposes for the risk of HIV infection in MSM. MSM are also prone for HPV-induced anal carcinoma (Palefsky et al. 1998).

Key Questions

- What are the common sexually transmitted infections and their causative organisms in males?
- Describe the pathogenesis of genital herpes infection.
- Write the different stages of syphilis with their clinical manifestations.
- What are the diagnostic tests of syphilis infection?
- Describe the features of organism causing chancroid.
- What are the complications of granuloma inguinale infection?
- Write briefly the principles of management of genital warts.
- Describe the pathogenesis and management of human immunodeficiency virus infection.

References

Albritton WL. Biology of Haemophilus ducreyi. Microbiol Rev. 1989;53:377–89.

Ashley RL. Laboratory techniques in the diagnosis of herpes simplex infection. Genitourin Med. 1993;69:174–83.

Bai Y, Xue H, Wang K, Cai L, Qiu J, Bi S, et al. Covalent fusion inhibitors targeting HIV-1 gp41 deep pocket. Amino Acids. 2013;44(2):701–13.

Benedetti J, Corey L, Ashley R. Recurrence rates in genital herpes after symptomatic first-episode infection. Ann Intern Med. 1994;121:847–54.

Berger RE. Acute epididymitis. Sex Transm Dis. 1981;8:286–9.

Bignell C, Unemo M. 2012 European guideline on the diagnosis and treatment of gonorrhoea in adults. Int J STD AIDS. 2013;24:85–92.

Burd EM. Human papillomavirus laboratory testing: the changing paradigm. Clin Microbiol Rev. 2016;29:291–319.

Carter JS, Kemp DJ. A colorimetric detection system for Calymmatobacterium granulomatis. Sex Transm Infect. 2000;76:134–6.

Cherneskey MA. Chlamydia trachomatis diagnostics. Sex Transm Infect. 2002;78:232–4.

Corey L, Adams HG, Brown ZA, Holmes KK. Genital herpes simplex virus infections: clinical manifestations, course, and complications. Ann Intern Med. 1983;98:958–72.

D'Aunoy R, von Haam E. General reviews: venereal lymphogranuloma. Arch Pathol. 1939;27:1032–82.

Das K, Arnold E. HIV-1 reverse transcriptase and antiviral drug resistance. Part 1. Curr Opin Virol. 2013;3(2):111–8.

Deeks SG. The end of AIDS: HIV infection as a chronic disease. Lancet. 2013;382(9903):1525–33.

Dyson N, Howley PM, Munger K, Harlow E. The human papilloma virus-16 E7 oncoprotein is able to bind to the retinoblastoma gene product. Science. 1989;243:934–7.

Engelkens HJ, ten Kate FJ, Vuzevski VD, van der Sluis JJ, Stolz E. Primary and secondary syphilis: a histopathological study. Int J STD AIDS. 1991;2:280–4.

Grünewald K, Desai P, Winkler DC, Heymann JB, Belnap DM, Baumeister W, et al. Three-dimensional structure of herpes simplex virus from cryo-electron tomography. Science. 2003;302:1396–8.

Hart G. Donovanosis. Clin Infect Dis. 1997;25:24–30.

Hatch TP. Developmental biology. In: Stephens RS, editor. Chlamydia: intracellular biology, pathogenesis and immunity. Washington, DC: ASM Press; 1999. p. 29–67.

Holmes KK, Handsfield HH, Wang SP, Wenworth BB, Turck M, Anderson JB, et al. Etiology of nongonococcal urethritis. N Engl J Med. 1975;292:1199–205.

Howles JK. Synopsis of clinical syphilis. St.Louis: The C. V. Mosby Co; 1943.

Hurwitz DJ, Pincus L, Kupper TS. Imiquimod: a topically applied link between innate and acquired immunity. Arch Dermatol. 2003;139:1347–50.

Jacobs Jr NF, Kraus SJ. Gonococcal and nongonococcal urethritis in men: clinical and laboratory differentiation. Ann Intern Med. 1975;82:7–12.

Kallstrom H, Liszewski MK, Atkinson JP, Jonsson AB. Membrane cofactor protein (MCP or CD46) is a cellular pilus receptor for pathogenic Neisseria. Mol Microbiol. 1997;25:639–47.

Lafferty WE, Coombs RW, Benedetti J, Critchlow C, Corey L. Recurrences after oral and genital herpes simplex virus infection. Influence of site of infection and viral type. N Engl J Med. 1987;4(316):1444–9.

Larsen SA, Steiner BM, Rudolph AH. Laboratory diagnosis and interpretation of tests for syphilis. Clin Microbiol Rev. 1995;8:1–21.

Lewis DA. Chancroid: clinical manifestations, diagnosis, and management. Sex Transm Infect. 2003;79:68–71.

Lieberman-Blum SS, Fung HB, Bandres JC. Maraviroc: A CCR5-receptor antagonist for the treatment of HIV-1 infection. Clin Ther. 2008;30(7):1228–50.

Maartens G. HIV infection: epidemiology, pathogenesis, treatment and prevention. Lancet. 2014;384:258–71.

Merz AJ, So M. Interactions of pathogenic Neisseriae with epithelial cell membranes. Annu Rev Cell Dev Biol. 2000;16:423–57.

Metifiot M, Marchand C, Pommier Y. HIV integrase inhibitors: 20 years landmark and challenges. Adv Pharmacol. 2013;67:75–105.

Mindel A, Tovey SJ, Timmins DJ, Williams P. Primary and secondary syphilis, 20 years' experience. 2. Clinical features. Genitourin Med. 1989;65:1–3.

Norris SJ, Larsen SA. Treponema and other host-associated spirochetes. In: Murray PR, Baron EJ, Pfaller MA, Tenover FC, Yolken RH, editors. Manual of clinical microbiology. 6th ed. Washington, D.C: ASM Press; 1995. p. 636–51.

Okagaki T. Impact of human papillomavirus research on the histopathologic concepts of genital neoplasms. Curr Top Pathol. 1992;85:273–307.

Oriel JD. Natural history of genital warts. Br J Vener Dis. 1971;47:1–13.

Orle KA, Gates CA, Martin DH, Body BA, Weiss JB. Simultaneous PCR detection of Haemophilus ducreyi, Treponema pallidum, and herpes simplex virus types 1 and 2 from genital ulcers. J Clin Microbiol. 1996;34:49–54.

Palefsky JM, Holly EA, Ralston ML, Jay N. Prevalence and risk factors for human papillomavirus infection of the anal canal in human immunodeficiency virus (HIV)-positive and HIV-negative homosexual men. J Infect Dis. 1998;177:361–7.

Parekh BS, Ellenberger D, Westerman L, Yang C, Nkengasong JN. Human immunodeficiency virus (HIV) infections. In: Unemo M (editor in chief), Ballard R, Ison C, Lewis D, Ndowa F, Peeling R, editors. Laboratory diagnosis of sexually transmitted infections, including human immunodeficiency virus 2013. WHO Document Production Services, Geneva; 2013. p.155–166.

Pund ER, Greenblatt RB. Specific histology of granuloma inguinale. Arch Pathol. 1937;23:224–9.

Rahman MU, Hudson AP, Schumacher Jr HR. Chlamydia and Reiter's syndrome (reactive arthritis). Rheum Dis Clin North Am. 1992;18:67–79.

Reeves WC, Corey L, Adams HG, Vontver LA, Holmes KK. Risk of recurrence after first episodes of genital herpes: relation to HSV type and antibody response. N Engl J Med. 1981;305:315–9.

Richens J. The diagnosis and treatment of donovanosis (granuloma inguinale). Genitourin Med. 1991;67:441–52.

Satterwhite CL, Torrone E, Meites E, Dunne EF, Mahajan R, Ocfemia MC, et al. Sexually transmitted infections among US women and men: prevalence and incidence estimates , 2008. Sex Transm Dis. 2013;40(3):187–93.

Schachter J, Dawson CR. Human chlamydial infections. Littleton: PSG Publishing; 1978. p. 45–62.

Scheffner M, Werness BA, Huibregtse JM, Levine AJ, Howley PM. The E6 oncoprotein encoded by human papillomavirus types 16 and 18 promotes the degradation of p53. Cell. 1990;63:1129–36.

Singh A, Preiksaitis J, Ferenczy A, Romanowski B. The laboratory diagnosis of herpes simplex virus infections. Can J Infect Dis Med Microbiol. 2005;16:92–8.

Stamm WE, Guinan ME, Johnson C, Starcher T, Holmes KK, McCormack WM. Effect of treatment regimens for Neisseria gonorrhoeae on simultaneous infection with Chlamydia trachomatis. N Engl J Med. 1984;310:545–9.

Stokes JH, Beerman H, Ingraham NR. Modern clinical syphilology. Philadelphia: The W. B. Saunders Co; 1944.

Uyeda CT, Welborn P, Ellison-Birang N, Shunk K, Tsaouse B. Rapid diagnosis of chlamydial infections with the MicroTrak direct test. J Clin Microbiol. 1984;20:948–50.

Watsford SD, Alderman LW. Granuloma inguinale. Med J Aust. 1953;2:50–2.

Wensing AM, van Maarseveen NM, Nijhuis M. Fifteen years of HIV protease inhibitors: raising the barrier to resistance. Antiviral Res. 2010;85(1):59–74.

WHO case definitions of HIV for surveillance and revised clinical staging and immunological classification of HIV-related disease in adults and children. Geneva, World Health Organization; 2007. http://www.who.int/hiv/pub/guidelines/HIVstaging150307.pdf. Accessed on 4 Aug 2016.

Workowski KA, Bolan GA. Sexually transmitted diseases treatment guidelines, 2015. MMWR. 2015;64(3):1–134.

zur Hausen H. Papillomaviruses and cancer: from basic studies to clinical application. Nat Rev Cancer. 2002;2:342–50.

Male Reproductive Cancers

28

Kunhi Parambath Haresh and Rony Benson

Learning Objectives

Prostate, testicular and penile cancer

- Aetiology, symptoms and signs
- Diagnostic evaluation
- Staging
- Principles of management

28.1 Introduction

Malignancies of the male reproductive system includes carcinoma of the prostate, testis and penis. Adenocarcinoma of prostate is the most common cancer of the male reproductive system. Prostate cancer mainly affects elderly men. Germ cell tumors are the most common malignancies involving the testis. Tumors of the penis are rare among the male reproductive cancers.

K.P. Haresh, MD, DNB, MNAMS (✉) • R. Benson, MD
Department of Radiotherapy, All India Institute of Medical Sciences, New Delhi, India
e-mail: drkpharesh@gmail.com; ronybenson@gmail.com

A. Kumar, M. Sharma (eds.), *Basics of Human Andrology*,
DOI 10.1007/978-981-10-3695-8_28

28.2 Prostate Cancer

Carcinoma of the prostate is the most common cancer of the male reproductive tract and mainly affects elderly males. Prostate cancer is considered the second most common malignancy among males in the world (Globocan 2012). The incidence increases significantly with increasing age (Zhou et al. 2016). Most common histological variety is adenocarcinoma and accounts for more than 95% of all cases. Rare malignant tumors include small-cell carcinoma and sarcomas of the prostate (Mazzucchelli et al. 2008).

28.2.1 Aetiology

Increasing age is the most important risk factor for prostate cancer, and it is rare below the age of 40 years. Increase in the life expectancy has led to an increased prevalence of prostate cancer. Increase in body weight and body mass index has also been proposed as risk factors for development of prostate cancer (Lynch et al. 2014). Positive family history of prostate cancer is a known risk factor especially if first-degree relatives are involved (Mai et al. 2010). The exact gene for development of familial prostate cancer is yet to be identified, though single nucleotide polymorphisms have been reported to be associated with it (Zhang et al. 2016). The role of dietary factors in the development of prostate cancer is also controversial with some reports mentioning the association of low vitamin D levels to the development of prostate cancer (Tuohimaa et al. 2004). There is a clear role for androgen in the development of prostate cancer. Androgens bind to receptors and this complex translocate to the nucleus for DNA binding and activation of genes that control cell division (Coffey 1993). Studies have reported the use of anti-androgens such as 5 alpha-reductase inhibitors for the chemoprevention of prostate cancer (Musquera et al. 2008).

Five to ten percent of all prostate cancers are hereditary, and a number of heritable genetic changes have been identified to contribute to a person's risk of developing prostate cancer. BRCA2 gene has been associated with an increased risk for prostate cancer. Other genes, such as BRCA1 and HOXB13, have been seen associated with some cases of hereditary prostate cancer.

28.2.2 Symptoms and Signs

The symptoms associated with prostate cancer vary widely and may overlap with symptoms of benign prostatic hyperplasia (BPH). Most of the patients with early-stage prostate cancer and screening-detected patients are asymptomatic except if associated with benign prostatic hyperplasia (Imai et al. 1988). As the prostatic urethra lies within the prostate gland, any enlargement of the prostate gland produces urinary symptoms including increased frequency of urination, increased urination at night and difficulty in initiating urination. The symptoms overlap with the symptoms of BPH and are not specific to prostatic cancer though haematuria is more commonly associated with prostatic cancer (Ramyil et al. 2008).

Per-rectal examination may reveal prostate enlargement with nodules palpable on the prostate. Depending on the spread of prostatic cancer, the patient may present with features of bone pain and tenderness in the bones (Muralidharan and Smith 2013). Involvement of the spinal cord may produce features of spinal cord compression in some patients with lower limb weakness and urinary and rectal symptoms (Helweg-Larsen and Sørensen 1994).

28.2.3 Diagnostic Evaluation

The baseline investigations must include complete blood counts, liver, kidney function tests and prothrombin time before a biopsy is taken. Serum PSA must be done in all patients at baseline and must be done preferably before any manipulation is done to prostate. Usually, a level of <4.0 ng/mL is considered normal. The half-life for serum PSA is 2.2–3.2 days (Oesterling 1991). The sensitivity of PSA may be improved by using free PSA, in patients with a total PSA between 4 and 10 ng/ml. Patients with free PSA greater than 25% of total are more likely to have a non-malignant cause than a cancer, making a biopsy unnecessary. If free PSA is less than 10% of total, the patient is more likely to have prostate cancer.

Transrectal ultrasound (TRUS) is a very useful investigation for the diagnosis of prostate cancer. It helps in tumor staging of prostate cancer and documents the prostate volume and is a very sensitive investigation. TRUS also helps in taking guided biopsy from suspicious areas of the prostate. Magnetic resonance imaging (MRI) is also very helpful imaging. Prostate cancer is well appreciated on T2-weighted image where the tumor appears hypointense compared to the normal prostatic parenchyma (Penzkofer and Tempany-Afdhal 2014). The advantage of MRI over TRUS is that it helps in nodal staging (Taneja 2004). A computed tomography (CT) scan may be used for nodal staging when MRI is not available, but it is not a good imaging modality for early-stage prostate cancer.

Biopsy of the prostate is necessary for the diagnosis and for getting the Gleason's score of the patient (Shah and Zhou 2016). Biopsy helps in the final histological confirmation of the adenocarcinoma as other histologies may be present, whose diagnosis may need the alteration of treatment. The pattern may range from pattern 1 being the most differentiated, with discrete glandular formation, to pattern 5, which is the most undifferentiated. The pathologist looks for the two most common tumor growth patterns. The Gleason's score is summarised as the sum of both, e.g. (4 + 3 = 7). In some cases a tertiary pattern may also be seen which needs to be mentioned in the biopsy report. Bone scan must be done in patients with a PSA of more than 20 or Gleason's score more than 7 or in T3 tumors.

The guidelines are given for screening of prostate cancer in asymptomatic individuals; given by the American Society of Clinical Oncology (ASCO) are most widely approved. It is recommended to do screening only in men with a life expectancy more than 10 years. Serum prostate specific antigen (PSA) testing must be

Table 28.1 The American Joint Committee on Cancer 7th edition staging of prostatic cancer

T staging	N staging	M staging
T1 – neither palpable nor visible on imaging T2 – tumor limited to the prostate T3 – spread beyond capsule or involvement of seminal vesicles T4 – spread to surrounding organs	N0 – negative regional lymph node N1 – positive regional lymph node	M1 – positive for metastasis either to the non-regional lymph node, bone or other visceral organs

used for prostate cancer screening in these men. The risk for overdiagnosis must be matched with the available resources for treatment before routinely implementing any screening strategy.

28.2.4 Staging

The most widely accepted system of staging prostate cancer is given by the American Joint Committee on Cancer (AJCC). The staging of cancer includes status of tumor, node and metastasis (TNM).

The AJCC 7th edition staging of prostate cancer is summarised in Table 28.1.

28.2.5 Principles of Management

The management of prostate cancer depends on various factors such as age at diagnosis, TNM staging, serum PSA levels and Gleason's score. Another important factor that is important in deciding treatment of patients is the expected life expectancy.

28.2.5.1 Role of Surgery

Radical prostatectomy with or without lymph node dissection is the standard surgery for carcinoma of the prostate. Laparoscopic surgery in a good oncological centre may provide similar outcomes as with that of open surgery (Kurpad et al. 2016). The need for lymph node dissection has been controversial. A pelvic lymph node dissection may be recommended in high-risk prostate cancer patients (Safir et al. 2015).

28.2.5.2 Role of Radiotherapy

Radiotherapy may be used as a radical treatment, postoperative treatment or palliative treatment in patients with prostatic cancer. Radiotherapy alone may be a treatment option in patients with low-risk prostatic adenocarcinoma. The advantage of radiotherapy over surgery is that it is safe, is noninvasive and can be administered even in patients with medical comorbidities. In metastatic disease, radiotherapy may be used for palliation of local symptoms as well as pain of bone metastasis (Cameron et al. 2014).

28.2.5.3 Role of Androgen Deprivation Therapy

The importance of androgen in the development of prostatic adenocarcinoma led to the use of androgen deprivation therapy (ADT) as a possible therapeutic option (Grossmann et al. 2013). The addition of ADT to radiation therapy has been found to confer survival advantage in patients with stage T3, PSA levels >20 ng/mL and Gleason's score ≥8 or high-risk prostate cancer. ADT may be started 2–3 months before the starting of radiotherapy as it reduces the volume of prostate and helps in easier radiation planning. ADT may be used in adjuvant setting in postoperative cases with lymph node-positive disease. In patients with metastatic disease, ADT may be the first-line therapy and is associated with good disease control and minimal side effects (Shafi et al. 2013).

Surgical orchidectomy or medical orchidectomy may be done in these patients, with the advantage of surgical orchidectomy being one-time procedure and cheaper. Medical orchidectomy with GnRH analogues is another option with similar efficacy but requires repeated injections.

28.2.5.4 Role of Chemotherapy

Mitoxantrone, a type II topoisomerase inhibitor, which disrupts DNA synthesis, was the initial drug used in prostate cancer, and now it has been replaced by docetaxel plus prednisolone. Docetaxel acts by disrupting the microtubular network in cells by promoting the formation of stable microtubules.

28.2.5.5 Follow-Up

Follow-up after radical treatment should include a serum PSA every 3–6 months and digital rectal examination every 6–12 months for 5 years and then yearly thereafter.

28.3 Testicular Cancer

Testicular cancer is the odd one among the male genitourinary cancers in that it relatively affects younger people with the median age of diagnosis ranging from 20 to 40 years (Stevenson and Lowrance 2015). A patient presenting with a testicular mass after the fifth decade usually has lymphoma and not a germ cell tumor although spermatocytic seminomas can occur in this age group.

28.3.1 Aetiology

The risk of testicular tumor has been found to be slightly higher in white men than people with darker complexion (Spitz et al. 1986). Cryptorchidism or undescended testis has been found to be a risk factor for the development of testicular germ cell tumors, and prophylactic orchiopexy has been shown to reduce the incidence of germ cell tumor if performed before puberty (Toppari et al. 2014). If the testis is found nonfunctional, then an orchidectomy may be necessary. Patients with Klinefelter syndrome and those with a history of mumps orchiditis are also at an increased risk of development of testicular tumors (De Sanctis et al. 2013).

There have been reports that up to 50% of the testicular cancers have some abnormal gene inherited from parents. The genes associated with testicular cancer include TERT, DMRT1 and ATF7IP. In future, identification of these genes might help in early diagnosis as well as the development of newer more effective treatments for testicular cancers.

28.3.2 Symptoms and Signs

The patient classically presents with a painless swelling of the testis which is progressively increasing in size. Sometimes the patient may have pain in the scrotum or pain radiating to the lower abdomen. Spread of germ cell tumor to the para-aortic lymph node may produce backache. Production of hormones like beta human chorionic gonadotropin (β-hCG) can also produce symptoms like gynaecomastia. Spread of the germ cell tumor to the lung may produce breathing difficulties, and spread to the brain causes central nervous system symptoms such as headache, nausea, vomiting, seizures and focal neurological deficit. Examination of the abdomen is also very important to rule out para-aortic lymphadenopathy.

28.3.3 Tumor Subtypes

The various histological subtypes of testicular cancers are given in Table 28.2. Germ cell tumors (GCT) constitute the major histological variety of testicular tumors. GCT can be broadly classified as seminomatous and non-seminomatous germ cell tumors. Other rare subvarieties include interstitial/stromal tumors which include Leydig cell tumors and Sertoli cell tumors, lymphomas and leukemic infiltrates to the testis (Looijenga 2014).

28.3.4 Diagnostic Evaluation

The baseline investigations must include complete blood counts and liver and kidney function tests. The initial investigation for any patient with testicular swelling should be testicular ultrasound which helps in differentiating cystic and solid areas (Islam et al. 2015). Ultrasound also helps in staging of testicular tumors. Tumor markers must be done in all patients and must include serum lactate dehydrogenase (LDH), β-hCG and alpha-fetoprotein (AFP) (Reiter et al. 2015). AFP is never raised in pure seminomas but may have moderately raised β-hCG.

Table 28.2 Various histological subtypes of testicular tumors

1. Germ cell tumors (95%) A. Seminomatous B. Non-seminomatous Embryonal cell carcinoma Choriocarcinoma Yolk sac tumor Teratoma
2. Interstitial cell tumors/sex cord/stromal tumors Leydig cell tumors Sertoli cell tumors Gonadoblastoma Granulosa cell tumors
3. Miscellaneous testicular neoplasms Epidermoid cyst Adenocarcinoma of rete testis
4. Lymphoma
5. Leukemic infiltration
6. Metastases

Table 28.3 Tumor markers in testicular tumors

	hCG	AFP	LDH
Nature	Glycoprotein	Serum protein	Enzyme
Cancers in which marker is raised	Choriocarcinoma	Embryonal carcinoma Yolk sac tumor Combined tumor Hepatocellular carcinomas	Germ cell tumors Lymphomas
Normal level	<5 IU/Ml	<16 ng/ml	105–333 IU/L
Half-life	22 h	5 days	1 day

β-hCG is classically raised in choriocarcinomas, while AFP is raised in yolk sac tumors. LDH usually indicates the bulk of the disease. A summary of the various tumor markers and its significance in testicular tumors is summarised in Table 28.3.

A contrast-enhanced computed tomography (CECT) of the abdomen should be done to look for any para-aortic or pelvic lymph node (Brunereau et al. 2012). A chest X-ray should be done in all patients. A CECT of the chest should be done in patients with positive para-aortic lymph node or those with abnormal chest X-ray. An MRI of the brain should be done in patients with central nervous system symptoms, and a bone scan should be done in all patients with clinical suspicion of bone metastasis.

A trans-scrotal fine needle aspiration cytology or biopsy should never be done in a patient with suspected testicular tumors. This is because it leads to alteration in the lymphatic drainage and increases the risk of both local recurrence and aberrant

Table 28.4 AJCC 7th edition staging of testicular cancer

T staging	N staging	M staging
T1: tumor confined to the testicle; may invade into the albuginea T2: tumor with involvement of tunica vaginalis or the presence of angio-lymphatic invasion T3: spermatic cord involvement T4: scrotal involvement	N1: single or multiple lymph node <2 cm N2: single/multiple node, 2–5 cm N3: any node >5 cm	M1: positive for metastasis A: non-regional nodal/lung metastasis B: other metastases

lymph nodal recurrence. A high inguinal orchidectomy is usually diagnostic and curative in testicular tumor.

28.3.5 Staging

The AJCC staging is the most widely accepted staging system for testicular cancers. It is divided into three stages. Stage I includes tumor limited to the testis without nodal involvement, while nodal involvement is characterised as stage II. Stage III represents metastatic spread of the testicular tumors. The AJCC 7th edition staging of testicular tumor is summarised in Table 28.4.

28.3.6 Principles of Management

The management of testicular tumors depends on the histology and stage grouping of the patient. The patient after proper investigations must undergo a high inguinal orchidectomy, and the tumor marker level must be documented after surgery.

Any seminomatous tumor with elevated AFP must be treated like non-seminomatous germ cell tumor. It is common for large seminomatous tumors to have a residual tumor after chemotherapy. The patient diagnosed as non-seminomatous germ cell tumor after a high inguinal orchidectomy must be properly evaluated, and the tumor marker level must be documented after surgery. The adjuvant treatment depends on tumor stage and level of tumor markers.

28.3.6.1 Role of Surgery

High inguinal orchidectomy is both diagnostic and therapeutic in germ cell tumors. Any trans-scrotal surgery or biopsy must be avoided as it may lead to alteration in lymphatic drainage and aberrant lymphatic drainage.

Biopsy of the contralateral testis may be indicated in patients with contralateral atrophic testis or cryptorchid testis or if there is abnormality in ultrasound. Surgery may also have a role in resection of residual tumor after chemotherapy in both seminomatous and non-seminomatous germ cell tumors.

28.3.6.2 Role of Radiotherapy

The role of radiotherapy is more in seminomatous germ cell tumor than non-seminomatous tumor (Wilder et al. 2012). Seminoma is a very highly radio- and chemosensitive tumor.

28.3.6.3 Role of Chemotherapy

Both seminomatous and non-seminomatous germ cell tumors are considered as chemosensitive, and platinum is the most effective drug (Kawai 2011). The main mechanism of action of platinum drugs is by formation of cross link with the purine bases on the DNA. The other effective drugs are bleomycin and etoposide. The mechanism of action of bleomycin includes inhibition of DNA synthesis and generation of activated oxygen-free radical species, which causes DNA breaks. Etoposide is a topoisomerase II inhibitor which prevents religation of the DNA strands and thus causes DNA breaks.

28.3.6.4 Follow-Up

The follow-up strategy for patients varies with the adjuvant strategy planned in patients. Patients kept on active surveillance require more frequent follow-up with contrast-enhanced computed tomography of the abdomen and pelvis every 3–6 months for the first year and then annually. Tumor markers may be repeated every 2 months and chest X-ray repeated every 6 months for the first year for patients kept on active surveillance. The patients who received adjuvant chemotherapy or radiotherapy require CECT of the abdomen and pelvis annually.

28.4 Penile Cancer

Penile cancer is the rarest among the cancers of the male reproductive system and can involve the penile tissue or prepuce. Penile cancer usually occurs in elderly male with a median age at diagnosis of around 60 years. Squamous cell carcinomas account for around 95% of the cancers of the penis with the rest being small-cell carcinoma and melanomas. The cancer usually occurs in the prepuce and the glans of the penis with shaft being a rare site.

28.4.1 Aetiology

Human papilloma virus (HPV) has been clearly associated with the development of penile cancers with squamous cell carcinoma particularly common. It is estimated that about one third of all penile cancers are caused usually by HPV 16 and 18 subtypes (Anic and Giuliano 2011). The presence of genital wart may also be an increased risk factor for development of penile cancers. Genital hygiene is another very important factor that has been seen associated with penile cancer, which may be due to increased infection rate and due to accumulation of smegma.

Infection with human immunodeficiency virus (HIV) has also been seen to be a risk factor associated with the development of penile cancer in men (Silverberg et al. 2009). Some premalignant conditions are also associated with carcinoma of the penis, namely, leukoplakia and balanitis xerotica obliterans (Minhas et al. 2010). Leukoplakia of the penis appears as whitish spots over the penis and is usually associated with phimosis where foreskin of the penis cannot be retracted. Balanitis xerotica obliterans of the penis may present with pruritus or burning sensation over the penis. Topical treatment is usually sufficient for balanitis xerotica obliterans.

28.4.2 Symptoms and Signs

Penile cancer may present as a non-healing ulcer over the penis and may or may not be associated with pain. There may also be a subtype that presents as a proliferative growth over the penis. The patient may also have fowl smelling discharge or bleeding from the ulcerated growth. The patient may have urinary complaints if the disease has spread up to the urethra to cause obstruction and blood in urine. Spread of the tumor to the nodes may produce swelling in the groin.

Once suspected of penile cancer, the patient should be clinically examined, and the induration surrounding the growth should be noted, and the inguinal area must be closely examined for the presence of any palpable nodes. The patient must also be looked for any abdominal lymphadenopathy, hepatomegaly and bony tenderness.

28.4.3 Diagnostic Evaluation

The baseline investigations must include complete blood counts, liver and kidney function test and prothrombin time. Biopsy of the lesion should be done to confirm the presence of malignancy and its subtype. Ultrasound of the penis may help in the proper tumor staging, while MRI in addition to the staging of the primary tumor helps in the nodal staging also (Singh et al. 2005).

Fine needle aspiration cytology of any suspicious node may be done as about half of the enlarged inguinal nodes may be infective and thus help in accurate staging especially if the patient is planned for organ preservation with radiotherapy. Any patient with a palpable inguinal lymph node should undergo an MRI/computed tomography (CT) of the pelvis and abdomen to look for pelvic and intra-abdominal disease. A chest X-ray should be done in all cases, and bone scan should be done if clinical suspicion or if alkaline phosphatase is raised.

28.4.4 Staging

The widely accepted staging system for penile cancers is the AJCC-TNM staging. In the AJCC staging, the tumor staging is as per invasion of the corpora and the urethra. The AJCC 7th edition staging of penile cancer is summarised in Table 28.5.

Table 28.5 AJCC 7th edition staging of penile cancer

T staging	N staging	M staging
T1 – tumor involving the subepithelial connective tissue T2 – tumor involving the corpus spongiosum/cavernosum T3 – tumor invading the urethra T4 – tumor invading contiguous structures	N1 – single inguinal lymph node N2 – multiple or bilateral inguinal lymph node N3 – extranodal extension/ pelvic lymph node	M1 – positive for metastasis

28.4.5 Principles of Management

The management of penile cancer depends on the clinical stage, histology and desire of the patient for organ preservation. Patients with T1 node-negative disease have a wide variety of organ preservation strategies including surgery (Micali et al. 1996). Surgery for patients with T2 disease onwards may require at least a partial penectomy, but organ preservation can be achieved in selected patients using radiotherapy.

28.4.5.1 Local Therapy

Topical 5-flurouracil (5-FU) and imiquimod may be tried in patients with carcinoma in situ of the penis. 5-FU acts via inhibition of the enzyme of thymidylate synthase, thereby inhibiting DNA synthesis. Imiquimod acts via activating interferon alpha (IFN-α), interleukins and tumor necrosis factor alpha (TNF-α) which stimulates natural killer cells and cytotoxic T lymphocytes.

28.4.5.2 Role of Surgery

Surgery for penile cancer includes surgery to the primary tumor and the nodes. For early T1 and T2 disease, wide local excision with circumcision or partial penectomy may be adequate (Schoeneich et al. 1999). Total penectomy may be needed in some T2 patients and almost all T3 patients but has the disadvantage of organ compromise (Magoha 1995). Surgery for inguinal lymph node usually involves an inguinal lymph node dissection. Pelvic lymph node dissection may be required in patients with more than two lymph nodes positive in inguinal area (Misra et al. 2004; Mosconi et al. 2005).

28.4.5.3 Role of Radiotherapy

Radiotherapy may be used as a part of organ preservation approach for penile cancer (Gerbaulet and Lambin 1992; Antunes et al. 2007). The main advantage of using radiotherapy over surgery in penile cancer is penile preservation which can be achieved with radiotherapy. Elective and involved nodal irradiation may be done as indicated.

28.4.5.4 Role of Chemotherapy

Platinum remains the cornerstone for any chemotherapy regimens in penile cancer. The main mechanism of action of platinum drugs is by formation of cross link with the purine bases on the DNA. Other agents active in the carcinoma penis are methotrexate and taxanes. Methotrexate inhibits the enzyme dihydrofolate reductase, thus inhibiting DNA synthesis. Taxanes act via promoting the formation of stable microtubules there by inhibiting cell division.

28.4.5.5 Follow-Up

The patient may be kept on clinical follow-up after completion of treatment with follow-up every 3 months for the first 2 years, then every 6 months till 5 years and then annually. Imaging of the abdomen and pelvis with computed tomography may be done only in N2/N3 disease.

Key Questions

- What are the risk factors of prostate cancer?
- Describe the role of hormone therapy in the management of prostate cancer.
- What are the various histological types of testicular tumors?
- Discuss the management options for seminoma testis.
- Describe the aetiological factors associated with development of penile cancer.
- What are the organ preservation approaches available for treatment of penile cancer?

References

AJCC- Americal Joint Committee on Cancer. https://cancerstaging.org/Pages/default.aspx. Accessed 1 May 2016.

Anic GM, Giuliano AR. Genital HPV infection and related lesions in men. Prev Med. 2011;53(Suppl 1):S36–41.

Antunes AA, Dall'Oglio MF, Srougi M. Organ-sparing treatment for penile cancer. Nat Clin Pract Urol. 2007;4(11):596–604.

Brunereau L, Bruyère F, Linassier C, Baulieu J-L. The role of imaging in staging and monitoring testicular cancer. Diagn Interv Imaging. 2012;93(4):310–8.

Cameron MG, Kersten C, Guren MG, Fosså SD, Vistad I. Palliative pelvic radiotherapy of symptomatic incurable prostate cancer – a systematic review. Radiother Oncol. 2014;110(1):55–60.

Coffey DS. The molecular biology of the prostate. In: Lepor H, Lawson RK, editors. Prostate disease. Philadelphia: Saunders; 1993. p. 28–56.

De Sanctis V, Fiscina B, Soliman A, Giovannini M, Yassin M. Klinefelter syndrome and cancer: from childhood to adulthood. Pediatr Endocrinol Rev. 2013;11(1):44–50.

Gerbaulet A, Lambin P. Radiation therapy of cancer of the penis. Indications, advantages, and pitfalls. Urol Clin North Am. 1992;19(2):325–32.

Globocan. Estimated cancer incidence, mortality and prevalence worldwide in 2012. 2012. http://globocan.iarc.fr/Default.aspx. Accessed 1 May 2016.

Grossmann M, Cheung AS, Zajac JD. Androgens and prostate cancer; pathogenesis and deprivation therapy. Best Pract Res Clin Endocrinol Metab. 2013;27(4):603–16.

Helweg-Larsen S, Sørensen PS. Symptoms and signs in metastatic spinal cord compression: a study of progression from first symptom until diagnosis in 153 patients. Eur J Cancer. 1994;30A(3):396–8.

Imai K, Zinbo S, Shimizu K, Yamanaka H, Kumasaka F, Sato Z. Clinical characteristics of prostatic cancer detected by mass screening. Prostate. 1988;12(3):199–207.

Islam SR, Rahman S, Boksh Z, Pasha K, Tanim SA. Testicular tumor – a review article. Mymensingh Med J. 2015;24(1):207–14.

Kawai K. Current understandings of chemotherapy for testis cancer. Gan To Kagaku Ryoho. 2011;38(1):54–8.

Kurpad R, Woods M, Pruthi R. Current status of robot-assisted radical cystectomy and intracorporeal urinary diversion. Curr Urol Rep. 2016;17(6):42.

Looijenga LHJ. Testicular germ cell tumors. Pediatr Endocrinol Rev. 2014;11(Suppl 2):251–62.

Lynch BM, Friedenreich CM, Kopciuk KA, Hollenbeck AR, Moore SC, Matthews CE. Sedentary behaviour and prostate cancer risk in the NIH-AARP diet and health study. Cancer Epidemiol Biomarkers Prev. 2014;23(5):882–9.

Magoha GA. Management of carcinoma of the penis: a review. East Afr Med J. 1995;72(9):547–50.

Mai PL, Wideroff L, Greene MH, Graubard BI. Prevalence of family history of breast, colorectal, prostate, and lung cancer in a population-based study. Public Health Genomics. 2010;13(7–8):495–503.

Mazzucchelli R, Lopez-Beltran A, Cheng L, Scarpelli M, Kirkali Z, Montironi R. Rare and unusual histological variants of prostatic carcinoma: clinical significance. BJU Int. 2008;102(10):1369–74.

Micali G, Innocenzi D, Nasca MR, Musumeci ML, Ferraú F, Greco M. Squamous cell carcinoma of the penis. J Am Acad Dermatol. 1996;35(3 Pt 1):432–51.

Minhas S, Manseck A, Watya S, Hegarty PK. Penile cancer – prevention and premalignant conditions. Urology. 2010;76(2 Suppl 1):S24–35.

Misra S, Chaturvedi A, Misra NC. Penile carcinoma: a challenge for the developing world. Lancet Oncol. 2004;5(4):240–7.

Mosconi AM, Roila F, Gatta G, Theodore C. Cancer of the penis. Crit Rev Oncol Hematol. 2005;53(2):165–77.

Muralidharan A, Smith MT. Pathobiology and management of prostate cancer-induced bone pain: recent insights and future treatments. Inflammopharmacology. 2013;21(5):339–63.

Musquera M, Fleshner NE, Finelli A, Zlotta AR. The REDUCE trial: chemoprevention in prostate cancer using a dual 5alpha-reductase inhibitor, dutasteride. Expert Rev Anticancer Ther. 2008;8(7):1073–9.

Oesterling JE. Prostate specific antigen: a critical assessment of the most useful tumor marker for adenocarcinoma of the prostate. J Urol. 1991;145(5):907–23.

Penzkofer T, Tempany-Afdhal CM. Prostate cancer detection and diagnosis: the role of MR and its comparison with other diagnostic modalities – a radiologist's perspective. NMR Biomed. 2014;27(1):3–15.

Ramyil VM, Dakum NK, Liman HU, Udeh E. The management of prostatic haematuria. Niger J Med. 2008;17(4):439–42.

Reiter MJ, Costello JE, Schwope RB, Lisanti CJ, Osswald MB. Review of commonly used serum tumor markers and their relevance for image interpretation. J Comput Assist Tomogr. 2015;39(6):825–34.

Safir IJ, Lian F, Alemozaffar M, Master VA. Surgery for high-risk prostate cancer and metastatic prostate cancer. Curr Probl Cancer. 2015;39(1):33–40.

Schoeneich G, Perabo FG, Müller SC. Squamous cell carcinoma of the penis. Andrologia. 1999;31(Suppl 1):17–20.

Shafi AA, Yen AE, Weigel NL. Androgen receptors in hormone-dependent and castration-resistant prostate cancer. Pharmacol Ther. 2013;140(3):223–38.

Shah RB, Zhou M. Recent advances in prostate cancer pathology: gleason grading and beyond. Pathol Int. 2016;66(5):260–72.

Silverberg MJ, Chao C, Leyden WA, Xu L, Tang B, Horberg MA. HIV infection and the risk of cancers with and without a known infectious cause. AIDS. 2009;23(17):2337–45.

Singh AK, Saokar A, Hahn PF, Harisinghani MG. Imaging of penile neoplasms. Radiographics. 2005;25(6):1629–38.

Spitz MR, Sider JG, Pollack ES, Lynch HK, Newell GR. Incidence and descriptive features of testicular cancer among United States whites, blacks, and Hispanics, 1973–1982. Cancer. 1986;58(8):1785–90.

Stevenson SM, Lowrance WT. Epidemiology and diagnosis of testis cancer. Urol Clin North Am. 2015;42(3):269–75.

Taneja SS. Imaging in the diagnosis and management of prostate cancer. Rev Urol. 2004;6(3):101–13.

Toppari J, Rodprasert W, Virtanen HE. Cryptorchidism – disease or symptom? Ann Endocrinol. 2014;75(2):72–6.

Tuohimaa P, Tenkanen L, Ahonen M, Lumme S, Jellum E, Hallmans G, et al. Both high and low levels of blood vitamin D are associated with a higher prostate cancer risk: a longitudinal, nested case-control study in the Nordic countries. Int J Cancer. 2004;108(1):104–8.

Wilder RB, Buyyounouski MK, Efstathiou JA, Beard CJ. Radiotherapy treatment planning for testicular seminoma. Int J Radiat Oncol Biol Phys. 2012;83(4):e445–52.

Zhang Y, Li P, Xu A, Chen J, Ma C, Sakai A, et al. Influence of a single-nucleotide polymorphism of the DNA mismatch repair-related gene exonuclease-1 (rs9350) with prostate cancer risk among Chinese people. Tumor Biol. 2016;37(5):6653–9.

Zhou CK, Check DP, Lortet-Tieulent J, Laversanne M, Jemal A, Ferlay J, et al. Prostate cancer incidence in 43 populations worldwide: an analysis of time trends overall and by age group. Int J Cancer. 2016;138(6):1388–400.

Part VIII

Regulation of Male Fertility and Sexual Health

Male Contraception

29

N.K. Lohiya, Imtiyaz Alam, and A.S. Ansari

Learning Objectives

- Sites targeted for contraception
- Barrier method
- Vas-based methods
  Vasectomy
  Reversible inhibition of sperm under guidance
  Intra-vas device
- Hormones, drugs, and herbs as contraceptives
- Potential methods of contraception

29.1 Introduction

According to World Health Organization (WHO) report, 122 million planned pregnancies occur each year worldwide. In addition to this, 87 million pregnancies are reported to be unintended, and 46 million pregnancies are terminated medically

N.K. Lohiya, PhD, FNASc, FAMS, FIAES, FEMSI (✉) • I. Alam, MSc • A.S. Ansari, PhD
Centre for Advanced Studies, Department of Zoology, University of Rajasthan, Jaipur 302004, India
e-mail: lohiyank@gmail.com; imtiyazalam@hotmail.co.uk; abdulsansari@yahoo.com

A. Kumar, M. Sharma (eds.), *Basics of Human Andrology*,
DOI 10.1007/978-981-10-3695-8_29

(WHO 2005). Surprisingly, in 50% of unintended pregnancies, some contraceptive measures are used. This shows convincing failure rate of available contraceptives. While there are numerous forms of contraception available, it is clear that there is no single method which fulfills the requirement of an ideal contraceptive. It is estimated that male contraceptives account for 37% of all contraceptive measures. Currently, among all available male contraceptives, condoms, abstinence, withdrawal, and vasectomy are believed to be successful. However, these methods are reported to be unpleasant, doubtful, and unacceptable for many users. The requirement of a safe, effective, and reversible male contraception is, thus, an essential requirement.

29.2 Sites Targeted for Male Contraception

The primary targets for male contraception are testis, epididymis, and ductus deferens. Intervention in testis has potential endocrine consequences. Damage in DNA during spermatogenesis may lead to teratogenic effects. The blood-testis barrier may also impose immunogenic problems. Consequently, attention has been given to epididymis and ductus deferens as viable targets for male contraception. Sperm has the ability to bind and fertilize egg following its maturation or capacitation.

Capacitation involves functional changes by localized luminal microenvironment of the epididymis and post-testicular remodeling of sperm plasma membrane by addition, removal, or alteration of cell surface molecules. Epididymis is a potential target for contraception. The ductus deferens is also a potential site for intervening sperm transportation with minimum hormonal and systemic interference. Besides epididymis and ductus deferens, there are other methods which target hormonal cascade, immune system, and sperm-egg binding mechanism (see Fig. 29.1).

29.3 Barrier Method

Barrier methods such as condoms have been used since ancient times; the initial use was mostly to provide protection from sexually transmitted infections (STIs). Further, the method also attracted attention as a means of family planning. Majority of condoms are made up of latex rubber and tested for water leakage and tensile strength. It has been suggested that condoms used with or without nanoxynol-9 (N-9), a spermicidal agent, show no difference in the rate of urogenital gonorrhea and chlamydial infection (Roddy et al. 2002). Despite its huge acceptance in couples, errors in condom use such as breakage and slippage are quite common.

29.4 Vas-Based Methods

29.4.1 Vasectomy

Vasectomy is an effective method of contraception. It is regarded as a permanent method of contraception. However, reversal has been performed by microsurgical

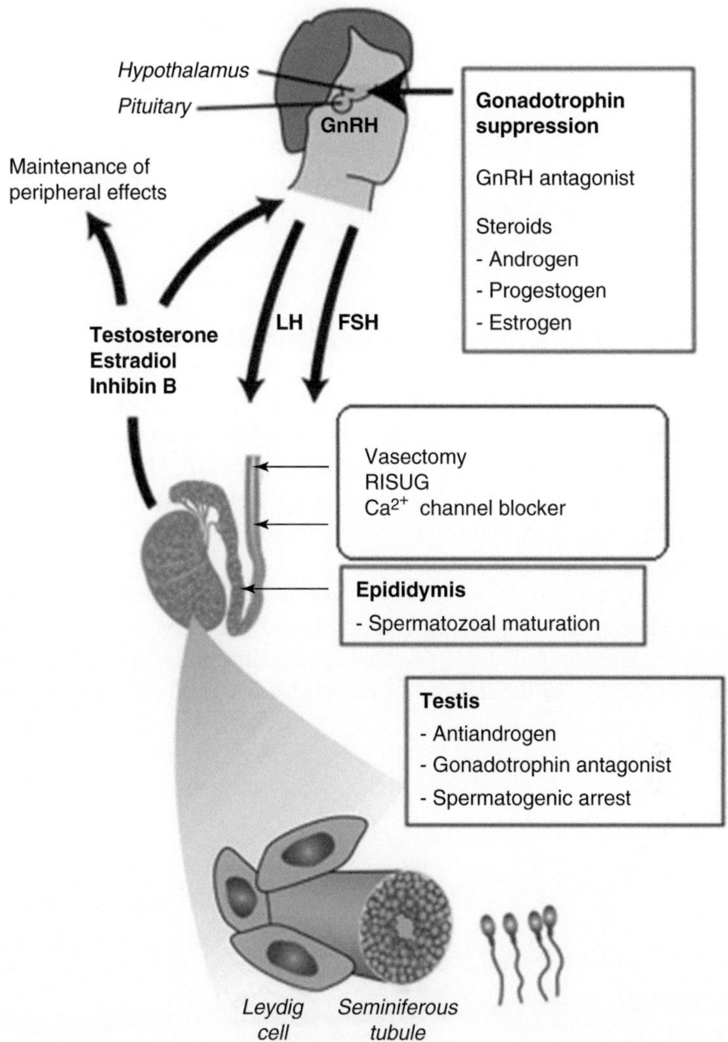

Fig. 29.1 Various sites targeted for male contraception. *LH* luteinizing hormone, *FSH* follicle-stimulating hormone, *GnRH* gonadotropin-releasing hormone, *RISUG* reversible inhibition of sperm under guidance, Ca^{2+} calcium

techniques (Sharlip 1993). It is considered completely safe with minimum morbidity and almost no mortality. However, it is accepted reluctantly due to cultural and social factors.

The method includes a no-scalpel approach which is performed under local anesthesia. Around 15 mm length of vas deferens is removed for the ease of vasovasostomy, if indicated later. Noninvasive methods of vasectomy have been used to reduce complications related to surgery and to increase its acceptance. Recently, noninvasive laser thermal technique for vas occlusion in canine has been reported (Cilip et al. 2012). High-resolution and superficial imaging of vas deferens is done by optical coherence tomography (OCT), while high-frequency ultrasound (HFUS) provides deeper imaging of vas deferens. Both OCT and high HFUS are potential imaging techniques for accurate confirmation of vas occlusion during noninvasive laser vasectomy. After vasectomy, up to 6% of men usually desire to undergo vasectomy reversal. For these men, patency can be achieved by recanalization of vas deferens or surgical sperm retrieval along with assisted reproduction. Nevertheless, it remains difficult for surgeons to accurately counsel men about the chance of patency after vasectomy reversal.

29.4.2 Reversible Inhibition of Sperm Under Guidance

Reversible inhibition of sperm under guidance (RISUG) is a new injectable male contraceptive, presently undergoing extended phase III clinical trials in India, being coordinated by the Indian Council of Medical Research (ICMR), New Delhi. The polymer is made by adding 40 g of styrene and maleic anhydride in ethyl acetate under nitrogen atmosphere and irradiated at the dose of 30–40 rad/sec with a total 0.2–0.24 megarad dose.

There are two approaches of vas occlusion by RISUG – partial occlusion and complete occlusion (see Fig. 29.2). Hypothetically, the polyelectrolytic nature of the polymer induces an imbalance of charges on sperm plasmalemma that leads to deactivation of sperm in the lumen of the vas (Kumar et al. 2007). However, few

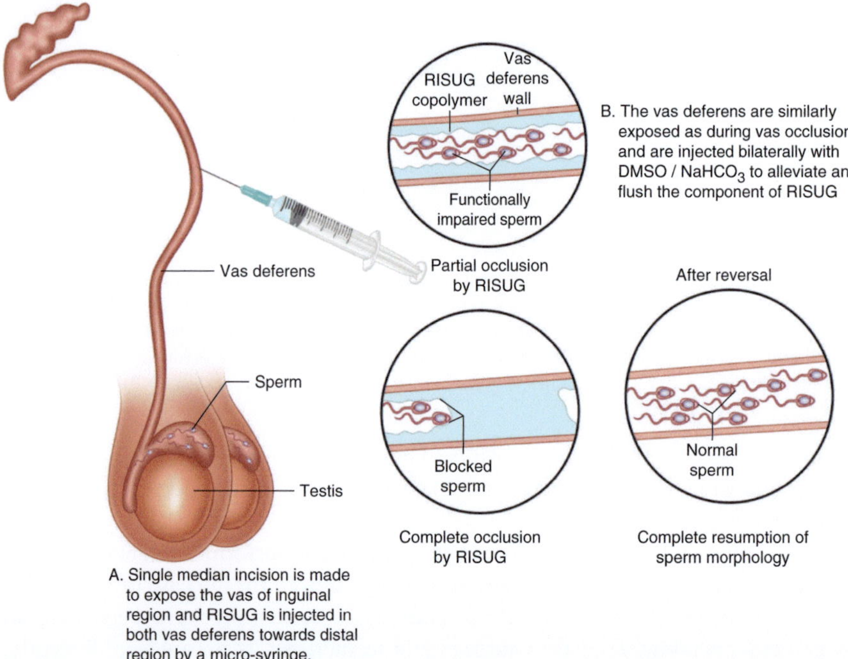

Fig. 29.2 The concept of RISUG injection and its reversal. *DMSO* dimethyl sulfoxide, *RISUG* reversible inhibition of sperm under guidance, *$NaHCO_3$* sodium bicarbonate

other mechanisms of action have also been proposed, e.g., pH lowering and formation of styrene-maleic anhydride (SMA) agglomerate (Guha 1996).

Inhibition of fertility has been observed following 15 days of vas occlusion in rats with structural abnormalities of sperm as indicated by damage in acrosomal membrane, segmented column loss, deformities in neck centriole, degenerated mitochondrial sheath, and axoneme with membrane loss in midpiece and tail (Lohiya et al. 2010). Clinical trials on humans also indicate 100% sterility following vas occlusion with RISUG; however, duration to achieve complete sterility varies greatly.

In an extended phase III clinical trial, 141 healthy adult male subjects received prefilled RISUG injections containing 60 mg SMA dissolved in 120 μl dimethyl sulfoxide (DMSO). Analyses of seminal characteristics and accessory gland function were carried out in 25 subjects for a period of 6 months of RISUG injection. All subjects showed impairment of sperm motility and increased percentage of sperm abnormalities like bent and coiled tail, amorphous, tapering head, etc. All azoospermic samples showed presence of immature germ cells. A significant lowered level of neutral α-glucosidase was evident in all subjects. It was concluded that the presence of immature germ cells and occasional sperm or both types of cells in the semen were due to partial patency of the vas deferens.

Reversal of vas occlusion by RISUG has been reported in animal models. RISUG/SMA dissolves at higher pH, becomes unstable at the pH ranging 8–9, and detaches from the vas epithelium. SMA dissolves in DMSO and sodium bicarbonate ($NaHCO_3$) solutions. Reversal studies have been conducted successfully in monkeys by means of noninvasive method where the SMA depot is broken into pieces and pushed toward the urethra by manual manipulations such as palpation, percutaneous electrical stimulation, forced vibratory movements, suprapubic percussion, and per-rectal digital massage. This reversal procedure, however, is not found feasible in humans where vas deferens is located deeply in the scrotum. Therefore, DMSO and $NaHCO_3$ have been used for reversal studies.

Following vas occlusion with RISUG, the vaginal smear shows several morphological aberrations in voided sperm as indicated by head-tail separation, damaged acrosome, and bent tail. Following reversal with DMSO, the morphology of sperm and fertility returned to normal range within 90 days (Lohiya et al. 2010). Recently, relative reversal suitability of both solvents (DMSO and $NaHCO_3$) following RISUG-induced long-term (360 days) vas occlusion has been investigated in albino rats (Ansari et al. 2016). The reversal study with $NaHCO_3$ indicates an early recovery of fertility as compared to reversal with DMSO.

29.4.3 Intra-Vas Device

Studies related to use of copper (Cu) intra-vas device (IVD) has shown successful results in animals (see Fig. 29.3). Cu IVD inserted in male rhesus monkeys exhibited significant reduction of fertility up to 13 months. Following 13 months of post-treatment recovery in contraceptive effect, rise in sperm count and motility has been recorded (Kapur et al. 1984).

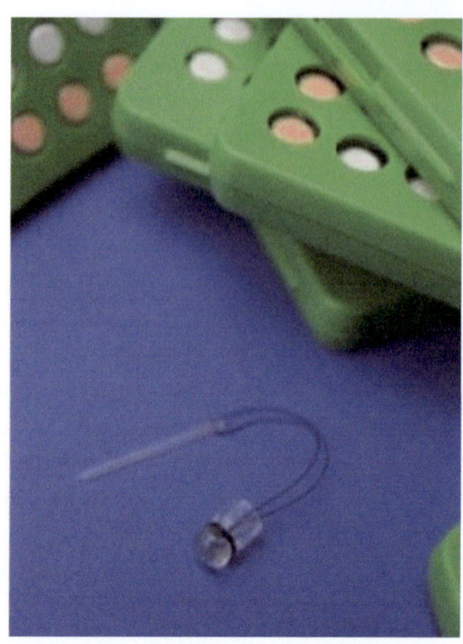

Fig. 29.3 Intra-vas device

Other types of IVDs such as nano-copper complex/polymer composites, solid (rod) and open (tube) prototype gold-platinum, nano-SiO(2)-copper complex cross-linking polymer composites, etc., have also been studied extensively in animal models (Chen et al. 2010).

29.5 Hormonal

29.5.1 Combination Protocol

Investigations on hormonal male contraception with testosterone alone and/or with gonadotropin suppressive agents have been attempted. Weekly intramuscular injections of testosterone-enanthate are administered for steady suppression of gonadotropins.

For more stable and long-term release of testosterone into the blood circulation, a new injectable formulation of testosterone-undecanoate in tea seed oil has been developed (Chen et al. 1991; Li et al. 1994). A long-term administration of testosterone-undecanoate in monkeys and humans has shown better fertility control as compared to testosterone-enanthate.

29.5.2 Desogestrel

Desogestrel is a progestogen which gets converted to an active agent-etonogestrel (Hasenack et al. 1986). A study has shown that administration of 300 μg oral desogestrel daily in combination with 50 mg testosterone weekly induces infertility (Wu et al. 1999).

Testosterone pellets in combination with higher doses of oral desogestrel have been used. The higher-dose regimen induces azoospermia in all subjects (Kinniburgh et al. 2002).

29.5.3 Dimethandrolone Undecanoate

Dimethandrolone (DMA) or 7α, 11β-dimethyl-19-nortestosterone is a known artificially synthesized androgen (Cook and Kepler 2005). In men, the 17β-undecanoic acid ester of DMA (DMAU) is an oral active agent used for androgen therapy. In body, the liberation of biologically active DMA is due to the cleavage of the 17β-ester bond. 19-norandrogens are considered as impeded progestin agonists (Reel et al. 1979). DMA also possess progestational activity that makes it suitable as hormonal contraceptive for men.

It has been shown that DMA and DMAU when administered in castrated rats enhance weights of accessory reproductive glands and decreases levels of gonadotropins.

29.5.4 Gonadotropin-Releasing Hormone (GnRH)

The metabolic effects can be predicted by steroidal negative feedback mechanism on the secretion of gonadotropins. By interference with the GnRH activity on gonadotrophs, the gonadotropin secretion can be inhibited. Side effect of hypogonadism can be overcome by administration of exogenous testosterone. GnRH agonists such as nafarelin and buserelin desensitize GnRH receptors, thereby decreasing gonadotropin release. Use of GnRH agonists along with testosterone has not been found satisfactory (Behre et al. 1992).

The GnRH antagonists inhibit activity of GnRH on the gonadotroph. Few drawbacks of antagonists are histamine-like allergic reactions at the site of injection, requirement for daily injection, difficulties, and expense of manufacture. Nevertheless, the clinical studies that have been carried out showed promising results, and new antagonists have less side effects. GnRH antagonists can result in azoospermia in larger number with less metabolic impact than the progestogen-based regimens.

29.6 Drugs

29.6.1 Nifedipine

Nifedipine, a dihydropyridine (see Fig. 29.4), is a nonhormonal contraceptive which acts by blocking Ca^{2+} channels and modifying concentration of cholesterol in sperm membrane (Benoff et al. 1994).

The ion channel specific to sperm flagella is the cation channel of sperm (CatSper) which causes sperm hyperactivation by regulating Ca^{2+} influx. Disruption of CatSper

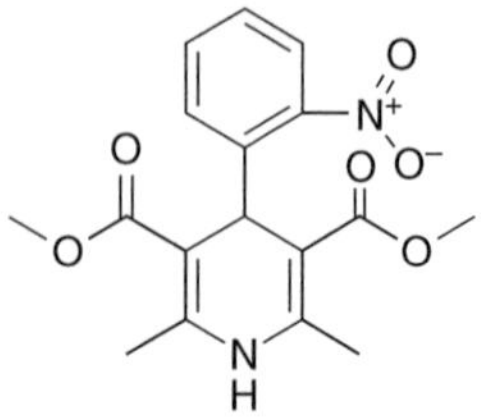

Fig. 29.4 Nifedipine

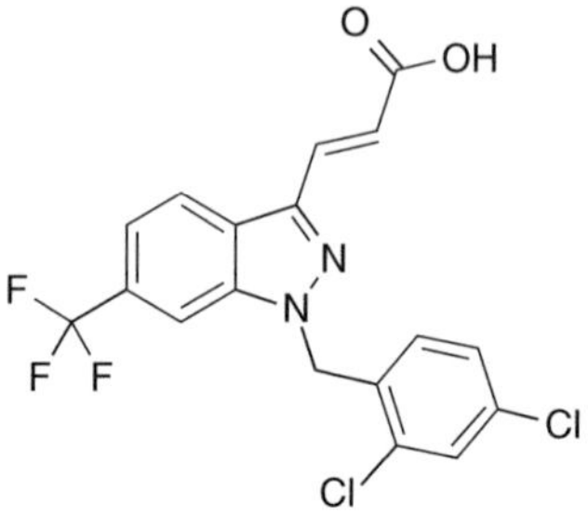

Fig. 29.5 Gamendazole

gene causes lack of sperm motility leading to sterility (Babcock 2007). Whether nifedipine acts on CatSper is not clearly known. Side effects associated with nifedipine include headache, heart burn, dizziness, nausea, and edema. These types of male contraceptives directed toward blocking sperm-specific ion channels hold a bright future as the effect would be restricted to testis only.

29.6.2 Gamendazole

Gamendazole is an indazole carboxylic acid derived from lonidamine (LND; see Fig. 29.5).

Reversible and complete inhibition of spermatogenesis can be achieved by the administration of indazole carboxylic acids in variety of animal models (Coulston et al. 1975; De Martino et al. 1981). Its use has been limited due to side effects such as muscular pain, testicular pain, nausea, vomiting, somnolence, and enhanced levels of liver enzymes. However, at the dose of 25 mg/kg, inhibition of spermatogenesis in 50% treated animals has been observed without any side effects. Gamendazole produces more extensive loss of spermatids than LND. Both LND and gamendazole cause loss of spermatogenic cells from the germinal epithelium.

29.6.3 Adjudin

1-(2,4-Dichlorobenzyl)-1H-indazole-3-carbohydrazide, commonly known as adjudin, is another analogue of indazole-carboxylic acid which has been developed as a nonhormonal male contraceptive (see Fig. 29.6).

Fig. 29.6 Adjudin

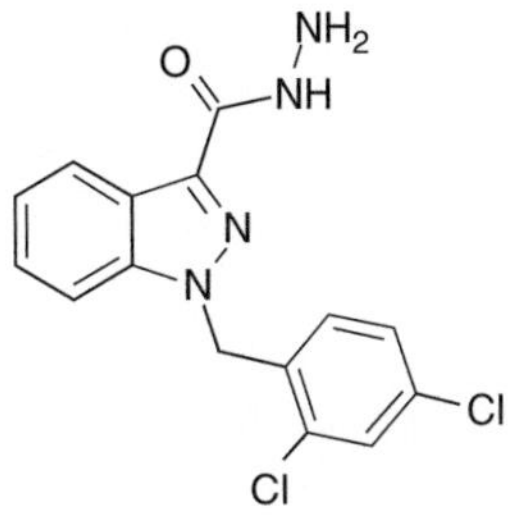

Fig. 29.7 Phenoxybenzamine

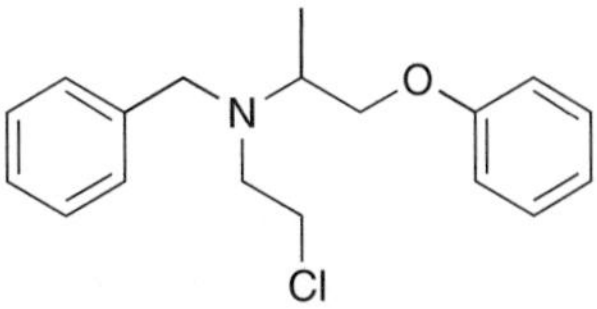

Adjudin acts by disturbing adhesion between germ cells and Sertoli cells. Adhesion between cells in other organs remains unchanged (Cheng et al. 2001). The complete mechanism of action of adjudin is not known. However, in testis, the principal target of adjudin appears to be apical ectoplasmic specializations present between elongated spermatids and Sertoli cells. The effect of adjudin is so prompt that within 6–7 h of treatment, 50% of seminiferous tubules show degenerative changes followed by damage to secondary spermatocytes and spermatids.

29.6.4 Phenoxybenzamine

Phenoxybenzamine (PBZ; see Fig. 29.7) or Dibenzyline has nonselective and irreversible alpha-antagonist properties. Administration of PBZ at doses up to 20 mg/day results in aspermia. Continued treatment of this chemical results in reversible inhibition of ejaculation due to paralyzing effects on vas deferens, ampulla, and ejaculatory ducts. However, long-term administration does not affect seminal characteristics. The withdrawal of treatment leads to reversal of these effects with occurrence of normal ejaculation. This drug is also useful for improvement of symptoms of premature ejaculation in men.

29.7 Herbal

29.7.1 *Carica papaya* Seeds

The seed products of *Carica papaya* have been used as potential male contraceptive in laboratory animals. Administration of aqueous oral suspension of ripe papaya seed induces sterility in male albino rats and rabbits (Vyas and Jacob 1984). *Carica papaya* seeds extract induces gradual degradation of germinal epithelium, Sertoli cells, and Leydig cells.

Administration of chloroform extract of seeds of *Carica papaya* shows total inhibition of sperm motility without any adverse effects (Lohiya and Goyal 1992). Chloroform extract possesses mild estrogenic properties. Preclinical investigations in monkeys reveal that chloroform extract results in gradual decrease in sperm concentration and motility. The sperm function tests and semen biochemical parameters such as fructose, Ca^{2+}, and glycerylphosphorylcholine (GPC) show significant reduction. Ultrastructure of the testis shows necrotic changes. These altered parameters are reversible following withdrawal of treatment (Panneerdoss 2004). No human trials have been reported; hence, much work is required to elucidate its potential as herbal male contraception.

29.7.2 *Tripterygium wilfordii*

The Chinese herbal medicine *Tripterygium wilfordii* is used in treatment of glomerulonephritis, arthritis, psoriasis, and other autoimmune diseases (see Fig. 29.8).

The glycosides of *Tripterygium wilfordii* show reversible sperm motility impairment in humans. Studies show that multi-glycosides extracted from this plant result in infertility in rats (Qian et al. 1986). A series of six male antifertility diterpene epoxides have been isolated from *Tripterygium wilfordii* such as triptolide, tripdiolide, triptolidenol, tripchloride, 16-hydroxytriptolide, and T7/19. The compounds cause defects in spermiogenesis and sperm along with exfoliation and inhibition of basic nuclear protein turnover of late spermatid, delayed spermiation, and various sperm abnormalities. Human trials show similar effects in small group

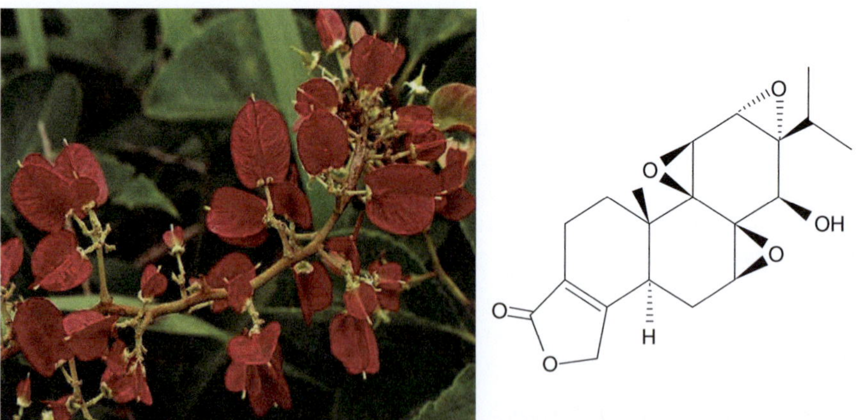

Fig. 29.8 *Tripterygium wilfordii* and structure of triptolide

of men. Administration of isolated triptolide results in decrease in sperm motility and fertility followed by inhibition of spermatogenesis during long-term treatment in rats (Qian et al. 1995).

29.7.3 Gossypol

Gossypol is a phenol derived from plant *Gossypium herbaceum* (see Fig. 29.9). Gossypol is found to be effective with 99% efficacy which has been assessed by sperm profiling in human trials. Further study on gossypol has been stopped due to its major side effect, hypokalemia, low therapeutic index, and uncertain recovery.

Gossypol trial has shown limited azoospermia cases with minor to serious side effects such as fatigue, gastrointestinal symptoms, decreased libido, dizziness, and hypokalemic paralysis (Qian et al. 1972). Low-dose gossypol acetic acid in combination with desogestrel/ethinyl estradiol and testosterone-undecanoate leads to damage of epididymal sperm (Yang et al. 2004).

29.7.4 Oleanolic Acid

Oleanolic acid is a triterpenoid present in medicinal plants such as *Phytolacca americana*, *Eugenia jambolana* (see Fig. 29.10), *Syzygium* species, garlic, etc. Oleanolic acid has hepatoprotective, antiviral, and antitumor properties. It inhibits spermatogenesis by decreasing forward sperm motility, which is restored within 14 days of its withdrawal (Mdhluli and van der Horst 2002). No serious side effect is noted with the use of oleanolic acid as contraceptive in Wistar rats (Rajasekaran et al. 1988).

Fig. 29.9 Plant of *Gossypium herbaceum* and structure of gossypol

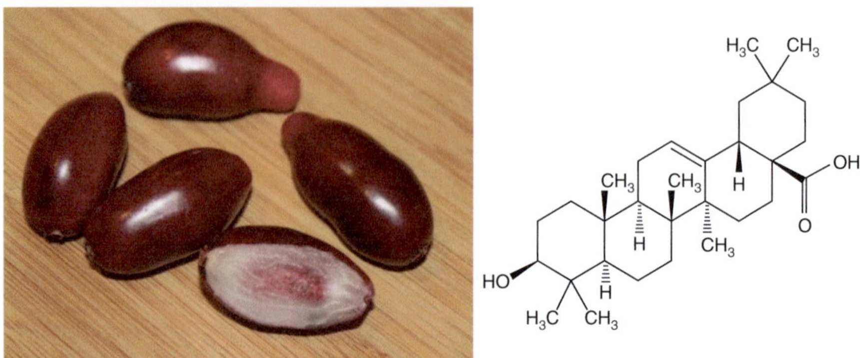

Fig. 29.10 *Eugenia jambolana* and structure of oleanolic acid

29.8 Potential Methods

29.8.1 Chromatin Remodeling

Approximately 4% of genes encoded by mammalian genome are expressed during spermatogenesis in male germ cells. The proteins encoded by these spermatogenesis-specific genes have been identified. One such protein is bromodomain-containing protein (BRDT). BRDT is a chromatin-associated protein which binds to histone molecules and shows expression in pachytene and diplotene spermatocytes, as well as round spermatids (Shang et al. 2007). Deletion of BRDT1 gene causes sterility in homozygous hypomorphic male mice, while single nucleotide polymorphism of BRDT causes oligozoospermia and azoospermia (Aston et al. 2010). Use of bromodomain antagonist JQ1 against BRDT inhibits spermatogenesis. JQ1 mediates its action on meiotic and postmeiotic male germ cells severely affecting spermatogenesis and sperm motility. Thus, JQ1 inhibition of BRDT shows how chromatin remodeling can be used for contraception.

29.8.2 Genomics

Interfering normal epididymal function by using molecular inhibitors of proteins required for post-testicular maturation of sperm can be another target for contraceptive intervention. This can be achieved by suitable selection of the proteins with the help of epididymal transcriptional profiles and list of genes expressed along the length of epididymis (Johnston et al. 2005).

29.8.3 Miglustat

Miglustat, an N-butyl-deoxynojirimycin (NB-DNJ), is an inhibitor of glucosylceramide synthase enzyme given for treating type I Gaucher disease. Miglustat also affects sperm morphology and motility in epididymis causing infertility. NB-DGJ, another alkylated imino sugar, has similar effects (Van der Spoel et al. 2002). However, unalkylated imino sugar, deoxynojirimycin (DNJ), has no such effect (Suganuma et al. 2005). This suggests that the antifertility effect of imino sugars is provided by some unique biochemical properties of the alkyl group. Because miglustat cleared all safety tests for its use in humans to treat Gaucher disease, it seems to be an effective birth control pill for males. However, some studies did not find any inhibitory effect of miglustat on spermatogenesis (Amory et al. 2007). Therefore, more studies are needed to comment on its antifertility potential.

29.8.4 Immunocontraception

Immunocontraception involves the use of body defense to disrupt critical aspects of reproductive process which provides long-term, but temporary protection against pregnancy. Active immunization of mice with anti-ACTL7a antibodies causes agglutination of mouse and human sperm in vitro thereby affecting fertility (Fu et al. 2012). Immunization of monkeys with high-titer antibodies against eppin, present on sperm surface in the ejaculate, causes alteration in sperm motility and results in complete infertility, whereas discontinuation of immunization reverses fertility (O'Rand et al. 2007). Immunization against sperm surface protein semenogelin is also known. Semenogelin has role in sperm mobilization as its proteolytic cleavage into biologically active fragments during ejaculation liquefies the semen. Thus, immunization against semenogelin causes defects in its degradation affecting sperm motility. Cysteine-rich secretory protein 2 (CRISP2) is a testicular sperm protein proposed to be involved in gamete fusion. Anti-CRISP2 sera significantly inhibit the fertilizing ability of sperm (Weigel Muñoz et al. 2012). The two epitopes in CatSper1 have much similarity between mouse and human and may be used for fertility regulation in humans (Li et al. 2012). Limitations of immunocontraception include latent time period needed to reach effective immunity, lack of specificity, and individual variations in response to immunization. These kinds of immune-specific proteins identified and translated into pharmaceutical products hold huge potential as immunocontraceptives.

29.8.5 Indenopyridines

Indenopyridines are antihistaminic compounds which also have anti-spermatogenic effects in many species. High doses of indenopyridine (2.5–5.0 mg/kg) induce infertility in rats. However, the dose of indenopyridine-inducing infertility varies in different species. Indenopyridine (CDB-4022) primarily targets Sertoli cells and decreases the production of inhibin and androgen-binding proteins which are the markers of Sertoli cell function. There is reduction in seminiferous tubule fluid secretions in addition to a series of detrimental ultrastructural changes such as mitochondrial swelling, vacuolization, loss of cytoplasm, and disorganization of Sertoli cell tight junctions (Hild et al. 2007). These changes impair Sertoli cell function causing loss of germ cells from seminiferous epithelium, ultimately leading to infertility. An associated decrease in serum testosterone shows that CDB-4022 might also affect Leydig cell function via Sertoli cells. CDB-4022 mediated abolition of spermatogenesis and fertility is irreversible unless endogenous testosterone is suppressed by gonadotropin-releasing hormone (GnRH) agonist or antagonist; the mechanism is yet to be identified.

Key Questions

- Describe the principle of RISUG as contraceptive method.
- What is the mechanism of action of nifedipine as a male contraceptive?
- Write briefly the basis of using chromatin remodeling as male contraceptive method.
- Discuss the principle of immunocontraception.

References

Amory JK, Muller CH, Page ST, Leifke E, Pagel ER, Bhandari A, et al. Miglustat has no apparent effect on spermatogenesis in normal men. Hum Reprod. 2007;22(3):702–7.

Ansari AS, Hussain M, Khan SR, Lohiya NK. Relative suitability of DMSO and $NaHCO_3$ for reversal of RISUG® induced long-term contraception. Andrology. 2016;4(2):306–13.

Aston KI, Krausz C, Laface I, Ruiz-Castané E, Carrell DT. Evaluation of 172 candidate polymorphisms for association with oligozoospermia or azoospermia in a large cohort of men of European descent. Hum Reprod. 2010;25(6):1383–97.

Babcock DF. Wrath of the wraiths of CatSper3 and CatSper4. Proc Natl Acad Sci U S A. 2007;104(4):1107–8.

Behre HM, Nashan D, Hubert W, Nieschlag E. Depot gonadotrophin-releasing hormone agonist blunts the androgen-induced suppression of spermatogenesis in a clinical trial of male contraception. J Clin Endocrinol Metab. 1992;74(1):84–90.

Benoff S, Cooper GW, Hurley I, Mandel FS, Rosenfeld DL, Scholl GM, et al. The effect of calcium ion channel blockers on sperm fertilization potential. Fertil Steril. 1994;62(3):606–17.

Chen ZD, Liu JD, Kong ZL, Huang YZ, Jiang RG, Ren FJ, et al. Clinical studies on therapeutical efficacy of TU in the treatment of patients with sexual dysfunction and male infertility. J Zhejiang Med Univ. 1991;20:126–9.

Chen ZL, Huang XB, Suo JP, Li J, Sun L. The contraceptive effect of a novel filtering-type nanocopper complex/polymer composites intra-vas device on male animals. Int J Androl. 2010;33(6):810–7.

Cheng CY, Silvestrini B, Grima J, Mo MY, Zhu LJ, Johansson E, et al. Two new male contraceptives exert their effects by depleting germ cells prematurely from the testis. Biol Reprod. 2001;65(2):449–61.

Cilip CM, Allaf ME, Fried NM. Application of optical coherence tomography and high-frequency ultrasound imaging during noninvasive laser vasectomy. J Biomed Opt. 2012;17(4):046006-1-5.

Cook CE, Kepler JA. 7alpha,11beta-Dimethyl-19-nortestosterone: a potent and selective androgen response modulator with prostate-sparing properties. Bioorg Med Chem Lett. 2005;15(4):1213–6.

Coulston F, Dougherty WJ, LeFevre R, Abraham R, Silvestrini B. Reversible inhibition of spermatogenesis in rats and monkeys with a new class of indazole-carboxylic acids. Exp Mol Pathol. 1975;23:357–66.

De Martino C, Malcorni W, Bellocci M, Floridi A, Marcante ML. Effects of AF 1312 TS and lonidamine on mammalian testis. A morphological study. Chemotherapy. 1981;27 Suppl 2:27–42.

Fu J, Wang Y, Fok KL, Yang D, Qiu Y, Chan HC, et al. Anti-ACTL7a antibodies: a cause of infertility. Fertil Steril. 2012;97(5):1226–33.

Guha SK. Contraceptive for use by a male US. 1996; Patent 54880705.

Hasenack HG, Bosch AM, Käär K. Serum levels of 3-keto-desogestrel after oral administration of desogestrel and 3-keto-desogestrel. Contraception. 1986;33(6):591–6.

Hild SA, Reel JR, Dykstra MJ, Mann PC, Marshall GR. Acute adverse effects of indenopyridine, CDB-4022, on the ultrastructure of Sertoli cells, spermatocytes and spermatids in rat testes: comparison to the known Sertoli cell toxicant, di-n-pentylphthalate (DPP). J Androl. 2007;28(4):621–9.

Johnston DS, Jelinsky SA, Bang HJ, DiCandeloro P, Wilson E, Kopf GS, et al. The mouse epididymal transcriptome: transcriptional profiling of segmental gene expression in the epididymis. Biol Reprod. 2005;73(3):404–13.

Kapur MM, Mokkapati S, Farooq A, Ahsan RK, Laumas KR. Copper intravas device (IVD) and male contraception. Contraception. 1984;29(1):45–54.

Kinniburgh D, Zhu H, Cheng L, Kicman AT, Baird DT, Anderson RA. Oral desogestrel with testosterone pellets induces consistent suppression of spermatogenesis to azoospermia in both Caucasian and Chinese men. Hum Reprod. 2002;17:1490–501.

Kumar S, Chaudhury K, Sen P, Guha SK. Quantitative analysis of surface micro-roughness alteration in human spermatozoa using atomic force microscopy. J Microsc. 2007;227(Pt 2):118–23.

Li H, Ding X, Guo C, Guan H, Xiong C. Immunization of male mice with B-cell epitopes in transmembrane domains of CatSper1 inhibits fertility. Fertil Steril. 2012;97(2):445–52.

Li MH, Guo SB, Fang RY, Zhang YP. Pharmacological and clinical effects of testosterone undecanoate. N Drugs Clin Remedies. 1994;13:279–81.

Lohiya NK, Goyal RB. Antifertility investigation on the crude chloroform extract of Carica papaya Linn. Seeds in male albino rats. Indian J Exp Biol. 1992;30:1051–5.

Lohiya NK, Suthar R, Khandelwal A, Goyal S, Ansari AS, Manivannan B. Sperm characteristics and teratology in rats following vas occlusion with RISUG® and its reversal. Int J Androl. 2010;33(1):e198–206.

Mdhluli MC, van der Horst G. The effect of oleanolic acid on sperm motion characteristics and fertility of male Wistar rats. Lab Anim. 2002;36(4):432–7.

O'Rand MG, Widgren EE, Wang Z, Richardson RT. Eppin: an epididymal protease inhibitor and a target for male contraception. Soc Reprod Fertil Suppl. 2007;63:445–53.

Panneerdoss S. Preclinical investigation on the contraceptive efficacy and toxicological evaluation of the Carica papaya seed product in langur monkey Presbytis entellus entellus, PhD. Thesis, University of Rajasthan, Jaipur. 2004.

Qian SZ, Hu JH, Ho LX, Sun MX, Huang YZ, Fang JH. The first clinical trail of gossypol on male. In: Turner P, editor. Clinical pharmacology and therapeutics. Proceeding of first world conference on Clinical Pharmacology & Therapeutics, McMillan; 1972. p. 489–492.

Qian SZ, Zhong CQ, Xu N, Xu Y. Antifertility effect of Tripterygium wilfordii in men. Adv Contracept. 1986;2:253–4.

Qian SZ, Xu Y, Zhang JW. Recent progress in research on Tripterygium: a male antifertility plant. Contraception. 1995;51:121–9.

Rajasekaran M, Bapna JS, Lakshmanan S, Ramachandran Nair AG, Veliath AJ, Panchanadam M. Antifertility effect in male rats of oleanolic acid, a triterpene from Eugenia jambolana flowers. J Ethnopharmacol. 1988;24(1):115–21.

Reel JR, Humphrey RR, Shih YH, Windsor BL, Sakowski R, Creger PL, et al. Competitive progesterone antagonists: receptor binding and biologic activity of testosterone and 19-nortestosterone derivatives. Fertil Steril. 1979;31(5):552–61.

Roddy RE, Zekeng L, Ryan KA, Tamoufé U, Tweedy KG. Effect of nonoxynol-9 gel on urogenital gonorrhea and chlamydial infection: a randomized controlled trial. JAMA. 2002;287(9):1117–22.

Shang E, Nickerson HD, Wen D, Wang X, Wolgemuth DJ. The first bromodomain of Brdt, a testis-specific member of the BET sub-family of double-bromodomain-containing proteins, is essential for male germ cell differentiation. Development. 2007;134(19):3507–15.

Sharlip ID. What is the best pregnancy rate that may be expected from vasectomy reversal? J Urol. 1993;149:1469–71.

Suganuma R, Walden CM, Butters TD, Platt FM, Dwek RA, Yanagimachi R, et al. Alkylated imino sugars, reversible male infertility-inducing agents, do not affect the genetic integrity of male mouse germ cells during short-term treatment despite induction of sperm deformities. Biol Reprod. 2005;72(4):805–13.

van der Spoel AC, Jeyakumar M, Butters TD, Charlton HM, Moore HD, Dwek RA, et al. Reversible infertility in male mice after oral administration of alkylated imino sugars: a nonhormonal approach to male contraception. Proc Natl Acad Sci U S A. 2002;99(26):17173–8.

Vyas DK, Jacob D. Effect of Papaya (Carica papaya) seeds on the reproductive structure and fertility of the male rabbits. Indian Zool. 1984;8:105–8.

Weigel Muñoz M, Ernesto JI, Bluguermann C, Busso D, Battistone MA, Cohen DJ, et al. Evaluation of testicular sperm CRISP2 as a potential target for contraception. J Androl. 2012;33(6):1360–70.

World Health Organization. The World Health Report 2005: Make every mother and child count. Geneva; 2005: 243.

Wu FCW, Balasubramanian R, Mulders TMT, Coelingh-Bennink HJT. Oral progestogen combined with testosterone as a potential male contraceptive: additive effects between desogestrel and testosterone enanthate in suppression of spermatogenesis, pituitary-testicular axis, and lipid metabolism. J Clin Endocrinol Metab. 1999;84:112–22.

Yang ZJ, Ye WS, Cui GH, Guo Y, Xue SP. Combined administration of low dose gossypol acetic acid with desogestrel/mini-dose ethinylestradiol/testosterone undecanoate as an oral contraceptive for men. Contraception. 2004;70:203–11.

Yoga and Sexual Health

30

Manjit Singh and Amit Kumar

Learning Objectives
- Asanas, mudras and bandhas for sexual health
- Pranayama and meditation for sexual disorders

30.1 Introduction

The chapter is based on our experience with Yoga, Ayurveda and other alternative systems of Indian medicine and their positive impact on sexual health. Consciousness or the inner awareness exists within our body in two forms: static and dynamic. There is a concept of a nonanatomical-subtle energy body (SEB) and a non-physiological-subtle energy generator (SEG). SEG is a sort of dynamic consciousness and is also called *Mool Prana*, *Anhad Naad*, *Kundalini*, etc.

The SEB pulsates, expands or contracts with breathing. The pattern of breathing and thought can regulate the activity of the SEG and, therefore, the energy levels of the SEB. SEG has no fixed location in the body; it moves and stabilises with the shift of inner consciousness or attention. If one concentrates in the forehead, SEG

M. Singh, MD (✉)
Safdarjung Hospital and Vardhman Mahavir Medical College, New Delhi, India
e-mail: drmanjitsinghji@yahoo.com

A. Kumar, MA, DYED
Department of Orthopaedics, All India Institute of Medical Sciences, New Delhi, India

A. Kumar, M. Sharma (eds.), *Basics of Human Andrology*,
DOI 10.1007/978-981-10-3695-8_30

gets located in the forehead, and if one shifts the attention to the heart, the SEG gets into the heart.

Pattern of breathing is well known to be regulated by the state of mind. Mental agitation, anger, anxiety, etc. induce tachypnoea and psychogenic hyperventilation. On the other hand, a peaceful and relaxed mind slows the rate of breathing. Yogis have noticed that by changing the pattern of breathing, one can make the mind peaceful.

If we breathe from the right nostril, SEG shifts to the right side of the body which is dominated by a nonanatomical energy entity, the *pingala* or solar channel (*nadi*). If one breathes through the left nostril, SEG shifts to the left side which is dominated by *Ida* or lunar channel. Right nostril breathing has a *heating* effect, whereas left side breathing has a *cooling* effect on our mind and body. According to a study, right nostril breathing increases oxygen consumption by 37%, left nostril breathing by 24% and alternate nose breathing by 18% (Telles et al. 1994). The study suggests that this is possibly due to an increased central sympathetic discharge. However, there is also increase in galvanic skin resistance during left nostril breathing, suggesting a reduced peripheral sympathetic tone; this is not so during both or only right nostril breathing. It has been also shown that there is increase in blood glucose level by right nostril breathing, whereas left nostril breathing lowers it (Backon 1988). Rhythmic breathing and *Sudarshan Kriya*, a kind of pranayama, in cancer patients improve cell-mediated immunity (Kochupillai et al. 2005). We, in our patients, have found that if a patient has chronic pain in the left side of the body, the same side nostril breathing for few months improves the pain. The same is the case with the right side.

According to Yoga, diseases are caused by imbalance in flow of energy in the subtle body. Thus, whenever there is increase in subtle energy in any part of the body, blood circulation and metabolism improve. This increase of energy in SEB also improves our local immunity.

During sleep, SEG rejuvenates energy in SEB. The SEG can be experienced through various meditational techniques, and in the end, it sublimates and unites with the static consciousness, the state of *Samadhi*.

Effect of yogic practices has been observed in sexual disorders such as hypersexuality, premature ejaculation and erectile dysfunction. One of the studies have shown that the practice of certain breathing exercises (pranayamas), gestures (mudras) and physical postures (asanas) for 1–2 months is as effective as a dose of 20–60 mg of fluoxetine capsules in prolonging the latency of ejaculation (Dhikav et al. 2007). The details of their list of yogic practices can be seen in their remarkable paper. We have mentioned our own protocol here.

These psychophysical practices include yogic interventions on the body, breath and mind:

- *Body*: asanas , bandhas and mudras
- *Breath*: pranayamas
- *Mind*: meditation/concentration

Under each group, we shall first describe the general and then specific practices, general practices to open energy channels by removing blocks in them and specific practices which redirect energy into specific areas and channels.

30.2 Asanas

Asanas refer to specific body positions and are usually named after animal postures. Asanas open energy channels and psychic centres as described in the Yogic texts. Yogic asanas help in controlling mind and subtle energies within the body. Asanas involve muscular and physiological aspects. Names of the useful asanas are given here. Their description is beyond the scope of this book. One can learn these asanas from Yoga experts.

Surya Namaskar or salutation to sun consists of a set of 12 asanas in a stepwise manner. Some other useful asanas are pawanmuktasana (fart release postures), sarvangasana (legs up posture), halasan (hand plough posture), dhanurasana (bow posture), bhadrasana (gentle posture), mayurasana (peacock posture), bhujangasana (cobra posture), paschimottanasana (sitting with legs stretched out and head and torso bent over the knees posture), vajrasana (thunderbolt posture) and matsyasana (fish posture).

30.3 Bandhas

Bandhas refer to tightening or locking. The bandhas lock subtle life forces in a particular area and redirect their flow into *Sushumna nadi* (channel) in the spinal canal for spiritual awakening. Hatha Yoga Pradipika, a fifteenth-century textbook of yogasanas by Swami Sva-aatma-raam, deals with bandhas and mudras together. Following bandhas are practised to increase the regional blood flow and metabolism.

30.3.1 Mulabandha

Sit in cross-legged posture, say, sukhasana (see Fig. 30.1). Focus the awareness at perineal region. Contract this region by pulling up the muscles of pelvis floor and then relaxing them for some time (see Fig. 30.2). Then, slowly contract this region and hold the contraction. Continue to breathe normally. This helps in sublimation of sex desire and hence helps in hypersexuality, premature ejaculation and also erectile dysfunction.

30.3.2 Uddiyanabandha

Cross-legged posture stabilises the base of the body and keeps the spine erect. Sit cross-legged with spine erect in siddhasana or sukhasana (see Fig. 30.1). Breathe in deeply through nostrils and exhale through mouth emptying lungs to the maximum possible extent. Now, hold the breath as long as you can and contract the abdominal wall inward and upward as long as possible without strain. Then, release the abdominal lock and continue normal breathing. This is useful in hypersexuality and premature ejaculation in association with practice of moolbandha.

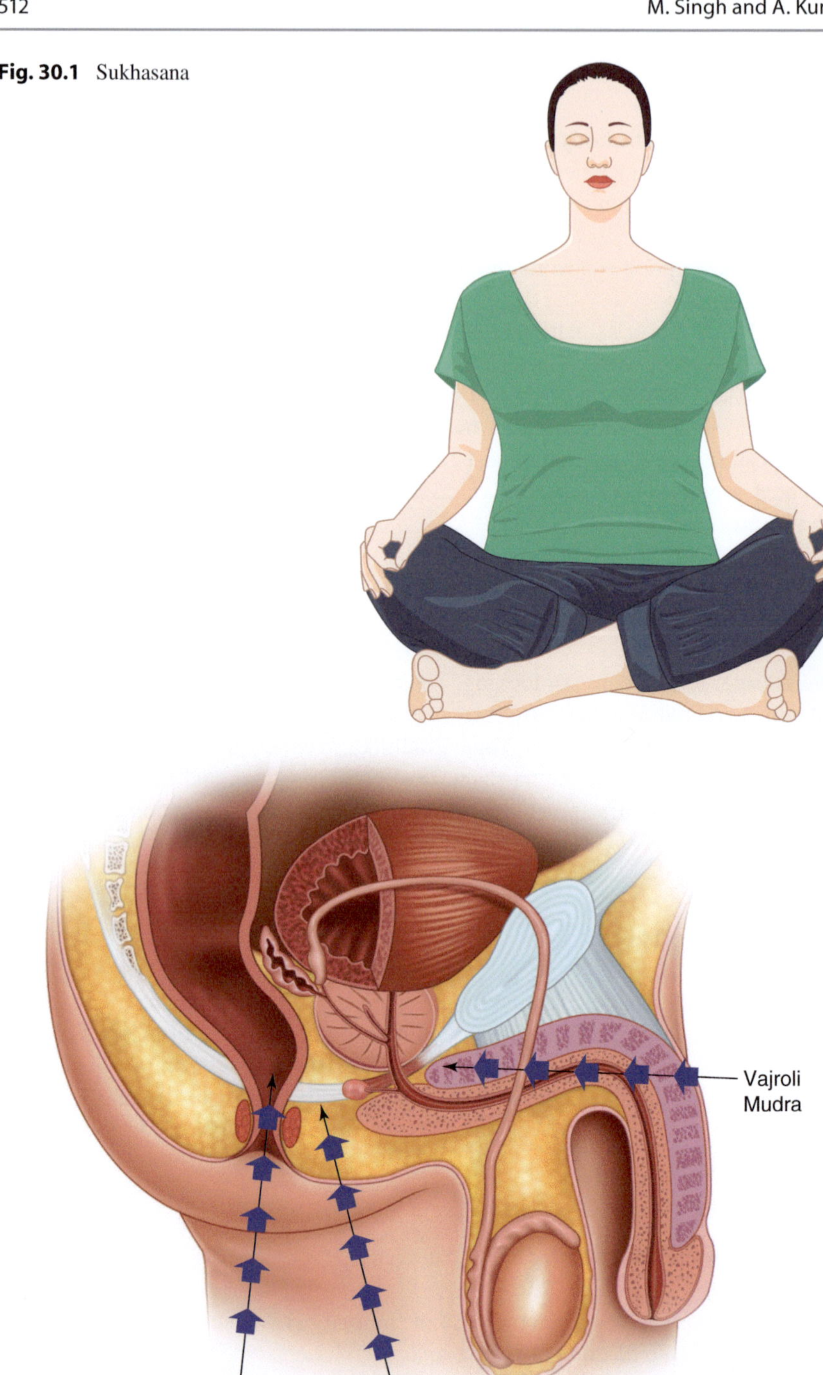

Fig. 30.1 Sukhasana

Fig. 30.2 Schematic diagram showing the concentration points in different bandhas and mudras

30.4 Mudras

Mudras are described as psychic, emotional, devotional and aesthetic gestures. These are cultivations of specific psychological association with a certain physical gesture. Mudras such as Ashwini and Vajroli can be practised.

30.4.1 Ashwini Mudra

Sit in sukhasana (see Fig. 30.1). Contract the sphincter muscle of the anus for a few seconds, then relax, continue contraction and relaxation for a few minutes (see Fig. 30.2). Repeat the practice for 5 min at a time and then repeat the process three or four times a day.

30.4.2 Vajroli Mudra

Sit in sukhasana with the head and spine straight (see Fig. 30.1). Take the awareness to the urethra at the root of the penis (see Fig. 30.2). Inhale and hold the breath; and simultaneously try to draw the urethra upwards as if one is trying to hold an intense urge to urinate. Testicles in men should also move slightly upwards with this contraction. Hold the contraction as long as comfortable and exhale while releasing the contraction and relax. Adepts in Vajroli can withhold ejaculation as long as they want or may not ejaculate at all.

30.5 Pranayama

Holding breath during sexual intercourse may prolong the ejaculation and helps achieving good orgasm. Commonly practiced pranayama exercises are as follows:

30.5.1 Anulom-Vilom

Sit in a sukhasana (see Fig. 30.1). Close the right nostril with the thumb and breathe in through the left nostril and exhale through the right nostril (see Fig. 30.3). Now, inhale from the right nostril and exhale from the left nostril. Repeat this practice for some time. Breath holding is not required.

30.5.2 Suryabhedana

Sit in sukhasana (see Fig. 30.1). Close the left nostril and breathe in through the right nostril and hold the breath for some time. Now, breathe out from the left nostril. Repeat this process for some time. This pranayama exercise is commonly practised for erectile dysfunction.

30.5.3 Chandrabhedana

Sit in sukhasana (see Fig. 30.1). Close the right nostril and breathe in through the left nostril and hold the breath for some time (see Fig. 30.3). Now, breathe out from the right nostril. Repeat this process for some time. This pranayama exercise is commonly practised for premature ejaculation.

30.5.4 Bhastrika

At the end of above pranayamas, Bhastrika is performed. Sit in sukhasana keeping the head and spine straight (see Fig. 30.1). Take a deep breath in and out forcefully through the nose. Keep repeating this process for some time. This is very powerful

Fig. 30.3 Right nostril breathing as in *Anulom-Vilom and Chandrabhedana*

pranayama. It improves energy flow in the subtle body. To remind, these are to be learnt from a yoga instructor.

30.6 Meditation

Meditation is forbidden in depression and psychotic disorders. Meditation should be learnt from a Guru than an instructor. Meditation reduces anxiety and mental stress which actually may lead to sexual disorders. Different types of meditations that can be practised are as follows:

30.6.1 Anapan

Sit in a sukhasana (see Fig. 30.1). Keep your awareness on your incoming and outgoing breath on the tip of your nose as long as possible.

30.6.2 Vipasana

Sit in sukhasana (see Fig. 30.1). Observe thoughts and emotions moving in the mind as a neutral observer, without analysing or judging the moral content.

30.6.3 Kriya Yoga

During inhalation, attention is taken to the perineal area between the root of the penis and the anus. During exhalation, attention is taken away from the perineal area to the crown of the head. This practice is repeated twenty times, twice daily.

30.6.4 Raja Yoga

Sit in sukhasana (see Fig. 30.1). Mind/attention is focussed in the forehead area in the centre of the two eyebrows. This yoga is practised in cases of premature ejaculation.

In premature ejaculation during the sexual act, one may try to relax the perineal area, and attention should be shifted above to the area in the forehead.

References

Backon J. Changes in blood glucose level induced by differential forced nostril breathing, a technique which affects brain hemisphericity and autonomic activity. Med Sci Res. 1988;16:1197–9.

Dhikav V, Karmarkar G, Gupta M, Anand KS. Yoga in premature ejaculation: a comparative trial with Fluoxetine. J Sex Med. 2007;4:1726–32.

Kochupillai V, Kumar P, Devender S, Aggarwal D, Bhardwaj N, Bhutani M, et al. Effect of rhythmic breathing (Sudarshan Kriya and Pranayama) on immune functions and tobacco addiction. Ann N Y Acad Sci. 2005;1056:242–52.

Swami Sva-atmaa-raam. Hatha Yoga Pradipika. 15th Century CE. Several English translations are available.

Telles S, Nagarathna R, Nagendra HR. Breathing through a particular nostril can alter metabolism and autonomic activities. Indian J Physiol Pharmacol. 1994;38:133–7.

Further Reading

Saraswati SS. Asana pranayama mudra bandha. Munger: Yoga Publications Trust; 1996.

Opinions

'*Written explicitly, the book will be a useful guide for the postgraduates, undergraduates, nursing students and also for lay persons*'.

PN Dogra, MS, MCh, DSc (Hon)/Professor and Chairperson, Department of Urology, All India Institute of Medical Sciences, New Delhi, India

'*The book has a holistic approach and is truly commendable*'.

KK Deepak, MD, PhD, FAMS/Professor and Chairperson, Department of Physiology, All India Institute of Medical Sciences, New Delhi, India

'*It is a remarkable and unique book in the field of andrology which fulfills a definite void. A must read for all those dealing with male infertility*'.

JB Sharma, MD, FRCOG, FAMS/Professor of Obstetrics and Gynaecology, All India Institute of Medical Sciences, New Delhi, India

A. Kumar, M. Sharma (eds.), *Basics of Human Andrology*,
DOI 10.1007/978-981-10-3695-8

Index

A. Kumar, M. Sharma (eds.), *Basics of Human Andrology*,
DOI 10.1007/978-981-10-3695-8

H

R

S

T

MIX
Papier aus verantwortungsvollen Quellen
Paper from responsible sources
FSC® C105338

If you have any concerns about our products,
you can contact us on
ProductSafety@springernature.com

In case Publisher is established outside the EU,
the EU authorized representative is:
Springer Nature Customer Service Center GmbH
Europaplatz 3, 69115 Heidelberg, Germany

Printed by Libri Plureos GmbH
in Hamburg, Germany

Printed by Books on Demand GmbH, Norderstedt / Germany

Printed by Books on Demand GmbH, Norderstedt / Germany